Nanoscale Insights into Ion-Beam Cancer Therapy

Andrey V. Solov'yov
Editor

Nanoscale Insights into Ion-Beam Cancer Therapy

Springer

Editor
Andrey V. Solov'yov
MBN Research Center at Frankfurter
Innovationszentrum Biotechnologie
Frankfurt am Main
Germany

and

A.F. Ioffe Physical-Technical Institute
Saint Petersburg
Russia

ISBN 978-3-319-43028-7 ISBN 978-3-319-43030-0 (eBook)
DOI 10.1007/978-3-319-43030-0

Library of Congress Control Number: 2016950401

Printed on acid-free paper

This Springer imprint is published by Springer Nature
The registered company is Springer International Publishing AG
The registered company address is: Gewerbestrasse 11, 6330 Cham, Switzerland

Preface

This book provides a unique and comprehensive overview of state-of-the-art understanding of the molecular and nanoscale processes that play important roles in ion-beam cancer therapy. It covers experimental design and methodology, and reviews the theoretical understanding of the processes involved. It offers the reader an opportunity to learn from a coherent approach about the physics, chemistry and biology relevant to ion-beam cancer therapy, a growing field of important medical application worldwide. The book describes phenomena occurring on different time, spatial and energy scales relevant to the radiation damage of biological targets and ion-beam cancer therapy from the molecular (nano) scale up to the macroscopic level. It illustrates how ion-beam therapy offers the possibility of excellent dose localization for treatment of malignant tumours, minimizing radiation damage in normal tissue whilst maximizing cell-killing within the tumour, offering a significant development in cancer therapy. The full potential of such therapy can only be realized through the understanding of the physical, chemical and biological mechanisms, on a range of time and space scales that lead to cell death under ion irradiation. This book describes how the recently developed multiscale approach, unifying all the experimental and theoretical expertise available in the field, leads to greater insight at the nanoscopic and molecular level into radiation damage of biological targets induced by ion impact. The book is intended for the master and Ph.D. students and specialists in the areas of physics, chemistry, biology and medicine related to ion-beam therapy, radiation protection, biophysics, radiation nanophysics and chemistry, atomic and molecular physics, condensed matter physics, and the physics of interaction of charged particles with matter. The most important features and benefits of the book are in the systematic description of the inclusive multiscale approach for the description of complex and highly interdisciplinary processes behind the ion-beam cancer therapy and its key components. This theoretical approach, being stretched from the atomistic level up to the biological scale, is demonstrated to be in the excellent agreement with experimental observations.

Ion-beam cancer therapy (IBCT, or hadron therapy) represents an effective method for providing high-dose delivery into tumours, thereby maximizing the probability of killing the cancer cells whilst simultaneously minimizing the radiation damage to surrounding healthy tissue [1–3]. Despite its high cost, proton-beam therapy is widely spread around the world with over 65 operational centres[1]. In ten European and Asian centres, patients are irradiated with carbon ions. Nonetheless, the full potential of these therapies can only be realized by achieving a better understanding of physical, chemical and biological mechanisms, over a range of time and space scales, that lead to cell inactivation under ion radiation.

The damaging effect of ionizing radiation has been known for many years. It has been commonly accepted that high-energy tracks formed by α, β, and γ radiation and atomic ions ionize cell components along the track, thereby leading to various dissociation channels and to the formation of damaging radicals. This has led to intensive research on the study of the mechanisms for the formation of such radicals and the fragmentation pattern of biomolecules by photons, electrons and ions. Such fundamental data underpins the study of radiation protection and the development of biomedical uses of different radiation, generally called radiotherapy, for treatment, of tumoural diseases in particular. The next generation of radiotherapy may be based on hadron therapy [4] and in particular ion-beam therapy. To date the development of ion-beam therapy has been based on empirical rather than phenomenological or *ab initio* scientific methods [5]. The emergence of the "RADAM" [5] and then "NanoIBCT" [6] communities has played an important role in attracting physicists, chemists and biologists into the field to tackle a plethora of scientific questions raised by the technological advances in this field.

The majority of biological effects of ion beams are associated with the process of ionization of the medium by traversing ions. It is commonly accepted that secondary electrons, ejected by ionization, are mainly responsible for DNA damage, either breaking the DNA strands directly, or reacting with molecules of tissue, producing free radicals and other reactive species. Macroscopically, the advantages of using ion beams compared to photons are related to the presence of a Bragg peak in the depth–dose distribution, where the production of secondary electrons is maximized. This localizes irradiation effects deep in tissue thus increasing the treatment efficiency and reducing side effects by sparing neighbouring healthy tissue. However, the mechanisms involved in radiation damage at the nanoscale and molecular level are still a subject of fundamental multidisciplinary research.

In 2010–2014, the European Concerted Research Action, COST Action MP1002: "Nano-scale insights in ion beam cancer therapy (Nano-IBCT)" was devoted to acquiring a deeper understanding of radiation-induced damage with ions on the nanoscopic and molecular level [6]. This endeavour clustered around the multiscale approach to the physics of radiation damage with ions [1, 7], designed to achieve a quantitative understanding of the physical, chemical and biological effects that take place on a wide range of spatial, temporal and energy scales. The COST Action combined European experimental and theoretical expertise in several topics including nuclear reactions and electromagnetic processes during the propagation of

---

[1]As of May 2016 [4].

ion beams in tissue, primary ionization in the medium (water and biological molecules), direct damage and production of secondary species (secondary electrons, radicals, holes), propagation of secondary species and their interaction with DNA, and radiobiological scale effects.

Action was formally launched in December 2010 and since then has brought together more than 300 experts from different disciplines (physics, chemistry, biology, etc.) drawn from more than fifty different institutions including hadron therapy centres and medical institutions. The Action also engaged with colleagues working in countries outside the EU, including Canada, Australia, Japan, India, China and the USA. Two-thirds of those participating were early-career researches and a quarter were postgraduate students half of which were young female researchers. The Action also supported more than 100 short-term scientific missions between different institutions and countries, which resulted in more than 200 publications in high-impact journals. Within the framework of the COST Action Nano-IBCT three major conferences (held in Caen, France October 2011, Sopot, Poland May 2013 and Boppard, Germany 2013, see conference photos in Figs. 1–3) and 12 workshops were organized. These conferences provided the opportunity to review the progress in the field of radiation damage to biomolecular systems and how such knowledge can be applied to the development of new cancer therapies. For further details, see the Action's website [6], the special issue of the Journal of Physics: Conference Series [8], and the topical issues of the European Physical Journal D: Atomic, Molecular, Optical and Plasma Physics [9, 10].

The main objective of the Action was to address the basic scientific questions which underpin the nanoscopic and molecular mechanisms associated with ion-beam cancer therapy. In particular, the following goals were pursued:

- Understanding the unique features of ion irradiation on the molecular level, e.g. site and bond selectivity, clustered damage, local temperature and chemical effects. Some of the open questions are related to the ratio of direct/indirect damage, the mechanisms leading to double strand break (single- or multiple electron-induced fragmentation) and the elucidation of possible lethal effects that are not present during photon irradiation.
- To establish comprehensive databases of recommended values for all the major processes involved in IBCT: ion and electron interaction cross sections, energy loss in biologically relevant systems, etc. This objective implies an important experimental and theoretical effort to determine differential and integral cross sections, both elastic and inelastic, for low incident energies.
- To yield a quantitative prediction of dose distribution and molecular damage generated by the passage of ions through cells, for example determining the rate and type of DNA single and double strand breaks, as well as complex damage sites.
- To develop a multiscale approach for the quantitative analysis of radiobiological effects and therapy planning, tested at different levels with experiments; also including reliable estimates of the relative biological effectiveness (RBE) for different ions.

Fig. 1 Conference photo, 1st Nano-IBCT Conference, 2–6 October, 2011, Caen, France

Fig. 2 Conference photo, 2nd Nano-IBCT Conference, 20–24 May, 2013, Sopot, Poland

Fig. 3 Conference photo, 3rd Nano-IBCT Conference, 27–31 October, 2014, Boppard, Germany

- To develop a new low-energy particle track simulation method based on the distribution functions derived from evaluated experimental and theoretical cross sectional data and energy loss providing information on energy distribution and induced damage at the molecular level.

The COST Action was organized through five Working Groups (WGs) each dedicated to a specific topic. WG1: Ion propagation, WG2: Primary ionization in the medium, direct damage and production of secondary species, WG3: Propagation of secondary species, WG4: Electron attack on DNA, WG5: Radiobiological scale effects. These working groups form the pillars upon which the research plan for a multiscale approach to the physics of ion-beam cancer therapy was based [1].

This book summarizes the main research achievements of the COST Action Nano-IBCT. It provides a comprehensive overview of state-of-the-art understanding of the molecular and nanoscale processes that play significant role in ion-beam cancer therapy. It covers experimental design and methodology, and reviews the theoretical understanding of the processes involved. It is based on the reviews written by the teams of experts devoted to the essential aspects of the multiscale scenario of the complex cascade of physical, chemical and biological processes involved into the ion-beam cancer therapy [1]. The topics of the reviews and their interrelationship arise naturally from the action aims, its research and organizational structure, as well as from the follow-up research developments.

Chapter "Multiscale Physics of Ion-Beam Cancer Therapy" by E. Surdutovich and A.V. Solov'yov is the most comprehensive review of the multiscale approach to the physics of radiation damage with ions. The approach allows one to predict survival probabilities for cells irradiated with ions based on the series of phenomena that take place on a variety of scales in time, space, and energy. The scenario of biodamage starting from ion entering tissue is the basis for an analytic synthesis of microscopic effects that comprise the macroscopic coefficients of the linear-quadratic model describing survival probabilities. The latter are calculated for both aerobic and hypoxic conditions at a variety of linear energy transfers. The oxygen enhancement ratio is obtained as a byproduct of these calculations. The calculated survival curves are compared with experiments on different cell lines and ready for medical applications.

Chapter "Propagation of Swift Protons in Liquid Water and Generation of Secondary Electrons in Biomaterials" by P. de Vera, R. Garcia-Molina, and I. Abril introduces a proper description of the propagation of a swift proton beam through biomaterials, accounting for the energy deposited as well as the geometrical evolution of the beam as a function of the target depth and nature, that is a crucial issue in proton therapy. For this purpose, simulation is a very adequate tool, since the most relevant interactions that take place between the projectile and the target constituents (electrons and nuclei) can be conveniently accounted for in a controlled manner. For this purpose, in this chapter an overview and relevant results for hadron therapy are presented which were obtained using the code SEICS (Simulation of Energetic Ions and Clusters through Solids). This approach combines Monte Carlo and molecular dynamics, to follow in detail the motion and energy deposition of swift protons through targets of hadron therapeutic interest, mainly liquid water. The main interactions considered in our study are of elastic nature (affecting mainly the projectiles direction) and inelastic processes (leading to either nuclear reactions or electronic energy loss). The performance of the code, as well as the quality of its main input, namely the stopping force for proton beams in liquid water (which is the main tissue constituent), are benchmarked by comparing the results of the simulations with available experimental proton energy spectra as a function of the detection angle after traversing a micrometric water jet. The excellent agreement with experiments validates the SEICS code, which can be used then to study several problems of interest for proton therapy, including the calculation of depth−dose curves and lateral dose profiles, the energy evolution of the proton beam along the target, as well as the production of secondary electrons at the Bragg peak in relevant biomaterials.

Chapter "Monte Carlo-Based Modeling of Secondary Particle Tracks Generated by Intermediate- and Low-energy Protons in Water" by A. Verkhovtsev, P. Arce, A. Munoz, F. Blanco, and G. Garcia gives an overview of recent developments in the Monte Carlo-based modelling of the interaction of ionizing radiation with biologically relevant systems. Several track structure codes, such as Geant (GEometry ANd Tracking), Geant4-DNA and LEPTS (Low-Energy Particle Track Simulation), are described. Main features, areas of application and current limitations of each tool are discussed. A special attention is focused on the energy range covered by primary and secondary charged particles and on the type of interactions included in the simulation. A recent development of LEPTS is presented, aimed at the simulation of full slowing down of protons in water together with all molecular processes involving particles. The utilized approach allows one to study radiation effects on the nanoscale in terms of the number and the type of induced molecular processes. Development of new tools for the simulation of biologically relevant materials opens the way for a more realistic, physically meaningful description of radiation damage in living tissue.

Chapter "Ion Collisions with Biomolecules and Biomolecular Clusters" by P. Rousseau and B.A. Huber describes the recent progress which has been made in experimental studies of ion collisions with biomolecular systems, either in form of isolated biomolecules in the gas phase or as clusters containing up to several tens of biomolecules. Most of the work has been performed with projectiles which play an important role in ion beam cancer therapy applications as protons or multiply charged ions of carbon and oxygen. The biomolecular targets are characterized by an increasing complexity and include water molecules, nucleobases, nucleosides and nucleotides, as well as amino acids and protein segments. Other complex targets are heterogeneous clusters containing biomolecular systems which are embedded in a water environment. After an introduction to ion–molecule collisions using C60 fullerene as a model system, ionization and charge transfer processes as well as ion-induced fragmentation studies are reviewed. Finally the effect of the environment considering clusters of biomolecules including hydrated systems is discussed.

Chapter "Dissociative Electron Attachment to Biomolecules" by I. Bald, R. Curik, J. Kopyra, and M. Tarana discusses the biomolecular damage caused by reactions induced by low-energy electrons (<20 eV). In this energy regime electrons can efficiently decompose molecules such as DNA or DNA building blocks by Dissociative Electron Attachment (DEA). Experiments on single DNA building blocks have been performed in the gas phase revealing that DEA can proceed with remarkable site selectivity. Low-energy electron-induced DNA strand breakage is typically investigated using plasmid DNA in the condensed phase. Very recently, a pronounced dependence of electron-induced DNA strand breakage on the nucleotide sequence was found using different experimental approaches suggesting that at least part of the observed strand breaks are due to initial electron attachment to the nucleobases. Currently, a strong research focus is on the fundamental understanding of DEA to therapeutically administered radiosensitizers. In the near future DEA to novel potential radiosensitizers will be explored, and the electron-induced damage of biomolecules within complex environments has to be investigated. Considerable attention has been paid to the theoretical research of the DEA in the context of the DNA damage. With respect to this, the theoretical part of the chapter reviews all the computational approaches that have been used to study DEA to biomolecules over the last decade. These approaches are divided into two classes. The first class consists of electronic structure methods studying the transient negative ions formed by electrons captured by the neutral building blocks of the DNA. Approaches dealing with the complicated nuclear dynamics of the DEA to biomolecules form the second class explored in this chapter.

Chapter "Photoprocesses with Biomolecules in the Gas Phase" by P. Bolognesi and L. Avaldi reviews the basic processes in molecules of biological interest induced by the absorption of VUV and soft X-rays. The study of excitation, ionization and dissociation in the gas phase on the one hand provides detailed information on the electronic structure and geometry that determine the functioning of these molecules in macroscopic systems and, on the other hand, sheds light on the microscopic effects of radiation damage in living cells.

Chapter "Irradiation-Induced Processes with Atomic Clusters and Nanoparticles" by A. Verkhovtsev, A.V. Korol, and A.V. Solov'yov gives an overview of theoretical and computational studies of physical phenomena manifesting themselves in photon, electron and ion collisions with atomic clusters and nanoparticles (NPs). The emphasis is made on ion and electron scattering as well as photoabsorption of metal NPs which are of current interest in application in cancer treatments with ionizing radiation. Although the number of reports on dose enhancement and radio sensitization due to metal NPs has been rapidly increasing during the past years, physical mechanisms of enhanced production of secondary electrons and reactive species due to sensitizing NPs are still a debated issue and require thorough investigation. In this chapter the essential role of collective electron excitations in the formation of electron emission spectra of metal clusters and NPs is elucidated. These effects appear also in other types of nanoscale systems, such as carbon-based NPs. A number of recent Monte Carlo-based studies devoted to the investigation of radiosensitization and dose enhancement effects for proton irradiation combined with metal NPs are also briefly discussed.

Chapter "On the Quantum Description of Irradiation Dynamics in Systems of Biological Relevance" by P.M. Dinh, L. Bouessel du Bourg, C.-Z. Gao, Bin Gu, L. Lacombe, M. McAllister, M. Smyth, G. Tribello, M. Vincendon, J. Kohanoff, P.-G. Reinhard, L. Sanche, and E. Suraud discusses the two main products of ionizing radiation in biological tissues, namely electrons and radicals. Numerous secondary electrons are generated by ionization in the molecules in the vicinity of DNA and are produced with a mean energy about 10 eV. These low-energy electrons can lead to DNA strand breaks via dissociative electron attachment and other mechanisms. The modelling of these phenomena requires, on the one hand, an explicit quantum description of the electrons of the target molecule (typically, a subunit of a DNA strand), and on the other hand, a realistic account of the DNA environment. This chapter reviews theoretical and computational approaches that have allowed studies of electron dynamics (excitation, ionization, transport and localization) in systems of biological interest.

Chapter "Multiscale Modelling of Molecular Processes for Biomedical and Nanotechnology Applications with MBN Explorer" by A.V. Solov'yov introduces MesoBioNano Explorer (MBN Explorer), a software package for the advanced multiscale simulations of complex molecular structure and dynamics and highlights some of its biomedical and nanotechnology applications. MBN Explorer has many unique features, a wide range of applications in Physics, Chemistry, Biology, Material Science, and in related Industries. It is suitable for classical molecular dynamics, Monte Carlo and relativistic dynamics simulations of a large range of molecular systems of different kind, such as nano- and biological systems, nanostructured materials, composite/hybrid materials, gases, liquids, solids and various interfaces, with the sizes ranging from atomic to mesoscopic. MBN Explorer can be exploited together with MBN Studio, a specially developed graphical user interface, visualizer, and analytic toolkit.

Chapter "Thermo-Mechanical Damage of Biomolecules Under Ion-Beam Radiation" by P. de Vera, N.J. Mason, E. Surdutovich, and A.V. Solov'yov reviews the recent studies of new pathways of the ion-induced radiation damage. The prediction of the relative biological effectiveness of ion beams requires the quantification of all the biomolecular damage processes involved in the interaction of energetic ions with biological media. Traditionally, the damage pathways have been classified as direct or indirect, the former being related to the direct action of the secondary electrons produced along the ion path with DNA molecules, while the latter are referred to the damage produced by the other chemical species generated, mainly free radicals. However, the development over the last years of the multiscale approach to ion-beam cancer therapy has revealed the contribution of a new damage mechanism, not present in conventional therapy with photons or electrons: the thermo-mechanical DNA damage arising from the development of shock waves on the nanometer scale around the swift ion path. The present chapter explains the theoretical framework in which this effect is predicted and reviews the work performed over the last years to try to understand the role of this damage pathway in the mechanisms of ion-beam cancer therapy.

Chapter "Predictive Assessment of Biological Damage Due to Ion Beams" by A. Verkhovtsev, E. Surdutovich, and A.V. Solov'yov presents recent achievements in validation of the Multiscale Approach (MSA) to the physics of radiation damage with ions. An analytical recipe for the assessment of biological damage, developed using the phenomenon-based MSA, has been applied to numerous experiments, where survival curves were obtained for different cells and irradiation conditions. Contrary to other, in essence empirical methods for evaluation of macroscopic effects of ionizing radiation, the MSA predicts the biodamage based on the physical effects related to ionization of the medium, transport of secondary particles, chemical interactions, thermo-mechanical pathways of biodamage, and heuristic biological criteria for cell survival. An extensive comparison with experimental data for cell survival probability demonstrates the validity of the MSA to predict the macroscopic effects of ionizing radiation through an understanding of biological damage at the nanoscale. The analysis performed allows us to conclude that the biodamage can be accurately predicted in both aerobic and hypoxic conditions. Therefore, we anticipate this method to give great impetus to the practical improvement of ion-beam cancer therapy and the development of more efficient treatment protocols.

Chapter "New Research in Ionizing Radiation and Nanoparticles: The ARGENT Project" by M. Bolsa Ferruz, V. Ivošev, K. Haume, L. Ellis-Gibbings, A. Traore, V. Thakare, S. Rosa, P. de Vera, V.-L. Tran, A. Mika, D. Boscolo, S. Grellet, A. Verkhovtsev, B.A. Huber, K.T. Butterworth, K.M. Prise, F.J. Currell, N.J. Mason, J. Golding, E. Scifoni, G. Garcia, F. Boschetti, F. Lux, O. Tillement, C. Louis, K. Stokbro, A.V. Solov'yov, and S. Lacombe gives an overview of ARGENT (Advanced Radiotherapy, Generated by Exploiting Nanoprocesses and Technologies), an ongoing international Initial Training Network project, supported by the European Commission. The project, bringing together world-leading researchers in physics, medical physics, chemistry, and biology, aims to train 13 early-stage researchers whose research activities are linked to understanding and exploiting the nanoscale processes that drive physical, chemical and biological effects induced by ionizing radiation in the presence of radiosensitizing nanoparticles. This research is at the forefront of current practices and involves many experts from the respective scientific disciplines. In this chapter, the research topics covered by the ARGENT project are briefly reviewed.

Chapter "Biophysics Modeling to Optimize Ion Beam Cancer Therapy" by M. Beuve discusses the ion-beam cancer therapy (IBCT) and its optimization. The optimization of treatments by IBCT relies on modelling to simulate the transport of the incident ions (and the secondary particles) into patients, and, to predict the biological effects induced by all these particles. Considering the complexity of biological systems, multiscale approaches seem necessary to build the bridge between the primary physical and chemicals events and the consequences for patients both in healthy tissues and tumours. After a brief history of IBCT in France, this chapter presents models used to estimate the probability of tumour control by IBCT, showing the importance of predicting the survival of biological cells to complex irradiation. Then, follows a presentation and analysis of models predicting cell survival to irradiation with ions, including: the procedure developed in Japan for cancer treatments with passive beams; the microdosimetry models TDRA and MKM, and, the MMKM, a modified version of MKM used for active beam in Japan; the Katz models and the LEM, which is presently used by the European centres of therapy with carbon ions. Then, as perspectives, modelling based on nanodosimetry are addressed with a focus on the NanOx™ model.

Chapter "Treatment Planning Systems and Hadron Therapy Practice in France" by L. De Marzi, A. Patriarca, A. Mazal, J.-L. Habrand describes briefly the history of particle therapy development in France and discusses the technical and clinical aspects of proton-beam treatment planning, as many similarities exist between proton and ion therapy. This chapter includes a summary of the physics and approximations used in proton dose algorithms, including the impact of accelerator and nozzle modelling, a description of conventional delivery approaches such as passive scattering or pencil beam scanning, immobilization specificities and the need for accurate imaging of patient geometry. The issues of neutron generation, risk of second cancers, and radiobiological effectiveness (RBE) of protons will also be discussed. As several of these aspects are common to proton and ion therapy, one section of this chapter will be devoted to the differences between these techniques, especially the biological effects of radiation. Finally, recent developments and perspectives in the planning process will be presented.

Concluding, the COST Action Nano-IBCT played a very important role in the foundation of a strong European Nano-IBCT community, which inherited and broadened the traditions of the initial RADAM network. Ideas that emerged during the Nano-IBCT COST Action led to many research collaborations including the establishment of the current ARGENT programme exploring nanoparticles as radiosenistizers (www.itn-argent.eu).

Acknowledgements

We are grateful to the support of COST Action MP1002 "Nano-scale insights in ion beam cancer therapy" and FP7 ITN-ARGENT.

Frankfurt, Germany Andrey V. Solov'yov

References

1. E. Surdutovich, A. Solov'yov, Eur. Phys. J. D **68**, 353 (2014)
2. D. Schardt, T. Elsässer, D. Schulz-Ertner, Rev. Mod. Phys. **82**, 383 (2010)
3. M. Durante, J. Loeffler, Nat. Rev. Clin. Oncol. **7**, 37 (2010)

4. Particle therapy co-operative group. http://www.ptcog.ch/index.php/facilities-in-operation. Accessed May 2016
5. I. Baccarelli, F. Gianturco, E. Scifoni, A. Solov'yov, E. Surdutovich, Eur. Phys. J. D **60**, 1 (2010)
6. Cost action nano-ibct. http://mbnresearch.com/project-nanoibct
7. A. Solov'yov, E. Surdutovich, E. Scifoni, I. Mishustin, W. Greiner, Phys. Rev. **E79**, 011909 (2009)
8. B.A. Huber, Ch. Malot, A. Domaracka, A.V. Solov'yov, 1st Nano-IBCT Conference 2011 Radiation damage of biomolecular systems: nanoscale insights into-ion beam cancer therapy. J. Phys.: Conf. Ser. **373**, 011001 (2012)
9. M.A. Smialek, P. Limao-Vieira, N.J. Mason, A.V. Solov'yov, Radiation Damage of Biomolecular Systems: Nanoscale insights into Ion Beam Cancer Therapy. 2nd Nano-IBCT Conference. Editorial for Topical issue Nanoscale insights into Ion-Beam Cancer Therapy. Eur. Phys. J. **D68**, 312 (2014)
10. E. Surdutovich, G. Garcia, N.J. Mason, A.V. Solov'yov, Nano-scale processes behind Ion-Beam Cancer Therapy. Editorial for Topical issue Nano-scale processes behind Ion-Beam Cancer Therapy. Eur. Phys. J. **D70**, 86 (2016)

Contents

Multiscale Physics of Ion-Beam Cancer Therapy

Eugene Surdutovich and Andrey V. Solov'yov

Abstract This is the most comprehensive review of the multiscale approach to the physics of radiation damage with ions. The approach allows one to predict survival probabilities for cells irradiated with ions based on the series of phenomena that take place on a variety of scales in time, space, and energy. The scenario of biodamage starting from ion entering tissue is the basis for an analytic synthesis of microscopic effects that comprise the macroscopic coefficients of the linear-quadratic model describing survival probabilities. The latter are calculated for both aerobic and hypoxic conditions at a variety of linear energy transfers. The oxygen enhancement ratio is obtained as a byproduct of these calculations. The calculated survival curves are compared with experiments on different cell lines and ready for medical applications.

1 Introduction

The physics and chemistry of radiation damage caused by irradiation with protons and heavier ions has recently become a subject of intense interest because of the use of ion beams in cancer therapy [1–5]. Ion-beam cancer therapy (IBCT) was first realised in the 1950s as proton-beam therapy after being suggested by Robert R. Wilson in 1946 because of the favourable shape of the depth-dose distribution due to the fundamental difference in the energy deposition profile between charged projectiles and photons. For ions, this shape is characterised by the Bragg peak, which is a sharp maximum in the dose deposition close to the end of their trajectories. This peak cor-

E. Surdutovich (✉)
Oakland University, Rochester, MI, USA
e-mail: surdutov@oakland.edu

A.V. Solov'yov
MBN Research Center at Frankfurter Innovationszentrum Biotechnologie,
Altenhöferallee 3, 60438 Frankfurt am Main, Germany
e-mail: solovyov@mbnresearch.com

A.V. Solov'yov
A.F. Ioffe Physical-Technical Institute,
Polytekhnicheskaya Ul. 26, 194021 Saint Petersburg, Russia

A.V. Solov'yov (ed.), *Nanoscale Insights into Ion-Beam Cancer Therapy*,
DOI 10.1007/978-3-319-43030-0_1

responds to the peak in the linear energy transfer (LET) and we will refer to the latter as the Bragg peak. Due to this key feature, IBCT allows a delivery of high doses into tumours, maximising cancer cell destruction, and simultaneously minimising the radiation damage to surrounding healthy tissue. The effectiveness of radiation with ions depends on the choice of ions; it can be described by three factors: the peak value of LET, the proximal plateau value of LET, and the size of a tail distal to the peak. Since the LET is proportional to the square of charge of the projectile, ions heavier than protons are expected to be more effective; however, the increase of LET in the plateau region and the increasing size of the tail hinder the usage of heavier ions and, as a result, carbon ions, besides protons, are the most clinically used modality [3, 4]. Despite its high cost, proton-beam therapy is widely spread around the world with over 60 operational centres.[1] More proton centres are under construction. Although heavy ion therapy was adopted in the 1990s, there are only 10 clinical centres (in China, Germany, Italy, and Japan) where carbon ions are used [6].

The Bragg peak occurs because the inelastic cross sections of interactions of projectiles with the molecules of the medium increase up to the maximum values as the speed of the projectile decreases. As a result, the deposition of destructive energy to the tissue per unit length of the ion's path is maximised within 1 mm of the ion's trajectory. The location of the Bragg peak depends on the initial energy of the ions. Typical depths for carbon ions (in liquid water representing tissue) range from about 2.5 to 28 cm as the initial energy ranges from 100 to 430 MeV/u [4, 7–11]. Hence, a deeply-seated tumour can be scanned with a well-focused pencil beam of ions with minimal lateral scattering.

Over the past 20 years, technological and clinical advances of IBCT have developed more rapidly than the understanding of radiation damage with ions. Although an empirical approach has produced exciting results for thousands of patients thus far, many questions concerning the mechanisms involved in radiation damage with ions remain open and the fundamental quantitative scientific knowledge of the involved physical, chemical, and biological effects is, to a significant extent, missing. Indeed, the series of works that elucidated the importance of low-energy (below ionisation threshold) electrons appeared in ca. 2000, while the treatment of patients at GSI[2] started in 1997. The dominant molecular mechanism of a double strand break (DSB), the most important DNA lesion [12, 13], still remains unknown. Even the significance of the relation of DNA damage (including DSBs) compared to the damage of other cellular components to the cell inactivation or sterilisation is not entirely clear. This list can be continued. Besides IBCT, the mechanisms of biodamage due to irradiation with heavy ions have attracted attention in regards to radioprotection from galactic cosmic rays, especially during potential long-term space missions [4].

Over many decades of using radiation with photons, vast data relating the radiation damage to deposited dose were accumulated. These data are currently used to describe the biological damage due to ions [4]. Nonetheless, there are substantial qualitative and quantitative differences between the effects of ions and photons

[1] As of March 2016 [6].

[2] Gesellschaft für Schwerionenforschung, Darmstadt, Germany.

on tissue. The first difference is in the localisation of the dose distribution for ions distinguished from the mostly uniform dose distribution for photons. This feature reveals itself longitudinally (along the ion's path) as the Bragg peak. Radially (with respect to the ion's path) it shows up as the sharply decreasing (within several tens of nm) radial dose distribution, while the average distance between adjacent ions in clinically used beams are several hundreds of nm.

The second difference is a consequence of the first. Secondary particles such as electrons, free radicals, etc., produced as a result of the interaction (ionisation and excitation) of ions with the medium, emerge at the location of the Bragg peak in much larger number densities than those produced by photons, and their distribution is also non-uniform. These secondary particles largely cause the biological damage, and in order to assess the damage, it is important to distinguish the biological effects of the locally deposited dose and the local number density of secondary particles. In other words, the (radial) dose is not the only characteristic that determines the biological damage. For instance, clustered damage, deemed to be more lethal than isolated damage, can be caused by several low-energy electrons, which are not associated with a large dose deposition. This qualitatively and quantitatively changes the effect of the radiation [1, 4, 14].

There are also differences in the chemical interactions related to a different balance between free electrons, free radicals, and other agents for ions versus photons. These differences, for example, affect the resistivity of cells to radiation and thus are quite important for the assessment of radiation damage. Finally, the Bragg peak leads to thermomechanical effects, which stem from the non-uniformity of the radial dose deposition.

One of the most important questions in the foundation of science devoted to radiation damage with ions is the question about molecular mechanisms leading to DNA damage, or more generally, biodamage. While "whether the biodamage leads to cell sterilisation?" is a biological question, the question about the mechanisms of biomolecular damage belongs to the realms of physics and chemistry. The role of low energy (sub-15 eV) electrons has been especially emphasised in Refs. [15–18]. A number of quantum effects, such as dissociative electron attachment (DEA), formation of electronic and phononic polarons, are discussed in the context of the interaction of these electrons with biomolecules. DEA is deemed to be the leading mechanism for DNA single strand breaks (SSBs) at low energies, while a number of ideas, including the action of Auger electrons, in relation to the mechanism of double strand breaks (DSBs) has been suggested [17, 19]. The Auger effect along with intermolecular Coulombic decay (ICD) are discussed not only in relation to the mechanism of DSBs, but also as important channels for production of secondary electrons, especially in the presence of nanoparticles as sensitizers [20]. Still more understanding is needed for the interaction of electrons of higher energies.

This chapter is devoted to the overview of the main ideas of the multiscale approach to the physics of radiation damage that has been designed with the goal of developing knowledge about biodamage at the nanoscale and molecular level and finding the relation between the characteristics of incident particles and the resultant biological damage [1, 21]. This approach is unique in distinguishing essential phenomena relevant to radiation damage at a given time, space, or energy scale and assessing the resultant damage based on these effects. The significance of understanding the fundamental mechanisms of radiation damage in order to exploit this knowledge for practical applications has inspired the European COST Action [22], which supported collaborations of physicists, chemists, and biologists, studying these phenomena both theoretically and experimentally. Even though a number of questions remain unanswered, at this point the goal of the multiscale approach has been achieved, since the "big" picture of relation of physical effects to the biological consequences has been painted.

The multiscale approach (MSA) was formulated and then elaborated upon, as different aspects of the scenario were added in a series of works [10, 11, 14, 21, 23–32]. Its name emphasizes the fact that important interactions involved in the scenario happen on a variety of temporal, spatial, and energy scales. Temporal and spatial scales are schematically shown in Fig. 1 (adopted from Ref. [30]). Even though the name of the approach emphasises these variety of scales, most of the methods devoted to assessment of biological damage are multiscale as well [33–38]. What singles this approach is its inclusiveness with respect to effects that happen on all scales relevant to the radiation damage. From the very beginning, the approach was formulated as, phenomenon-based and was aimed at elucidating the physical, chemical, or biological effects that are important or dominating on each scale in time, space, and energy. The practical goal of the MSA is the calculation of survival probabilities for cells irradiated with ions. These survival probabilities are directly related to the relative biological effectiveness (RBE) [3, 4, 39, 40], one of the key integral characteristics of the effect of ions compared to that of photons. The RBE is defined as a ratio of doses of photons and different projectiles leading to the same biological effect, such as killing a given percentage of cells in an irradiated region. This is why the calculation of survival probabilities is so important. The path to the calculation of survival

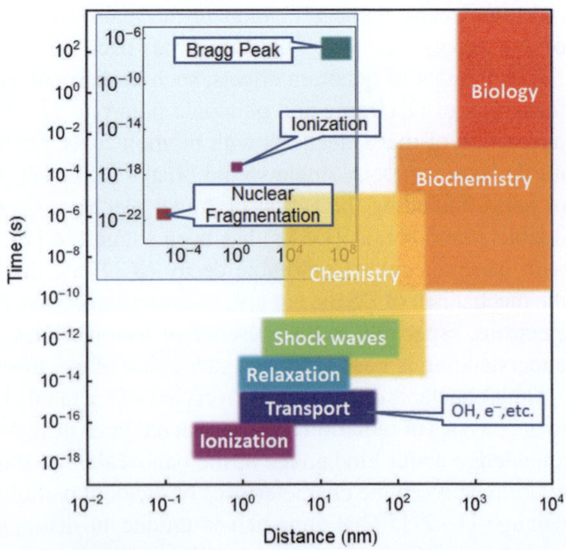

Fig. 1 Features, processes, and disciplines, associated with radiation therapy shown in a space-time diagram, which shows approximate temporal and spatial scales of the phenomena. The history from ionization/excitation to biological effects on the cellular level are shown in the main figure and features of ion propagation are shown in the *inset*

probabilities has been marked out in Ref. [30], but this is a recent Ref. [32], where the calculations were successfully compared with a series of experiments and the linear-quadratic model coefficients were calculated on the basis of the MSA. The oxygen enhancement ratio (OER), which compares the biological action of given projectiles to that at different aerobic or hypoxic conditions of irradiated targets, is calculated as a byproduct since the survival probabilities are calculated at different conditions.[3]

This chapter is organised in the following way. Section 2 briefly discusses the existing approaches to the calculation of RBE and explains the lack of satisfaction with a style and tools used in these calculations that inspired the authors to think of a different approach. In Sect. 3, the scenario of radiation damage with ions is described. Section 4 is devoted to the ion's transport in the medium. The secondary electron production, discussed therein as well brings the reader to Sect. 5, where the transport of secondary electrons and other reactive species is analysed and applied to targeting biomolecules. Section 6 explores thermomechanical effects, which play an important role in the MSA. Section 7 is devoted to the evaluation of the probability for an irradiated cell to survive based on the calculation of clustered DNA damage. First, the probability of clustered lesions is calculated. Then a criterion for DNA lesion lethality is introduced. Then the survival curves are calculated. This is where the whole MSA is synthesised into predictive quantitative scheme, which is later formulated as a recipe for the assessment of radiation damage with ions. This is followed with conclusions. The paper by and large discusses carbon ions as projectiles. However, there are several exceptions. In Sect. 4.2.2 protons are discussed as projectiles, since the calculations for carbon ions have not yet been accomplished. In Sect. 6, a wide variety of ions is discussed since LET at the Bragg peak is used as a parameter. Finally, in Sects. 7.1 and 7.2, the examples of calculations are discussed for protons and α-particles in order to compare these calculations with experiments.

2 Other Models for the Assessment of Radiation Damage with Ions

The expertise for the assessment of radiation damage historically comes from the times when x-rays (photons) were used as the only projectiles. For x-rays, the dose distribution is practically uniform. Therefore, it is not accidental that the dose has been chosen as the main parameter for prediction of radiation damage. Treatment plans have to deliver certain doses to certain locations in order to achieve the desired results.[4]

A survival curve is the dependence of the probability of cell survival on the absorbed dose of radiation. On one side, it relates the goal with the means, i.e., it

[3]More detail on the OER calculations can be found in the Chapter by A.V. Verkhovtsev, E. Surdutovich and A.V. Solov'yov

[4]The optimisation related to reducing dose deposition in healthy regions and treatment partitioning is left aside.

predicts the dose that is necessary in order to achieve cell deactivation with a desired probability. On the other side, it allows comparing different modalities (photons, protons, heavier ions, etc.) and thus allows one to optimize the choice of therapy. This comparison is achieved via the calculation of the ratio of doses of different projectiles necessary to achieve the same probability of cell survival. The ratio of the dose due to photons to that for other projectiles is called the relative biological effectiveness (RBE).

A vast majority, if not all, of other existing models that "calculate" the survival curves are based on an empiric formula,

$$- \ln \Pi_s = \alpha d + \beta d^2, \tag{1}$$

where d is the dose and α and β are coefficients. Several features of these curves have been discussed, one of which is the ratio of α/β. If this ratio is large, the survival curve is "steep" and more like a straight line; if it is small then it is a "shouldered" curve. A series of models suggested since 1955 [33, 36, 39, 41–44] provided a phenomenological explanation to this dependence and developed approaches to the calculation of RBE. The coefficients α and β depend on the kind of cells, on the cell cycle, on the access of oxygen to the irradiated cells and other factors.

For many practical purposes, an experimentally obtained curve, given by Eq. (1) is *sufficient* information for the evaluation of radiation damage, and it has been used for many years for treatment planning and optimization. Atomic or molecular interactions are not mentioned in those models; these and more information are hidden in the purely empirical coefficients α and β. Since the dose distribution is uniform, it is possible to solve all practical problems without atomic/molecular physics, since it brings up too many difficult questions involving interactions with biomolecules that seem irrelevant as compared to the biological unknowns related to repair mechanisms.

Particle projectiles change this picture. As was shown above, the dose distribution around each particle's path is highly nonuniform. The track structure and the consequent damage are much more complicated. A solution to this problem was suggested by the Katz approach in which the radial dose distribution is calculated and related to the inactivation of sub-cell-nucleus targets [45–48]. The quality factor of radiation was introduced in order to relate the survival curve parameters to a given type of radiation, differentiating between track types, inactivation modes, the structural complexity of targets, etc. The eventual goal of the Katz model was to calculate the RBE. Nevertheless, the biological relation of the radial dose distribution with the cell survival probability was done based on the survival curves for x-rays, without analyzing particular physical processes. The empiric survival curves provide the information such as characteristic γ-ray dose or coefficients α and β.

The Local Effect Model (LEM), developed at GSI, calculates the RBE assuming that the biological effect of radiation is entirely determined by the spatial distribution of the radial dose inside the cell nucleus. It relates the response of biological systems, following ion irradiation, to the corresponding response after x-ray irradiation [4, 49]. Corrections for the quality of damage was included in a later version of the LEM [4].

This model operates on the schematic level using Eq. (1) with empirical coefficients α and β. The LEM solves technical problems related to the optimization of treatments, leaving no place for ab initio approaches and physical, chemical, or biological effects in general; even a consideration of DNA lesions such as DSBs were only included in the LEM recently [50].

Most known approaches to the calculation of RBE in a sense start from the survival curve described using coefficients α, β, or other parameters such as characteristic dose. Then they point at possible biological effects that can explain the described features of the survival curves. Many of these approaches, such as microdosimetric-kinetic model (MKM) [41, 42] or molecular models [39] are more mathematical (the physics is defined by the words "dose" or "LET").

The MSA takes the opposite direction. It starts from physics and chemistry and moves in the direction of biological effects. The scenario of biodamage starting from ion entering tissue is the basis for an analytic synthesis of microscopic effects that comprise the macroscopic coefficients of the linear-quadratic model describing survival probabilities. These calculations, discussed in Sect. 7.2, are now verified experimentally [32] and demonstrate the fulfillment of the goal of relating the physical parameters to biological outcomes and completion of a microscopic theory.

3 Multiscale Scenario of Radiation Damage

Radiation damage due to ionizing radiation is initiated by the ions incident on tissue. Initially, they have energy ranging from a few to hundreds of MeV. In the process of propagation through tissue they lose their energy in the processes of ionization, excitation, nuclear fragmentation, etc. Most of the energy loss of the ion is transferred to tissue.[5] Naturally, radiation damage is associated with this transferred energy, and the dose (i.e., deposited energy density) is a common indicator for the assessment of the damage [1, 4, 39]. The profile of the LET along the ion's path is characterised with a plateau followed by a sharp Bragg peak. The position of this peak depends on the initial energy of the ion and marks the location of the maximum radiation damage. In the process of radiation therapy, a tumour is being "scanned" with the Bragg peak both laterally and longitudinally. The active longitudinal scanning is achieved by changing of initial energy of projectiles, while passive is achieved by using scatterers of variable thickness in front of the target[6] in order to deposit a large dose to the target and spare healthy tissues surrounding it.

However, the deposition of large doses in the vicinity of the Bragg peak does not explain how the radiation damage occurs, since projectiles themselves only interact with a few biomolecules along their trajectory and this direct damage is only a small fraction of the overall damage. It is commonly understood that the secondary

[5]The only part that is not transferred is emitted as radiation. This part, in the case of ions interacting with tissue, is deemed to be insignificant.

[6]The longitudinal scanning produces the so-called spread-out Bragg peak (SOBP).

electrons and free radicals produced in the processes of ionization and excitation of the medium with ions are largely responsible for the vast portion of the biodamage.

Secondary electrons are produced during a rather short time of 10^{-18}–10^{-17} s following the ion's passage. The energy spectrum of these electrons has been extensively discussed in the literature [10, 11, 51–53] and the main result (relevant for this discussion) is that most secondary electrons have energy below 50 eV (more than 80 % for 0.3 MeV/u ions[7]) and only a few (less than 10 % for 0.3 MeV/u-ions) have energy higher than 100 eV. Moreover, this is true for a very large range of ion energy. This has several important consequences. First, the ranges of propagation of these electrons in tissue are rather small, around 10 nm [54]. Second, the angular distribution of their velocities as they are ejected from their original host, and as they scatter further, is largely uniform [35]; this allows one to consider their transport using a random walk approach [14, 19, 21, 55, 56].

The next time scale 10^{-16}–10^{-15} s corresponds to the propagation of secondary electrons in tissue. These electrons (which start with about 45–50 eV energy) are called ballistic. In liquid water, the mean free paths of elastically scattered and ionizing 50-eV electrons are about 0.43 and 3.5 nm, respectively [35]. This means that they ionize a molecule after about seven elastic collisions, while the probability of second ionization is small [10]. Thus, the secondary electrons are losing most of their energy within first 20 collisions and this happens within 1–1.5 nm of the ion's path [28]. After that they continue propagating, elastically scattering with the molecules of the medium until they get bound or solvated electrons are formed. It is important to notice that these low energy electrons remain important agents for biodamage since they can attach to biomolecules like DNA causing dissociation [18, 57]. The solvated electrons may play an important role in the damage scenario as well [13, 58, 59].

Additionally, the energy lost by electrons during the previous stage in the processes of ionization, excitation and electron-phonon interaction is transferred to the medium. As a result of this relaxation, the medium within about a 1–1.5-nm cylinder (for ions not heavier than iron) around the ion's path becomes very hot [25, 28]. This cylinder is referred to as the hot cylinder. The pressure inside this cylinder increases by a factor of about 10^3 compared to the pressure in the medium outside the cylinder. This pressure builds up by about 10^{-14}–10^{-13} s and it is a source of a cylindrical shock wave [27]. This shock wave propagates through the medium for about 10^{-13}–10^{-11} s. Its relevance to the biodamage is as follows. If the shock wave is strong enough (the strength depends on the distance from the ion's path and the LET), it may inflict damage directly by breaking covalent bonds in a DNA molecule [28]. Besides, the radial collective motion that takes place during this time is instrumental in propagating the highly reactive species such as hydroxyl radicals, just formed solvated electrons, etc. to a larger radial distance (up to tens of nm) thus increasing the area of an ion's impact.

The assessment of the primary damage to DNA molecules and other parts of cells due to the above effects is done within the MSA. This damage happens within

[7] This value corresponds to the kinetic energy of ions near the Bragg peak.

10^{-5} s from the ion's passage and consists of various lesions on DNA and other biomolecules. Some of these lesions may be repaired by the living system, but some may not and the latter may lead to cell sterilisation. The scenario described above is illustrated in Fig. 2 (adopted from Refs. [30, 60]).

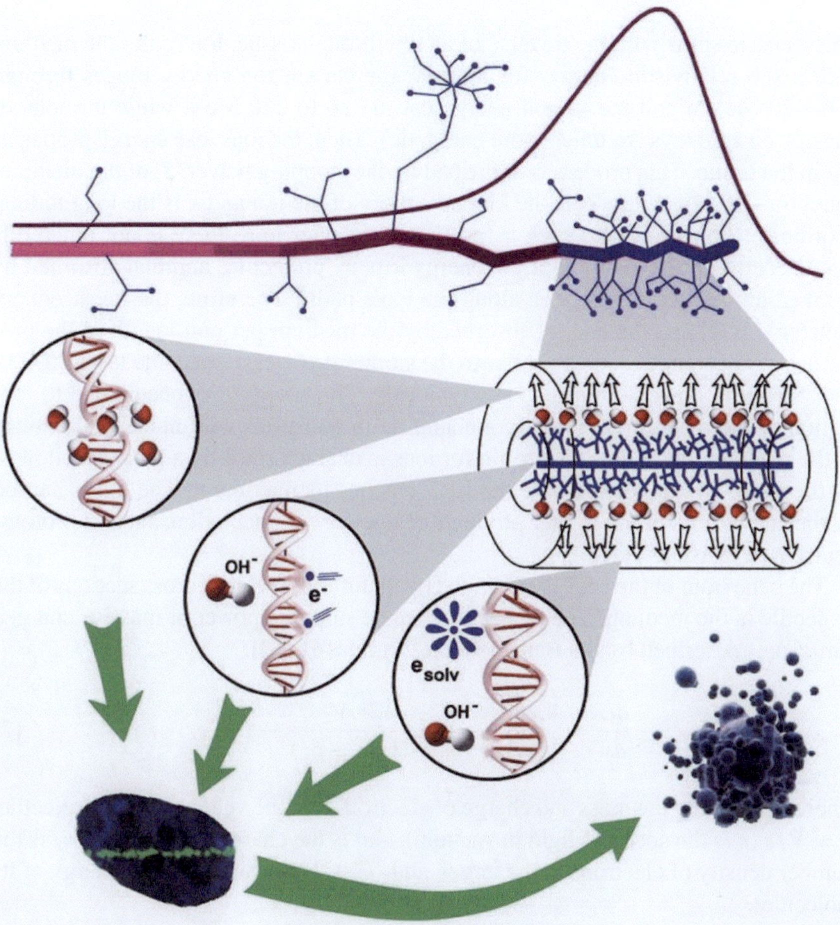

Fig. 2 The scenario of biological damage with ions. Ion propagation ends with a Bragg peak, shown in the *top right corner*. A segment of the track at the Bragg peak is shown in more detail. Secondary electrons and radicals propagate away from the ion's path damaging biomolecules (*central circle*). They transfer the energy to the medium within the *hot cylinder*. This results in the rapid temperature and pressure increase inside this cylinder. The shock wave (shown in the expanding *cylinder*) due to this pressure increase may damage biomolecules by stress (*left circle*), but it also effectively propagates reactive species, such as radicals and solvated electrons to larger distances (*right circle*). A living cell responds to all shown DNA damage by creating foci (visible in the *stained cells*), in which enzymes attempt to repair the induced lesions. If these efforts are unsuccessful, the cell dies; an apoptotic cell is shown in the *lower right corner*

4 Propagation of Ions in Tissue and Primary Ionization of the Medium

4.1 The Main Characteristics of Ion's Propagation in the Medium

The scenario starts with the traverse of an ion through tissue. Ions enter the medium with a sub-relativistic energy (for therapy, the carbon ion energy ranges through 100–420 MeV/u and the proton energy can be up to 250 MeV, while the ions of galactic cosmic rays are much more energetic). Then, the ions lose energy propagating in the tissue. This process is described by the stopping power, S, of the medium, equal to $-dE/dx$, where E is the kinetic energy of the ion and x is the longitudinal coordinate. For projectiles such as protons or heavier ions, there is not much difference between the location of the energy loss by projectiles and that absorbed by the medium *longitudinally*, i.e., along the ion's path.[8] Therefore, the linear energy transfer (LET), i.e., the energy absorbed by the medium per unit length of the projectiles's trajectory becomes similar to the stopping power. Hence, the terms "LET" and "stopping power" are used synonymously. The energy loss occurs due to ionization of the medium, nuclear fragmentation in collisions with nuclei, excitations of the medium, etc. The LET profile for ions is characterized by a plateau followed by the sharp Bragg peak, where the LET reaches its maximum. The tail is caused by the energy loss of the lighter products of nuclear fragmentation, such as protons, neutrons, α-particles, etc.

The behaviour of the LET is explained by features of inelastic cross sections of the projectile in the medium. The Bragg peak in the stopping power of massive charged particles is described by the Bethe–Bloch formula [61–63].

$$-\frac{dE}{dx} = \frac{4\pi n_e z^2 e^4}{mV^2}\left[\ln\frac{2mV^2}{\langle I\rangle(1-\beta^2)} - \beta^2\right], \tag{2}$$

where m and e are the mass and charge of electron, V is the velocity of the projectile, $\beta = V/c$ (c is the speed of light in vacuum), $z|e|$ is the charge of projectile, n_e is the number density of electrons in the target, and $\langle I\rangle$ is the mean excitation energy of its molecules.

This formula provides the dependence of the stopping power on the energy of the ion and practically depends on a single parameter, the mean excitation energy. This parameter for liquid water is chosen empirically somewhere between 70 and 80 eV [9, 64]. The use of such a non-physical parameter is sufficient for the calculations of the position of the Bragg peak and its shape, and Eq. (2) is used in many Monte Carlo (MC) simulations [9] for that purpose. This parameter, however, hides all physical processes such as ionization and excitation of the medium, even though

[8]This is so because the energy is mostly transferred to electrons and other secondary particles, whose longitudinal ranges are many times smaller than the characteristic scale of x.

these same processes are important for the understanding of the scenario of radiation damage. Therefore, a different approach has been used. In Refs. [10, 11, 21], the singly-differentiated (with respect to the secondary electron energy) ionization cross sections of water molecules in the medium have been employed as a physical input. This uncovers the physics integrated in the empirical parameter and allows not only describing the features of the Bragg peak, but also obtaining the energy spectrum of secondary electrons, which are very much involved in subsequent radiation damage.

4.2 *Singly-Differentiated Cross Sections of Ionization*

The total ionization cross section, σ_t, differentiated with respect to secondary electron kinetic energy, W, i.e., singly-differentiated cross section (SDCS) is the main quantity in our analysis. Besides the kinetic energy of secondary electrons and the properties of water molecules, the SDCS depends on the velocity V of the projectile and its charge, $z|e|$.

4.2.1 Calculation of the SDCS Using a Parametric Semiempirical Approach

In Refs. [10, 11, 65], the semi-empirical Rudd's expression [66] for the calculation of SDCS has been used. This analytic expression, containing a number of parameters, is a combination of the experimental data and calculations within the plane-wave Born approximation and other theoretical models [66]. Since this model was developed for non-relativistic protons, it had to be modified to include heavier ions at relativistic velocities. The original SDCS is given in the following form [66]:

$$\frac{d\sigma_t}{dW} = z^2 \sum_i \frac{4\pi a_0 N_i}{I_i} \left(\frac{I_0}{I_i}\right)^2 \times \frac{F_1(v_i) + F_2(v_i)\omega_i}{(1+\omega_i)^3 \left(1 + \exp(\alpha(\omega_i - \omega_i^{\max})/v_i)\right)}, \tag{3}$$

where the sum is taken over the electron shells of the water molecule, $a_0 = 0.0529$ nm is the Bohr radius, $I_0 = 13.6$ eV, N_i is the shell occupancy, I_i is the ionization potential of the shell, $\omega_i = W/I_i$ is the dimensionless normalised kinetic energy of the ejected electron, v_i is the dimensionless normalised projectile velocity given by

$$v_i = \sqrt{\frac{mV^2}{2I_i}}. \tag{4}$$

When $V \ll c$, $V = \sqrt{\frac{2E}{M}}$ (where M is the mass of a projectile), and, hence $v_i = \sqrt{\frac{m}{M}\frac{E}{I_i}}$. When V approaches c, the definition of v_i, given by (4), holds, however, the projectile's velocity V is given by βc, where $\beta^2 = 1 - 1/\gamma^2 = 1 - (Mc^2/(Mc^2 + E))^2$, and γ is the Lorentz factor of the projectile.

Functions F_1 and F_2 in (3) are given by

$$F_1(v) = A_1 \frac{\ln(1+v^2)}{B_1/v^2 + v^2} + \frac{C_1 v^{D_1}}{1 + E_1 v^{D_1+4}}, \tag{5}$$

and

$$F_2(v) = C_2 v^{D_2} \frac{A_2 v^2 + B_2}{C_2 v^{D_2+4} + A_2 v^2 + B_2}. \tag{6}$$

The fitting parameters $A_1 \ldots E_1, A_2 \ldots D_2$, and α depend on the medium. In Ref. [66], they are given for water vapour. The comparison of positions of Bragg peaks for different initial carbon ion energies with those measured in experiments provided sufficient material for refitting of these parameters for liquid water medium [11]. These parameters are listed in Table 1 [11]. The cut-off energy $\omega^{\max}$ is given by

$$\omega_i^{\max} = 4v_i^2 - 2v_i - \frac{I_0}{4I_i}, \tag{7}$$

where the first term on the right-hand side represents the free-electron limit, the second term represents a correction due to electron binding, and the third term gives the correct dependence of the SDCS for $v_i \ll 1$ [66]. For $v_i \gg 1$, Eq. (3) should asymptotically approach the relativistic Bethe–Bloch formula (2). This is accomplished when F_1, given by (5), is replaced by the following expression,

$$F_1(v) = A_1 \frac{\ln(\frac{1+v^2}{1-\beta^2}) - \beta^2}{B_1/v^2 + v^2} + \frac{C_1 v^{D_1}}{1 + E_1 v^{D_1+4}}. \tag{8}$$

Indeed, the asymptotic behaviour of (8) at $v \gg 1$ is given by

Table 1 Fitting parameters and ionization energies for three outer and two inner shells (symbols $1a_1$, $2a_1$, $1b_1$, $3a_1$, and $1b_2$ represent corresponding molecular orbitals), of water molecules in a liquid water environment [11]

Shells	Ionization energies (eV)	A_1	B_1	C_1	D_1	E_1	A_2	B_2	C_2	D_2	α
Outer: $1b_1$, $3a_1$, $1b_2$	10.79, 13.39, 16.05	1.02	82	0.5	−0.78	0.38	1.07	14.5	0.61	0.04	0.64
Inner: $2a_1$, $1a_1$	32.3, 539.0	1.25	0.5	1.0	1.0	3.0	1.1	1.3	1.0	0.0	0.66

$$\frac{A_1}{v^2}\left[\ln\left(\frac{v^2}{1-\beta^2}\right)-\beta^2\right], \tag{9}$$

which, after being substituted to Eq. (3) and the understanding that $\frac{dE}{dx} \sim \sum_i \int (W + I_i)\frac{d\sigma_i}{dW} dW$, leads to Eq. (2). The correction of Eq. (8) reveals itself as an increase of the cross section at high energies.

4.2.2 Calculations of SDCS Based on the Energy-Loss Function

An alternative method has been used in Ref. [29], where the dielectric formalism based on the experimental measurements of the energy-loss function (ELF) of the target medium, Im $(-1/\varepsilon(\mathscr{E}, q))$, where $\varepsilon(\mathscr{E}, q)$ is the complex dielectric function, and $\hbar q$ and $\mathscr{E}$ are the momentum and energy transferred in the electronic excitation, respectively [67, 68]. This formalism allows obtaining the SDCS not only for liquid water but for a real biological medium containing sugars amino acids, etc. If the ELF is experimentally known, many-body interactions and target physical state effects are naturally included in these calculations.

According to that formalism, the macroscopic (nonrelativistic) SDCS for ionization of the electronic shell i is given by,

$$\frac{d\sigma_i(W,\mathscr{E})}{dW} = \frac{e^2}{n\pi\hbar^2}\frac{Mz^2}{E}\int_{q_-}^{q^+}\frac{dq}{q}\mathrm{Im}\left[\frac{-1}{\varepsilon(q, I_i + W)}\right]_i, \tag{10}$$

where $q_\pm = \sqrt{2M}(\sqrt{E} \pm \sqrt{E - \mathscr{E}})$. Equation (10) can be used for different charged projectiles by properly taking into account their charge state, or for electrons by introducing an exchange term in the integrand and imposing the correct integration limits.

Since Eq. (10) requires the contribution of each electronic shell of the target to its ELF, and the latter is usually measured for all the excitations and ionizations of the electronic system in the optical limit ($q = 0$), the algorithm for obtaining the data at $q > 0$ and splitting this ELF into different electronic shells is needed in addition to the experimentally measured ELF.

The optical ELF for bioorganic condensed compounds and liquid water are rather similar and can be parameterized with a single-Drude function [69]

$$\mathrm{Im}\left[\frac{-1}{\varepsilon(q=0,\mathscr{E})}\right] = \frac{a(Z)\mathscr{E}}{\left(\mathscr{E}^2 - b(Z)^2\right)^2 + c(Z)^2\mathscr{E}^2}, \tag{11}$$

where $a(Z)$, $b(Z)$, and $c(Z)$ are the functions of the mean atomic number of the target Z, corresponding to the height, position, and width [69]. While $b(Z)$ and $c(Z)$ are parametric functions, $a(Z)$ is obtained by imposing the f-sum rule [70], linked to the number of electrons in the target, Z, also accounting for the contribution from the inner shells, as explained in Ref. [69]. Using this approach the ELF of an arbitrary bioorganic compound can be estimated, even in the case where no experimental data exist. A wide variety of extension algorithms for extrapolation of optical-ELF to $q > 0$ are available due to extensive research [71]. In Ref. [29] a simple quadratic dispersion relation introduced by Ritchie and Howie [72], with its parameters for liquid water [71], has been used.

The issue of splitting of the ELF into contributions from different shells has been studied for liquid water [73, 74], providing parameterizations of the ELF split in ionization and excitation arising from each different shell. In Ref. [29] a specially designed approximation has been applied to split the ELF for biomolecules. In order to describe the outer-shell ionization of biomolecules, the mean value of their binding energies, $\bar{I}$, is calculated.[9] It is then assumed that the outer-shell electrons will be ionized if the transferred energy satisfies $\mathscr{E} > \bar{I}$. Then, the ejected electron energy is $W = \mathscr{E} - \bar{I}$. In Ref. [29], SDCSs are calculated and compared with other calculations and experiments for protons interacting with water, adenine, and benzene.

The total ionization cross sections (TICS) can also be estimated for different biomolecules relevant for IBCT [29]. For example, in Fig. 3a [29] the macroscopic TICSs are calculated for proton impact in five representative biological materials relevant for cancer therapy: liquid water, dry DNA ($C_{20}H_{27}N_7O_{13}P_2$), protein, lipid,

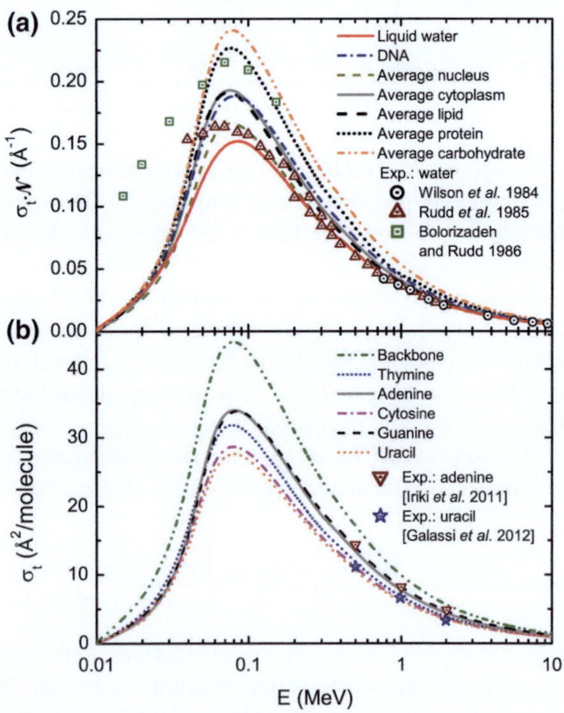

Fig. 3 **a** Calculated macroscopic TICS for proton impact in liquid water, DNA, protein, lipid, and cell nucleus. **b** Calculated microscopic TICS for proton impact in the DNA components adenine, cytosine, guanine, thymine, and sugar-phosphate backbone. Symbols represent experimental data [29]. $\mathscr{N}$ is the molecular density of the target

[9]The relevant data are available for some biological molecules, such as the DNA bases and the sugar-phosphate backbone [75] and some amino acids [76] and others [77].

and the cell nucleus. Their atomic compositions and densities can be found in the ICRU Report 46 [78] and other sources, and a reasonable value of their mean binding energies can be estimated from the values of their molecular components, such as the water molecule, DNA bases and backbone, and amino acids [73, 75, 76]. The experimental data for water vapour [79–81] are also shown. They agree well with the calculations of Ref. [29] above 100 keV, where the first Born approximation is applicable without further corrections. From these results, it is plausible that all the biological targets different from water have a larger ionization probability than water. One can also see that the TICS of a cell nucleus is only slightly larger than that of liquid water, and that protein has a slightly larger TICS than the rest of the biomaterials.

In Fig. 3b [29, 30] the microscopic TICS per molecule for proton impact in the DNA molecular components, such as adenine, cytosine, guanine, thymine, and sugar-phosphate backbone are shown. Their atomic composition can be easily found in the literature, and their mean binding energies were estimated from quantum chemistry calculations [75, 76]. Also shown are experimental data at high energies for adenine [82], which are in excellent agreement with the predictions of Ref. [29]. This method allows one to estimate the ionization probability of each constituent of the DNA molecule, which gives important information on the sensitivity of each one to radiation damage. According to these results, the DNA backbone is the most probable part of the DNA to be ionized by proton impact (a similar behaviour was previously observed for electron impact in Ref. [75]; also, recent theoretical estimates [83] point towards sugar-phosphate C–O bond cleavage due to interaction with low energy electrons) and, between bases, adenine and guanine are the most sensitive to proton impact ionization. This fact could have important implications in the DNA damage, since it seems that single or double strand breaks could be more probable than base damage, or that regions of the DNA with a bigger concentration of adenine or guanine would be more likely damaged by radiation than other parts of the genome, attending to direct ionization effects.

Much more information can be obtained with this method, such as the number of emitted electrons, the average energy of electrons, SDCS and TICS for other biological targets and projectiles. This model, using little input information and physically motivated approximations, can provide useful information about the ion impact ionization of a huge number of relevant biological targets, for which data are lacking, both experimentally and theoretically. This model can be easily extended to ions heavier than protons, in different charge states, as well as to electron impact ionization, by introducing appropriate corrections, such as the description of the electronic structure of the ion, or exchange and relativistic corrections for electrons.

4.3 The Position of the Bragg Peak

The stopping cross section, defined as

$$\sigma_{\rm st} = \sum_i \int_0^\infty (W + I_i) \frac{d\sigma_{t,i}}{dW} dW, \tag{12}$$

where the sum is taken over all electrons of the target, gives the average energy lost by a projectile in a single collision, which can be further translated into energy loss within an ion's trajectory segment, dx:

$$S(E) = -\frac{dE}{dx} = n\sigma_{st}(E), \tag{13}$$

where n is the number density of molecules of the medium. This quantity is known as the stopping power [39, 64]. As was discussed above in Sect. 4.1, for ions this quantity is similar to the linear energy transfer (LET).

The LET found from Eq. (13) is a function of the kinetic energy of the ion rather than the ion's position along the path in the medium. The dependence of LET (and other quantities) on this position, however, is more suitable for cancer therapy applications. Integrating inverse LET, given by (13), yields[10]

$$x(E) = \int_E^{E_0} \frac{dE'}{|dE'/dx|}, \tag{14}$$

where E_0 is the initial energy of the projectile. We obtain the correspondence between the position of the ion along the path and its energy. This allows one to obtain all quantities of interest in terms of x rather than E. The depth dependence of the average LET (stopping power S) as a function of x is shown in Fig. 4 [30]. The calculations of the LET include the effects that were discussed above, such as SDCS calculated using semi-empirical parametrisation (3), modified for relativistic energies (8) with the use of the effective charge described below in Sect. 4.4. The effect of energy straggling due to multiple ion scattering, described in the Sect. 4.5 is also taken into account. This effect explains why the height of the Bragg peak decreases with the increasing initial energy of ions and thus increasing depths of the corresponding Bragg peaks. The contribution of non-ionization processes, such as excitation of neutral molecules, are also included in these calculations. In order to accomplish this, the excitation cross sections for proton projectiles [85] were scaled using the ratio of the effective charges for carbon and proton at a given energy E.

In Fig. 4, our calculated LET is compared with the experimental results [7]. As can be seen from the figure, the experimental dots at the Bragg peak are systematically lower than the calculated curve, the difference being due to in the nuclear fragmentation component, which has not been included in the analytical calculations. It is

[10]This is known as the continuous slowing down approximation (CSDA) range [84].

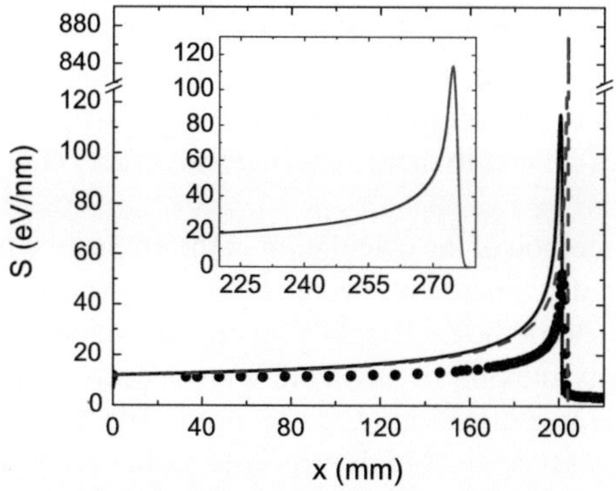

Fig. 4 The dependence of the LET on depth with the Bragg peak, plateau, and tail for carbon ions in liquid water. The calculations (*solid line*) are done for ions with the initial energy of 330 MeV/u and with use of Eqs. (12) and (13). Experimental results [7] for the same energy are shown with *dots*. The *dashed line* depicts the LET dependence without the effect of energy straggling. In the *inset*, two almost *coinciding curves* show the agreement between the analytical calculations and MC simulations [9] for 420 MeV/u carbon ion projectiles with straggling being included

feasible to include it, as has been done in Ref. [86] for protons, if the appropriate fragmentation cross sections are known.

As confirmed by MCHIT MC simulations [9], nuclear fragmentation reactions become important for heavy-nuclei beams and deeply-located tumours. For example, both experimental data [7] and MCHIT calculations [9] indicate that more than 40 % of primary 200 MeV/u $^{12}C^{6+}$ nuclei undergo fragmentation before they reach the Bragg peak position, and this fraction exceeds 70 % for a 400 MeV/u $^{12}C^{6+}$ beam. As a result of nuclear reactions the beam is attenuated. New projectiles such as protons, neutrons, and α-particles are formed. Since these particles are lighter than the incident ions, after fragmentation they carry a larger portion of the energy and their penetration depths are larger than that of the original ions [7]. This results in a tail after the Bragg peak also seen in Fig. 4.

4.4 Charge Transfer Effect

The incident ions are usually stripped of all electrons, but as they slow down they pick electrons off and their charge reduces. The dependence of the charge of ions on their velocity has been suggested by Barkas [87], where the following empirical formula for the effective charge, z_{eff}, is introduced,

$$z_{eff} = z(1 - \exp(-125\beta z^{-2/3})), \tag{15}$$

where z is the charge of the stripped ion. This formula is a result of studies of energy loss of ions in emulsions. More detailed descriptions of charge transfer effects have become available recently [88]. These studies allow one to not only estimate the effective charge of the ion, but also find its fluctuations. These fluctuations are important since LET increases proportionally as z^2 and if LET becomes large enough, qualitative differences related to thermomechanical effects may become substantial (see Sect. 6 below).

Regardless of the method of the calculation of the effective charge, in order to find the stopping power and estimate the secondary electron spectra (in the first approximation) z in Eq. (3) should be replaced by an effective charge z_{eff} which decreases with decreasing energy making the ionization cross section effectively smaller. In Refs. [10, 11] the parameterization (15) was used. The effective charge given by this expression slowly changes at high projectile velocity, but rapidly decreases in the vicinity of the Bragg peak. As a result, charge transfer significantly affects the height of the Bragg peak, and only slightly shifts its position towards the projectile's entrance. This happens because the stopping cross section as a function of velocity has a sharp peak as velocity decreases. At the same time σ_{st} is proportional to z_{eff}^2. If the latter decreases with decreasing V, the Bragg peak shifts towards the direction of the beam's entrance to the tissue. For instance, with the account for charge transfer, for carbon ions the Bragg peak occurs at $E = 0.3$ MeV/u rather than at $E = 0.1$ MeV/u.

4.5 The Effect of Ion Scattering

It will become clear below, in Sects. 5.6.3 and 7.2, that tracks of ions emerging from clinically used accelerators do not interfere, i.e., the effects of a single ion do not spread far enough to reach the area affected by adjacent ions. Therefore, it is usually sufficient to study a single ion interacting with tissue and then combine these effects relating the action of the beam with the dose. Even though the Bragg peak is a feature of every ion's LET, each peak cannot be observed separately. Since each of the projectiles in the beam experiences its own multiple scattering sequence, peaks for different ions occur at a slightly different spatial location and only the Bragg peak, averaged over the whole beam, is observed experimentally. Therefore, in order to compare the shape of the Bragg peak with experiments, the whole ion beam should be considered.

In Ref. [10], the Bragg peak for an ion beam was obtained via introduction of the energy-loss straggling due to ion scattering. The energy straggling, described by a semi-analytical model [89], is given by

$$\left\langle \frac{dE}{dx}(x) \right\rangle = \frac{1}{\lambda_{str}\sqrt{2\pi}} \int_0^{x_0} \frac{dE}{dx}(x') \exp\left[-\frac{(x'-x)^2}{2\lambda_{str}^2}\right] dx', \quad (16)$$

where x_0 is a maximum penetration depth of the projectile and $\lambda_{str} = 0.8$ mm is the longitudinal-straggling standard deviation computed by Hollmark et al. [90] for a carbon ion of that range of energy. The Bragg peak shown in Fig. 4 was calculated using Eq. (16).

4.6 Energy Spectra of Secondary Electrons

The most important effect that takes place during the propagation of the ion in tissue is the ionization of the medium. This is how, when, and where the secondary electrons, the key player in the scenario of radiation damage, are produced. The information, required for the understanding of phenomena related to secondary electrons, is the number of electrons produced per unit length of the ion's trajectory and their energy distribution. This section is devoted to the analysis of the electron energy distributions obtained from ionization cross sections discussed above.

The emission of electrons in collisions of protons with atoms and molecules has been under theoretical and experimental investigation for decades [53, 66, 91, 92]. The quantity of interest is the probability to produce N_e secondary electrons with kinetic energy W, in the interval dW, emitted from a segment Δx of the trajectory of a single ion at the depth x corresponding to the kinetic energy of the ion, E. This quantity is proportional to the singly-differentiated cross ionization section (SDCS),[11] discussed in Sect. 4.2.

$$\frac{dN_e(W,E)}{dW} = n\Delta x \frac{d\sigma_t}{dW}. \tag{17}$$

where n is the number density of molecules of the medium (for water at standard conditions $n \approx 3.3 \times 10^{22}\,\text{cm}^{-3}$). Equation (17) relates the energy spectrum of secondary electrons to the SDCS regardless of the method, by which the latter are obtained.

One important characteristics that can be obtained from the SDCS is the average energy of the secondary electrons, $\langle W \rangle$, which is given by

$$\langle W \rangle (E) = \frac{1}{\sigma_t} \int_0^{\infty} W \frac{d\sigma_t}{dW} dW. \tag{18}$$

The dependence of the average energy of electrons on the energy of the projectile, given by the result of integration (18) for liquid water medium is shown in Fig. 5 [30]. Notice, that this figure is different from similar figures of Refs. [10, 21], where the calculations were done with parameters for water vapour. This dependence indicates that the energy of secondary electrons is somewhere below 50 eV for the whole range of the ion's energy and it levels out as the energy of projectiles increases. There are

[11] The SDCS are integrated over full solid angle of electron emission.

Fig. 5 Average energy of secondary electrons produced as the result of impact ionization as a function of kinetic energy (per nucleon) of $^{12}C^{6+}$ ions

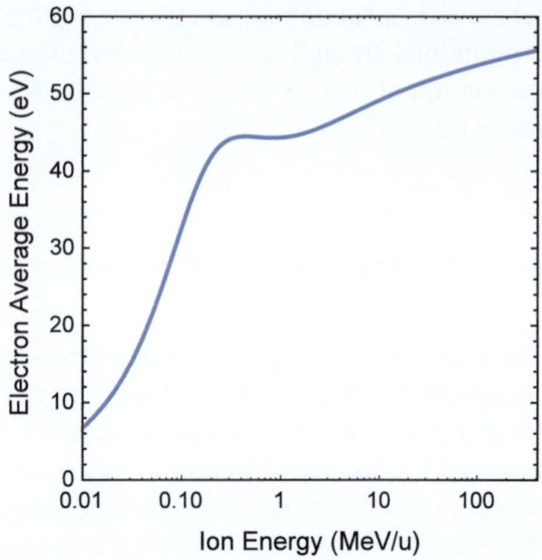

several consequences from this. First, since the dependence of $\langle W \rangle$ on the ion's energy E on a relevant range of projectile energies (0.3–400 MeV/u) is weak for the large range of the ion's energy, the number of produced secondary electrons is largely proportional to the value of LET, more precisely to the electronic component of the LET, S_e, that excludes nuclear stopping. Indeed, if the ion is destroyed in a nuclear collision, ionization due to its debris should be discussed instead; if it survives then its ionizing capabilities do not change too much, unless it slows down considerably; then, its stopping power may change correspondingly. Second, the expression for $\langle W \rangle$ is independent of the charge of the projectile, e.g., the difference between, say, protons and iron ions is in their values of S_e, i.e., in the number of secondary electrons, but not in their relative energy spectra. Therefore, the difference between the effects of these different ions will be in the number of secondary electrons produced by these ions per unit length of path. Third, most of the secondary electrons are capable of ionizing just one or two water molecules; thus, there is no significant avalanche ionization effect [10]. This can be explained by a simple estimate. Since the average energy of secondary electrons in the vicinity of the Bragg peak is about 40 eV (somewhat below this value), the maximum average energy that can be transferred to the next generation secondary electron is just $(40 \text{ eV} - I_i)/2$, which is about 15 eV for the outermost electrons, an energy barely enough to cause further ionization. Finally, what is of crucial importance for the consideration of the next scale of electron propagation is that at sub-50 eV energies, the electrons' cross sections are nearly isotropic [35, 93] and it is possible to use the random walk approximation in order to describe their transport [14, 21, 55, 56]. This transport is described in the next section.

5 Transport of Secondary Electrons and Reactive Species in Ion Tracks

The next stage of the scenario is related to secondary electrons ejected from the molecules of the medium as a result of ionization. As has been discussed above, most of these electrons have energies below 50 eV. They are called ballistic electrons until their energy becomes sufficiently small and coupling with phonons, recombination, and other quantum processes start dominating their transport. While the electrons are ballistic, their interactions with molecules can be described as a sequence of elastic and inelastic collisions. Many works, by and large using MC simulations, describe the transport of ballistic electrons. They are known as track structure codes [35]. Some of them describe chemical reactions in the medium including production of radicals and their propagation. However, regardless of how sophisticated these codes are, they do not contain the whole physical picture as will be shown below. In this section, a rather simple analytical approach is applied to the description of the propagation of ballistic electrons and its results are compared to MC simulations. It is also demonstrated how to make sense of radiation damage based on these calculations.

The main mechanism of radiation damage by ballistic electrons is inelastic collisions with targets. A target in this discussion is a biomolecule, such as DNA. Therefore, the probability of biodamage is a combination of the number of electrons (or other secondary particles) colliding with a given segment of a biomolecule and the probability of a certain inelastic process on impact. The first part is described by the fluence of electrons or other particles on the target. Fluence is the integral of the flux of particles (the number of particles hitting a part of the target's surface per unit time) over the entire time after the ion's passage and over the surface of the target. In general, the fluence depends on the distance of the target from the ion's path and its geometrical orientation. It will be shown that this part can be calculated analytically with accuracy, sufficient for understanding the scenario of radiation damage. The second part, i.e., the probability of a certain inelastic process on impact, is more difficult to assess mainly because of the diverse variety of possible processes. However, there are plenty of data that allows one to make reasonable quantitative estimates for this probability.

We start with the analysis of transport of secondary electrons. We also consider the production and transport of reactive species such as solvated electrons and radicals. Then the fluence for relevant configurations is calculated. It will be shown that important characteristics of the track structure such as radial dose can also be calculated as a byproduct of transport analysis. The random walk approach [31, 94] used for these problems allows one to make simple analytical calculations. The main requirement for the use of this approach is that the elastic and inelastic scattering of secondary electrons is isotropic. The anisotropy in the angular dependence of the cross sections for sub-50-eV electrons appears to be insignificant [35]. As was noted above, more than 80 % of secondary electrons satisfy this condition and only for less than 10 % of δ-electrons with energies higher than 100 eV is this condition violated significantly. The effects of δ-electrons will be discussed in Sect. 5.7. In Sect. 5.4, the production and transport of radicals, whose role in radiation damage is quite substantial, will be discussed. Later on, in Sect. 7.1 their effect will be incorporated into the calculation of the probability of complex damage.

This section is largely devoted to the transport of sub-50-eV electrons. Moreover, unless specifically stated to the contrary, these secondary electrons are produced by carbon ions in the vicinity of the Bragg peak in liquid water. At this part of the ion's trajectory, while a 0.3-MeV/u carbon ion passes 1 μm along the path, a typical radius within which the secondary electrons propagate is about 2–3 nm [31]. This allows one to assume that the electron diffusion is cylindrically symmetric with respect to the ion's path. The electronic component of the LET, S_e, remains nearly constant along this 1 μm of ion's path described by the coordinate ζ. Therefore, the number of ejected secondary electrons per unit length $\frac{dN_e}{d\zeta}$ is independent of ζ. A typical elastic mean free path of sub-50-eV electrons ranges between 0.1 and 0.45 nm [35, 93]. Since the scale along the Bragg peak is measured in tens of μm, while the radial scale is only tens of nm, therefore one can assume ζ to be ranging from $-\infty$ to $+\infty$.

5.1 *Random Walk of Secondary Electrons Originating on the Ion's Path*

A large number of secondary electrons is produced in the immediate vicinity of the ion's path. The secondary electrons are ejected as a result of ionizations of the molecules of the medium by projectiles. The production of electrons on the ion's path is described by the number of ionization events per length of the ion's path, $dN_e/d\zeta$. The three-dimensional diffusion of these electrons is described by the following equation,

$$\frac{\partial n_{ep}(\mathbf{r},t)}{\partial t} = D_e \nabla^2 n_{ep}(\mathbf{r},t) - \frac{n_{ep}(\mathbf{r},t)}{\tau_e}, \tag{19}$$

where $n_{ep}(\mathbf{r},t)$ is the average number density of electrons at a location $\mathbf{r}$ (this vector connects points of origin of electrons on the path and their observation), D_e is the diffusion coefficient for ballistic electrons in the medium, given by $vl/6$, where v is the electron's velocity and l is their elastic mean free path in the medium; τ_e is the average lifetime. An additional subscript "p" indicates that these electrons originate from the ion's path.

Equation (19) with boundary conditions at the origin and infinity has a well-known solution [30, 94]

$$\frac{dn_{ep}(t,r)}{d\zeta} = \frac{dN_e}{d\zeta}\left(\frac{1}{4\pi D_e t}\right)^{3/2} \exp\left(-\frac{r^2}{4D_e t} - \frac{t}{\tau_e}\right), \tag{20}$$

where r is the length of $\mathbf{r}$. The transport of secondary electrons is cylindrically symmetric with respect to the ion's path [21, 30]. Therefore, Eq. (20) can be integrated over ζ to obtain the number density for electrons originating at the ion's path,

$$n_{ep}(t,\rho) = \frac{dN_e}{d\zeta}\frac{1}{4\pi D_e t}\exp\left(-\frac{\rho^2}{4D_e t} - \frac{t}{\tau_e}\right). \tag{21}$$

The flux of these electrons, $d\Phi_{ep}$, through an element of a surface $d\mathbf{A}$ is given by

$$d\Phi_{ep} = \int_{-\infty}^{\infty}\left(d\mathbf{A}\cdot\nabla D_e\frac{dn_{ep}(t,r)}{d\zeta}\right)d\zeta, \tag{22}$$

where the integration is done only over ζ, but not $d\mathbf{A}$.

5.2 *Transport of Particles Produced as a Result of Ionization of the Medium by Secondary Electrons*

Many of secondary electrons produced on the ion's path have a sufficient energy for ionization of molecules of the medium. These ionizations take place at different distances from the path. After these ionizations, both the newly ejected electrons and those that ionize the molecules have lower energies and, therefore, shorter elastic mean free paths. Then, the ionized molecules are likely to become sources of free radicals, which also originate away from the path. Both of these processes can be described by the coupled transport equations,

$$\begin{aligned}\frac{\partial n_{ep}(\mathbf{r},t)}{\partial t} &= D_e\nabla^2 n_{ep}(\mathbf{r},t) - \sum_i \frac{n_{ep}(\mathbf{r},t)}{\tau_{ei}},\\ \frac{\partial n_i(\mathbf{r},t)}{\partial t} &= D_i\nabla^2 n_i(\mathbf{r},t) + \frac{n_{ep}(\mathbf{r},t)}{\tau_{ei}} - \frac{n_i(\mathbf{r},t)}{\tau_i},\end{aligned} \tag{23}$$

where index "i" marks the a specific kind of newly formed reactive species. The first of Eqs. (23) describes the random walk of secondary electrons ejected by the ion. This equation can be solved the same way as (19) to give

$$n_{ep}(t,\rho) = \frac{dN_e}{d\zeta}\frac{1}{4\pi D_e t}\exp\left(-\frac{\rho^2}{4D_e t} - \frac{t}{\tau}\right), \tag{24}$$

where $1/\tau = \sum 1/\tau_{ei}$.

Then, the rest of Eqs. (23) become diffusion equations with sources, and they can be solved analytically as well with a use of the Green's functions, which are given by,

$$G_i(t-t',\mathbf{r}-\mathbf{r}') = \left(\frac{1}{4\pi D_i(t-t')}\right)^{3/2} \times \exp\left(-\frac{(\mathbf{r}-\mathbf{r}')^2}{4D_i(t-t')} - \frac{t-t'}{\tau_i}\right). \quad (25)$$

The solutions are:

$$n_i(t,\mathbf{r}) = \int G_i(t-t',\mathbf{r}-\mathbf{r}')\frac{n_{ep}(t',\mathbf{r}')}{\tau_{ei}}dt'd\mathbf{r}'. \quad (26)$$

Each of these solutions requires four integrations, which are rather bulky, but doable. In order to simplify these integrations, it is advisable to use cylindrical coordinates with ζ-axis along the ion's path. The expression looks as,

$$n_i(t,\mathbf{r}) = \frac{1}{\tau_{ei}}\frac{dN_e}{d\zeta}\int\left(\frac{1}{4\pi D_i(t-t')}\right)^{3/2} \times \exp\left(-\frac{(\mathbf{r}-\mathbf{r}')^2}{4D_i(t-t')} - \frac{t-t'}{\tau_i}\right) \times \frac{1}{4\pi D_e t'}\exp\left(-\frac{\rho'^2}{4D_e t'} - \frac{t'}{\tau_{ei}}\right)dt'd\mathbf{r}', \quad (27)$$

where

$$(\mathbf{r}-\mathbf{r}')^2 = \rho^2 - 2\rho\rho'\cos(\phi-\phi') + \rho'^2 + (\zeta-\zeta')^2. \quad (28)$$

and $d\mathbf{r}' = \rho' d\rho' d\phi' d\zeta'$. Since ζ' dependence only appears in the exponential of (27) after substituting (28), this integration in infinite limits can be done first; and it is equal to $\sqrt{4\pi D_i(t-t')}$. Similarly, the dependence on the azimuthal angle also only appears in the exponential of (27) after substituting (28), and

$$\int \exp\left(\frac{2\rho\rho'\cos(\phi-\phi')}{4D_i(t-t')}\right)d\phi' = 2\pi I_0\left(\frac{\rho\rho'}{2D_i(t-t')}\right), \quad (29)$$

where I_0 is a Bessel function. After this, two more integrals are remaining,

$$n_i(t,\mathbf{r}) = \frac{1}{\tau_{ei}}\frac{dN_e}{d\zeta}\int\frac{1}{8\pi D_i D_e(t-t')t'} \times \exp\left(-\frac{\rho^2+\rho'^2}{4D_i(t-t')} - \frac{t-t'}{\tau_i} - \frac{\rho'^2}{4D_e t'} - \frac{t'}{\tau_{ei}}\right) I_0\left(\frac{\rho\rho'}{2D_i(t-t')}\right)dt'\rho' d\rho'. \quad (30)$$

First, the integral over ρ' is of the kind

$$\int_0^\infty \exp\left(-\frac{\rho'^2}{a^2}\right) I_0\left(\frac{\rho'}{b}\right) \rho' d\rho' = \frac{a^2}{2} \exp\left(\frac{a^2}{4b^2}\right). \tag{31}$$

After substituting the expressions for a and b into (31) and (30) becomes

$$n_i(t,\rho) = \frac{1}{4\pi\tau_{ei}} \frac{dN_e}{d\zeta} \int_0^t \frac{1}{D_e t' + D_i(t-t')} \\ \times \exp\left(-\frac{\rho^2}{4(D_e t' + D_i(t-t'))} - \frac{t-t'}{\tau_i} - \frac{t'}{\tau_{ei}}\right) dt'. \tag{32}$$

The last integration is not analytic, but it can be readily done numerically. Thus, n_i is obtained as a function of time and the distance from the axis.

The contribution to the flux due to ∇n_i can also be calculated as a function of ρ and t. All these calculations include parameters $dN_{ep}/d\zeta$, τ_i, and τ_{ei}.

5.3 Transport of Electrons Ejected by Secondary Electrons

Many of the ejected secondary electrons have enough energy to ionize water molecules in the medium. Indeed, more than 65 % of these electrons have energies higher than the ionization potential of water molecules. The second "wave" of ionization (i.e., ionization by secondary electrons that were ejected by ions) has been discussed [10], but only from the point of view of remaining energy, and it was concluded that the third wave is insignificant. For some applications, it is sufficient to consider the diffusion of secondary electrons leaving the second wave aside [55]. However, Sect. 5.2 suggests an analytic method for accounting of the second generation of electrons.

The formation and transport of the second wave ionization can be described by Eq. (23),

$$\frac{\partial n_1(\mathbf{r},t)}{\partial t} = D_1 \nabla^2 n_1(\mathbf{r},t) - \frac{n_1(\mathbf{r},t)}{\tau_1}, \\ \frac{\partial n_2(\mathbf{r},t)}{\partial t} = D_2 \nabla^2 n_2(\mathbf{r},t) \\ + 2\frac{n_1(\mathbf{r},t)}{\tau_1} - \frac{n_2(\mathbf{r},t)}{\tau_2}. \tag{33}$$

Here, index "1" marks secondary electrons of the first generation. Their energy is taken to be equal to be the average energy of secondary electrons formed in the vicinity of the Bragg peak, i.e., ~45 eV [65]. Hence, $D_1 = vl/6 = 0.265\ \text{nm}^2\,\text{fs}^{-1}$ and $\tau_1 = l_{ion}/v = 0.64$ fs (all mean free path data are taken from Refs. [93, 95]. In the process of ionization, these electrons lose ionization energy to the medium and share the rest with electrons of the second generation. We assume the remaining energy of first-generation electrons and the energy acquired by the second-generation electrons to be equal, and we refer to both of these kind of electrons as second-generation electrons. The factor of two that appears in front of the second term on the r.h.s. of the second equation of (33) formalises this assumption. Index "2" in this equation corresponds to the second-generation electrons; their energy is about 15 eV (as will be shown below) and, correspondingly, $D_2 = 0.057\ \mathrm{nm}^2\,\mathrm{fs}^{-1}$ and $\tau_2 = 15.3\,\mathrm{fs}$. The solutions of the above equations are obtained similarly to those of (23),

$$\begin{aligned} n_1(t,\rho) &= \frac{dN_e}{d\zeta}\frac{1}{4\pi D_1 t}\exp\left(-\frac{\rho^2}{4D_1 t}-\frac{t}{\tau_1}\right), \\ n_2(t,\rho) &= 2\frac{1}{4\pi\tau_1}\frac{dN_e}{d\zeta}\int_0^t \frac{1}{D_1 t' + D_2(t-t')} \\ &\times \exp\left(-\frac{\rho^2}{4(D_1 t' + D_2(t-t'))}-\frac{t-t'}{\tau_2}-\frac{t'}{\tau_1}\right)dt'. \end{aligned} \tag{34}$$

From the solutions (34) it follows [31] that a few fs after the ion's passage, all secondary electrons lose energy and the number density of secondary electrons is by and large given by $n_2(\rho, t)$. With time the distribution becomes a little broader, but the main effect is the exponential decrease with time. As a result of these decrease, the so-called pre-solvated electrons are formed. A pre-solvated stage of electrons is a transition stage between low-energy ballistic electrons and a relatively stable compound of electrons with water molecules known as solvated electrons. This transition takes about 1 ps.

The formation of pre-solvated electrons can be found from the conservation of electrons corresponding to (33),

$$\frac{\partial n_{aq}(\mathbf{r},t)}{\partial t} = \frac{n_2(\mathbf{r},t)}{\tau_2}, \tag{35}$$

where the subscript "*aq*" corresponds to pre-solvated electrons. In (35), the diffusion and all chemical terms are dropped since the diffusion, $D_{aq} = 4.5\times10^{-6}\,\mathrm{nm}^2\,\mathrm{fs}^{-1}$ [96] and chemical reactions are happening on a much longer scale. Equation (35) can be integrated over time to get the initial distribution of pre-solvated electrons. During the time of formation of solvated electrons, pre-solvated electrons may interact with nearby biomolecules, but, since their diffusion coefficient is much smaller compared to those of ballistic electrons, their location does not change considerably compared to the position of last inelastic event. Therefore, the distribution obtained by the integration of (35) is the initial radial distribution of solvated electrons. This distribution is shown in Fig. 6 [31]. Later on they slowly diffuse and react with other solvated electrons and hydroxyl radicals to form stable OH^- [96]. Chemical reactions dominate the diffusion and, as will be discussed below, given that diffusion is the only mechanism for the transport of solvated electrons, they are unlikely to leave the track.

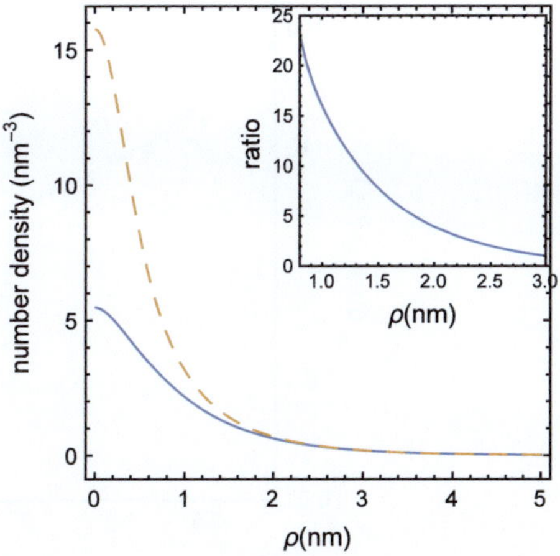

Fig. 6 Number density of pre-solvated electrons (*solid line*) and initial distribution of hydroxyl radicals (*dashed line*) as a function of distance from the ion's path at 50 fs when the transport of ballistic electrons is over. The ratio of chemical annihilation of hydroxyl radicals to their diffusion (discussed at the end of Sect. 5.4) as a function of the distance from the ion's path is shown in the *inset*

It is also possible to assess the energy deposition density (dose), $\varepsilon(\rho, t)$, in the medium by ions and secondary electrons. In order to do this, we assume that the average energy, $\bar{w}$, is deposited to the medium with each ionization. Moreover, since the first-generation electrons start with about 45 eV and $\bar{w} \approx 15$ eV, the second-generation electrons energy is also approximately $(45 - 15)/2 = 15$ eV. Then, the rate of energy deposition is proportional to the rate of inelastic events,

$$\frac{\partial \varepsilon(\rho, t)}{\partial t} = \bar{w} \left(\frac{dN_e}{d\zeta} \delta^{(2)}(\rho) \delta(t) + \frac{n_1(\rho, t)}{\tau_1} + \frac{n_2(\rho, t)}{\tau_2} \right). \tag{36}$$

The first energy deposition occurs right at the ion's path where the molecules are ionized by the ion; this corresponds to the first term on the right hand side of (36), where δ's are the corresponding δ-functions. The second deposition (second term on the r.h.s.) is the ionization by secondary electrons at the end of their ionization mean free paths. Finally, the third deposition is due to remaining energy loss due to excitation of molecules by the electrons of second generation. After that, the electrons enter the pre-solvated stage.

The time integration of (36) gives the dependence of the radial dose on time. The radial dose distributions at times 5, 10, 20, and 50 fs are shown in Fig. 7. At small radii, the distribution is due to primary ionization and it does not change with time. At larger radii, the dose slowly increases because of energy loss by the second generation of electrons. The shown results are obtained with $\bar{w} = 16.5$ eV, which corresponds to the normalization

$$\int_0^{\infty} \int_0^{t_r} \frac{\partial \varepsilon(\rho, t)}{\partial t} dt 2\pi \rho d\rho = S_e, \tag{37}$$

where $S_e = 890$ eV/nm is the linear energy transfer (LET) of a *single ion* at the Bragg peak for carbon ions [10] and t_r is the time by which most of electrons stop being

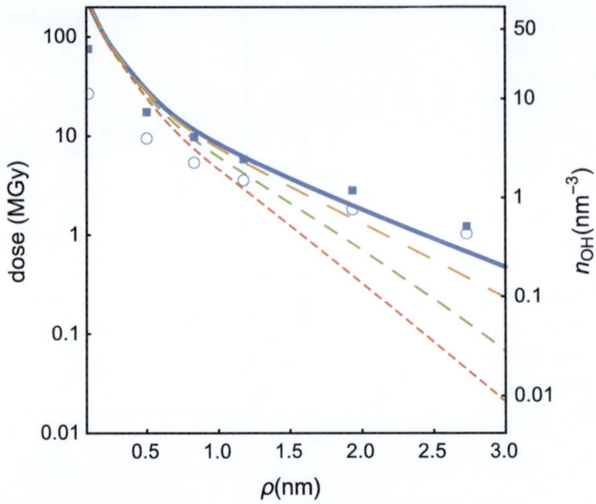

Fig. 7 The radial dose (in MGy) distributions as functions of the distance from the ion's path at times 50, 20, 10, and 5 fs are shown with *solid* and *dashed lines* with a diminishing dash size, correspondingly. The *solid line* also represents the initial radial distribution of hydroxyl radicals. The corresponding labels are shown on the *right side of the frame*. *Solid squares mark* the radial dose data for 2-MeV/u [97] multiplied by the factor of four. *Empty circles* represent the radial dose by a single generation of secondary electrons calculated in Ref. [55]

ballistic (this time is taken to be 50 fs). In Fig. 7 [31], the results of integration of (36) are compared with the radial dose of Ref. [97]. While the shapes of these distributions are alike, the absolute values are different by the factor of about four. Part of this disagreement (factor of about 2 [10]) can be explained by the fact that in [97], the radial dose presented for 2-MeV/u, i.e., proximal of the Bragg peak for carbon ions. The remaining factor can be due to the energy straggling present in Ref. [97], but absent for a single ion data of (36). Also, while the solid line represents the radial dose at 50 fs, in Ref. [97] there is no information about the time. This is typical for Monte Carlo simulations, but since they are compared with experimental data the time corresponding to dots is likely to be on a ps scale.

The results shown in Fig. 7 may be used as a starting point for shock wave development [28, 30], since the radial dose distribution evolved by the time $\leq$50 fs determines the temperature distribution around the ion's path following from the process of energy relaxation, in which the energy stored in electronic excitations is transferred into vibrational excitations and then to translational degrees of freedom. The characteristic time for the decay of an electronic excitation in Na clusters is estimated to be about 0.4 ps [98] and it is likely to be several times longer for the liquid water, which is consistent with the analysis performed in Ref. [25].

Curiously, that the solid curve in Fig. 7 gives the upper estimate of pressure near the ion's path. Indeed, the energy $U(\rho)$ inside a cylinder of radius ρ and length L, coaxial with the ion's path, is given by

$$U(\rho) = \int_0^{\rho} \varepsilon(\rho')2\pi\rho' d\rho' L, \tag{38}$$

where $\varepsilon(\rho)$ is the radial dose given by

$$\varepsilon(\rho) = \int_0^{t_r} \frac{\partial\varepsilon(\rho, t)}{\partial t} dt. \tag{39}$$

Then, the force, normal to the surface of this cylinder is given by

$$F(\rho) = -\frac{\partial U(\rho)}{\partial \rho} = \varepsilon(\rho)2\pi\rho L, \tag{40}$$

and the pressure $P(\rho)$ is

$$P(\rho) = \frac{F(\rho)}{2\pi\rho L} = \varepsilon(\rho). \tag{41}$$

The units MGy used in Fig. 7 correspond to GPa. At one point, $\rho = 2$ nm, this pressure can be compared to that assessed in Ref. [25]. Figure 7 gives the value of 1.8 GPa, which is by a factor of 1.8 larger than the estimate of Ref. [25].

Finally, the radial dose has been studied in Ref. [55] using a random walk approximation. In that work, we obtained a reasonably good comparison of shapes with Ref. [97] for intermediate distances. Only one generation of secondary electrons was used, but the parameters, such as mean free path and the relaxation time were chosen close to those for the second generation of this work. Therefore, the results obtained in Ref. [55] remain meaningful; they are shown in Fig. 7 with circles. They do not include term $\bar{w}\frac{dN_e}{d\zeta}\delta^{(2)}(\rho)$ that comes from primary ionization events, and, hence, the dose in the region adjacent to the ion's path is considerably lower than that obtained by the integration of (36) over time. Besides, the value of $\bar{w}$ used to obtain the absolute value of radial dose according to Ref. [55] is 33 eV, since only one generation is involved in radial energy transport.

5.4 *Transport and Production of Hydroxyl Radicals*

Hydroxyl among other radicals plays the most important role in the DNA damage [13, 99]. Therefore, we are considering the production and transport of these radicals in more detail. In the liquid water environment irradiated with ions such as carbon, the hydroxyl radicals are primarily produced as a consequence of ionization of water molecules [13],

$$H_2O^+ + H_2O \rightarrow H_3O^+ + OH^{\cdot}, \tag{42}$$

and by excited water molecules,

$$H_2O^* \to H^{\cdot} + OH^{\cdot}. \tag{43}$$

Both of these processes can be initiated by either ions or secondary electrons that ionize and excite water molecules. The time scales play an important role in further analysis. The ionization is happening in 10^{-17} s and secondary electrons diffuse by several nm within several fs. However, $OH^{\cdot}$ is produced through (42) and (43) only on a ps scale, i.e., after the electron transport is over and pre-solvated electrons are formed. Neither H_2O^+, nor H_2O^* are remaining at rest during this time; they are diffusing away from the path. Since their diffusion coefficients are of the order of $10^{-6}\,\mathrm{nm^2\,fs^{-1}}$, they diffuse by only about $\sqrt{D\tau_{form}} \approx \sqrt{10^{-6}10^3} = 3 \times 10^{-2}$ nm away from places where the inelastic event (ionization or excitation of water molecule) has taken place. This distance is too small, therefore, we can consider the loci of inelastic events to be the initial loci of hydroxyl radicals.

The distribution of inelastic events was analysed in Sect. 5.3 in Eq. (36), where the rate of energy deposition was obtained. In the first approximation, we can assume that each inelastic event produces a hydroxyl radical, then the equation, similar to (36) for production of radicals is as follows,

$$\frac{\partial n_{\mathrm{OH}}(\rho,t)}{\partial t} = \frac{dN_e}{d\zeta}\delta^{(2)}(\rho)\delta(t) + \frac{n_1(\rho,t)}{\tau_1} + \frac{n_2(\rho,t)}{\tau_2}. \tag{44}$$

Similarly to (36), it can be integrated over time to obtain the initial number density of hydroxyl radicals, $n_{\mathrm{OH}}(\rho)$. It is shown with a solid line in Fig. 7 with labels on the right side of the graph. The results of the integration of second and third terms are shown with a dashed line in Fig. 6 to compare the distributions of hydroxyl with solvated electrons at larger distances.

Thus, with the above assumptions, each ionization on the ion's path produces one ~45-eV electron, which after another ionization loses energy and produces two ~15-eV electrons, which then become two solvated electrons. These two ionization events together with final energy loss by two 15-eV electrons produce four hydroxyl radicals. However, the initial radial distributions of solvated electrons and hydroxyl are different; the number density of hydroxyl radicals strongly dominates that of solvated electrons near the path and these number densities are nearly the same at $\rho > 1.5$ nm. Now, let us analyse the further evolution of hydroxyl.

The ballistic electrons are gone by 50 fs after the ion's passage; this time much shorter than that of formation of hydroxyl. Therefore, by the time hydroxyl radicals are formed, there are no more active sources since both ionization and excitation processes are over. All reactions that eradicate hydroxyl are second order reactions with respect to perturbations in water; these are

$$\begin{aligned} OH^{\cdot} + H^{\cdot} &\to H_2O, \\ OH^{\cdot} + e^-_{aq} &\to OH^-, \\ OH^{\cdot} + OH^{\cdot} &\to H_2O_2, \end{aligned} \tag{45}$$

and others [96]. This means that all equations such as (19) and (23) acquire additional terms on the right hand side that account for chemical reactions, but lose terms like $n_{\rm OH}(\mathbf{r},t)/\tau_{\rm OH}$, since hydroxyl does not react with water molecules and is stable by itself. For example, Eq. (19) becomes

$$\frac{\partial n_{\rm OH}(\mathbf{r},t)}{\partial t} = D_{\rm OH}\nabla^2 n_{\rm OH}(\mathbf{r},t) - \sum k_{{\rm OH}i} n_{\rm OH} n_i, \tag{46}$$

where $k_{{\rm OH}i}$ is the rate constant corresponding to the reaction of species "i" with hydroxyl. This is a nonlinear equation, and we start with comparing all terms on the right hand side of (46).

First, let us consider the hydroxyl radicals that are formed due to ionizations or excitations by ions. Since the diffusion coefficient for hydroxyl is so small (2.8×10^{-6} nm^2 fs^{-1}) compared to electrons discussed above, the first term in Eq. (46) (for carbon ions in the vicinity of Bragg peak) is of the order of 10 if we consider the estimated diffusion radius to be 0.03-nm. The second term is equal to $k_{\rm OH,OH} n_{\rm OH}^2$, since hydroxyl by itself is the dominant reagent in this region. With $k_{\rm OH,OH} = 1.0 \times 10^{-5}$ nm^3 fs^{-1} [13, 96, 99], this term exceeds 200 nm^{-3} fs^{-1}, since the concentration near the path is so high. This concentration decreases by the factor of 100 within 2 ps as most of the hydroxyl becomes involved in the formation of peroxide. Therefore, the hydroxyl, formed by direct ion's action by and large does not exit the a sub-nm region around the ion's path.

The situation with hydroxyl formed by secondary electrons is somewhat different, since it is formed with a much smaller concentration and the diffusion term may be comparable with that of peroxide formation. The equation of interest can be written as,

$$\frac{\partial n_{\rm OH}(\rho,t)}{\partial t} = D_{\rm OH}\frac{1}{\rho}\frac{\partial}{\partial \rho}\left(\rho \frac{\partial n_{\rm OH}(\rho,t)}{\partial \rho}\right)$$
$$-k_{{\rm OH},e_{aq}} n_{\rm OH}(\rho,t) n_{aq}(\rho,t) - k_{\rm OH,OH} n_{\rm OH}^2(\rho,t). \tag{47}$$

This equation includes two relevant chemical reactions from (45). In the beginning, at $t \approx 1$ps, the number density of hydroxyl, $n_{\rm OH}(\rho,t)$, exceeds that of solvated electrons, $n_{aq}(\rho,t)$, as can be seen in Fig. 6. The reaction constant $k_{{\rm OH},e_{aq}}$ is larger than $k_{\rm OH,OH}$ by the factor of 4.2 [13, 96, 99]. Therefore, the reaction terms in (47) dominate the diffusion term in the whole domain of ρ. The most conservative estimate of the ratio of reaction to diffusion terms in (47) is shown in the inset of Fig. 6. It is apparent that terms become comparable only at $\rho \geq 2.5$ nm. Only a small (less than 10 %) fraction of hydroxyl diffuses to these distances.

Therefore, we can summarise that in the first approximation, which can be improved by introduction of probabilities of formation of hydroxyl radicals, the initial distribution of $\rm OH^{\cdot}$ is shown in Fig. 7. These radicals would react with biomolecules such as DNA if the latter appear to be in the track. However, if the collective transport with a shock wave does not take place, the radicals react with each other and with solvated electrons and by and large do not leave the track. Other species, such as H_2O_2 and OH^- may propagate on larger distances and may be detected outside tracks.

Concluding, the random walk approximation, applied to two generations of secondary electrons showed that most of these electrons do not spread beyond the 2-nm cylinder around the ion's path. Pre-solvated electrons, important agents of strand breaks in DNA molecules, are also formed within this small region. Their small diffusion coefficients do not allow them to be transferred far enough since they readily react with hydroxyl radicals abundant in the same region. The latter also react with themselves and their high number density makes this (nonlinear) reaction much more important than their (linear) diffusion. Thus, we can conclude that the diffusion mechanism does not allow most of the reactive species to leave the few-nm track radius. Carbon ions near the Bragg peak were used in the above example. The effect of containment of reactive species in the track will be stronger for heavier ions and weaker for lighter ones, since the concentration of reactive species is much smaller.

Alternatively, another transport mechanism, i.e., collective transport in a shock wave, see Sect. 6 should be studied both theoretically and experimentally in order to understand whether the reactive species actually propagate to larger distances. This question is very important for the assessment of radiation damage with ions [30]. The radial dose distribution obtained in Fig. 7 gives the initial conditions for the development of cylindrical shock waves.

5.5 Calculation of the Fluence of Secondary Electrons

Now we return to Eq. (22)

$$d\Phi_{ep} = \int_{-\infty}^{\infty} \left(d\mathbf{A} \cdot \nabla D_e \frac{dn_{ep}(t, r)}{d\zeta} \right) d\zeta. \tag{48}$$

It gives the number of secondary electrons passing through an area $d\mathbf{A}$ located at a distance ρ from the ion's path, as is shown in Fig. 8 [30] per unit time. These electrons originate from a segment $d\zeta$ of the ion's path and the integration is done over ζ, but not over $d\mathbf{A}$. Vector $\mathbf{r}$ connects the element $d\zeta$ with $d\mathbf{A}$.

The next step is the integration of (48) over time in order to calculate the total number of particles incident on $d\mathbf{A}$, equal to the fluence multiplied by dA. In order to do this, we change variables from t to the number of steps by secondary electrons k using $\bar{v}t = kl$. We rewrite Eq. (48), substituting (20), and switching from variable t to k as

$$dN_A(\mathbf{r}) = \int_t d\Phi_{ep} dt = \int_{-\infty}^{\infty} d\zeta d\mathbf{A} \cdot \mathbf{n_r} \frac{dN_e}{d\zeta} \int_{r/l}^{\infty} dk \\ \times \frac{r}{2k} \left(\frac{3}{2\pi k l^2} \right)^{3/2} \exp\left(-\frac{3r^2}{2kl^2} - k\frac{l}{l_i} \right), \tag{49}$$

where $l_i = v\tau_e$ is the inelastic mean free path.

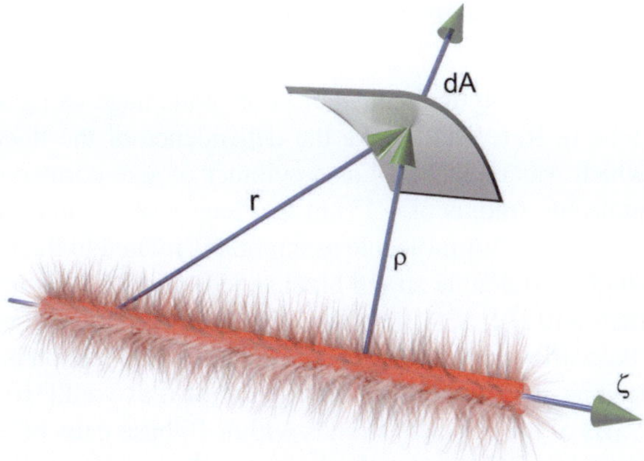

Fig. 8 Geometry for the general calculation of fluence through a segment of surface $d\mathbf{A}$. The ion path is along the axis

A lower limit, r/l, introduced in Eq. (49), corresponds to the earliest time a particle emitted at zero time can reach the segment dA located a distance r away from the point of emission. It is translated to minimal number of mean free paths required to pass this distance. This issue is discussed in more detail in Ref. [55].

After integrating over k, the number of particles incident on the whole target, $\mathscr{F}(\rho)$, can be calculated as the integral over the surface of the target:

$$\mathscr{F}(\rho) = \int_A dN_A(\mathbf{r}) = \int_A d\mathbf{A}\cdot\mathbf{n_r}\frac{dN_e}{d\zeta}d\zeta\int_{r/l}^{\infty}dk \\ \times\frac{r}{2k}\left(\frac{3}{2\pi kl^2}\right)^{3/2}\exp\left(-\frac{3r^2}{2kl^2}-k\frac{l}{l_i}\right). \quad (50)$$

That is, the number of secondary particles incident per unit area (fluence) integrated over the area of the target yields the number of incident particles per target. This quantity is an integrated fluence and below it is referred to as fluence. Then, the average fluence is found by dividing this number by the lateral area of the target. Strictly speaking, the fluence given by Eq. (50) depends on more variables than just the distance between the target and the ion's path. These variables include the elastic and inelastic mean free paths of secondary electrons and more geometrical parameters. The mean free paths corresponds to some energy between zero and 50 eV and thus energy averaging is achieved. After this averaging, the energy of electrons is assumed to be constant. In different works [14, 19, 21, 55, 56] this averaging was done according to the particular physical problem. However, it is the dependence on the distance ρ, kept in Eq. (50), that remains important for calculations of radiation damage. The application of this method to specific geometries that were considered in some of these works are demonstrated below.

5.6 Targeting a Twist of DNA with Secondary Electrons

The first analytical calculation of biodamage using a random walk approach was done in Ref. [21], where the dependence of the fluence through a twist of DNA, which was represented as a cylinder of size corresponding to one twist of a DNA molecule (radius of 1.15 nm and length of 3.4 nm), was calculated. A choice of a twist of a DNA molecule as a target is related to the types of DNA damage, such as single and double strand breaks (SSB and DSB) which are widely discussed in the literature [12, 13, 100]. The DSB is a severe lesion, which can still be repaired, but its contribution to the probability of cell sterilisation is significant, especially if other lesions are produced in the vicinity of it as well [101]. The DSB is defined as two SSBs of the opposite strands within 10 base pairs of each other, i.e., within a single twist of a DNA molecule.

The probability of an SSB or a DSB in a given twist is related to the fluence of secondary electrons produced by the passing ion. Therefore, a natural problem of calculating the fluence of these electrons through a cylinder enwraping the twist was among the first to be considered [21, 30]. Here, we only show the results of these calculations. They are presented in Fig. 9 [30] for two different orientations of the cylinder containing a DNA twist; a perpendicular orientation is shown in the inset.

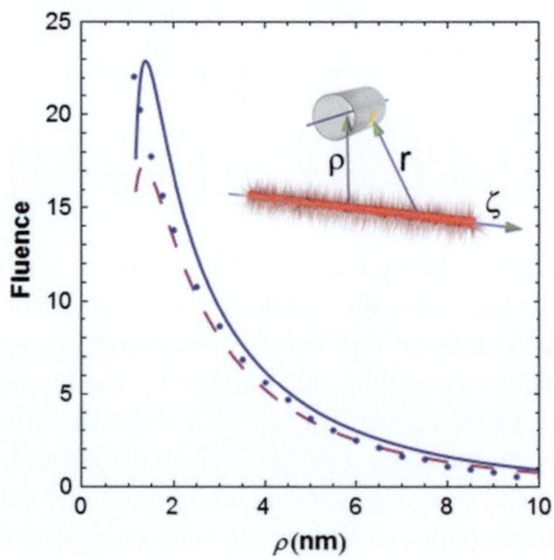

Fig. 9 Fluences of secondary electrons produced by a single $^{12}C^{6+}$ ion in the vicinity of a *Bragg peak* through a *cylinder enwraping* a DNA twist are shown with respect to the distance of the *cylinder* from the ion's path. Two different orientations (*parallel* and *perpendicular*) are shown as well as MC simulations for the perpendicular case. In the *inset*, the geometry for the calculation of fluence through a (*perpendicular*) cylinder enwraping a DNA twist is shown

5.6.1 Calculation of the Number of SSBs per Single Ion

An estimate of the number of SSBs per unit length of the ion's trajectory can be obtained assuming that this number is proportional to the number of secondary electrons incident on a given twist of a DNA molecule. For example, for a straight segment of length $d\zeta$ of the ion's path, the number of SSBs is given by the integral,

$$\frac{dN_{SSB}}{d\zeta} = \Gamma_e \int_0^\infty \mathscr{F}_e(\rho) n_t 2\pi\rho d\rho, \tag{51}$$

where Γ_e is the probability that an electron incident on a DNA twist induces a SSB and n_t is the number density of DNA twists (i.e., cylinders). Since the spatial dependence of n_t is unknown, it is reasonable (in the first approximation) to assume that it is constant. The fluence $\mathscr{F}_e(\rho)$ for carbon ions at the Bragg peak, obtained in Sect. 5.6, can be substituted in the integral (51). This gives us an estimate of

$$\frac{dN_{SSB}}{d\zeta} = n_t \sigma_{\rm SSB}, \tag{52}$$

where $\sigma_{\rm SSB} = \Gamma_e \int_0^\infty \mathscr{F}(\rho) 2\pi\rho d\rho = \Gamma_e \times 1.1 \times 10^3\ {\rm nm}^2$. The value of $\sigma_{\rm SSB}$ is obtained using a simple diffusion model that contains two parameters, the elastic and inelastic mean free paths, l and l_i (assumed to be the same for all electrons). The third input in this value is $\frac{dN_e}{d\zeta}$. This number can be calculated using ionization cross sections discussed in Sect. 4.2. However, the number calculated from Eq. (52) only includes the electrons of the first generation. The calculation can be improved if two generations are taken into account. Since the low-energy electrons produced in the latter are important for biodamage, the number $\frac{dN_e}{d\zeta}$ and, therefore, both the fluence and $\sigma_{\rm SSB}$ are underestimated by a factor of about two [10].

The probability of the production of a SSB by an electron incident on a twist of a DNA molecule, Γ_e, is included in $\sigma_{\rm SSB}$ and appears in Eq. (51) as well as in (52). This probability can be estimated as the cross section for breaking an important covalent bond that leads to a SSB multiplied by the number of such bonds in a single DNA twist and divided by the lateral area of this DNA segment, represented above by a cylinder. However, the cross section for breaking a covalent bond is energy-dependent and the energy of secondary electrons varies from zero to about 50 eV. At low energies (below the ionization threshold) the cross section is deemed to be that of DEA, i.e., resonant attachment of the secondary electron to the molecule (formation of temporary negative ion) followed by dissociation (SSB). At higher energies of impact electrons, the cross sections contributing to Γ_e are defined by the ionization cross sections provided that the formation of a cation leads to a strand break. These processes are being studied theoretically and experimentally [18, 59, 83, 102–104]. Their typical cross sections vary, but the cross section for a SSB as a consequence of DEA for 1–3 eV electrons is about 2 nm^2 [105, 106] per plasmid DNA. According to assumptions of Ref. [106], the probability of an SSB can be obtained by dividing this cross section by the area of a flat plasmid ($\sim 2.2 \times 10^3$ nm^2). This probability can be seen as a product of a probability that incident electron hits a given DNA twist, the number of twists in a plasmid, and Γ_e. Substituting the numbers from Ref. [106] yields

$$\frac{2\,\mathrm{nm}^2}{2.2\times 10^3\,\mathrm{nm}^2}=\frac{6\,\mathrm{nm}^2}{2.2\times 10^3\,\mathrm{nm}^2}\times 320\times \Gamma_e, \tag{53}$$

where 320 is the approximate number of twists in a plasmid used in Ref. [106] and 6 nm^2 is the average cross sectional area of a DNA twist. Equation (53) gives an estimate $\Gamma_e \approx 10^{-3}$. For higher energy electrons, ionization of a DNA molecule does not necessarily lead to a SSB and many pathways are being discussed. Nevertheless, reported SSB yields at higher electron energies are of the same order (if not higher) as those for low-energy electrons [18].

Another estimate for Γ_e can be obtained by multiplying a cross section of a rupture of a covalent bond with an incident electron by the number of bonds per DNA twist whose break leads to a SSB, and dividing this product by the lateral area of the twist. In this estimate, $\sigma_b \times 36/30\,\mathrm{nm}^2$, the cross section is the most questionable component. According to different studies, it can vary from 10^{-5} nm^2 to 10^{-1} nm^2 depending on the electron's energy and conditions such as the chemical environment of the molecule [18, 102]. For instance, the above experiments [18, 105, 106] were done in vacuum with dry plasmid molecules retaining only few adjacent water molecules. For DNA molecules in a natural environment, these cross sections may be much larger [58]. The above cross section variation gives a range for Γ_e from 10^{-4} to 10^{-1}.

5.6.2 Calculation of the Number of DSBs

The estimate of the number of DSBs is more ambiguous than that of SSBs. This is mainly due to the lack of understanding of the mechanism of producing this lesion. Many works [15, 17] suggest that a DSB is the result of the action of a single electron that dissociatively attaches to a DNA molecule. The dissociative attachment is considered to be an important pathway of SSBs at very low energies and in about one out of five such incidences, a DSB takes place due to the interactions with the debris of a SSB [17]. Alternatively, DSBs can be due to two separate SSBs on opposite strands. This may be possible if the number density of secondary electrons is high enough. It is also possible that double ionization events play a significant role [19]. Such events create a high local number density of low energy electrons at a considerable distance from the ion's path and if this occurs in the vicinity of a DNA twist, at least two of the three electrons involved in a double ionization event may be incident on the same twist. This depends on the values of the cross sections for double ionization. The probability of ICD-effects[12] on DNA and water molecules adjacent to it may also be an important factor [19, 108].

[12]ICD is a type of non-radiative relaxation process, similar to the Auger effect, except in the case of the ICD the extra electron is emitted by the neighbouring molecule [107].

Regardless of the pathway for DSBs, for a given ion in a given medium, the ratio of yields of DSBs and SSBs (per unit length of ion trajectory) is fixed and *dose independent* unless tracks of different ions interact. Indeed, each ion's track is determined by the type of ion and an increased dose just means an increase in the density of ion tracks. Only after some critical value of dose is reached, do the tracks start overlapping. Only then can the dependencies of yields of SSBs and DSBs on dose become not proportional to each other. These conditions are not being observed in the analysed experiments or in therapy,[13] however, if laser-driven ion beams are used [109], track interaction effects may become important.

Therefore, the DSB yield can be calculated as a sum of two terms, the first of which represents the events where SSBs are converted to DSBs and the second accounts for DSBs due to separate electrons. In order to calculate the second term, the average number of SSBs per twist, $\mathscr{N}_e$, can be introduced as[14]

$$\mathscr{N}_e = \Gamma_e \mathscr{F}_e(\rho). \tag{54}$$

Then, with the probability of a DSB due to two separate electrons in this twist is given by $\frac{1}{4}\mathscr{N}_e^2 \exp[-\mathscr{N}_e]$, the DSB yield is

$$\frac{dN_{DSB}}{d\zeta} = \lambda n_t \sigma_{\rm SSB} + \frac{n_t}{4}\int_0^\infty \mathscr{N}^2 \exp[-\mathscr{N}]2\pi\rho d\rho, \tag{55}$$

where λ is a fraction of SSBs converted to DSBs, i.e., the number of DSBs due to the action of a single electron. The second term in the DSB yield is given by the integration over the volume similar to Eq. (51).

At this point the phenomenon-based approach can be related to experiments. If real tissue is irradiated, one can only find the percentage of cells surviving. If this value is measured as a function of dose, the survival curve is obtained as a result. Many interactions on sub-cellular, cellular, or even at the organismic level may affect the survival curve. In in vitro experiments on cell cultures, elimination of some of these interactions allows, e.g., synchronizing cell cycles, control over the environment, etc. Still, there are no direct ways of relating cell sterilisation to, e.g., DSBs produced by secondary electrons. Therefore, the comparison with experiments on DNA molecules irradiated with ions is the most appropriate.

5.6.3 Comparison with Experiments on Plasmid DNA

Of all the experiments investigating DNA molecules irradiated with ions, the study of plasmid DNA is the most valuable, since there are reasonably reliable ways to distinguish the intact molecules from those with a SSB and from those with a DSB.

---

[13] In this section only effects of secondary electrons are discussed. The situation may be different when radicals are included.

[14] This number is a part of the integrand of Eq. (51).

Another important feature is that the effects of DNA damage observed in these experiments are not affected by the biological effects of repair that take place in living cells. This allows for a more pure comparison.

An undisturbed plasmid is a closed loop of a supercoiled DNA molecule [110]. This loop contains a given number of base pairs, e.g., in experiments described in Ref. [111] plasmid DNA pBR322 irradiated with carbon ions contains 4361 bp. The characteristic size of this molecule is about 100 nm [110]. A disk shape (with a radius of 50 nm and thickness of 1 nm) for a supercoiled plasmid is assumed for all calculations along with a uniform chromatin distribution. Therefore, the cross sectional area is $A_p = 7.8 \times 10^3$ nm^2, the volume is $V_p = 7.8 \times 10^3$ nm^3, and the average twist number density is $n_t = 436.1/V_p = 5.6 \times 10^{-2}$ nm^{-3}.

If such a molecule experiences a SSB, it becomes "circular" or just a loop without the supercoil structure. A DSB makes the plasmid "linear" since both of its strand are broken. These structural conformations can be distinguished using electrophoresis or high-performance liquid chromatography [18, 111]. This allows the measuring of SSB and DSB yields experimentally. In one of the experiments described in Ref. [111], plasmid DNA was dissolved in a 600 mmol/l solution of mannitol in water. Mannitol serves as a radical scavenger so their contribution to DNA damage may be neglected. This is adequate for the theoretical treatment (Sects. 5.6.1 and 5.6.2), which only includes secondary electrons.

The results of experiments of Ref. [111] are shown in Fig. 10 [30] with dots. They represent the probabilities for two outcomes after an irradiation with carbon ions at the spread-out Bragg peak. The first outcome (open squares) is for the plasmid to become open circular (not supercoiled), associated with a SSB. There is a reported

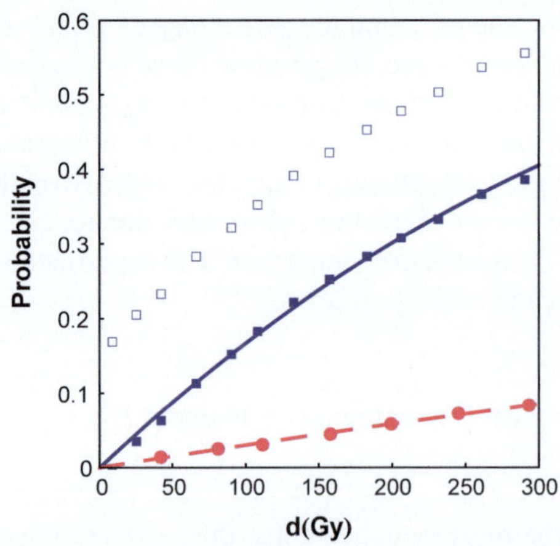

Fig. 10 Probabilities for SSBs and DSBs induced in plasmid DNA by secondary electrons as a function of dose. Dots correspond to experiments [111]: *open squares* to the original SSBs, *filled squares* to the "cleaned" SSBs, and *filled circles* to DSBs. Calculated probabilities are shown with *lines*. The *solid line* corresponds to the probability of SSBs calculated using Eq. (58). The *dashed line* depicts the probability for DSBs calculated using Eq. (60)

problem with the quality of the data resulting in the probability corresponding to SSBs not starting from zero at a zero dose [111]. This means that some plasmids are either not supercoiled to begin with or appear as such in the electrophoresis. This probability remains elevated by about the same value throughout the dose range. In order to compare these data with our calculations, the zero-level probability of the SSB yield was subtracted in order to "clean" the data. These data points are shown with filled squares. The second outcome is for the plasmid to become linear, associated with a DSB and is shown with filled circles. These probabilities (filled squares and circles) are monotonically increase with dose with the SSB dependance being slightly non-linear. In order to explain these data using the MSA, let us start with the dose dependance. When a beam of carbon ions is incident on the plasmid solution, there is a dose-dependent probability that ν ions will traverse through a plasmid. This probability is given by the Poisson distribution:

$$P_\nu = \frac{N_{ion}^\nu}{\nu!} \exp\left[-N_{ion}\right], \tag{56}$$

where N_{ion} is the average number of ions passing through the cross sectional area of a plasmid, A_p. The average number of ions passing through this area is equal to the ratio of this area to the average area per ion. The average area per ion, $\mathscr{A}$, can be calculated if a uniform distribution of ions in the beam is assumed. Then the dose is equal to the LET (which is associated with the average for the Bragg peak stopping power due to ionization processes, $\bar{S}_e$) divided by the average area per ion, i.e., $d = \frac{\bar{S}_e}{\mathscr{A}}$. Then, N_{ion} is given by:

$$N_{ion} = \frac{A_p}{\mathscr{A}} = \frac{A_p}{\bar{S}_e} d. \tag{57}$$

In Ref. [111], the average LET over the spread-out Bragg peak, $\bar{S}_e$ is 189 ± 15 eV/nm. This includes energy straggling effect along the ion's trajectory. Then, the number of SSBs that are likely to be induced in a plasmid, i.e., SSB yield per plasmid is given by the sum,

$$Y_{SSB,e} = \frac{dN_{SSB}}{d\zeta} \bar{x}_p \sum_{\nu=1}^{\infty} \nu P_\nu, \tag{58}$$

where $\bar{x}_p$ is the average length of an ion's path through the plasmid. The subscript "e" indicates that this yield is only due to secondary electrons. Each term of this sum is a product of the number of SSBs per unit length of trajectory of a single ion, the length of this trajectory through the plasmid and the number of ions traversing the plasmid. The average length of a trajectory, $\bar{x}_p$, for larger objects, such as cell nucleus, can be evaluated geometrically. For plasmid, it is assumed that an ion traverses a chromatin fibre twice and on this ground, $\bar{x}_p \approx 4$ nm. If exactly ν ions pass through the plasmid, this length is multiplied by ν. This is the first term in the sum of Eq. (58). Then the factor P_ν gives the probability that ν ions are passing through the target. Hence, the whole sum multiplied by $\bar{x}_p$ determines the average length of tracks through the plasmid.

The sum in Eq. (58) does not include interactions of different ions that could occur if trajectories of two or more ions are so close that the same twist of a DNA molecule could be hit with electrons originating from the different ions. The probability of such an interference can be estimated. Since the range of 50-eV electrons in liquid water is about 10 nm, the two ion's trajectories must be within 20 nm, for the interference to occur. Then the estimate is obtained from Eqs. (56) and (57) with $\nu = 2$ and $A_p = \pi \times 10^2$ nm^2. For the maximal dose of 300 Gy used in Ref. [111] the resulting probability is 5×10^{-6}. This number is very small compared to the probability that one ion will pass through the plasmid at this dose (equal to 0.3) or even that two ions will pass through it (equal to 0.02). Therefore, the interference term in Eq. (58) can be neglected.

The only term of Eq. (58) that depends on dose is P_ν, therefore the dose dependence of the yield is contained in the sum $\sum_{\nu=1}^{\infty} \nu P_\nu$. This dependence is not unique for the yield of SSBs. The same sum appears in all calculations, provided that the damage due to each ion is localised in its track and the tracks do not interfere. The dependence of this sum on dose is asymptotically exponential at large values of N_{ion}. This means that on a semi-logarithmic plot the dose dependence will be asymptotically a straight line. This will be seen below in the analysis of survival curves in Sect. 7.2.

The numbers relevant to the experiments of Ref. [111], such as $A_p = 7.8 \times 10^3$ nm^2 and $\bar{S}_e = 189$ eV/nm, substituted to Eq. (57) give $N_{ion} = 2.6 \times 10^{-4} d$ with the dose in Gy. This means that even at the highest dose of 300 Gy used in Ref. [111] $N_{ion} \ll 1$. However, Ref. [111] gives the dose dependence of the *probability* of a SSB per plasmid rather than yield. This probability is obtained from the Poisson statistics,

$$P_{SSB,e} = Y_{SSB,e} \exp\left[-Y_{SSB,e}\right] + \frac{1}{4} Y_{SSB,e}^2 \exp\left[-Y_{SSB,e}\right], \qquad (59)$$

where the first term corresponds to a single SSB in the plasmid DNA and the second term corresponds to two SSBs on the same strand. The first term is just a probability of one SSB produced based on the average number of SSBs per plasmid (yield). The second term is one half of the probability that two SSBs are produced. The fit of Eq. (59) to the probability of the SSB dependence on dose, shown in Fig. 10, gives $\frac{dN_{SSB}}{d\zeta} \approx 2.0$ nm^{-1}. Comparing this with Eq. (52) and taking $n_t = 5.6 \times 10^{-2}$ nm^{-3}, we obtain an estimate for $\Gamma_{SSB} = 0.03$, which is within the range discussed in Sect. 5.6.1.

Now the comparison for DSBs can be made. Similar to Eq. (58), the number of DSBs induced in a plasmid (a DSB yield per plasmid) is given by the sum,

$$Y_{DSB,e} = \frac{dN_{DSB}}{d\zeta}\bar{x}_p \sum_{\nu=1}^{\infty} \nu P_\nu, \tag{60}$$

and the probability of a DSB per plasmid is given by,

$$P_{DSB} = Y_{DSB,e} \exp\left[-Y_{DSB,e}\right]. \tag{61}$$

A fit of Eq. (61) to the probability of the DSB dependence on dose (for Ref. [111]) gives $\frac{dN_{DSB}}{d\zeta} = 0.3\ \text{nm}^{-1}$. Then, comparing this with Eq. (55) and taking $n_t = 5.6 \times 10^{-2}\ \text{nm}^{-3}$, we obtain an estimate for $\lambda = 0.15$, which is in reasonable agreement with the values between 0.1 and 0.2 for different electron energies [15, 17, 18]. Thus, the comparison of our model for the effect of secondary electrons with the results of Ref. [111] for a plasmid DNA solution in the presence of radical scavengers is reasonable.

Some comments regarding these calculations should be made. First, as has been noted in Sect. 5.6.1, the number of secondary electrons is underestimated. This happens because in our calculations only the electrons ejected by ions were included, missing those ejected in the process of secondary ionization by electrons. The correction for this number will increase fluence, but will not affect the dose dependence. Since the actual fluence will then be larger (by the factor of about two [10]), Γ_{SSB} will be smaller. The second issue is that the treatment of a supercoiled plasmid as an object with uniformly distributed chromatin may be a little far-fetched. Also, if a plasmid suffers a single strand break, its size increases by a factor larger than two and then it may be a target for another ion. Nevertheless, the comparison that was just made is quite reasonable and encouraging for further steps in the assessment of radiation damage.

5.7 Accounting for δ-Electrons

The effects due to secondary electrons with energies of 100 eV and above or the so-called δ-electrons should be discussed separately. These particles cannot be included in the diffusion model because their cross sections are strongly peaked in the forward direction, their mean free paths exceed 1 nm and they lose their energy ionizing the medium and are capable of producing a number of extra electrons and creating a cluster-damage site. In order to estimate corrections due to δ-electrons, several quantities pertinent to these particles need to be analysed.

The first is the mean free path. According to Ref. [35], both the elastic and inelastic mean free paths of 100-eV electrons are about 1 nm. If such an electron is ejected in the most likely direction according to the binary interaction model [66], about

$70°$, this electron will start losing energy within 1 nm of the ion's trajectory. Even if it produces more electrons than a sub-50-eV electron, they will not spread much further than them. Because of the kinematic limit, for an ion in the Bragg peak region, energies of ejected secondary electrons are below 0.7 keV. These electrons with elastic mean free path of about 4.5 nm are emitted in the forward direction, and it can be shown that the maximal distance between the first collision and the ion's path is 1.6 nm and it is reached by the electrons of energies 400–500 eV. There is no way that further transport can carry further generations of electrons far beyond the 10 nm distance off the ion's path. In addition, the probability of producing a 400-eV secondary electron is only about 0.02 of that producing a 50-eV electron. Therefore, even though δ-electrons are not included in the random walk approach, the location of their effect is by and large overlapped with that of sub-50 eV secondary electrons. The number of electrons ejected as the consequence of ionization by δ-electrons can be estimated from energy conservation and these electrons have already been effectively included in the random walk, since $dN_e/d\zeta$ was obtained from the value of the stopping power, S_e.

Still another possibility exists for δ-electrons to affect the discussed scenario. If a much more energetic electron, i.e., with energy larger than 20 keV, then with the mean free path of the order of 100 nm, it can cause damage elsewhere. Moreover, these electrons cannot be ejected in the Bragg peak region, since the required ion energy must be over 9 MeV. The probability of such events is very small; it is less than that of emitting a 50-eV electron by a factor over 10^6. Therefore, this possibility is realised so rarely that it can be neglected.

6 Thermomechanical Effects

Thus far, the energy loss by incident ions, the transport of produced secondary particles, and the radiation damage induced by these particles have been discussed. The transport, described by diffusion or MC simulations, is that of the ballistic electrons, radicals, etc. in a static medium. This transport does not include the whole physical picture because propagating secondary particles transfer the energy further, making the medium hot and dynamic.

Energy relaxation in the medium has been studied in Ref. [25], where the inelastic thermal spike model was applied to liquid water irradiated with carbon ions. This model has been developed to explain track formation in solids irradiated with heavy ions and it studies the energy deposition to the medium by swift heavy ions through secondary electrons [112–122]. In this model, the electron-phonon coupling (strength of the energy transfer from electrons to lattice atoms) is an intrinsic property of the irradiated material.

The application of the inelastic thermal spike model to liquid water predicted that the temperature increases by 700–1200 K inside the hot cylinder by 10^{-13} s after the ion's traverse [25]. However, within this model, only coupled (between electrons and atoms of the medium) thermal conductivity equations are solved, while the further dynamics of the medium is missing. This dynamics is the consequence of a rapid pressure increase inside the hot cylinder around the ion's path up to 10^4 atm, while the pressure outside of it is about atmospheric. Since the medium is liquid, this pressure difference prompts rapid expansion, resulting in a shock wave, which has been analysed in Refs. [27, 28, 123, 124].

The shock waves discussed below have not yet been discovered experimentally. Therefore, their influence on the biodamage, also discussed in this section is somewhat precursory. Nevertheless, some of these effects were included in the calculation of clustered damage and survival curves in Sects. 7.1 and 7.2 in order to emphasise the potential importance of these effects and, therefore, for experimental efforts devoted to their discoveries.

6.1 Hydrodynamic Expansion on the Nanometre Scale

The problem of the expansion of the medium driven by the high pressure inside the hot cylinder is in the realm of hydrodynamics and it has been thoroughly analyzed in Ref. [27]. It has been shown that the expansion is cylindrically symmetric. If the ratio of pressures inside and outside of the hot cylinder is high enough, as happens for large values of LET, the cylindrical expansion of the medium is described as a cylindrical shock wave, driven by a "strong explosion" [125]. For an ideal gas, this condition holds until about $t = 1$ ns, but in liquid water the shock wave relaxes much sooner. In Ref. [123] the molecular dynamics simulations of liquid water expansion showed that the shock wave weakens by about 0.5 ps after the ion's passage.

The solution of the hydrodynamic problem describing the strong explosion regime of the shock wave, as well as its mechanical features and limitations, are very well described in Refs. [125–127]. In Ref. [27], the solution for the cylindrical case has been reproduced and analyzed in order to apply it for the nanometre-scale dynamics of the DNA surroundings. In this section, only the results pertinent to the further discussion of biodamage are presented.

The self similar flow of water and heat transfer depend on a single variable, ξ. This variable is a dimensionless combination of the radial distance, ρ, from the axis, i.e., the ion's path, the time t after the ion's passage, the energy dissipated per unit length along the axis, which is equal to the LET per ion, S_e, and the density of undisturbed water, $\rho = 1$ g/cm^3. This combination is given by

$$\xi = \frac{\rho}{\beta\sqrt{t}}\left[\frac{\rho}{S_e}\right]^{1/4}, \tag{62}$$

where β is a dimensionless parameter equal to 0.86 for $\gamma = C_P/C_V = 1.222$ [27] (where C_P and C_V are molar heat capacity coefficients at constant pressure and volume, respectively). The radius and the speed of the wave front are given by

$$R = \rho/\xi = \beta\sqrt{t}\left[\frac{S_e}{\rho}\right]^{1/4} \tag{63}$$

and

$$u = \frac{dR}{dt} = \frac{R}{2t} = \frac{\beta}{2\sqrt{t}} \left[\frac{S_e}{\rho} \right]^{1/4}, \tag{64}$$

respectively. It is also worthwhile to combine Eqs. (63) and (64) and obtain the expression of the speed of the front in terms of its radius R,

$$u = \frac{\beta^2}{2R} \left[\frac{S_e}{\rho} \right]^{1/2}. \tag{65}$$

Using Eq. (65), pressure P at the wave front can be obtained as

$$P = \frac{2}{\gamma + 1} \rho u^2 = \frac{1}{\gamma + 1} \frac{\beta^4}{2} \frac{S_e}{R^2}. \tag{66}$$

Then, one can solve the hydrodynamic equations in order to obtain the expressions for speed, pressure, and density in the wake of the shock wave, i.e., behind the wave front [27].

The following intriguing questions have been raised in Refs. [25, 27]. What can such a shock wave do to biomolecules such as DNA located in the region of its propagation through the medium; can it cause biodamage by mechanical force? The forces acting on DNA segments were predicted to be as large as 2 nN, which is more than enough to break a covalent bond, causing a strand break; however, these forces are only acting for a short time and it remained unclear whether this is sufficient to cause severe damage to DNA molecules. The other question is: how significant can the transport due to the collective flow of this expansion be compared to the diffusion of secondary particles?

6.2 *Transport of Reactive Species by the Radial Collective Flow*

The study done in Ref. [27] suggests that a considerable collective radial flow emerges from the hot cylinder region of medium. The maximal mass flux density carried by the cylindrical shock wave is given by $\rho_f u$, where $\rho_f = \frac{\gamma+1}{\gamma-1}\rho$ is the matter density on the wave front. This expression is proportional to u and its substitution from Eq. (65) yields that the mass flux is proportional to the $\sqrt{S_e}$. This flux density is inversely proportional to radius ρ and is linear with respect to the $\sqrt{S_e}$. It sharply drops to zero in the wake of the wave along with the density. A sharp rarefaction of the volume in the wake of the wave follows from the results of Ref. [27]. This is the effect of cavitation on a nanometer scale and due to this effect the water molecules of the hot cylinder along with all reactive species formed in this cylinder are pushed out by the radial flow. Such a mechanism of propagation of reactive species, formed within the hot cylinder, is competitive with the diffusion mechanism, studied in MC simulations done using track structure codes [35].

Intriguingly, the cylindrical shock wave accomplishes the transfer of reactive species such as hydroxyl and solvated electrons, which play important roles in chemical DNA damage [13, 18, 58] much more effectively than the diffusion mechanism. Indeed, the time at which the wave front reaches a radius ρ can be derived from Eq. (63) as it is equal to $(\rho^2/\beta^2)\sqrt{\rho/S_e}$. This time has to be compared to diffusion times, which can be estimated for different reactive species as ρ^2/D, where ρ is the distance from the ion's path and D is the corresponding diffusion coefficient. The ratio of these times is equal to $(\sqrt{\rho/S_e})D/\beta^2$. For all relevant species, the diffusion coefficient is less than 10^{-4} cm^2/s [96]. Therefore, the above ratio is less than $10^{-3}/\sqrt{S_e(\mathrm{keV/nm})}$, which is much less than unity even for protons. For instance, for carbon ion projectiles, the wave front reaches 5 nm from the path in 2.8 ps after the ion's traverse, while hydroxyl radicals reach the same distance via the diffusion mechanism in about 9 ns, a more than 3000 times longer time. In fact, the lifetime of hydroxyl free radicals is shorter than 5 ns [13, 39, 96], therefore the shock wave transport may be the only means to deliver hydroxyl radicals to distances farther than 3.5 nm of the ion's path.

The collective flow is expected to play a significant role in the transport of reactive species at values of LET that are large enough to produce a shock wave, even if this wave is not sufficiently strong to cause covalent bond ruptures. The analysis shows that even at small values of LET, typical for the plateau region in the LET dependence on depth (well before the Bragg peak), a shock wave is formed; however it damps and becomes acoustic at radii under 10 nm. At $S_e = 0.9$ keV/nm shock waves propagate further than 10 nm.

Thus, the effects following the local heating of the medium in the vicinity of an ion's path are quite striking. The MD simulations of a shock wave on a nanometer scale, initiated by an ion propagating in tissue-like medium, demonstrate that such a wave generates stresses, capable of breaking covalent bonds in a backbone of a DNA molecule located within 1.5 nm from the ion's path when the LET exceeds 4 keV/nm and this becomes the dominating effect of strand breaking at[15] $S_e > 5$ keV/nm. The LET of ∼4–5 keV/nm corresponds to the Bragg peak values for ions close to Ar and heavier in liquid water. Besides the dramatic effects at such high values of LET, it was found that weaker shock waves produced by carbon ions or even protons transport the highly reactive species, hostile to DNA molecules, much more effectively than diffusion.

Presently, shock waves on nm scale have not been discovered experimentally. Such a discovery would shift a paradigm in the understanding of radiation damage due to ions and would require re-evaluation of the relative biological effectiveness. The collective transport effects for all ions and direct covalent bond breaking by shock waves for ions heavier than argon are important factors in the scenario. These effects will also have to be considered for high-density ion beams, irradiation with intensive laser fields, and other conditions prone to causing high gradients of temperature and pressure on a nanometer scale.

---

[15]These values correspond to conservative estimates ($\varepsilon_0 = 3$ eV) [28]. They may be much lower if the actual thresholds appear to be smaller [58].

7 Estimation of Radio-Biological Effects

The essence of results obtained in Sects. 5.6.1 and 5.6.2 is that for a given ion, the numbers of SSBs and DSBs per unit length of the ion's path can be calculated. Or, alternatively, for a given DNA twist, the probability of the above lesions can be calculated. However, those calculations are still far from predicting whether the cell containing a given segment of DNA molecule will be inactivated or survive. This question is largely in the realm of biology, because of a variety of biological mechanisms, which are activated following the creation of a lesion. Nearby proteins are engaged in DNA repair and may or may not be successful. Such an activity is marked by the appearance of the so called foci that can be observed experimentally [128–130]. These protein foci remain visible (the cells must be stained beforehand) until the repair is finished. If a lesion cannot be repaired the cell containing this DNA molecule is likely to be inactivated. There is a plethora of biological studies directed at determining the probabilities of a successful DNA repair depending on the extent of the damage [101].

It is established that a simple SSB is most likely to be fixed within minutes after this lesion is produced. DSBs can also be fixed, however, with smaller probability and there is also a chance that its repair (e.g., the non-homologous end joining (NHEJ) type of DSB repair [101, 131]) may not be successful. The probability of repair is even smaller for multiply-damaged sites per target containing clustered DNA lesion or complex DNA damage. A clustered DNA lesion is defined as the number of DNA lesions, such as DSBs, SSBs, abasic sites, damaged bases, etc., that occur within about two helical turns of a DNA molecule so that, when repair mechanisms are engaged, they treat a cluster of several of these lesions as a single damage site [100, 101, 132–136]. In this section, we will explain the criterion for lethality of damage that was suggested in Ref. [30] and then implement it for the calculation of survival curves. This criterion is based on the idea that among different DNA lesions caused by interaction with reactive species the multiply damaged sites with a sufficient complexity may not be repaired. Namely, it was postulated [30] that a lesion combined of a DSB and at least two other simple lesions such as SSB within two DNA twists is lethal. Then in Ref. [32] this criterion was applied and justified for a number of cell types.

In regard to irradiation with heavy charged particles, the key assumption adopted in the MSA following Refs. [134, 135, 137] is that the leading cause of cell inactivation is the complexity of nuclear DNA damage. However, this may change in different conditions, e.g., when biodamage takes place in presence of sensitising nanoparticles [20, 138]. Indeed, it is currently established that radiosensitising nanoparticles delivered to cells are preferentially localised outside cell nuclei [139]. Therefore, the damage of other organelles may become more important.

7.1 Assessment of the Complex DNA Damage

When a DSB is induced due to secondary electrons, as discussed in Sects. 5.6.2 and 5.6.3, there is a substantial probability (between 0.1 and 0.2 for plasmid DNA) for a DSB to occur as a result of the interaction of the molecule with a single electron. However, it is difficult to expect a clustered DNA damage site to be caused by a single electron or another secondary particle, since the distance between lesions in such a site can be too large (more than 5 nm). Therefore, in Refs. [14, 26], the complexity of DNA damage has been quantified by defining a cluster of damage as a damaged portion of a DNA molecule by several independent agents, such as secondary electrons or radicals. Then, it is reasonable to expect that the probability that the electrons or radicals induce clustered damage is related to the fluence of these agents on a given DNA segment in the same sense as the probabilities of other types of lesions, such as SSBs or DSBs, as discussed in Sects. 5.6.2 and 5.6.3.

Therefore, it is natural to start with the calculation of the number of clustered damage sites, produced by an ion, per unit length of its trajectory, $\frac{dN_C}{d\zeta}$, similar to what was done for SSBs and DSBs in Eqs. (51) and (55). DNA molecules are on the surface of nucleosomes and the latter is modelled as a cylinder of radius 5.75 nm. Then, an element of its lateral surface that enwraps two twists of a DNA molecule serves as a target for secondary electrons and radicals. This segment of the surface is 2.3 nm wide (along the axis of the cylinder) and 6.8 nm long (along the cylinder's circumference). For definiteness, this cylinder is taken to be perpendicular to the ion's path, its axis to be at a distance ρ from the path, and the cylinder to be symmetric with respect to the plane containing the path and vector $\boldsymbol{\rho}$.

First, the fluence of secondary electrons on such a target, $\mathscr{F}_c$, using Eq. (50) has to be calculated. The geometry for this problem is similar to that for the calculation of the fluence on a cylinder perpendicular to the ion's path, considered in Sect. 5.6. The details for the calculation of fluence $\mathscr{F}_c(\rho)$ on a double-twist target are given in Ref. [30]. This number multiplied by Γ_e gives the average number of SSBs for a given DNA segment due to secondary electrons.

Then, the fluence for free radicals and other reactive species, $\mathscr{F}_r(\rho)$, is calculated. This fluence includes the physics pertinent to the transport of reactive species, such as the relaxation of ionisation energy in the medium and the (predicted) cylindrical shock wave around the ion's path [30, 32]. In order to calculate the average number of SSBs and similar lesions for a given DNA segment due to reactive species, $\mathscr{N}_{\rm r}(\rho)$, fluence $\mathscr{F}_r(\rho)$ has to be multiplied by the probability of inducing a lesion per hit, Γ_r. Since neither the exact $\mathscr{F}_r(\rho)$, nor Γ_r are known, we assume the former to be uniform within a certain distance from the ion's path, and we introduce a convoluted function $\mathscr{N}_{\rm r}(\rho)$:

$$\mathscr{N}_{\rm r}(\rho) = \begin{cases} \mathscr{N}_{\rm r}, & \rho \le R_{\rm r} \\ 0, & \rho > R_{\rm r}, \end{cases} \tag{67}$$

where $R_{\rm r}$ is the effective distance for free radicals propagation which depends on the projectile's velocity and charge. We considered this value in the range between 5 and

10 nm [32]. A uniform distribution of radicals within a certain distance from the ion's path implies that the reactive species, formed in the nearest proximity to the path, are transported by a shock wave and their number density is nearly uniform inside the cylinder that enwraps the decayed shock wave [30]. The average probability $\mathcal{N}_{\rm r}$ for SSBs caused by free radicals to take place was estimated as 0.08 from the comparison of the experimental results [111] for plasmid DNA dissolved in pure water or in a scavenger-rich solution [30, 32].

The value of $\mathcal{N}_{\rm r}$ is also affected by environmental conditions of an irradiated target. In the case of hypoxic conditions, the value $\mathcal{N}_{\rm r}$ is reduced because the radical-induced damage may be repaired if oxygen is not present. The quantitatively correct description of the experimental survival probabilities of cells irradiated under hypoxic conditions was achieved by utilising the value $\mathcal{N}_{\rm r} = 0.04$ which is two times smaller than that in the aerobic environment; this corresponds to experimental data on the induction of DSBs and non-DSB clustered DNA lesions in mammalian cells at normal concentration of oxygen and at deep hypoxia [140]. Further work, however, is needed to explore, in more detail, how the probability of lethal lesion production by free radicals depends on the environmental conditions, e.g., at intermediate concentrations of oxygen [141].

Thus, the average number for multiple damage sites per target containing clustered damage at a distance ρ from the ion's path is given by

$$\mathcal{N}_c(\rho) = \mathcal{N}_e(\rho) + \mathcal{N}_r(\rho) = \Gamma_e \mathcal{F}_e(\rho) + \Gamma_r \mathcal{F}_r(\rho), \tag{68}$$

where the functions $\mathcal{N}_e(\rho)$ and $\mathcal{N}_r(\rho)$ define the average number of lesions like SSBs, base damages, abasic states, etc., induced by secondary electrons and other reactive species (free radicals, pre-solvated and solvated electrons, etc.), respectively.

Then, the criterion of lethality is introduced. From the analysis of survival curves, it has been assumed that a complex lesion that contains a DSB and at least two more simple lesions such as SSB, base damage, etc. is lethal[16] [30]. A cell that contains at least one lethal lesion becomes inactivated, i.e., it either dies, or becomes sterile. The assumption has been successfully tested on a number of cell lines in different conditions of irradiation [32]. Survival curves turned to be very sensitive to modifications to this criterion (e.g., no DSB, or DSB plus at least one simple lesion, or DSB plus at least three simple lesions). Therefore, in cases of disagreement, some corrections to other parameters were introduced, but this criterion has not been modified. So, according to this criterion, the probability of lethal damage, $P_l(\rho)$, is given by

$$P_l(\rho) = \lambda \sum_{\nu=3}^{\infty} \frac{\mathcal{N}_c^{\nu}}{\nu!} \exp\left[-\mathcal{N}_c\right]. \tag{69}$$

The sum starts from $\nu = 3$, which makes the minimum order of lesion complexity at a given site to be larger or equal to three. The factor λ, first introduced in Eq. (55),

[16]This corresponds to T3-DSB [101].

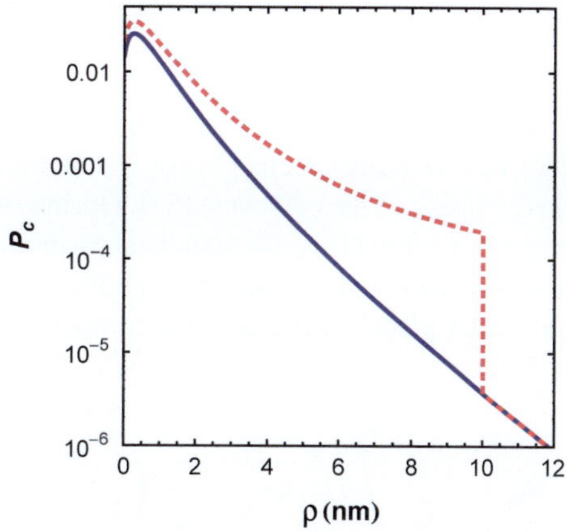

Fig. 11 The dependence of probabilities for complex damage to be induced by secondary electrons on the distance from the nucleosome to the ion's path. The *solid line* represents the order of complexity larger than three. The *dotted line* shows the inclusion of the effect of radicals uniformly distributed inside a 10-nm cylinder. This *curve* is plotted with parameters discussed in Sect. 7.2

indicates that one of the simple lesions is converted to a DSB. This implies that in the current model the DSBs occur via SSB conversion; in principle, other mechanisms can also be taken into account. The dependence of probability P_l on distance is shown in Fig. 11 [30]. The observed decrease at small ρ is again a geometric effect caused by the rapid decrease in the area.

The next step is the integration of inactivation probability over the volume of the cell nucleus with the number density of such sites in order to calculate the yield of lethal damages per segment of ion's trajectory (similarly to $\frac{dN_{SSB}}{d\zeta}$ and $\frac{dN_{DSB}}{d\zeta}$). This integral,

$$\frac{dN_l}{d\zeta} = n_s \int_0^{\infty} P_l(\rho) 2\pi\rho d\rho = n_s \sigma_l, \tag{70}$$

where n_s is the number density of sites, gives the number of clustered damage sites per unit length of the ion's trajectory. Equation (70) can be numerically integrated similar to Eqs. (51) and (55). For $\Gamma_e = 0.03$ corresponding to that found in Sect. 5.6.3, $\sigma_l = \int_0^{\infty} P_l(\rho) 2\pi\rho d\rho = 0.03\ \text{nm}^2$ (these calculations are done for alpha particles near the Bragg peak).

The number density of targets, n_{s}, is proportional to the ratio of base pairs accommodated in the cell nucleus to the nuclear volume, $n_{\text{s}} \sim N_{\text{bp}}/V_{\text{n}}$. The coefficient of proportionality takes into account that a target represents a double DNA twist comprising 20 bp [30]:

$$n_{\text{s}} = \frac{N_{\text{bp}}}{20\, V_{\text{n}}} = \frac{3\pi}{8} \times \frac{N_{\text{bp}}}{20\, A_{\text{n}}\, \bar{z}}, \tag{71}$$

where the cross sectional area, A_{n}, of the cell nucleus and its diameter, D_{n} are defined for given cells; an average length of ions' traverse through a nucleus is $\bar{z} = \pi D_{\text{n}}/4$. Taking into account the chromatin dynamics during the cell cycle and that diploid cells contain a double set of chromosomes, one gets the final expression for n_{s}:

$$n_s = \frac{1.67 \times 2}{20} \times \frac{3\pi}{8} \times \frac{N_g}{A_n\,\bar{z}} = \frac{\pi}{16} \times \frac{N_g}{A_n\,\bar{z}}, \tag{72}$$

where N_g is genome size (e.g., equal to 3.2 Gbp for human cells [142] and to 2.7 Gbp for Chinese hamster cells [143]). The factor 1.67 arises because of dependence of N_{bp} on the phase of the cell cycle [32]. During interphase, the number density n_s remains constant during G_1 phase, which takes about 1/3 of the total cell cycle duration (T_c) in human cells [144], but becomes doubled in the S and G_2 phase after DNA replication has taken place. Averaging the number density of DNA over the different phases, one gets

$$N_{bp} = \frac{2\,N_g}{T_c}\int_0^{T_c} f(t)\,dt = \frac{2\,N_g}{T_c}\left(\frac{1}{3}T_c + 2\times\frac{2}{3}T_c\right) = \frac{2\,N_g}{T_c}\times 1.67T_c = 3.33\,N_g\,. \tag{73}$$

7.2 Obtaining the Survival Curves

The assessment of RBE for ions, from the point of view of the MSA, starts from the calculation of survival curves for a given type of cell irradiated with a given type of ions at given conditions. This means that for a given type of cell and a given dose the probability of cell survival (or inactivation) has to be calculated. In the previous section, the probability of cell inactivation was related to the probability of inducing a nuclear DNA lesion of a certain complexity, so that it is unlikely to be repaired with proteins. In this section, the accomplishments of previous section are applied to the calculation of survival curves.

The effect of each ion can be treated independently from others as long as the average distance between the paths is considerably larger than the radii of tracks. Typical doses used in ion beam therapy are small [137] and the above condition is satisfied [30]. For a given type of cell, given dose, and LET, the average number of ions that traverse a cell nucleus can be calculated as $N_{ion} = A_n\,d/S_e$, where A_n is the cross sectional area of the cell nucleus and $S_e = |dE/dx|$ is a part of LET spent on ionization of tissue. This, similar to our experience with plasmid DNA in Sect. 5.6.3, gives us the dose dependence. The average number of lethal DNA lesions in the cell nucleus is given by the following expression, similar to Eqs. (58) and (60),

$$Y_l = \frac{dN_l}{d\zeta}\sum_{j=1}^{\infty} z_j\, j\, \mathscr{P}_j(d), \tag{74}$$

where the $\frac{dN_l}{d\zeta} z_j$ is the average number of lethal lesions per single ion's traverse and the sum $\sum_{j=1}^{\infty} z_j j\,\mathscr{P}_j(d) = \bar{z}\sum_{j=1}^{\infty} j\,\mathscr{P}_i(d)$ yields an average length of traverse of all ions passing through a cell nucleus for a given dose [30]. Probability $\mathscr{P}_j(d)$ for exactly j ions traversing the cell nucleus is found similarly to (56),

$$\mathscr{P}_j(d) = \frac{N_{\text{ion}}^j}{j!}\, e^{-N_{\text{ion}}} \tag{75}$$

At large values of N_{ion}, $\bar{z}$ becomes dose-dependent. For $N_{\text{ion}} \leq 10^2$ (in the current study this condition is satisfied), $\bar{z}$ is nearly constant and substitution of Eq. (75) into (74) yields a linear dependence of the number of lethal lesions per cell nucleus on dose:

$$Y_l = \frac{dN_l}{dx}\,\bar{z}\,N_{\text{ion}}(d) = \frac{\pi}{16}\sigma_l N_{\text{g}}\frac{1}{S_e}\,d\,. \tag{76}$$

Equation (76) gives the number of lethal lesions per cell nucleus. Therefore, the probability of cell inactivation, Π_l is obtained by subtracting the probability of zero lethal lesions occurrence from unity,

$$\Pi_s = 1 - \exp\left[-Y_l\right]. \tag{77}$$

Then, the probability of cell survival is given by unity less the probability of cell inactivation, i.e., by that second term of Eq. (77):

$$\Pi_{surv} = 1 - \Pi_l = \exp\left[-Y_l\right]. \tag{78}$$

In virtue of (77) and (76), the logarithm of Π_{surv},

$$\ln \Pi_{surv} = -Y_l = -\frac{\pi}{16}\sigma_l N_{\text{g}}\frac{1}{S_e}\,d, \tag{79}$$

is linear with dose.

Figure 12 [32] shows the survival curves for human adenocarcinomic A549 cells irradiated with protons and alpha-particles at different values of LET. The calculated curves (lines) are compared to the experimental data (symbols) on survival of the same cells in the same conditions. These survival curves are calculated using Eq. (79) and exhibit a good agreement with experiments. More examples are demonstrated

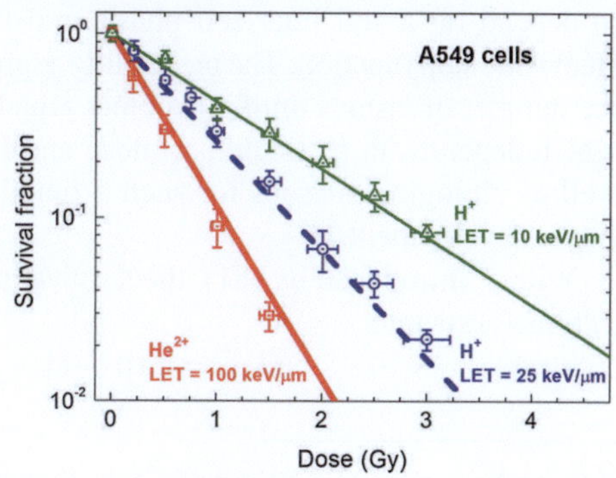

Fig. 12 Survival curves for A549 cells irradiated with protons for values of LET of 10 and 25 eV/nm and α particles with the LET at 100 eV/nm. *Solid line* is calculated using Eq. (79). The *dots* represent the experimental data [145, 146]

in aerobic and hypoxic conditions for a variety of LET values in Ref. [32]. Linear survival curves are observed in a large variety of cells and projectiles [32, 147–149]. From the model point of view this corresponds to the so-called single-hit model described in Ref. [39]. The enzymatic repair is included in the above model in the definition of the criterion of lethality. A change in the definition would change the slope of survival curves, but will not affect its linearity.

However, in some cases, there is an evidence that survival curves as a result of irradiation with ions can be "shouldered" consequent to successful damage repair; i.e., in the language of LQ model, Eq. (1), the coefficient β may be noticeably large [150]. In the framework of the MSA, this means that in these cases, some complex lesions can still be repaired. A possibility of repair leads to the deviation from a purely exponential behaviour of survival probability and the appearance of shoulders in survival curves, which have been observed in experiments for specific cells [149, 151]. An example for such curves for the repair-efficient Chinese hamster CHO cell line is considered in Ref. [32].

In these cases, the deviation from a purely exponential behaviour of survival probability can be explained by a possibility of repair of complex lesions, which were assumed to be lethal above. For this purpose, a biological parameter, the probability of a successful repair of a complex lesion, χ, is introduced and Eq. (78) transforms into

$$\Pi_{\rm surv} = e^{-Y_l} + \sum_{\mu=1}^{\infty} \chi^{\mu} \frac{Y_l^{\mu}}{\mu!} e^{-Y_l} = e^{-(1-\chi)Y_l}, \tag{80}$$

where each term in the sum represents the probability of exactly μ complex lesions to be induced multiplied by χ^{μ}, since all of these lesions must be repaired.

The probability of repair of a complex damage may depend on the cell's response to radiation, which involves specific biological mechanisms of damage repair [152]. Although the exact form of this dependence is unknown, the simplest function of probability, χ, can be introduced as a linear function of Y_l,[17]

$$\chi = (\chi_0 - \chi_1 Y_l)\, \Theta(\chi_0 - \chi_1 Y_l), \tag{81}$$

where the positive parameters χ_0 and χ_1 of the function of probability are likely to depend on a cell line, cell phase, and irradiation conditions, and $\Theta(x)$ is the Heaviside step function. The probability χ gradually approaches zero with increasing the number of lesions until it becomes equal to zero at a critical value, $\tilde{Y}_l = \chi_0/\chi_1$, which depends, in particular, on dose and LET. A study of these dependencies as well as biological reasons for such a functional dependence may be a goal of the future development.

With χ introduced by (81), the survival probability (80) below the critical value $\tilde{Y}_l$ transforms into

$$\Pi_{\rm surv} = \exp\left[-(1-\chi_0)Y_l - \chi_1 Y_l^2\right]. \tag{82}$$

[17] As can be seen from Eq. 80, constant χ does not make the survival curve shouldered.

Above the critical value, Eq. (78) remains valid. Thus, these equations explain the meaning of the critical value $\tilde{Y}_l$ as the transition point in the survival curve from the linear-quadratic to the linear regime. Examples of calculations and comparisons with experiments are given in Ref. [32].

The logarithm of $\Pi_{\rm surv}$ given by (82) can be written as

$$-\ln \Pi_{\rm surv} = (1-\chi)Y_l = Y_l - \Theta(\chi_0 - \chi_1 Y_l)\,(\chi_0 - \chi_1 Y_l)\,Y_l\,. \tag{83}$$

For $Y_l < \chi_0/\chi_1$, the survival probability in virtue of Eq. (76) is as follows,

$$-\ln \Pi_{\rm surv} = (1-\chi_0)\,\frac{\pi}{16}\sigma_l N_{\rm g}\,\frac{d}{S_e} + \chi_1 \left(\frac{\pi}{16}\sigma_l N_{\rm g}\right)^2 \frac{d^2}{S_e^2}\,. \tag{84}$$

This result provides the molecular-level justification of the empirical LQ parameters α and β for doses $d \leq \frac{16}{\pi}\frac{S_e}{\sigma N_g}\frac{\chi_0}{\chi_1}$:

$$\alpha = (1-\chi_0)\,\frac{\pi}{16}\sigma_l N_{\rm g}\,\frac{1}{S_e}, \qquad \beta = \chi_1 \left(\frac{\pi}{16}\sigma_l N_{\rm g}\right)^2 \frac{1}{S_e^2}\,. \tag{85}$$

At $Y_l > \chi_0/\chi_1$, i.e., for $d > \frac{16}{\pi}\frac{S_e}{\sigma N_g}\frac{\chi_0}{\chi_1}$, survival curves are linear,

$$-\ln \Pi_{\rm surv} = \frac{\pi}{16}\sigma_l N_{\rm g}\,\frac{d}{S_e}, \tag{86}$$

and the parameter α transforms into

$$\alpha = \frac{\pi}{16}\sigma_l N_{\rm g}\,\frac{1}{S_e}\,. \tag{87}$$

The calculated survival probabilities depend on the following numbers, which were either used as parameters or were determined from the comparison with experiments: the number of secondary electrons produced by the ion per unit length, $dN_e/d\zeta$, that can be calculated using Eq. (17) and dependent on LET, the probability for an electron incident on DNA to induce a SSB per one DNA twist, Γ_e, the fluence of radicals on a DNA twist, $\mathscr{F}_r$, or the number of produced radicals and the average radius of their propagation from the ion's trajectory that allow one to estimate it, the probability for radicals to induce SSBs, Γ_r, the number density of targets, n_s, (assumed to be uniform) in a given cell. The enzymatic repair has been taken into account by the introduction of the criterion of lesion lethality and (for some cases) the probability χ of complex lesion repair. Some of the parameters in this list can be found from the literature, estimated or calculated (rather accurately) for given media, cells, and projectiles. Some of them remain unknown for now, but further research may clarify their values. It is noteworthy that a complicated problem of the calculation of survival probabilities can be reduced to the search of several microscopic parameters that can be determined theoretically or experimentally.

We just demonstrated the capability of MSA to produce the parameters of the linear quadratic model in a rather universal way. A variety of survival curves are calculated using the same criterion of lethality and lesion probability values successfully compare to experiments. (See the chapter by A.V. Verkhovtsev, E. Surdutovich and A. V. Solov'yov of this book for more detail.) These calculations are done for different values of LET that correspond to protons, α-particles, or carbon ions in the vicinity of the Bragg peak, or away from it, in aerobic or hypoxic conditions. This is an important verification of the predictive power of the MSA and this opens the way to clinical implementations of this approach.

7.3 The Recipe for Obtaining the Survival Probabilities

In this section, the MSA is summarised in a recipe for a phenomenon-based assessment of radiation damage that results in the calculation of survival curves. This section connects all ideas discussed above together into an integrated approach to the prediction of survival probabilities. These survival curves can be compared to those for photons and the RBE can be calculated. Below, we show all necessary steps in order to calculate the survival probability for cells in an irradiated region. Let us imagine that a certain type of cell is irradiated with certain ions. First, it is desirable to know the composition of the medium. Cross sections of ionization, excitation, and nuclear fragmentation affect the shape of the LET-depth dependence. Section 4.2 gives recipes for determining some of these cross sections for water and more complex media. Second, using Eqs. (12–16), the LET dependence on energy and longitudinal coordinate can be obtained. This gives the location of the Bragg peak, its height, and other features of the LET curve. The conditions related to secondary particles should be assessed: the energy spectrum of secondary electrons, their average energy, and other features provide the grounds for inference on what methods can be used for the calculation of their transport and energy transfer. If the value of the LET (at least at some section of the ion's propagation) is higher than 4 keV/nm, shock wave effects may dominate the scenario of biodamage.

Third, the cells should be thoroughly investigated. Thus far in the MSA, the cell nucleus was by and large considered to be the main target, however there could be conditions in which other parts of a cell, such as the cell membrane, cytoplasm, mitochondria, and other organella are targets, whose damage may be lethal to the cell. If the nuclear DNA is the target, it is important to know how it is distributed. Any information on the structure of chromatin, its volume, shape, and other characteristics (which depend on the position in the cell cycle) is important for the description of the target.

Fourth, as soon as the target is described the fluence of secondary particles, such as secondary electrons and radicals, on this target is calculated. The random walk has been used in this and other works related to the MSA to describe transport of electrons, but it can be calculated using more sophisticated methods. Radicals diffuse through the medium as well, but they may also be carried by the collective flow of the shock wave. The DNA damage due to radicals depends on their production and transport, both of which depend on the LET. High temperatures inside the hot cylinder and consequent shock waves contribute to these processes. More research is needed first to verify the existence of shock waves on the nanometre scale and then to describe the transport of reactive species including chemical reactions. The ultimate effect of radicals depends on the hypoxic/aerobic conditions in the medium. If the effectiveness is known, the OER can be determined for the given conditions.

Fifth, the average number of DNA lesions of interest (that could be lethal) per unit length of the ion's trajectory should be calculated. This implies the integration and averaging of damaging effects in the radial direction with respect to the ion's trajectory. Then, knowing the size of the cell nucleus (or other target), the dose dependence of number traversing ions and lethal lesions yield using Eqs. (57) and (58) can be determined. The survival curve can then be calculated using Eq. (79) or Eq. (84), if the special repair properties are involved. The comparison of the survival curve with that for x-rays gives the RBE for a given location. This location is described with the value of the LET and the depth coordinate that corresponds to it. This means that these calculations predict the RBE (and OER) at the Bragg peak, plateau, and the tail of the LET-depth dependence. Then, if the tissue can be scanned to produce a spread-out Bragg peak, the calculations can be superimposed.

Enzymatic repair mechanisms play an important role for the overall damage assessment. They are included in several places. Mainly, their effectiveness determines the criterion for lesion lethality. The resultant survival curves will be based only on α coefficient, given by Eq. 87. Function χ for additional repair effects should only be introduced if there is an evidence that the survival curves for given cells are shouldered. In general, more biological input is necessary both for given calculations and for the further development of the MSA. However, there is a hope that in the application to radiation damage with ions and high LET, the uncertainty related to biological effects is less than at lower LET values. This hope has been strongly supported by a recent successful verification of the recipe with the above criterion of lethality on a number of cell lines across a wide range of LET and different oxygen conditions [32]. Also, advanced track structure codes allow to do modelling on the biological scale [153]. For the analytical MSA, it would be interesting to develop the biological input in relation to these models.

8 Conclusions and Outlook

Thus, the multiscale approach to the assessment of radiation damage with ions has been reviewed. It has been initiated less than ten years ago and now it is in the stage when a number of aspects can still be improved, but by and large it is completed. It has evolved from a curious desire to understand peculiarities of radiation damage with ions to a unique scientific theoretical predictive method for calculation of survival probabilities of cells irradiated with ions. The main difference of MSA from other approaches is the in-depth focusing on physical effects, and, therefore, it is referred to as a phenomenon-based approach. Even though a huge effort should be undertaken to implement it for therapy planning, its experimental verification [32] makes such implementation feasible.

Section 7.3 gives an example of how the logic of the MSA is applied to the assessment of survival probabilities. This is presented as a recipe, which may be employed in a clinical code. This recipe allows one to look into the mechanisms that determine these probabilities. Every element in the scenario (representing a certain scale) can be investigated independently and in most cases this can be done rather simply. The use of analytical or semi-analytical tools allows to understand the key effects on each scale. The main advantages of the MSA follow from its architecture, its fundamentality, and its versatility. The approach evaluates the relative contributions and significance of a variety of phenomena; it elucidates a complex multiscale scenario in sufficient detail and has a solid predictive power. It is structurally simple and inclusive, and allows for modifications and extensions by including new effects on different scales and improvements on the way.

The latest achievements of the MSA include the calculations of survival probabilities for different conditions with respect to oxygen concentration [32]. The OER has been obtained as a natural byproduct of calculation of survival probabilities. More detail on this can be found in the chapter by A.V. Verkhovtsev, E. Surdutovich and A.V. Solov'yov of this book. Further understanding of how to treat biological effects such as enzymatic repair has been obtained as well [32]. This success hints that the further development of the MSA can be in done in the biological direction. Therefore, further validation of the method on different biological targets can be suggested as the first area of its development. It will further clarify the parameters used currently, the use of repair coefficients, and may relate the latter to biological repair models.

The second area is related to the further understanding of the role of shock waves predicted in the process of the analysis of scenario of radiation damage. Thus far, these waves have not been observed, therefore we welcome experimental efforts in order to detect them. In addition to this, theoretical ideas of effects that can be observed experimentally and that can confirm the existence of shock waves. The shock waves may play a crucial role in the direct biodamage and production and transport of radicals.

The third area of development is related to the modification of the medium. The use of nanoparticles such as gold nanoparticles (GNP) as sensitizers has been discussed both theoretically and experimentally [20, 154], in order to boost the production of secondary electrons near the target and thus increase the RBE. The use of nanoparticles is considered for different modalities. Such a modification of the medium should be feasible within the MSA. The relevant cross sections of secondary electron production and their energy spectrum will define their effect on nearby biomolecules.

The fourth area is the modification of modality. Ions heavier than carbon require a better understanding of thermomechanical effects discussed above, since the shock waves initiated in the Bragg peak area would be more pronounced. The use of these ions may not necessarily be therapeutic. Rather, the understanding of the mechanisms of radiation damage at very high values of LET will help the assessment of the hazards of exposure to such ions during space missions or elsewhere. Also, the targets may not necessarily be biological, e.g., the assessment of radiation damage of electronics or other equipment can be done in a similar fashion. Another aspect, which can also be regarded as a modality modification, is the series of effects related to irradiation with ion beams produced by high-power lasers [109, 155]. In these conditions, the beam is much more dense and the tracks substantially interfere. The application of the MSA for the calculation of survival curves may be especially beneficial in this case.

The future development of the MSA will make a worthwhile tool for the assessment of radiation damage on the molecular level. While there is more work to be done to make it practical, its fundamental basis and depth related to atomic/molecular physics has become evident.

Acknowledgements We are grateful to R. Garcia-Molina, M. Niklas, John Posa, I.M. Solovyeva, I.A. Solov'yov, P. de Vera, and A.V. Verkhovtsev for the assistance with figures, important advice, and insight, Center for Scientific Computing of Goethe University, and the support of COST Action MP1002 "Nano-scale insights in ion beam cancer therapy" and FP7 ITN-ARGENT.

References

1. Surdutovich E, Solov'yov A (2012) J Phys Conf Ser 373:012001
2. Baccarelli I, Gianturco F, Scifoni E, Solov'yov A, Surdutovich E (2010) Eur Phys J D 60:1
3. Amaldi U, Kraft G (2007) J Radiat Res 48:A27
4. Schardt D, Elsässer T, Schulz-Ertner D (2010) Rev Mod Phys 82:383
5. Durante M, Loeffler J (2010) Nat Rev Clin Oncol 7:37
6. Particle therapy co-operative group. http://www.ptcog.ch/index.php/facilities-in-operation. Accessed May 2016
7. Haettner E, Iwase H, Schardt D (2006) Rad Protec Dosim 122:485
8. Sihver L, Schardt D, Kanai T (1998) Jpn J Med Phys 18:1
9. Pshenichnov I, Mishustin I, Greiner W (2008) Nucl Inst Methods B 266:1094
10. Surdutovich E, Obolensky O, Scifoni E, Pshenichnov I, Mishustin I, Solov'yov A, Greiner W (2009) Eur Phys J D 51:63
11. Scifoni E, Surdutovich E, Solovyov A (2010) Phys Rev E 81:021903
12. Chatterjee A, Holley WR (1993) Adv Radiat Biol 17:181
13. von Sonntag C (1987) The chemical basis of radiation biology. Taylor & Francis, London
14. Surdutovich E, Gallagher DC, Solov'yov AV (2011) Phys Rev E 84:051918
15. Sanche L (2005) Eur Phys J D 35:367
16. Boudaïffa B, Cloutier P, Hunting D, Huels MA, Sanche L (2000) Science 287:1658
17. Huels MA, Boudaïffa B, Cloutier P, Hunting D, Sanche L (2003) JACS 125:4467
18. Sanche L (2010) In: Greenberg M (ed) Radical and radical ion reactivity in nucleic acid chemistry. Wiley, New York, p 239
19. Surdutovich E, Solov'yov AV (2012) Eur Phys J D 66:206
20. McMahon S, Hyland W, Muir M, Coulter J, Jain S, Butterworth K, Schettino G, Dickson G, Hounsell A, O'Sullivan J, Prise K, Hirst D, Currell F (2011) Sci Rep 1:18
21. Solov'yov A, Surdutovich E, Scifoni E, Mishustin I, Greiner W (2009) Phys Rev E 79:011909
22. Cost action nano-ibct. http://mbnresearch.com/project-nanoibct. Accessed Feb 2016
23. Surdutovich E, Solov'yov A (2009) Europhys. News 40/2:21
24. Surdutovich E, Scifoni E, Solov'yov A (2010) Mutat Res 704:206
25. Toulemonde M, Surdutovich E, Solov'yov A (2009) Phys Rev E 80:031913
26. Surdutovich E, Yakubovich A, Solov'yov A (2010) Eur Phys J D 60:101

27. Surdutovich E, Solov'yov A (2010) Phys Rev E 82:051915
28. Surdutovich E, Yakubovich AV, Solov'yov AV (2013) Sci Rep 3:1289
29. de Vera P, Garcia-Molina R, Abril I, Solov'yov AV (2013) Phys Rev Lett 110:148104
30. Surdutovich E, Solov'yov A (2014) Eur Phys J D 68:353
31. Surdutovich E, Solov'yov A (2015) Eur Phys J D 69:193
32. Verkhovtsev A, Surdutovich E, Solov'yov A (2016) Sci Rep 6:27654
33. Frese MC, Yu VK, Stewart RD, Carlson DJ (2012) Int J Radiat Oncol 83:442
34. Semenenko V, Stewart R (2006) Phys Med Biol 51:1693
35. Nikjoo H, Uehara S, Emfietzoglou D, Cucinotta FA (2006) Radiat Meas 41:1052
36. Nikjoo H, Bolton C, Watanabe R, Terrisol M, O'Neill P, Goodhead D (2002) Radiat Prot Dosim 99:77
37. Friedland W, Jacob P, Bernhardt P, Paretzke H, Dingfelder M (2003) Radiat Res 159:401
38. Plante I, Cucinotta F (2010) Radiat Environ Biophys 49:5
39. Alpen EL (1998) Radiation biophysics. Academic Press, San Diego, London, Boston, New York, Sydney, Tokyo, Toronto
40. Hall EJ, Giaccia AJ (2012) Radiobiology for radiologist. Lippincott Williams & Wilkins, Philadelphia, Baltimore, New York, London
41. Hawkins R (1996) Int J Radiat Biol 69:739
42. Hawkins R (2009) Radiat Res 172:761
43. Goodhead D, Thacker J, Cox R (1993) Int J Radiat Biol 63:543
44. Goodhead D (2006) Radiat Prot Dosim 122:3
45. Butts JJ, Katz R (1967) Radiat Res 30:855
46. Katz R, Ackerson B, Homayoonfar M, Sharma SC (1971) Radiat Res 47:402
47. Korcyl M, Waligórski M (2009) Int J Radiat Biol 85:1101
48. Cucinotta F, Nikjoo H, Goodhead D (1999) Radiat Environ Biophys 38:81
49. Scholz M, Kraft G (1996) Adv Space Res 18:5
50. Friedrich T, Scholz U, Elsässer T, Durante M, Scholz M (2012) Int J Radiat Biol 88:103
51. Pimblott S, Siebbeles L (2002) Nucl Inst Methods B 194:237
52. Pimblott S, LaVerne J, Mozumder A (1996) J Phys Chem 100:8595
53. Pimblott S, LaVerne J (2007) Radiat Phys Chem 76:1244
54. Meesungnoen J, Jay-Gerin JP, Filali-Mouhim A, Mankhetkorn S (2002) Radiat Res 158:657
55. Surdutovich E, Solov'yov AV (2012) Eur Phys J D 66:245
56. Bug M, Surdutovich E, Rabus H, Rosenfeld AB, Solov'yov AV (2012) Eur Phys J D 66:291
57. Park Y, Li Z, Cloutier P, Sanche L, Wagner J (2011) Radiat Res 175:240
58. Smyth M, Kohanoff J (2012) J Am Chem Soc 134:9122
59. Becker D, Adhikary A, Sevilla M (2010) Charged particle and photon interactions with matter recent advances, applications, and interfaces. CRC Press, Taylor & Francis, Boca Raton
60. Surdutovich E, Solov'yov AV (2013) J Phys Conf Ser 438:012014
61. Bethe H (1930) Ann Phys 397:325
62. Bloch F (1933) Z Phys A Hadrons Nucl 81:363
63. Bloch F (1933) Ann Phys 408:285
64. Abril I, Garcia-Molina R, Denton C, Kyriakou I, Emfietzoglou D (2011) Radiat Res 175:247
65. Obolensky O, Surdutovich E, Pshenichnov I, Mishustin I, Solov'yov A, Greiner W (2008) Nucl Inst Methods B 266:1623
66. Rudd ME, Kim YK, Madison DH, Gay T (1992) Rev Mod Phys 64:441
67. Landau L, Lifshitz E, Pitaevskii L (1984) Electrodynamics of continuous media, vol 8, 2nd edn. Butterworth-Heinemann, Burlington
68. Lindhard J (1954) K Dan Vidensk Selsk Mat Fys Medd 28:8
69. Tan Z, Xia Y, Zhao M, Liu X, Li F, Huang B, Ji Y (2004) Nucl Instrum Methods Phys Res B 222:27
70. Altarelli M, Smith D (1974) Phys Rev B 9:1290
71. Garcia-Molina R, Abril I, Kyriakou I, Emfietzoglou D (2012) Chap. 15. In: Gómez-Tejedor GG, Fuss MC (eds) Radiation damage in biomolecular systems. Springer, Dordrecht
72. Ritchie RH, Howie A (1977) Philos Mag 36:436

73. Dingfelder M, Hantke D, Inokuti M, Paretzke H (1999) Radiat Phys Chem 53:1
74. Emfietzoglou D (2003) Radiat Phys Chem 66:373
75. Bernhardt P, Paretzke HG (2003) Int J Mass Spectrom 223–224:599
76. Peudon A, Edel S, Terrisol M (2006) Radiat Prot Dosim 122:128
77. Kim YK et al (2004) Electron-impact ionization cross section for ionization and excitation database (version 3.0). http://www.nist.gov/pml/data/ionization/index.cfm
78. White DR, Griffith RV, Wilson IJ (1992) Photon, electron, proton and neutron interaction data for body tissues. International Commission on Radiation Units and Measurements (ICRU 46), Bethesda, MD
79. Wilson WE, Miller JH, Toburen LH, Manson ST (1984) J Chem Phys 80:5631
80. Rudd M, Goffe T, DuBois R, Toburen L (1985) Phys Rev A 31:492
81. Bolorizadeh MA, Rudd ME (1986) Phys Rev A 33:888
82. Iriki Y, Kikuchi Y, Imai M, Itoh A (2011) Phys Rev A 84:052719
83. Simons J (2007) Adv Quantum Chem 52:171
84. Bichsel H (1988) Rev Mod Phys 60:663
85. Dingfelder M, Inokuti M, Paretzke H (2000) Rad Phys Chem 59:255
86. Garcia-Molina R, Abril I, de Vera P, Kyriakou I, Emfietzoglou D (2012) J Phys Conf Ser 373:012015
87. Barkas WH (1963) Nuclear research emulsions I. techniques and theory, vol 1. Academic Press, New York, London
88. Schiwietz G, Grande PL (2001) Nucl Instr Methods B 175–177:125
89. Kundrat P (2007) Phys Med Biol 52:6813
90. Hollmark M, Uhrdin J, Belkic D, Gudowska I, Brahme A (2004) Phys Med Biol 49:3247
91. Inokuti M (1971) Rev Mod Phys 43:297
92. Schmidt-Böcking H, Schmidt L, Weber T, Mergel V, Jagutzki O, Czasch A, Hagmann S, Doerner R, Demkov Y, Jahnke T, Prior M, Cocke C, Osipov T, Landers A (2004) Radiat Phys Chem 71:627
93. Tung C, Chao T, Hsieh H, Chan W (2007) Nucl Inst Methods B 262:231
94. Chandrasekhar S (1943) Rev Mod Phys 15:1
95. Nikjoo H, Uehara S, Wilson WE, Hoshi M, Goodhead DT (1998) Int J Radiat Biol 73:355
96. LaVerne J (1989) Radiat Phys Chem 34:135
97. Waligorski M, Hamm R, Katz R (1986) Nucl Tracks Radiat Meas 11:309
98. Gerchikov LG, Ipatov AN, Solov'yov AV, Greiner W (2000) J Phys B 30:4905
99. Nikjoo H, O'Neill P, Goodhead DT, Terrissol M (1997) Int J Radiat Biol 71:467
100. Ward J (1995) Radiat Res 142:362
101. Schipler A, Iliakis G (2013) Nucl Acid Res 41:7589
102. Fabrikant II, Caprasecca S, Gallup GA, Gorfinkiel JD (2012) J Chem Phys 136:184301
103. Becker D, Sevilla M (1993) In: Lett J (ed) Advances in radiation biology, vol 17. Academic Press, pp 121–180
104. Gianturco FA, Sebastianelli F, Lucchese RR, Baccarelli I, Sanna N (2008) J Chem Phys 128:174302
105. Sanche L (2012) In: Garcia G, Fuss MC (eds) Radiation damage in biomolecular systems. Springer
106. Panajotovic R, Martin F, Cloutier P, Hunting D, Sanche L (2006) Radiat Res 165:452
107. Cederbaum LS, Zobeley J, Tarantelli F (1997) Phys Rev Lett 79:4778
108. Mucke M, Braune M, Barth S, Förstel M, Lischke T, Ulrich V, Arion T, Becker U, Bradshaw A, Hergenhahn U (2010) Nat Phys 6:143
109. Bulanov SS, Brantov A, Bychenkov VY, Chvykov V, Kalinchenko G, Matsuoka T, Rousseau P, Reed S, Yanovsky V, Krushelnick K, Litzenberg DW, Maksimchuk A (2008) Med Phys 35:1770
110. Adamcik J, Jeon JH, Karczewski KJ, Metzler R, Dietler G (2012) Soft Matter 8:8651
111. Dang HM, Goethem MJV, Graaf ERVD, Brandenburg S, Hoekstra R, Schlathölter T (2011) Eur Phys J D 63:359
112. Toulemonde M, Dufour C, Meftah A, Paumier E (2000) Nucl Inst Methods B 166–167:903

113. Toulemonde M, Trautmann C, Balanzat E, Hjort K, Weidinger A (2004) Nucl Inst Methods B 216:1
114. Skupinski M, Toulemonde M, Lindeberg M, Hjort K (2005) Nucl Inst Methods B 240:681
115. Pawlak F, Dufour C, Laurent A, Paumier E, Perrière J, Stoquert JP, Toulemonde M (1999) Nucl Inst Methods B 151:140
116. Toulemonde M, Assmann W, Dufour C, Meftah A, Studer F, Trautmann C (2006) Mat. Fys. Medd. 52:263
117. Dammak H, Lesueur D, Dunlop A, Legrand P, Morillo J (1993) Radiat Eff Defect Sol 126:111
118. Mieskes HD, Assmann W, Grüner F, Kucal H, Wang ZG, Toulemonde M (2003) Phys Rev B 67:155414
119. Meftah A, Djebara M, Khalfaoui N, Toulemonde M (1998) Nucl Instr Methods B 146:431
120. Katin V, Martinenko Y, Yavlinskii Y (1987) Sov Tech Phys Lett 13:276
121. Toulemonde M, Assmann W, Trautmann C, Grüner F (2002) Phys Rev Lett 88:057602
122. Meftah A, Brisard F, Costantini J, Hage-Ali M, Stoquert J, Studer F, Toulemonde M (1993) Phys Rev B 48:920
123. Yakubovich AV, Surdutovich E, Solov'yov AV (2011) AIP Conf Proc 1344:230
124. Yakubovich AV, Surdutovich E, Solov'yov AV (2012) Nucl Instr Methods B 279:135
125. Landau L, Lifshitz E (1987) Fluid dynamics, vol 6, 2nd edn. Reed-Elsevier, Oxford, Boston, Johannesburg
126. Zeldovich Y, Raiser Y (1966) Physics of shock waves and high-temperature hydrodynamic phenomena, vol 1, Oxford, New York
127. Chernyj G (1994) Gas dynamics. Nauka, Moscow
128. Niklas M, Abdollahi A, Akselrod M, Debus J, Jäkel O, Greilich S (2013) Int J Radiat Oncol Biol Phys 87:1141
129. Jakob B, Scholz M, Taucher-Scholz G (2003) Radiat Res 159:676
130. Tobias F, Durante M, Taucher-Scholz G, Jakob B (2010) Mutat Res 704:54
131. Roos WP, Kaina B (2006) Trends Mol Med 12:440
132. Ward J (1988) Prog Nucleic Acid Res Mol Biol 35:95
133. Goodhead DT (1994) Int J Radiat Biol 65:7
134. Malyarchuk S, Castore R, Harrison L (2009) DNA Repair 8:1343
135. Malyarchuk S, Castore R, Harrison L (2008) Nucleic Acids Res 36:4872
136. Sage E, Harrison L (2011) Mutat Res 711:123
137. Amaldi U, Kraft G (2005) Rep Prog Phys 68:1861
138. Zhang XD et al (2015) Sci Rep 5:8669
139. McQuaid HN et al (2016) Sci Rep 6:19442
140. Stewart R et al (2011) Radiat Res 176:587
141. Tinganelli W et al (2015) Sci Rep 5:17016
142. Alberts B, Johnson A, Lewis J, Raff M, Roberts K, Walter P (2007) Molecular biology of the cell. Garland Science, Hamden, CT
143. Lewis N et al (2013) Nat Biotechnol 31:759
144. Morgan D (2006) The cell cycle: principles of control. New Science Press
145. Wéra AC, Riquier H, Heuskin AC, Michiels C, Lucas S (2011) Nucl Instrum Methods B 269:3120
146. Wéra AC, Heuskin AC, Riquier H, Michiels C, Lucas S (2013) Radiat Res 179:273
147. Heuskin AC, Michiels C, Lucas S (2013) Phys Med Biol 58:6495
148. Scholz M, Kellerer A, Kraft-Weyrather W, Kraft G (1997) Radiat Environ Biophys 36:59
149. Weyrather WK, Ritter S, Scholz M, Kraft G (1999) Int J Rad Biol 75:1357
150. Krämer M, Scifoni E, Wälzlein C, Durante M (2012) J Phys Conf Ser 373:012017
151. Usami N et al (2008) Int J Radiat Biol 84:603
152. Falk M, Lukasova E, Kozubek S (2012) In: Gómez-Tejedor GG, Fuss MC (eds) Biomolecular systems in radiation damage. Springer, New York
153. Friedland W, Kundrát P (2013) Mutat Res Genet Toxicol Environ 756:213
154. Zheng Y, Hunting DJ, Ayotte P, Sanche L (2008) Radiat Res 169:19
155. Bulanov SS, Esarey E, Schroeder CB, Bulanov SV, Esirkepov TZ, Kando M, Pegoraro F, Leemans WP (2016) Phys Plasmas 23:056703

Propagation of Swift Protons in Liquid Water and Generation of Secondary Electrons in Biomaterials

Pablo de Vera, Rafael Garcia-Molina and Isabel Abril

Abstract A proper description of the propagation of a swift proton beam through biomaterials, accounting for the energy deposited as well as the geometrical evolution of the beam as a function of the target depth and nature, is a crucial issue in proton therapy. For this purpose, simulation is a very adequate tool, since the most relevant interactions that take place between the projectile and the target constituents (electrons and nuclei) can be conveniently accounted for in a controlled manner. In this chapter an overview and relevant results for hadron therapy are presented of the simulations we have developed using the code SEICS (Simulation of Energetic Ions and Clusters through Solids), which combines Monte Carlo and Molecular Dynamics, to follow in detail the motion and energy deposition of swift protons through targets of hadron therapeutic interest, mainly liquid water. The main interactions considered in our study are of elastic nature (affecting mainly the projectile's direction) and inelastic processes (leading to either nuclear reactions or electronic energy loss). The performance of the code, as well as the quality of its main input, namely the stopping force for proton beams in liquid water (which is the main tissue constituent), are benchmarked by comparing the results of the simulations with available experimental proton energy spectra as a function of the detection angle after traversing a

P. de Vera
Department of Physical Sciences, The Open University,
Milton Keynes, England MK7 6AA, UK
e-mail: p.devera@qub.ac.uk

P. de Vera
MBN Research Center, 60438 Frankfurt am Main, Germany

P. de Vera
School of Mathematics and Physics, Queen's University Belfast,
BT7 1NN Belfast, Northern Ireland, UK

R. Garcia-Molina (✉)
Departamento de Física – Centro de Investigación en Óptica y Nanofísica,
Regional Campus of International Excellence "Campus Mare Nostrum",
Universidad de Murcia, 30100 Murcia, Spain
e-mail: rgm@um.es

I. Abril
Departament de Física Aplicada, Universitat d'Alacant, 03080 Alacant, Spain

A.V. Solov'yov (ed.), *Nanoscale Insights into Ion-Beam Cancer Therapy*,
DOI 10.1007/978-3-319-43030-0_2

micrometric liquid water jet. The excellent agreement with experiments validates the SEICS code, which we can use then to study several problems of interest for proton therapy, including the calculation of depth-dose curves and lateral dose profiles, the energy evolution of the proton beam along the target, as well as the production of secondary electrons at the Bragg peak in relevant biomaterials.

1 Introduction

The interaction of fast charged particles with biological materials is a topic of great current interest, due to its possible (beneficial or harmful) consequences on human tissues. The passage of energetic ions through the body initiates a cascade of physical, chemical and biological processes, which can produce the damage and, eventually, the death, of human cells. Such results can be considered either as negative (when the damage should be prevented) or positive (when it is desired), so a proper description and understanding of the involved phenomena is desirable, in each case, for radiation protection or for radiotherapeutic purposes.

While radiation protection against energetic ions is relevant to prevent human exposure in the context of nuclear reactors and space exploration, radiotherapeutic purposes find an emergent application in the technique known as ion beam cancer therapy (or hadron therapy) [1]. Contrary to conventional radiotherapy, where energetic photon or electron beams are used, having a rather homogeneous energy deposition profile in human tissues, ion beam cancer therapy exploits the unique characteristics of energetic ion beams, which present an inverse depth-dose profile, losing more energy as their kinetic energy decreases. This behaviour gives place to the appearance of the Bragg peak, i.e., a sharp maximum in the energy deposition profile near the end of the trajectories of the energetic ions. The depth (in the target) of the Bragg peak has to be carefully tuned. For that purpose, it is necessary to improve the knowledge of the interaction processes that take place when swift ion beams move through biological materials. This will allow a precise energy delivery in deep-seated tumours, while minimising the damage to surrounding healthy tissues, which is especially desirable for treating tumours close to sensitive areas, such as the brain. For ion beam cancer therapy, usually proton and carbon ion beams are used, although the majority of centres all around the world employ protons [2].

Energetic ions moving through condensed matter (solids and liquids) lose energy mainly by electronic excitations and ionisations. The main quantity describing this energy loss is the stopping power (or stopping force) [3]. Its accurate determination for biological materials and, especially, for liquid water (the main constituent of living tissues) is an important issue, since it will determine the precision with which the Bragg peak can be positioned and, hence, the accuracy of treatment plannings [4, 5]. However, electronic excitation and ionisation are just two of the multiple phenomena which occur since the ion begins its propagation in the body until the final biological damage.

Ion beam cancer therapy should be studied, indeed, within a multiscale approach [6], since processes in very different energy, spatial, and time scales take place.

Apart from electronic interactions, ion propagation is influenced by nuclear fragmentation reactions, which convert the initial projectiles in a new whole family of propagating secondary particles. The electronic interactions lead to the ejection of a vast quantity of secondary species, including secondary electrons and free radicals, which propagate in nano- and micrometric scales, inducing the physical and chemical mechanisms that account for the major part of the initial damage of the biomolecules, especially the nuclear DNA. Then, biological processes come into play, dealing with the damage and reparation of the biomolecules, which will lead to cell death if the latter mechanism is not effective.

Therefore, it is clear that an optimal use of ion beam cancer therapy requires knowledge on multiple phenomena, among which, from the physics point of view, electronic excitation and ionisation, and nuclear fragmentation reactions, play a relevant role. A convenient way of modelling ion beam interaction with biological materials, taking into account all these physical interactions, is numerical simulation, such as Monte Carlo and Molecular Dynamics. In the Monte Carlo codes, the detailed history of each projectile of the beam is followed by randomly drawing the coordinate at which each possible interaction (elastic or inelastic scattering, electron exchange between the projectile and the target, nuclear fragmentation reaction...) takes place, as well as the final result of the interaction, by using appropriate interaction probabilities, known as cross sections.

Depending on the kind of code, more or less detailed information will be obtained from the simulation. Track-structure (or event by event) codes are the most complete ones, in which the result of each interaction is described and followed in detail. This is particularly important for the emission of secondary electrons: in the track-structure codes, the energy and angle of each secondary electron is determined after each ionising collision, and then the electrons are also followed until stopped. Examples of such codes are KURBUC [7], PARTRAC [8], NOTRE DAME [9], MC4 [10], and EPOTRAN [11], among others [10].

As a consequence of being very detailed, these simulations are also very time consuming, and not very convenient when only macroscopic dose distributions are needed. In those cases, radiation transport (or condensed history) codes can be used, by grouping bunches of interactions that, by themselves, have a small effect on the projectile history, which greatly reduce computational time. This is done, for example, with the electronic interactions when the trajectories of the secondary electrons are not important, i.e., for the calculation of macroscopic dose profiles. Some examples of condensed history codes are FLUKA [12], MCNPX [13], SRIM [14], PENELOPE [15] or GEANT4 [16], among others [17]. Also, codes exist that can use the condensed history algorithm for the whole target, while calculating the track-structure just for the regions of interest, such as GEANT4-DNA [18] or LEPTS [19, 20], which enable more efficient multiscale simulations.

However, in order to obtain proper results from Monte Carlo codes, they have to be fed with appropriate input, i.e., with accurate values of the cross sections for each interaction process, notably for the electronic excitation in condensed phase targets, such as liquid water. In this context, the code SEICS (Simulation of Energetic Ions and Clusters through Solids) has been developed over the last years, for the description of swift ion propagation through a target by taking special care on its condensed matter nature [21]. The SEICS code employs accurate electronic stopping quantities, obtained from the dielectric formalism [22], by properly taking into account the stochastic energy loss of swift ions in condensed organic targets, such as liquid water [23], bone [24] or other biotargets [25–28]. Therefore, this code represents a good tool for the simulation of ion beam interactions with biomaterials.

The SEICS code will be briefly described in Sect. 2. After some initial considerations on the way the code works, in Sect. 2.1 we will explain in detail how the electronic energy loss of proton beams in liquid water is calculated, taking into account the stochasticity of the process through the use of the stopping power and the energy-loss straggling, as well as the charge state of the projectile. Then, the rest of relevant interactions in the problem will be described, namely elastic scattering (Sect. 2.2), electron capture and loss processes (Sect. 2.3), and nuclear fragmentation reactions (Sect. 2.4). In particular, we will show how, even recognizing the complexity of nuclear fragmentation reactions, they can be implemented in a quite simple way for proton beams.

After describing the code, and before studying several problems related to proton therapy, in Sect. 3 we will benchmark both the code and its main input, i.e., the electronic stopping quantities for protons in liquid water. This will be done by comparing the outputs from SEICS with the results of the experiments performed in micrometric liquid water jets by the Kyoto group [29, 30], which were done to determine the stopping power of liquid water for proton beams of intermediate energy. The excellent agreement of our simulated energy distributions with the corresponding measurements validate the performance of our code, as well as the accuracy of the stopping quantities we use.

Finally, several applications of the SEICS code to calculate useful quantities in proton therapy are presented in Sect. 4. We will start by calculating depth-dose curves of proton beams in liquid water, Sect. 4.1. Apart from the satisfactory comparison with available experimental data, we will see how the SEICS code can be used to evaluate the contribution of each individual interaction process (electronic energy-loss and straggling, elastic scattering, nuclear fragmentation reactions) to the total dose deposited in the target. Then, in Sect. 4.2 the depth-dose curves will be complemented with the calculation of lateral dose profiles. Here, the simulation results for the lateral aperture of the beam will be parameterised, which yields useful results for the analytical calculations of dose profiles. The energy distribution of protons along the Bragg curve will be discussed in Sect. 4.3, where important characteristics of the energy spectra of primary projectiles along the beam trajectory will be featured. This discussion is very relevant, since the energy distribution of primary projectiles governs the production of secondary electrons, which in turn will determine the microscopic track-structure of the incident radiation. In Sect. 4.4 we will discuss the generalisation of the dielectric formalism to obtain energy spectra of secondary electrons produced by proton beams, and we will use them, in conjunction with the energy distribution of the primary protons as a function of the depth, to determine the realistic energy distributions of secondary electrons produced at the Bragg peak region. Such results are useful to establish the initial conditions for track-structure simulations in realistic circumstances, where ultimate cellular damage is expected to depend on the number and energy of electrons produced around the Bragg peak region in several relevant biomaterials [31, 32].

2 Propagation of Energetic Ions Through Condensed Media

In order to properly describe the propagation of energetic ions through condensed media, we use the simulation code SEICS (Simulation of Energetic Ions and Clusters through Solids), which is based on a combination of Molecular Dynamics and Monte Carlo techniques to follow the motion of swift projectiles through the target [21, 33–35]. The code implements all the relevant interactions between ions and the target constituents, which are described in what follows. The energy loss by the projectile is mainly due to the electronic excitations and ionisations that it generates along its trajectory, which is accounted for by solving the projectile's equation of motion by considering a stopping force (obtained from the stopping power, i.e., the average energy loss per unit path length) whose fluctuations, due to the stochastic nature of the interaction, are accounted for by the energy-loss straggling; due to the high energies typically used in hadron therapy, relativistic corrections have been included in the kinematics of the projectiles. The multiple Coulomb scattering of the projectile with the target nuclei is the major responsible for the beam angular spreading. The electron capture and loss processes by the projectile determine its charge state, which, in turn, determines the stopping force. Finally, the nuclear fragmentation reactions modify the number of projectiles in the beam and generate new ones, affecting the general shape of the Bragg curve.

Whereas the slowing down of the projectile is managed through a standard Molecular Dynamics procedure, the rest of the processes are drawn by the Monte Carlo technique, where the probability distributions are obtained from the corresponding cross sections.

Besides, an effort has been made to have a more realistic description of the electronic excitation spectrum of the more relevant biological targets, putting especial interest in liquid water, as it is the main constituent of the living tissues, as well as the DNA molecular constituents.

Taking into account all the ingredients cited previously, the SEICS code dynamically follows the trajectory of each projectile in the target, providing its position, velocity and charge-state at any instant. The position $\mathbf{r}$ and velocity $\mathbf{v}$ of a projectile, with mass M_1 and atomic number Z_1, are obtained by numerically solving its equation of motion at discrete time intervals Δt. For this purpose we rewrite the velocity variant of Verlet's algorithm [36], taking into account the relativistic velocities of the projectiles:

$$\mathbf{r}(t+\Delta t)=\mathbf{r}(t)+\mathbf{v}(t)\Delta t+\frac{\mathbf{F}(t)}{2M_1}(\Delta t)^2\left[1-\left(\frac{v(t)}{c}\right)^2\right]^{3/2}, \tag{1}$$

$$\mathbf{v}(t+\Delta t)=\mathbf{v}(t)+\frac{\mathbf{F}(t)+\mathbf{F}(t+\Delta t)}{2M_1}\Delta t\left[1-\left(\frac{v(t)}{c}\right)^2\right]^{3/2}, \tag{2}$$

where **F** is the electronic stopping force, and c is the speed of light. The classical trajectory of the projectile is followed until it reaches a threshold energy E_{th}. We use $E_{\text{th}} \sim 250$ eV, although reducing this value has not practical consequences in the final depth-dose distributions.

The force that acts on the projectile is due to its inelastic collisions with the target electrons. This produces the so-called electronic stopping force, which depends on the projectile charge-state q and speed v. Due to the stochastic nature of the interaction with the target electrons there are fluctuations in the force sensed by the projectile. Then, in the simulation we use the modulus of the electronic stopping force felt by the projectile (with charge state q) from a Gaussian distribution with mean value S_q (the stopping power or mean value of the energy loss), and a standard deviation given by:

$$\sigma=\sqrt{\Omega_q^2/\Delta s}, \tag{3}$$

where $\Delta s = v\Delta t$ is the distance travelled by the projectile (with velocity v) in a time step Δt, and Ω_q^2 is the energy-loss straggling (the mean square deviation per unit path length of the energy-loss distribution) for a projectile with charge state q. According to the Box-Müller procedure to generate a Gaussian distribution [37], the electronic stopping force acting on the projectile is written as:

$$\mathbf{F}=-\left[S_q+(\Omega_q/\sqrt{\Delta s})\sqrt{-2\ln\xi_1}\cos(2\pi\xi_2)\right]\hat{v}, \tag{4}$$

with $\hat{v}$ being the unit vector of the instantaneous projectile velocity **v**. The symbols ξ_i refer to random numbers uniformly distributed between 0 and 1 [38], with the value of the subscript i $(=1, 2, \ldots,)$ denoting each time a random number ξ_i is used in the simulation.

The stopping power S_q and the energy-loss straggling Ω_q^2 of the projectile are the main input quantities in the simulation code. They are calculated in the dielectric framework, together with the MELF-GOS (Mermin Energy Loss Function – Generalised Oscillator Strength) model [39, 40], which has been developed to realistically represent the electronic excitation spectrum of any condensed media, being particularly convenient for biological targets, which are liquids or solids [25, 41]. To speed up the simulation, at higher projectile velocities ($v \geq 20$ a.u.; kinetic energies $T \geq 10$ MeV/u) we use the analytical relativistic Bethe formula for the stopping power [42],

$$S = \frac{4\pi e^4 Z_2 Z_1^2 \mathscr{N}}{v^2} \ln\left[\frac{2mv^2}{I(1-(v/c)^2)} - (v/c)^2\right], \quad (5)$$

where Z_2 is the atomic number of the target, $\mathscr{N}$ is its atomic or molecular density, m and e are the electron mass and charge, respectively, and I is the mean excitation energy of the target, which only depends on its electronic structure [43], and it is found from the MELF-GOS model [25, 44]. Also at high energies, the energy-loss straggling Ω_q^2 is evaluated from the Bohr straggling formula [3]:

$$\Omega_{\mathrm{Bohr}}^2 = 4\pi e^4 Z_2 Z_1^2 \mathscr{N}. \quad (6)$$

The SEICS code becomes a convenient tool to address different problems related to ion beam cancer therapy, since the propagation and energy deposition of the ions through liquid water and other biologically relevant targets can be described with high accuracy.

Althought a complete description of the SEICS code can be found in Refs. [21, 35], in what follows we summarise how the different interactions between a swift charged particle and the target constituents are implemented in it.

2.1 *Inelastic Energy Loss Processes*

The dielectric formalism is a convenient approach for describing the interaction of fast charged particles with the electrons of a condensed target [22, 45, 46]. It assumes a linear response of the electronic system to the perturbation induced by the electric field of the projectile, providing a connection between the dielectric function of the target (a macroscopic property) and the matrix elements of the electronic transitions (a microscopic quantity). In this case, all the possible electronic excitations of the system are properly accounted for, including many-body and physical-state effects coming from intermolecular interactions, so important in condensed matter.

For an ion with atomic number Z_1, mass M_1 and charge state q, which travels with kinetic energy T (velocity v) through a condensed medium characterised by a dielectric function $\varepsilon(k, \omega)$, the dielectric formalism provides the different moments of the energy-loss distribution due to inelastic collisions of the projectile with the target electrons. The stopping power S_q is given by the first moment of the distribution:

$$S_q(T) = \frac{M_1 e^2}{\hbar \pi T} \int_{\omega_-}^{\omega_+} \mathrm{d}\omega\, (\hbar\omega) \int_{k_-}^{k_+} \frac{\mathrm{d}k}{k} [Z_1 - \rho_q(k)]^2 \mathrm{Im}\left[\frac{-1}{\varepsilon(k,\omega)}\right], \quad (7)$$

where $\hbar k$ and $\hbar\omega$ represent, respectively, the momentum and energy transferred to the target in an inelastic collision. $\rho_q(k)$ is the Fourier transform of the electronic density of the projectile, which is described by the statistical model proposed by Brandt and Kitagawa [47]. The integration limits are $\hbar k_\pm = \sqrt{2M_1\left(2T - \hbar\omega \pm 2\sqrt{T(T-\hbar\omega)}\right)}$, whereas $\hbar\omega_- = 0$ (for metals) or $\hbar\omega_- = E_{\rm gap}$ (for insulators) and $\hbar\omega_+ \simeq 4Tm/M_1$, where m is the electron mass.

The energy-loss straggling, Ω_q^2, which accounts for fluctuations in the energy loss due to the stochastic nature of the inelastic collisions, is related to the second moment of the distribution. It can be obtained from the following expression:

$$\Omega_q^2(T) = \frac{M_1 e^2}{\hbar\pi T}\int_{\omega_-}^{\omega_+} \mathrm{d}\omega\,(\hbar\omega)^2 \int_{k_-}^{k_+}\frac{\mathrm{d}k}{k}[Z_1 - \rho_q(k)]^2 \mathrm{Im}\left[\frac{-1}{\varepsilon(k,\omega)}\right]. \tag{8}$$

The Brandt-Kitagawa model presupposes that the cloud of bound electrons is screening the projectile nucleus over a certain radius that depends on the projectile velocity. Therefore, target electrons that approach the projectile with impact parameters larger than this radius (that is, distant collisions) perceive the projectile as a point charge with charge q, irrespective of its internal structure. However, when the impact parameter is smaller than the radius (i.e., close collisions) the target electrons penetrate the screening cloud of the bound electrons of the projectile, sensing a partially screened potential corresponding to a projectile charge larger than q. An average over all the impact parameters will give the charge of the projectile. The advantage of this model is that it is possible to derive analytical expressions for the Fourier transform of the electronic charge density of the projectile, $\rho_q(k)$. For more details about the implementation of this model in the SEICS code see Ref. [28].

In Eq. (7) the target properties enter through the energy loss function (ELF), $\mathrm{Im}[-1/\varepsilon(k,\omega)]$, which characterises the electronic excitation spectrum of the material, since it expresses the probability of producing a target excitation or ionisation with energy and momentum transfers $\hbar\omega$ and $\hbar k$, respectively. A good description of the target ELF at any energy and momentum transfer, the so-called Bethe surface, is basic to obtain accurate values of the stopping power.

On the other hand, the electric field, $\mathscr{E}_{\mathrm{ind},q}$, induced by the projectile with charge q in the target, which is the responsible of its stopping, can also distort the electronic cloud of the projectile, polarising it. The contribution of this process to the stopping power is given by [34, 48]:

$$S_{\mathrm{pol},q}(T) = \frac{e^2 M_1 Z_1}{\pi T}\int_{k_-}^{k_+}\frac{\mathrm{d}k}{k}\rho_q(k)\int_{\omega_-}^{\omega_+}\mathrm{d}\omega\,\omega\,\mathrm{Im}\left[\frac{-1}{\varepsilon(k,\omega)}\right]\left[1-\cos\left(\omega d_q\sqrt{\frac{M_1}{2T}}\right)\right]. \tag{9}$$

Here, $d_q = \mu_q \mathscr{E}_{\mathrm{ind},q}(T)$ is the displacement of the centre of the electron cloud from its nucleus, where μ_q is the projectile polarisability; the self-induced electric field is given by:

$$\mathscr{E}_{\mathrm{ind},q}(T) = \frac{M_1 e}{\pi T}\int_{k_-}^{k_+}\frac{\mathrm{d}k}{k}[Z_1 - \rho_q(k)]\int_{\omega_-}^{\omega_+}\mathrm{d}\omega\,\omega\,\mathrm{Im}\left[\frac{-1}{\varepsilon(k,\omega)}\right]. \tag{10}$$

This extra contribution to the energy loss of the projectile due to its self-polarisation is also included in the SEICS code.

At this point the only quantity that remains to be determined is the energy loss function of the target at any energy and momentum transfer, that is over the whole Bethe surface, which is not a trivial task. Most of the experiments can measure the ELF at the optical limit ($k = 0$), therefore it is necessary to use methodologies for a proper extension over the whole momentum plane ($k \neq 0$). A discussion about the different extended optical data models currently used to calculate the ELF at any momentum transfers can be found in [41]. In what follows, we will present the MELF-GOS methodology developed by our research group to describe the target ELF over the Bethe surface.

2.1.1 Target Description: The MELF-GOS Model

The MELF-GOS (Mermin Energy Loss Function-Generalised Oscillator Strength) methodology is based on the use of Mermin dielectric functions [49] for the description of the outer-shell electron excitations, together with generalised oscillator strengths in the hydrogenic approach [50] for the description of the inner-shell electron excitations.

Due to the different response to the perturbation induced by the projectile of either the tight-bound outer shell or the inner-shell electrons, the description of both contributions to the ELF can be safely assumed to be independent. Therefore the ELF can be divided in two components:

$$\mathrm{Im}\left[\frac{-1}{\varepsilon\,(k,\omega)}\right] = \mathrm{Im}\left[\frac{-1}{\varepsilon\,(k,\omega)}\right]_{\mathrm{outer}} + \mathrm{Im}\left[\frac{-1}{\varepsilon\,(k,\omega)}\right]_{\mathrm{inner}} . \tag{11}$$

The inner-shell electron excitations present large binding energies, preserving their atomic character and they do not participate in the chemical bonds of the target. Therefore, they can be regarded practically as atomic electrons, being properly described by the generalised oscillator strength (GOS) in the hydrogenic approach [40, 43, 50]. This fact allows to use the Bragg rule [51] for determining the ELF of the inner shells for a compound target as the sum of their atomic constituents [28]:

$$\begin{aligned}\mathrm{Im}\left[\frac{-1}{\varepsilon\,(k,\omega)}\right]_{\mathrm{inner}} &= \mathcal{N}\sum_j \alpha_j \frac{\mathrm{ELF}_j(k,\omega)}{\mathcal{N}_j} \\ &= \frac{2\pi^2\mathcal{N}}{\omega}\sum_j \alpha_j \sum_{n\ell} \frac{\mathrm{d}f^j_{n\ell}(k,\omega)}{\mathrm{d}\omega}\Theta(\omega-\omega^j_{\mathrm{ionis},n\ell}) ,\end{aligned} \tag{12}$$

where $\mathcal{N}$ is the molecular density of the target, α_j, $\mathrm{ELF}_j(k,\omega)$ and $\mathcal{N}_j$ are, respectively, the stoichiometry coefficient, the ELF of the jth atom and the atomic density of the jth element, $\mathrm{d}f^j_{n\ell}(k,\omega)/\mathrm{d}\omega$ is the hydrogenic GOS corresponding to the (n,ℓ)-

subshell of the jth element, and $\hbar\omega^{j}_{\mathrm{ionis},n\ell}$ is the ionisation energy of the (n, ℓ)-subshell of the jth-element in the compound target. $\Theta(\ldots)$ represents the Heaviside step function.

Due to the complex structure of the ELF for real materials, it is suitable to describe the outer-shells contribution to the ELF, $\mathrm{Im}\left[-1/\varepsilon\left(k, \omega\right)\right]_{\mathrm{outer}}$, by a linear combination of Mermin-type ELF (MELF), which must be fitted to the experimental optical ELF ($k = 0$), namely,

$$\begin{aligned}\mathrm{Im}\left[\frac{-1}{\varepsilon\left(k=0, \omega\right)}\right]_{\mathrm{outer}} &= \mathrm{Im}\left[\frac{-1}{\varepsilon\left(k=0, \omega\right)}\right]_{\mathrm{exp}} \\ &= \sum_i A_i \mathrm{Im}\left[\frac{-1}{\varepsilon_{\mathrm{M}}\left(k=0, \omega; \omega_i, \gamma_i\right)}\right]\Theta\left(\omega-\omega_{\mathrm{th},i}\right), \qquad (13)\end{aligned}$$

with ε_{M} being the Mermin dielectric function [49], given by:

$$\varepsilon_{\mathrm{M}}(k, \omega) = 1 + \frac{(1+i\gamma/\omega)\left[\varepsilon_{\mathrm{L}}(k, \omega+i\gamma)-1\right]}{1+(i\gamma/\omega)\left[\varepsilon_{\mathrm{L}}(k, \omega+i\gamma)-1\right]/\left[\varepsilon_{\mathrm{L}}(k, 0)-1\right]}, \qquad (14)$$

where ε_{L} is the Lindhard dielectric function [22, 52, 53], which provides an analytic expression for the dielectric response function of a homogeneous free-electron gas, where plasmons are undamped electronic excitations. However, experimental evidence indicates that a strong damping mechanism exists at all k for most materials. To solve this problem, the Mermin dielectric function phenomenologically modifies the Lindhard dielectric function by including plasmon damping through phonon-assisted electronic transitions. It should be noted that at the optical limit the Mermin and the Drude ELF are equivalent.

In Eq. (13) the coefficients A_i account for the intensity of each resonance in the experimental energy loss spectrum (i.e., the intensity of their oscillator strengths), $\hbar\omega_i$ and $\hbar\gamma_i$ represent the position and width of each excitation and $\hbar\omega_{\mathrm{th},i}$ is a threshold energy, usually the band gap energy, below which electronic excitations are not possible.

In summary, the fitting parameters A_i, ω_i and γ_i of the optical ELF in Eq. (13) are chosen in such a way that it reproduces the main features of the experimental optical ELF and fulfils physically motivated restrictions, such as the Kramers-Kronig (KK) sum rule [25]:

$$\frac{2}{\pi}\int_0^\infty \mathrm{d}\omega\, \frac{1}{\omega}\mathrm{Im}\left[\frac{-1}{\varepsilon(k=0, \omega)}\right] = 1 - \frac{1}{n^2(0)} \qquad (15)$$

and the f-sum rule (or Thomas-Reiche-Kuhn sum rule):

$$Z_2 = \frac{m}{2\pi^2\mathcal{N}e^2}\int_0^\infty \mathrm{d}\omega\, \omega\mathrm{Im}\left[\frac{-1}{\varepsilon(k=0, \omega)}\right]. \qquad (16)$$

$n(0)$ represents the refractive index at the static limit, whereas Z_2 is the target atomic number (per molecule). The KK sum rule, Eq. (15), guarantees a good behaviour of the ELF at low energy transfers. The f-sum rule, Eq. (16), links the ELF to the number of target electrons per molecule that can be excited by the projectile; it must be fulfilled for a good behaviour of the ELF at intermediate and high energy transfers.

An important advantage of the MELF-GOS model is that the extension from the optical data to $k \neq 0$ is automatically achieved through the analytic properties of the Mermin dielectric function, therefore no extra extension algorithms have to be introduced.

In order to test the accuracy of the MELF-GOS model, we calculate the Bethe surface of liquid water and compare with experimental measurements from inelastic x-ray scattering spectroscopy (IXSS) [54, 55]. Figure 1 shows the experimental data for the ELF of liquid water from $k = 0$ to $k = 3.59$ a.u. (symbols), together with the calculations obtained by the MELF-GOS model (lines). The very good agreement does not happen for other extended optical-data models (see [41]), which means that the inclusion of the damping in the electronic excitations is indispensable. Therefore, by starting from the experimental ELF at the optical limit, the MELF-GOS methodology can realistically predict the excitation spectrum of liquid water (and any other target) over the entire plane of energy and momentum transfers.

The dielectric formalism, together with a realistic description of the electronic properties of the target by the MELF-GOS model, has been successfully applied to calculate electronic energy-loss quantities for ions and electrons in a large number of inorganic and biological materials (elemental and compounds) [23–26, 40, 48, 56–65], which compare satisfactorily well with available experimental data.

As we discussed in Sect. 2, the SEICS code needs as input data the stopping power S_q and the energy-loss straggling Ω_q^2 for each charge state q of the projectiles in the irradiated target. Now we apply this methodology to calculate these quantities for proton beams in liquid water, as it is the main constituent of living tissues.

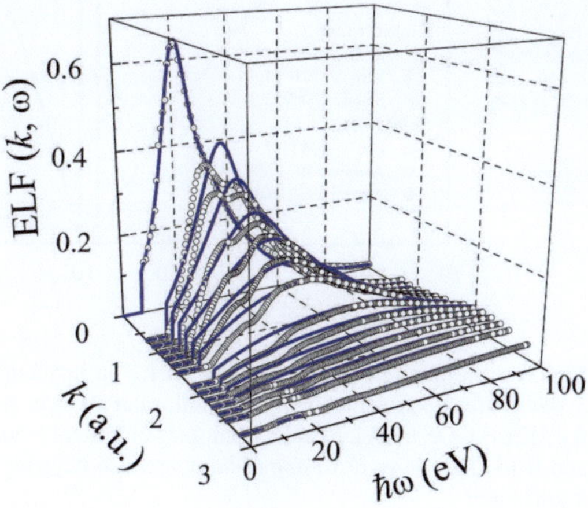

Fig. 1 Energy loss function (ELF) of liquid water as a function of energy $\hbar\omega$ and momentum transfer $\hbar k$. *Symbols* are experimental data from Hayashi [55] and *lines* represent the calculations from the MELF-GOS model

Before comparing the calculated energy-loss quantities with experimental data it is necessary to take into account that due to electron capture and loss processes by the projectile, the ion can be in different charge states q. The probability that the projectile reaches a given (equilibrium) charge state q at a given energy T is $\phi_q(T)$. Therefore, the stopping quantities are evaluated as:

$$S(T) = \sum_{q=0}^{Z_1} \phi_q(T)\, S_q(T)\,, \qquad \Omega^2(T) = \sum_{q=0}^{Z_1} \phi_q(T)\, \Omega_q^2(T)\,. \tag{17}$$

The equilibrium charge fractions $\phi_q(T)$ of the projectile, which depend on its energy and the target nature, are taken from a parameterisation to experimental data [66]. In Fig. 2 we show by solid lines the calculated stopping power S and energy-loss straggling Ω^2 for proton beams in liquid water as a function of the incident energy. S is compared with available experimental data (symbols) for liquid water [29, 30, 67] and ice [68–70]. The semiempirical results provided by the SRIM code [14] are depict by a grey dashed line, whereas the grey dash-dotted line corresponds to the stopping power collected in the ICRU Report 49 [71]. The predictions at high energies of all the models agree (among them and) with the newest experimental data in liquid water [67]. However, at energies aroud and lower than the stopping power maximum, the predictions of the models clearly disagree among them and depart from the available experimental data (although these are not for liquid water but for ice). Both SRIM and ICRU curves show a good agreement with the experimental data for ice, because they use a parameterisation to these available experimental data. We note that at low proton energies the inclusion of the electron charge-exchange

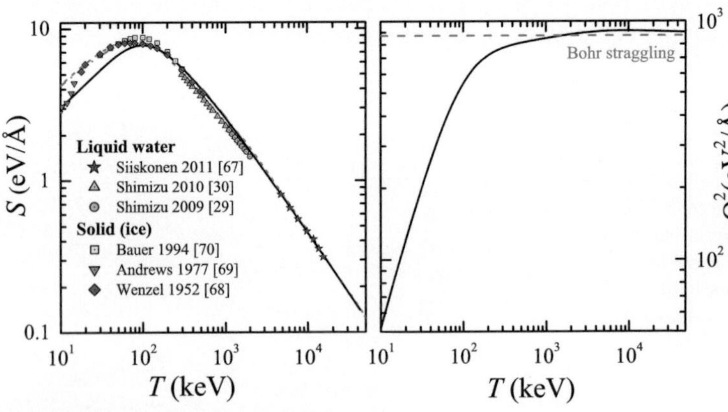

Fig. 2 **a** Stopping power of liquid water for an incident proton beam as a function of its energy. *Symbols* are experimental data for liquid water and ice. *Solid line* corresponds to calculations from the MELF-GOS model. Results from semi-empirical models, such as SRIM [14] (*grey dashed line*) and ICRU [71] (*grey dash-dotted line*), are also depicted. **b** Energy-loss straggling for protons in liquid water

process is essential to obtain suitable stopping power data. For a detailed discussion of the influence of several theoretical models in the stopping power calculations, the interested reader can consult Refs. [28, 41]. It is also interesting to note the differences between the experimental data for protons in liquid water at intermediate energies [29, 30] and the theoretical results. These experimental data rely on the measurement of the energy distributions of proton beams after traversing a liquid water jet, and on the interpretation of these energy distributions by Monte Carlo simulation. In Sect. 3 we will show that the stopping power at intermediate energy obtained in Refs. [29, 30] is questionable.

Moreover, at high projectile energies the SEICS code uses the Bethe stopping power (see Eq. (5)), where the mean excitation energy I of the material must be known, which only depends on the electronic structure of the target, and is defined in terms of the target ELF by the relation:

$$\ln I = \frac{\int_0^\infty \mathrm{d}\omega \, \ln(\omega)\, \omega \mathrm{Im}[-1/\varepsilon(0,\omega)]}{\int_0^\infty \mathrm{d}\omega \, \omega \mathrm{Im}[-1/\varepsilon(0,\omega)]} \, . \tag{18}$$

Therefore, accurate values of I are desirable, in particular for biological materials, where a few percent in the uncertainty of the I values might cause sizeable changes in the range and stopping maximum (i.e., the Bragg peak) of therapeutic ion beams, which have energies of the order of several hundreds of MeV/u. With the MELF-GOS model a value of the mean excitation energy 79.4 eV for liquid water is obtained. A comparison between different I values calculated from several models and several biological targets can be found in Ref. [28].

2.2 *Elastic Collisions*

The simulation code SEICS also considers multiple elastic scattering among the target nuclei and the projectile. These very frequent events modify the trajectory of the projectile providing its angular deflection (i.e., lateral spreading) and contribute to the energy-loss at low energies at the end of its travel, especially at the distal part of the Bragg peak, which affects the range of the projectile.

The elastic collisions are implemented in the code through a Monte Carlo algorithm [72, 73] that provides the projectile scattering angle and the corresponding elastic energy loss at each collision. The path length L_{el} of the projectile between two successive collisions with the target atoms is given by:

$$L_{\mathrm{el}} = -\frac{\ln \xi_3}{\sum_i \Lambda_{\mathrm{el},i}} \, , \tag{19}$$

where $\Lambda_{\mathrm{el},i}$ is the projectile inverse mean free path for having an elastic interaction with the i-atom of the target compound. Therefore $\sum_i \Lambda_{\mathrm{el},i}$ is the total macroscopic cross section for having an elastic collision with the target nuclei. We assume that each target atom is an effective scattering centre having a spherical volume with radius $r_0 = 1/(2\mathcal{N}^{1/3})$, where $\mathcal{N}$ is the target molecular density. To determine the type of the target atom that undergoes the collision with the projectile, we suppose that the collision probability with the i-atom type is proportional to the fractional contribution made by each atom to the total cross section [74]. The scattering between the projectile and the target atom is described by a screened Coulomb potential; here we use the universal interatomic potential [14] with the universal screening length. Therefore, the polar and the azimuthal scattering angles can be determined as well as the energy loss in the scattering process.

Summarising, the projectile direction of motion before the n-collision is defined by the polar angle Θ_{n-1} and the azimuthal angle Ψ_{n-1} in the laboratory frame of reference. The path length L_{el} until the next elastic collision is determined by using Eq. (19); after an elapsed time L_{el}/v, the n-collision takes place and we determine the i-type of target atom that is involved in the collision. Then, the scattering angles (polar and azimuthal), with respect to the direction of motion, are calculated and a new direction of motion, as well the modulus of the projectile velocity, in the laboratory frame of reference is obtained after the n-collision [73]. Once the velocity (modulus and direction) is known after the n-collision, the next collision takes place following the same steps that have been explained in the preceding paragraphs. A detailed description of the expressions used for the elastic scattering is presented in Ref. [35].

2.3 Charge Exchange Processes: Electron Capture and Loss

The capture and loss of electrons by the projectile, when it moves through the target, is included in the simulation code by a Monte Carlo procedure [34]. This affects the stopping force since it depends on the charge-state of the projectile.

The path length $L_{\text{C\&L}}$ of the projectile between two successive electronic capture or loss events is given by:

$$L_{\text{C\&L}} = -\frac{\ln \xi_7}{\Lambda_{\text{C}} + \Lambda_{\text{L}}}, \tag{20}$$

where Λ_{C} and Λ_{L} are the inverse mean free paths for electron capture and electron loss, respectively, both depending on the charge-state q of the projectile. We assume that the electron-loss cross section is proportional to both the geometrical cross section of the projectile and the number of its bound electrons. The inverse mean free path for electron capture, if multiple-electron processes are neglected, can be obtained from the equilibrium relation:

$$\Lambda_{\text{C}}(q+1 \rightarrow q) = \frac{\phi_q}{\phi_{q+1}} \Lambda_{\text{L}}(q \rightarrow q+1), \tag{21}$$

where ϕ_q and ϕ_{q+1} are the equilibrium fractions of the q and $q+1$ charge-states, respectively. For hydrogen projectiles:

$$\phi_0 + \phi_{+1} = 1 \,, \tag{22}$$

$$\phi_{+1} = \langle q \rangle \,, \tag{23}$$

where $\langle q \rangle$ is the average charge-state obtained through a fit to experimental data [66]. Since the probabilities of electron loss or electron capture by a projectile with charge-state q are proportional to the corresponding inverse mean free paths, it is possible to determine the new projectile charge-state from Eq. (23). In summary, according to this model we obtain the path length $L_{\mathrm{C\&L}}$ using Eq. (20); after an elapsed time $L_{\mathrm{C\&L}}/v$, either an electronic capture or loss event takes place determined according to their respective probabilities (see Ref. [35] for more details). We also take into account that electron capture and loss processes contribute to the energy loss of the projectile in an amount that is calculated according to the theoretical models outlined in Refs. [48, 75].

2.4 Nuclear Fragmentation Reactions

An accurate simulation of the propagation and energy deposition of swift protons in biological media at energies typically used in hadron therapy (several hundred of MeV) requires the inclusion of nuclear fragmentation reactions between primary protons and target nuclei. Complex processes occur in these collisions, which imply the excitation of the target nucleus, its fragmentation, the emission of secondary energetic particles (such as neutrons, photons, secondary protons or heavier particles), and the relaxation of the residual nucleus.

Nuclear fragmentation processes are included in the SEICS code assuming some pertinent simplifications. We consider that primary protons are removed from the beam according to their total non-elastic nuclear cross section, with a fraction of their residual energy being locally deposited.

According to the standard Monte Carlo procedure, the distance L_{frag} between two consecutive nuclear fragmentation collisions can be calculated as:

$$L_{\mathrm{frag}}(T) = -\lambda_{\mathrm{frag}}(T) \ln \xi_9 \,, \tag{24}$$

where λ_{frag} is the fragmentation mean free path given by:

$$\lambda_{\mathrm{frag}}(T) = \frac{A_2}{N_{\mathrm{A}}\, \rho\, \sigma_{\mathrm{frag}}(T)} \,, \tag{25}$$

with A_2 and ρ being, respectively, the total mass number (sum of the elemental constituents mass numbers) and the density of the compound target. N_{A} is the Avogadro number. $\sigma_{\rm frag}$ is the microscopic nuclear reaction cross section for the compound, i.e., the sum of the cross sections $\sigma_{{\rm frag},i}$ of its constituents weighted by the corresponding stoichiometric content α_i of each element:

$$\sigma_{\rm frag} = \sum_i \alpha_i \sigma_{{\rm frag},i} \; ; \tag{26}$$

the values of $\sigma_{{\rm frag},i}$ are chosen from recommended inelastic cross-sections compilations [76]. The fragmentation mean free path for protons in liquid water is shown in Fig. 3a as a function of the proton energy.

However, Eq. (24) is only valid when the mean free path does not change too much along the collision path. Since the change in the energy of the projectiles is large, due to the large values of the mean free path, we use an alternative algorithm to determine the distance at which the nuclear fragmentation reaction takes place. The formula for the exponential loss of particles in the beam is:

$$N(s + {\rm d}s) = N(s)\, {\rm e}^{-{\rm d}s/\lambda_{\rm frag}} \; , \tag{27}$$

where $N(s)$ is the number of remaining particles at a distance s, with ${\rm d}s$ being a differential path; $1 - N(s + {\rm d}s)/N(s)$ is a quantity between 0 and 1, which can be interpreted as the probability for a nuclear fragmentation to occur in this differential path ${\rm d}s$. A random number is sampled in each differential path and, if its value is less or equal than $1 - N(s + {\rm d}s)/N(s)$, then the primary proton disappears and deposits locally part of its energy. We consider that the energy transferred to neutral ejectiles (neutrons and gamma particles) escapes out of the treatment region and is lost [77], whereas the energy imparted to charged secondaries (protons, deuterons, tritons, alpha particles...) is deposited locally. Figure 3b shows the energy fraction $f_{\rm ejectile}$ transferred to different ejectiles, as recommended by the ICRU Report 63 [76].

3 Benchmarking the SEICS Code and the Experimental Stopping Power of Protons in Liquid Water

In this section we will benchmark the SEICS code (and one of its main ingredients, namely our calculated stopping power of liquid water) by comparing the simulated energy distribution of protons directed to a liquid water jet with the corresponding experimental data reported in Ref. [30].

One of the most relevant inputs for the simulation of the interaction of charged particles with condensed matter is the stopping power. Therefore an accurate determination of the stopping power of energetic protons in liquid water, as a subrogate of living tissues, is essential in hadron therapy [78, 79]. However only two sets of experimental data exist, due to the Kyoto group [30] and the Jyväskylä group [67]. Both are depicted in Fig. 2, together with theoretical curves as well as experimental stopping power of ice. Our calculated stopping power from the dielectric formalism

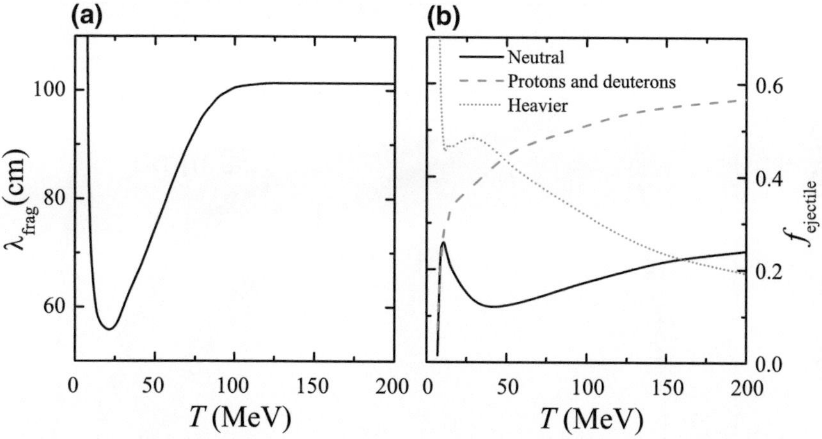

Fig. 3 **a** Nuclear fragmentation mean free path for protons in liquid water as a function of the incident energy, Eq. (25). **b** Fraction $f_{ejectile}$ of residual energy transferred from the incident proton to different ejectiles (neutral, protons and deuterons, and heavier particles) in liquid water, according to ICRU Report 63 data [76], as a function of the incident proton energy T

and the MELF-GOS model agrees very well with the data obtained by Siiskonen et al. [67]. However it is larger than the experimental data provided by Shimizu et al. [30]. This discrepancy can be understood from the way in which the experimental stopping powers were obtained by each group. Whereas the former [67] were derived directly from transmission measurements in a thin liquid water foil, the latter [30] was indirectly obtained though a fitting procedure from the energy spectrum at different exit angles of a proton beam after crossing a super-thin water jet.

The experimental stopping power data reported by the Kyoto group [29] were obtained by using the stopping power and the jet diameter as fitting parameters in the simulations performed with the GEANT4 code [16], until reaching satisfactory agreement with the experimental energy spectra. Figure 4 shows by symbols the experimental energy distributions for 2 MeV protons in liquid water at 10, 30 and 50 mrad; the GEANT4 simulations are depicted by dotted lines. In this figure, the energy of the ions after leaving the target, and when they reach the detector, is denoted by E.

However, these proton energy distributions after interaction with the liquid water jet can be satisfactorily reproduced using the SEICS code using the jet diameter as the only fitting parameter. We only need to take a 48.25 μm jet diameter (i.e., 3.5 % reduction compared to its nominal value) to obtain a very good agreement with the experimental distributions, being ever better than the comparison of GEANT4.

Proceeding in this manner we have validated the performance of the SEICS code for a suitable simulation of the motion and energy deposition of protons through a condensed target, and also, the reliability of our calculated stopping power values (solid curve in Fig. 2), which was obtained from the dielectric formalism and the MELF-GOS methodology. These results endorse what could be expected from the good agreement of our model-ELF for liquid water at several momentum transfers,

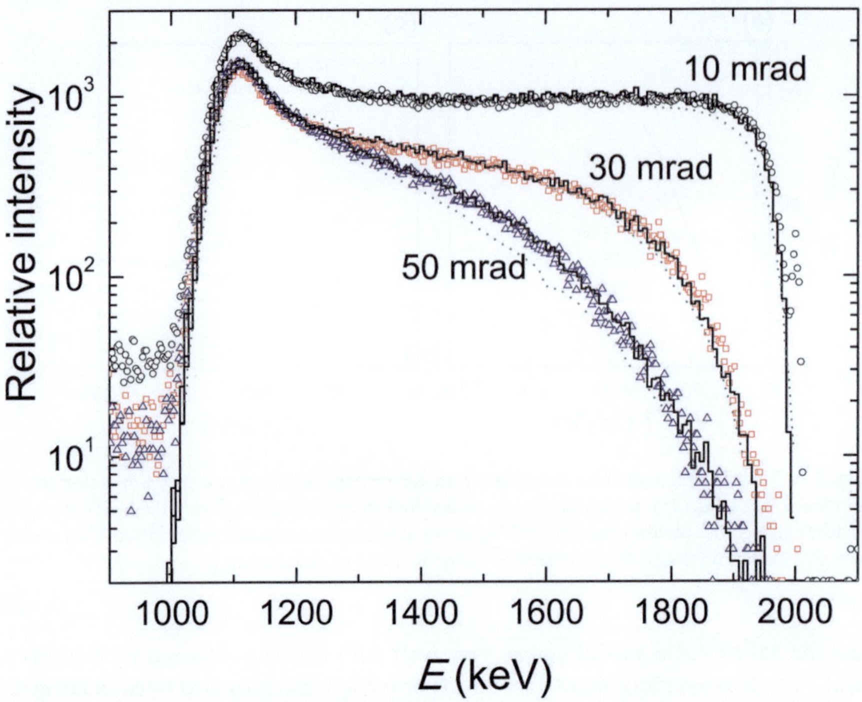

Fig. 4 Proton energy distributions at 10, 30 and 50 mrad after a 2 MeV proton beam interacts with a liquid water jet. *Symbols* correspond to experimental measurements [30]. *Dotted lines* are the distributions obtained with GEANT4, where the jet diameter and the liquid water stopping power had to be used as fitting parameters. *Solid lines* represent the results obtained with the SEICS code, where only the jet diameter was used as a fitting parameter, whereas the stopping power was the one provided by the dielectric formalism and the MELF-GOS methodology

which did agree very well with experimental measurements (Fig. 1), as well as the mean excitation energy $I = 79.4$ eV for liquid water predicted by the MELF-GOS method (see Eq. (18)), which is within the recent recommendation of 79.2 ± 1.6 eV given in Ref. [79].

4 Evaluation of Useful Quantities for Ion Beam Cancer Therapy

We will apply the simulation code SEICS to evaluate several quantities that are relevant in ion beam cancer therapy. We will focus our attention in the interaction of proton beams with liquid water, which is the main constituent of biological systems, and one of the most studied targets both theoretically and experimentally. The energy of the proton beam will be of the order of several hundred of MeV, as it is currently used in hadron therapy.

We will obtain the distribution of the dose deposited in the target as a function the depth (the Bragg curve), evaluating the influence of the different processes included in the code to describe the propagation of the projectiles through the target. The depth-dose curves will be compared to available experimental data.

From the detailed simulation it is also possible to extract the lateral dose distribution of the proton beams in the target, which is mainly due to elastic collisions. This can be quantified through the root mean square radius of the beam, which is parameterised as a function of the depth and energy of the incident projectiles.

Also, the energy distributions of the proton beam as a function of the depth in the target will be simulated, due to its significance in the generation of secondary electrons in the Bragg peak due to the proton impact.

4.1 Depth-Dose Distribution of Protons in Liquid Water

The precise knowledge of the energy deposited in the target by a proton beam as a function of the depth in the target (the depth-dose curve) is essential in proton therapy. This will allow a better control of the position of the Bragg peak, and of the energy deposited in it, as well as the energy deposited in the entrance-plateau region and distal part, where minimum damage is desired.

Due to its structure, the simulation code can be used to identify the role that each interaction process (elastic scattering, stopping power, energy-loss straggling, electron capture and loss, nuclear fragmentation reactions) has in its shape, since each process can be switched on and off in the simulations. Figure 5 represents the depth-dose profile of a 150 MeV proton beam in liquid water obtained with the SEICS code by removing in a controlled manner the contribution of different phenomena to the final result. The complete simulation, where all the processes (described in Sect. 2) are included is depicted by a solid line. Different processes have been eliminated from the simulation to analyse their contribution to the full simulation. Removing the energy-loss straggling (dotted line) has a big effect in the depth-dose curve, producing the appearance of an (unphysical) very sharp Bragg peak. Ignoring multiple elastic scattering (not shown in the figure) has a negligible influence on the depth-dose curve. Finally, a simulation without nuclear fragmentation reactions (dashed line) results in a broader (and slightly higher) peak and also in a reduction of energy deposited at the entrance region. Further information can be found for smaller proton incident energies in Refs. [21, 35, 80]. The choice of the stopping power used as input in the SEICS code (not shown in Fig. 5) mainly affects the range of the protons in the medium [41], and therefore the position of the Bragg peak [5, 78, 79, 81]. In all our calculations we use the electronic energy loss derived from the dielectric formalism and the MELF-GOS methodology, as described in Sect. 2.1.

The inclusion of the nuclear fragmentation reactions is an essential ingredient for projectiles having the typical energies used in hadron therapy, which is of the

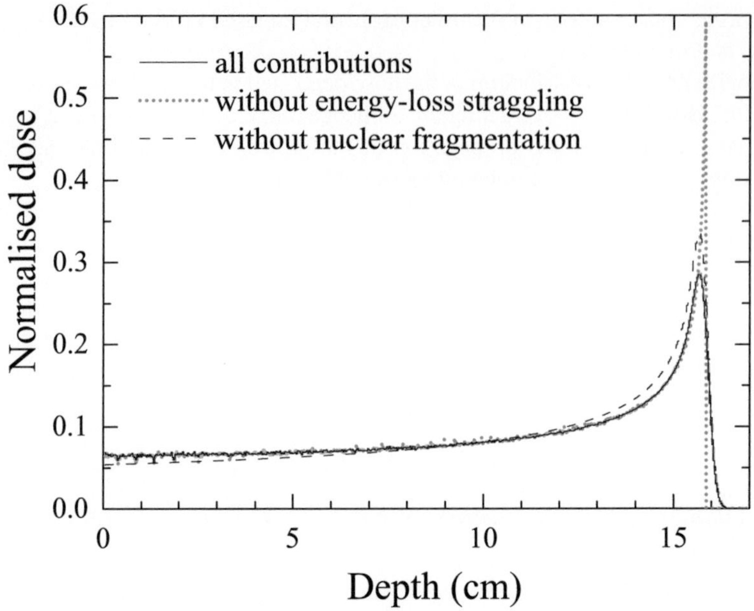

Fig. 5 Depth-dose distribution of a 150 MeV proton beam in liquid water simulated by the SEICS code (*solid line*). The relative contribution of different processes to the final result is considered by removing them from the simulation: without energy-loss straggling (*dotted line*), and without nuclear fragmentation (*dashed line*)

order of a few hundreds of MeV/u. We check our approach to implement the nuclear fragmentation reactions in the SEICS code comparing our results for depth-dose curves with available experimental data.

Figure 6 compares with experimental measurements [82] our simulated depth-dose curve for a 221.8 MeV proton beam in liquid water. Both curves are normalised to unit area. As it can be seen, the SEICS results, including the complete description of nuclear fragmentation reactions, perfectly agrees with the experiments at practically all the depths, except at the entrance of the target, where the simulation slightly overestimates the dose, as expected. In this figure we can also see the influence of the nuclear reaction model in the results. The dotted curve shows the simulated results when the nuclear fragmentation reactions are switched off. Clearly, it can not reproduce the experimental data. The dashed curve shows the results with nuclear reactions, but only removing the primary protons, without accounting for secondary particles, as it is done in some simpler approaches [83], which slightly improves the results without nuclear reactions. But the effect of secondary particles has to be taken into account, at least in an approximate manner as we have done, to obtain results in accordance with experimental data.

The good performance of our code can be further checked by comparing in Fig. 7 the simulated depth-dose curves with available experimental data for protons in liquid water at several energies. In all cases, there is an excellent agreement with the experimental data [82] for all the energies analysed.

It is noteworthy that the simple model we have used for nuclear fragmentation reactions allows a very nice comparison with experiments, without the necessity of implementing complex nuclear models.

4.2 *Lateral Spreading of the Depth-Dose Distribution*

Besides predicting with enough accuracy the depth-dose curves, in order to achieve high precision in hadron therapy treatment planning, another fundamental requisite is to know the lateral spreading of the beam with respect to the incident direction.

Multiple elastic scattering is the main responsible of the transversal deviations of the projectile trajectory, which results in the lateral spreading of the proton dose distribution, with the subsequent loss of precision in oncological treatment planning.

In what follows, we use the SEICS code to map the energy deposited in the target by the projectiles, both in the longitudinal and transverse directions of the incident beam.

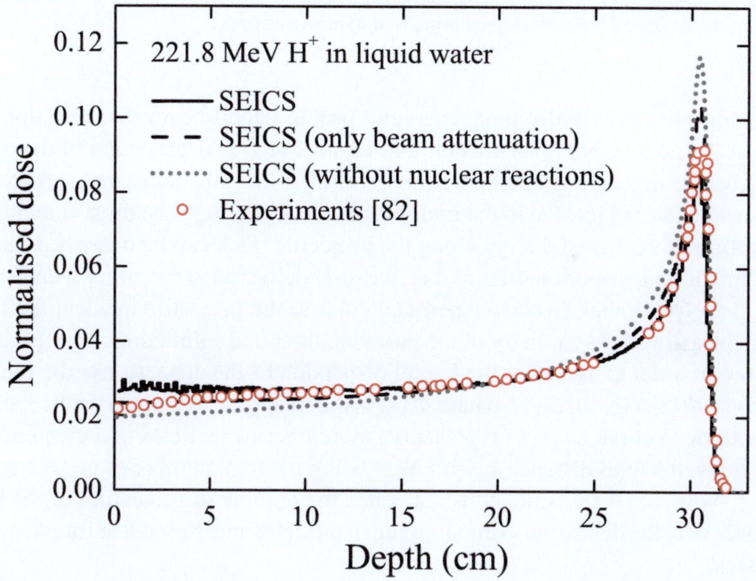

Fig. 6 Depth dose distribution in liquid water of a 221.8 MeV proton beam; experimental data are represented by *symbols* [82], whereas the *solid line* represents the results provided by the SEICS code. Simulations without nuclear reactions are depicted by a *dotted line* and with nuclear reactions but without including local energy deposition are shown by a *dashed line*. All the depth dose distributions are normalised to have unit area

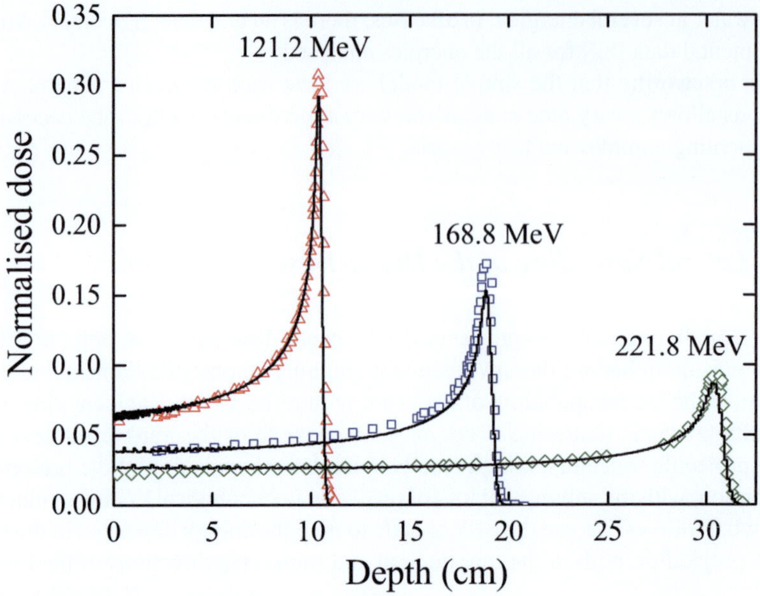

Fig. 7 Depth-dose curves of proton beams incident with different energies on liquid water. *Symbols* represent experimental data [82], whereas *solid lines* are the simulations obtained with the SEICS code. The depth-dose distributions are normalised to have unit area

We consider an initially monoenergetic proton pencil beam directed along the z-axis at therapeutic energies travelling in liquid water. A degradation of the proton energy occurs due mainly as a consequence of inelastic collisions, and a broadening of the beam takes place due to the multiple elastic scattering. The three-dimensional distribution of deposited energy along the projectile track can be described through the lateral dose distribution $\Phi(z, r)$, i.e., the dose delivered in the target at each depth z and at a given radial distance r perpendicular to the projectile incident direction. Due to the azimuthal symmetry of the problem, the x and y directions are equivalent, therefore in order to visualize the lateral dispersion of the dose we use the function $\Phi(z, x) = \Phi(z)N(z, x)/N(z)$, where $\Phi(z)$ is the total dose at a given depth z, that is, the depth-dose curve. $N(z, x)$ represents the number of particles at a given depth z and at a given lateral distance x, with $N(z)$ being the total number of projectiles at a depth z. Note that $N(z)$ is not constant, since the number of projectiles in the beam decreases with the depth due to the stopping of particles and the nuclear fragmentation processes.

In Fig. 8 we represent the simulated lateral distribution of the deposited dose for a 150 MeV proton beam in liquid water obtained with the SEICS code. It has to be noted the difference in the longitudinal and lateral scales, being the latter less than 10 % of the former, demonstrating the small lateral deflection of proton beams, which is one of their interesting features for treatment purposes. The larger concentration of deposited energy takes place along the beam entrance axis. However, due to multiple

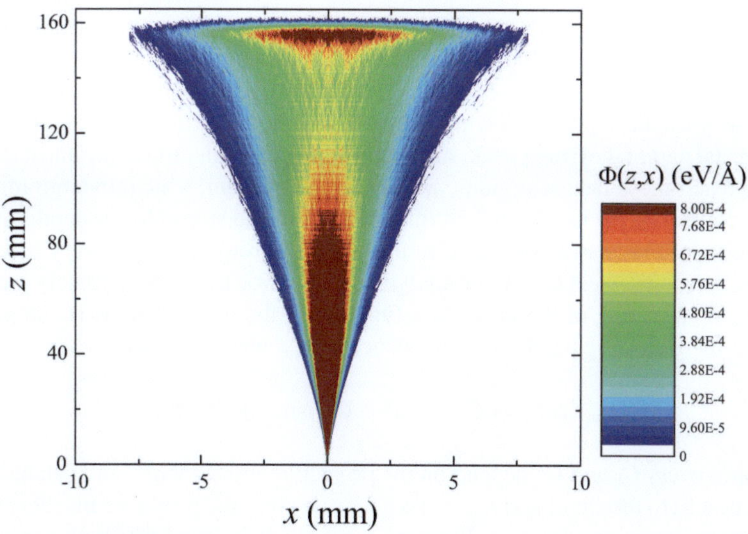

Fig. 8 Simulated lateral distribution of the dose deposited in liquid water by a 150 MeV proton beam, as a function of the depth z and the lateral distance x from the initial beam direction

elastic scattering there is a lateral spreading of the energy deposited that grows as the depth increases and reaches its maximum value around the Bragg peak. Because of this spread, the lateral dose presents two maxima along the initial projectile direction (i.e., the z axis), one at the Bragg peak depth, where the depth-dose curve is maximum, and another one at the very beginning of the protons track, where all the particles are concentrated in a small region because practically no elastic scattering has occurred yet. Although the total dose is smaller at the entrance plateau as compared to the Bragg peak region, the concentration of all the particles in small radial distances gives place to this maximum. Nonetheless, the integration of the lateral dose distribution recovers the expected shape of the integral depth-dose curve.

The lateral profile of the dose deposition pattern depends on the radial distribution of particles along the beam path, which can be represented by its root mean square radius $r_{\rm rms} = \langle r^2 \rangle^{1/2} = \sqrt{\langle x^2 \rangle + \langle y^2 \rangle}$, which is a function of the depth z. As the SEICS code follows the trajectories of the protons through the target, it is possible to evaluate $r_{\rm rms}$ at each depth, which indicates how much the incident particles scatter along their path. It has been shown [84] that our simulated results for $r_{\rm rms}$ of a proton pencil beam in liquid water agree with experimental data [85] as well as with an analytical model [86]. Although $r_{\rm rms}$ depends on the path traversed in the medium, on the atomic weight of the target and on the charge and energy of the incident projectiles, our simulations indicate that the quotient between the $r_{\rm rms}$ for a proton beam at the Bragg peak depth, $(r_{\rm rms})_{\rm max}$, and the Bragg peak depth, $z_{\rm max}$, is always around 3 %, that is $(r_{\rm rms})_{\rm max} \sim 0.03 z_{\rm max}$, independently of the initial proton energy in the range from 5 to 250 MeV for liquid water. For a 100 MeV proton beam, typically used in hadron therapy, the spread of the beam in liquid water at the Bragg peak is around 2 mm, whereas at 200 MeV it increases up to around 7 mm.

In the same proton energy range (5–250 MeV), our simulations predict that r_{rms} increases with the depth z following a parabolic dependence and suddenly falls down at the distal part of the Bragg peak, since only a few projectiles travel in almost straight line to reach these deeper regions, whereas most of them, which undergo multiple elastic collisions, deviate from their initial direction and stop at lower depths. As the simulation of the lateral dose distribution for each proton energy is very time consuming, it is convenient to have an analytical expression that approximately provides the lateral spreading of the beam as a function of the depth. This is accomplished through the following parabolic dependence of r_{rms} with the depth z:

$$r_{rms}(\mu m) = C_1\, z(\mu m) + C_2(\mu m^{-1})\, z^2(\mu m^2)\,. \tag{28}$$

The parameters C_1 and C_2 depend on the projectile initial energy, and can be found by the best fit to the simulated r_{rms}-curves for several initial proton beam energies T_0. These parameters are shown in Fig. 9 by symbols. It is found that both parameters follow a logarithmic behaviour with the initial proton energy T_0:

$$C_i = b_i\, [T_0(\mathrm{MeV})]^{a_i}\,, \tag{29}$$

where the constants a_i and b_i (the subscript i being associated to each C_i) are determined by the best fit to the C_i of the previous equation, shown by solid lines in Fig. 9. For protons in liquid water, these constants have the following values: $a_1 = -0.058$, $a_2 = -1.87$, $b_1 = 9.39 \times 10^{-3}$ and $b_2 = 1.56 \times 10^{-3}$.

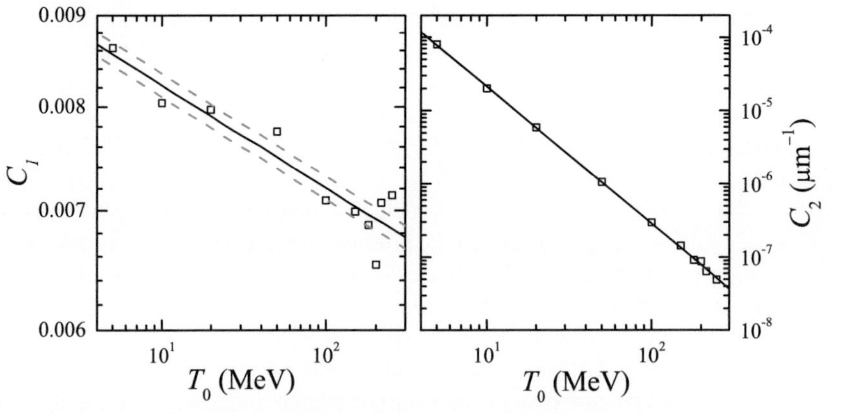

Fig. 9 Parameters C_1 and C_2 used to fit the r_{rms} of a proton pencil beam in liquid water to a parabola as a function of the initial proton energy. *Symbols* represent the results obtained with the SEICS code, *solid lines* are the least squares fits to these results and *dashed lines* represent their absolute error. In the case of C_2 the *lines* delimiting the error are indistinguishable from the *solid line*

As the lateral spread of the proton dose distribution due to multiple elastic scattering is one of the main causes of the spatial degradation of the Bragg peak, with the subsequent loss of precision required for a successful oncological treatment, a careful characterisation of the lateral spread of the dose deposited in a target of biological interest is of importance for improving the accuracy of clinical treatment planning in hadron therapy. The parameterisation we have obtained for $r_{\rm rms}$ as a function of the penetration depth z and the initial proton energy T_0 allows a quick calculation of $r_{\rm rms}$ for any proton energy and depth in liquid water.

4.3 Energy Distribution of the Proton Beam Along the Bragg Curve

Due to the stochastic nature of the interactions that take place between the projectile and the irradiated target, a monoenergetic distribution of the incident beam turns into an energy distribution as the particles of the beam move through the target. This distribution depends on the target nature, on the initial energy of the beam, and on the path travelled through the target. A detailed knowledge of the projectile energy distributions is very important since the generation of secondary electrons in the target due to proton impact strongly depends on the projectile energy. In this section we will study, by the simulation code SEICS, the energy distribution of a proton beam in liquid water and its evolution along the Bragg curve.

The SEICS code enables the calculation of the projectile energy distribution $\mathrm{d}N(T_0, z, T)/\mathrm{d}T$ at any depth z in the target of an initially T_0-monoenergetic proton beam, taking into account the interaction processes described previously (Sects. 2.1, 2.2, 2.3, and 2.4). The energy evolution of a proton beam incident with $T_0 = 150$ MeV in liquid water is shown in Fig. 10, where the mean energy of the beam (grey line) is depicted as a function of the depth, as well as the energy distribution of the proton beam (black lines) at several depths along the Bragg curve. To relate the Bragg peak with the proton beam energy distribution the depth-dose profile is also shown (dashed line). It can be seen how, while the mean energy of the beam decreases, the energy distribution has a Gaussian shape centred at the average energy, which broadens as the depth increases. At the Bragg peak depth, the distribution is centred around 15 MeV, and its full width at half maximum is practically 15 MeV wide.

In general, the proton energy distribution is rather broad around the Bragg peak, especially at its distal part, having a mean value and a full width at half maximum that always are $\sim 10\,\% T_0$, irrespectively of T_0, as we have found from several simulations. The physical origin of the broadening of the energy distribution is elucidated by switching on and off in the simulation code the different interaction phenomena. We find that the widening of the energy distribution is mainly due to the stochastic nature of the electronic interactions, accounted for through the electronic energy-loss straggling, while the multiple elastic scattering has an effect much less noticeable [21].

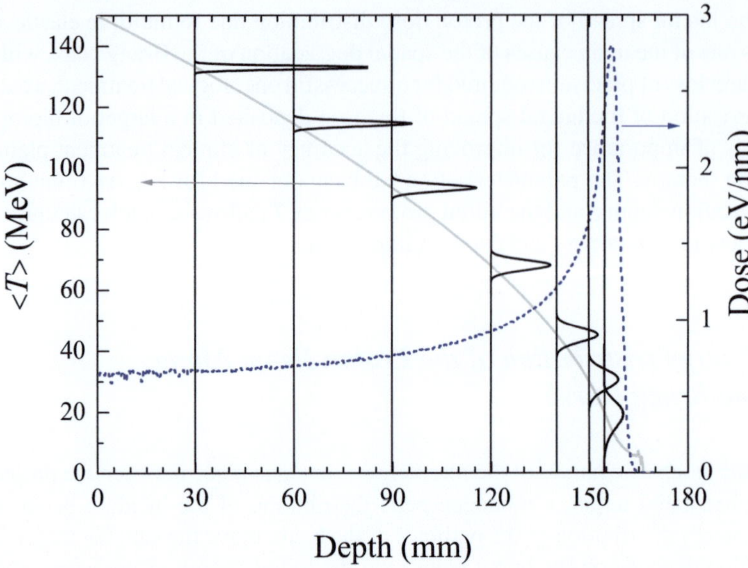

Fig. 10 (*Left axis*) Average energy $\langle T \rangle$ (*grey line*) of protons incident with 150 MeV in liquid water, as a function of the depth. The beam energy distribution at each depth is depicted by *black lines*. (*Right axis*) For comparison purposes, the depth-dose distribution is shown by a *dashed line*

We present in Fig. 11 the simulated energy distribution for a monoenergetic proton beam incident with $T_0 = 150$ MeV in liquid water at three depths around the Bragg peak: z_{80^-} is the depth corresponding to 80 % of the maximum dose before the Bragg peak, $z_{\max}$ is the depth where the maximum dose occurs (which corresponds to the Bragg peak), and z_{80^+} is the depth corresponding to 80 % of the maximum dose after the Bragg peak. The energy distributions are normalised to one incident projectile. It should be noted that when the beam moves through the target there is a reduction in the number of its projectiles due to nuclear reactions and to the energy loss. As it can be clearly seen in the Fig. 11, the proton energy spectrum in the region around the Bragg peak is very broad, especially at its distal part. We also found that the distributions are peacked at lower energies as the depth in the target increases.

4.4 Generation of Secondary Electrons in Biomaterials

Secondary electrons produced by ion impact play a central role in the multiscale picture of ion beam cancer therapy [87], since these electrons transport the energy lost by the projectile around its track at nanometre distances, giving place to a very sharp and intense radial dose distribution. This fact explains the increased radiobiological efficiency of ions compared to photons, for which the microscopic patterns of dose deposition are much more homogeneous [88, 89]. The secondary electrons generated

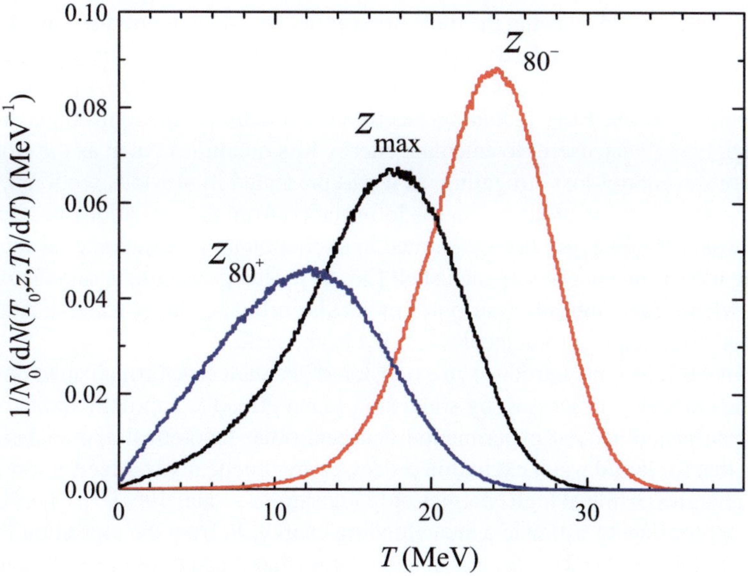

Fig. 11 Energy distributions of protons $\mathrm{d}N(T_0, z, T)/\mathrm{d}T$ for a monoenergetic beam incident with $T_0 = 150$ MeV on liquid water, at depths z_{80^-}, z_{max} and z_{80^+}, corresponding respectively to 80 % of the maximum dose before the Bragg peak, to the Bragg peak (maximum dose) and to 80 % of the maximum dose at the distal part of the Bragg peak. N_0 is the initial number of incident projectiles. The energy distributions are normalised to one incident projectile

along the proton track can also produce further ionisations, initiating an avalanche effect, leading to the energy transfer to sensitive biomolecular targets, such as DNA, lipids or proteins. Not only the number of emitted electrons is relevant, but also their energy spectrum, since although high energy electrons are capable of producing further ionisations, it has been shown that low energy electrons (below ionisation threshold) can also produce damage to biomolecules by dissociative electron attachment [31, 90].

There are several methodologies to find the energy distributions of secondary electrons generated by ion impact, going from very simple semiclassical models, such as the Rudd formula [91] or the Binary Encounter Approximation (BEA) [92], to complex and sophisticated time-consuming ab initio methods [93]. However, one of the limitations of these models is that they can only be applied with accuracy to some restricted ranges of energies, and specific projectile-target combinations. Therefore, it will be desirable to count on with a universal model applicable to a wide range of energies, any projectile-target system, and also to be simple, in order to be easily implemented in radiobiological models consuming a reasonable computing time.

This section is devoted to present a semiempirical model to calculate the energy distribution of secondary electrons generated by the impact of energetic proton beams in complex condensed biomaterials [94, 95].

A first idea could be using the dielectric formalism, since it provides an adequate framework to describe inelastic interactions of swift projectiles in condensed media. However in this formalism the target electronic response to external perturbations is embodied in the ELF, including electronic excitations and ionisations, which, although being very useful to calculate energy loss quantities (such as the stopping power or the energy-loss straggling, as it was presented in previous sections), complicates the separation of the ionisation information from the excitation one. Only for liquid water, the ELF has been separated in excitation and ionisation contributions, making use of the energies of each shell [96, 97]. This procedure is satisfactory for water, where there are only four outer molecular orbitals, but its generalisation for large and complex macromolecules is not a trivial task.

In what follows, we introduce an extension of the dielectric formalism to describe only the ionisation processes, by some simple but effective approximations, which have been tested against experimental data and other theoretical approaches. It is known that for liquid water excitation processes are practically confined at low transferred energies, while at high energies only ionisations remain [96, 97]. Therefore, it seems appropriate to estimate a mean binding energy, $\overline{B}$, from the ionisation thresholds of all the outer electronic shells, and assume that ionisations or excitations will only occur at energies above or below this threshold, respectively. This semiempirical method allows the calculation of ionisation cross sections for arbitrary biological materials within the dielectric formalism [94, 98].

Let us consider an energetic ion of mass M_1, atomic number Z_1 and charge q, moving with kinetic energy T in a medium characterised by its dielectric function $\varepsilon(k, \omega)$. Assuming that the outer shell electrons of the target can be characterised by a mean binding energy $\overline{B}$, then a secondary electron will be ejected with a kinetic energy $W = \hbar\omega - \overline{B}$. For each (n, ℓ)-subshell of each jth element, characterised by the ionisation energy $B^j_{\mathrm{ionis},n\ell}$, the secondary electrons will be ejected with energy $W = \hbar\omega - B^j_{\mathrm{ionis},n\ell}$. Then, the dielectric framework gives the following expression for the ionisation single differential cross section (ionisation SDCS), $\mathrm{d}\Lambda_{\mathrm{ionis}}/\mathrm{d}W$, or inverse mean free path, for the ejection of an electron with kinetic energy W:

$$\frac{\mathrm{d}\Lambda_{\mathrm{ionis}}(T, W, q)}{\mathrm{d}W} = \frac{e^2}{\pi\hbar^2}\frac{M_1}{T}\int_{\kappa_-}^{\kappa_+}\frac{\mathrm{d}k}{k}[Z_1 - \rho_q(k)]^2\,\mathrm{Im}\left[\frac{-1}{\varepsilon(k, W + \overline{B})}\right]_{\mathrm{outer}} \tag{30}$$
$$+ \frac{e^2}{\pi\hbar^2}\frac{M_1}{T}\sum_j \alpha_j \sum_{n\ell}\int_{K_-}^{K_+}\frac{\mathrm{d}k}{k}[Z_1 - \rho_q(k)]^2\,\mathrm{Im}\left[\frac{-1}{\varepsilon(k, W + B^j_{\mathrm{ionis},n\ell})}\right]_{\mathrm{inner},j,n\ell},$$

where $\kappa_\pm = (\sqrt{2M_1}/\hbar)(\sqrt{T} \pm \sqrt{T - W - \overline{B}})$, $K_\pm = (\sqrt{2M_1}/\hbar)(\sqrt{T} \pm \sqrt{T - W - B^j_{\mathrm{ionis},n\ell}})$, and α_j is the stoichiometry coefficient of each element j. Note that the key quantity to compute the ionisation cross section is the target ELF over the whole energy and momentum transfer, which is calculated by the MELF-GOS methodology (Sect. 2.1.1). So once the optical ELF of the target is known experimentally, the many-body interactions and the physical-state effects are naturally included in the subsequent calculations.

The mean binding energy of the target outer-shell electrons, $\overline{B}$, can be estimated from quantum chemistry calculations [99] for any biological media. Assuming a simple average of the binding energies of the outer-shell orbitals, the value for liquid water was set to 18.13 eV [94, 96], while for organic compounds it was assumed to be ~20 eV [98]. However, it is well known that the outer the shell is, the larger contribution to the ionisation cross section has. For this reason, we have re-estimated these energies, according to this fact, to be 13 eV for liquid water and ~ 15 eV for the rest of the biomaterials [100], and these are the values used for this work. The significance of this model lies in the possibility to calculate several radiological quantities such as single and total ionisation cross sections, as well as the number or the average energy of the emitted electrons, for any swift projectiles travelling in condensed biological targets, such as liquid water, DNA and its components, proteins, lipids, carbohydrates or cell constituents. It has been found that this approximation yields good results in comparison with available experimental data [94, 98, 101].

With the information obtained in the preceding paragraph, we can estimate the production of secondary electrons and its energy distribution along the whole Bragg curve, paying especial attention to depths around the Bragg peak, where the maximum energy deposition takes place. As we have presented in Sect. 4.3, an initial mono-energetic proton beam travelling through a medium, due to the stochastic inelastic interactions and the cumulative effect of the energy-loss straggling, develops a broad energy distribution, which depends on the initial proton energy and the travelled depth. In order to find realistic energy distributions of the secondary electrons generated by the projectile impact it is necessary to take into account the energy distribution of the proton beam, which can be reliably simulated by the SEICS code.

Once the energy distribution $\mathrm{d}N(T_0, z, T, q)/\mathrm{d}T$ at a depth z of the projectiles with charge q and initial kinetic energy T_0 is known, the realistic production of secondary electrons at any depth z generated by the proton beam must be calculated as a convolution between the ionisation inverse mean free path for a given energy T of the proton $\mathrm{d}\Lambda_{\text{ionis}}(T, W, q)/\mathrm{d}W$ (see Eq. (30)), and the energy distribution of the beam, namely:

$$\frac{\mathrm{d}\Lambda_{\text{ionis}}^{\text{conv}}(T_0, z, W)}{\mathrm{d}W} = \frac{1}{N_0}\int_0^{T_0} \mathrm{d}T \sum_{q=0}^{Z_1} \frac{\mathrm{d}N(T_0, z, T, q)}{\mathrm{d}T}\frac{\mathrm{d}\Lambda_{\text{ionis}}(T, W, q)}{\mathrm{d}W}, \tag{31}$$

where N_0 is the initial number of projectiles that bombard the target. A summation over all the charge states q must be performed to account for all the possible projectiles that can ionise the target.

In Fig. 12 we show the convoluted differential (in energy) ionisation cross section $\mathrm{d}\Lambda_{\text{ionis}}^{\text{conv}}(T_0, z, W)/\mathrm{d}W$ for a proton beam incident on liquid water with an initial energy $T_0 = 150$ MeV, as a function of the energy W of the emitted electrons and at depths z around the Bragg peak. The number of ejected electrons qualitatively follows the shape of the depth-dose curve, that is, the ionisation yield reaches a maximum at the Bragg peak (see Fig. 4). It is worth to notice that at each depth, the maximum number of secondary electrons is emitted at low energies, around 10 eV,

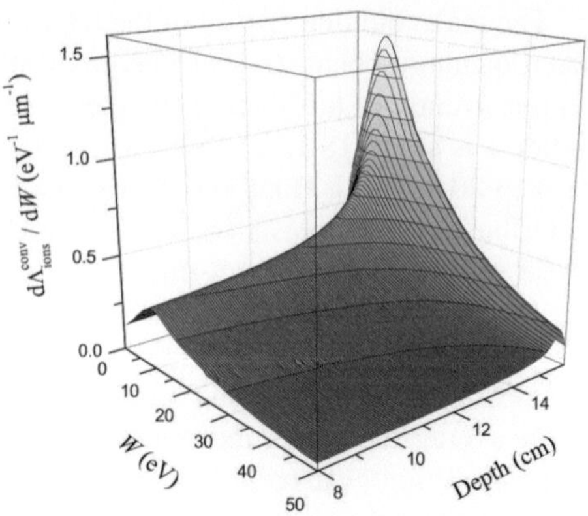

Fig. 12 Energy spectrum of secondary electrons generated by a 150 MeV proton beam in liquid water, as a function of the depth and the electron kinetic energy W. This result has been obtained from the convolution of the ionisation cross section and the proton energy distribution at each depth

with a sizeable increase in the Bragg peak region. This is particularly noteworthy due to the especial role played by low energy electrons in cellular damage [31, 32].

We analyse in more detail these convoluted energy distributions at the Bragg peak, z_{max}, in Fig. 13 for several relevant biomolecules. Curves in Fig. 13(a) show the SDCS for ionisation of liquid water by a proton beam incident with $T_0 = 150$ MeV, obtained by convolution with the proton energy distribution at z_{max} (solid line), as explained

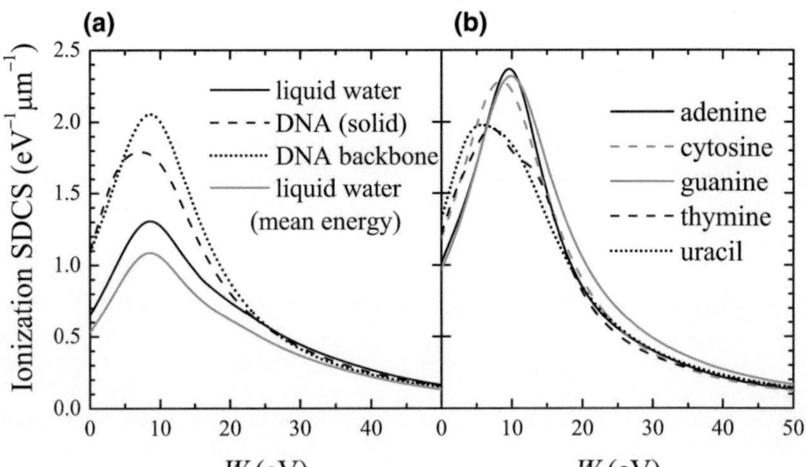

Fig. 13 **a** Convoluted energy distributions of secondary electrons (see text for more details), as a function of their ejection energy W, produced at the depth z_{max} corresponding to the Bragg peak by a 150 MeV proton beam incident on liquid water (*solid line*), DNA (*dashed line*) and backbone (*dotted line*); for comparison purposes, the energy distribution in liquid water obtained from the mean energy of the proton beam at the Bragg peak is also depicted (*grey solid line*). **b** Convoluted energy distributions of secondary electrons produced by at the Bragg peak by a 150 MeV proton beam in liquid water, but now considering that they are generated by adenide (*solid line*), cytosine (*grey dashed line*), guanine (*grey solid line*), thymine (*dashed line*) and uracil (*dotted line*)

previously, or when only considering the mean energy $\langle T \rangle$ of the beam at that depth (grey solid line). It can be seen that taking into account the energy spectrum of the protons increases the number of ejected secondary electrons in $\sim$20 % with respect to the mean proton energy calculation. This behaviour comes from the increased ionisation SDCS for the low energy protons [94] that appear in the distribution at Bragg peak.

Thus, the low energy tail of the proton energy distribution at each depth contributes largely to the ionisation SDCS. As low energy electrons are known to have an essential role in the damage of biological media due to dissociative electron attachment processes [31], a precise knowledge of the number and energy of the emitted electrons generated by the incidence of energetic ion beams in the biomaterials is of major relevance. Hence, it is indispensable to link the data obtained with the SEICS code for the energy distribution of the projectile at each depth with the energy distribution of the secondary electrons at each projectile energy in order to obtain accurate values for the initial conditions of the energy spectra, for the purpose of track-structure simulations of secondary electron transport. For these purposes, an increase of 20 % in the number of produced electrons might have noticeable consequences.

Figure 13a also depicts the ionisation SDCS for protons in DNA and its backbone calculated following the convolution method previously outlined, considering that the proton beam has been slowed down in liquid water. Although the Bragg curve is calculated in liquid water (the main medium where the incident projectiles propagate), the ionisation SDCS are obtained replacing at the Bragg peak depth the dielectric properties of liquid water by those of a different biological material, in order to determine the effect of the target composition on the electron production, which will determine the microscopic track-structure. The dashed and dotted lines in Fig. 13b correspond to the replacement of liquid water by solid DNA [102] and by the DNA sugar-phosphate backbone [99], respectively. The results for the rest of DNA/RNA bases (adenine, cytosine, guanine, thymine, and uracil) are shown by different lines in Fig. 13b. Their atomic compositions and densities can be found in Ref. [94]; however, here we have used a density of 1.35 g/cm^3 for the backbone, instead of 1 g/cm^3, to coincide with the DNA density. As it can be clearly seen, all the DNA/RNA components present a convoluted ionisation SDCS that is between 35 and 80 % larger than liquid water. This result indicates that the electron production can be substantially higher in a realistic biological target (e.g., cell nucleus) as compared to pure liquid water, which is currently used as a universal biological surrogate. In addition, the different molecular components of DNA/RNA could present somewhat different ionisation probabilities, although within a certain range, as it can be observed from the figure.

5 Summary and Conclusions

In this chapter, we have presented an overview of the procedures and results that can be obtained through a detailed simulation study of the propagation, energy deposition and electron production of energetic proton beams in materials of radiotherapeutic interest, especially liquid water. The outcomes of these simulations are relevant for ion beam cancer therapy purposes.

The study is conducted by using the SEICS code (Simulation of Energetic Ions and Clusters through Solids), which combines the Monte Carlo and Molecular Dynamics techniques to follow the motion of energetic ions through condensed matter. The main characteristics of SEICS have been explained in Sect. 2, where the procedure for treating the most relevant interactions between energetic protons and a condensed biological target have been featured, namely, electronic excitations, multiple elastic scattering, electron capture and loss, and nuclear fragmentation reactions.

The main interaction channel, affecting both the position and shape of the Bragg peak, i.e., the electronic excitation process, has been reviewed in Sect. 2.1. We have explained the main features of the dielectric formalism and the MELF-GOS method, which is particularly well suited for calculating the relevant electronic energy loss quantities in condensed matter (i.e., taking into account chemical and phase effects of the target). It has been shown how this methodology is capable of perfectly reproducing the experimental electronic excitation spectrum of liquid water reported in Refs. [54, 55]. The calculated stopping power of liquid water for proton beams has been compared to the available experimental data for liquid and solid water, finding a general good agreement.

The shape of the Bragg peak is also affected by the nuclear fragmentation reactions. We have implemented a simple method that incorporates this inelastic interaction into the simulation, based on the cross sections provided in Ref. [76], as explained in Sect. 2.4. The method accounts for the loss of primary protons according to the total nuclear cross sections, and then the residual energy of the charged secondary particles (protons and heavier ions) that are produced in the reaction is deposited locally.

Before applying the SEICS code, having as the main input the energy loss quantities calculated for liquid water with the dielectric formalism, to problems related to ion beam cancer therapy, we have benchmarked both of them (the code and the electronic energy loss) against experimental data in Sect. 3. Our simulations have reproduced the experimental energy distributions obtained at several angles when a 2 MeV proton beam irradiates a thin liquid water jet [29, 30]. We have found that it is possible to perfectly reproduce the experimental proton energy spectra by only reducing the jet diameter to 48.25 μm (3.5 % reduction compared to the nominal 50 μm diameter). Such a reduction is plausible, assuming a possible evaporation of the liquid water jet into vacuum. Therefore, we conclude that our calculated stopping power for protons in liquid water is accurate, and that the SEICS code works appropriately, so it can be used for studying problems related to ion beam cancer therapy.

This study has been presented in Sect. 4, where different characteristics of proton beams in liquid water have been discussed. The shape of the depth-dose (i.e., Bragg) curve, and the influence on it of the different interaction processes, was studied in Sect. 4.1. We showed how the stopping power determined the position of the Bragg peak, whereas the broadening of its shape was mainly due to the energy-loss straggling, with additional contribution from nuclear fragmentation reactions. The implementation of the nuclear fragmentation reactions, with the local energy deposition approximation, was checked against experimental data, finding a satisfactory agreement.

Regarding the multiple elastic scattering, it was shown to be responsible of the projectiles deviation from their initial direction, leading to the transversal dispersion of the beam, which results in a lateral spreading of the dose profile. We have studied this aspect in Sect. 4.2, where the root mean square radius of the transversal section of the proton beam has been studied in liquid water for different initial energies. We have found that it can be parameterised as a function of the incident energy, providing the corresponding parameters, which allows the calculation of the lateral dose analytically.

In Sect. 4.3, we have studied how the energy of the proton beam evolves along the Bragg curve. Due to the energy-loss straggling, the initially monoenergetic proton beam develops an energy distribution, which becomes rather wide at the Bragg peak depth. At this position, the mean energy of the distribution and the full width at half maximum are 10 % of the initial beam energy, as we have found from several simulations.

This distribution of the proton energy can affect the number and energy of secondary electrons produced at each depth, which has been studied in Sect. 4.4. The energy spectra of secondary electrons generated by a proton with a given energy, as calculated with the dielectric formalism, have been convoluted with the energy distributions of the primary protons, to produce realistic energy spectra of secondary electrons produced at each depth in the target. They have been compared with the spectra obtained when using instead the mean energy of the proton beam. The former (and more realistic) procedure leads to an increase of $\sim$20 % in the number of secondary electrons, compared to the latter. Finally, we have also calculated the ionisation single differential cross section for liquid water and other biological targets, namely DNA and its molecular components, i.e., the five DNA/RNA bases (adenine, cytosine, guanine, thymine, and uracil) and the sugar-phosphate backbone. By comparing these values, we have demonstrated that all these biomolecules present larger ionisation probabilities than liquid water. This is an important result for further studies of biodamage mechanisms at the nano- and microscale, which are largely related to the energy and number of electrons produced and propagating in realistic cellular components.

Acknowledgements Illuminating and fruitful discussions with many collaborators are acknowledged. Most of the research presented in this paper was developed under the warm and stimulating atmosphere of the COST Action MP 1002, Nanoscale Insights into Ion Beam Cancer Therapy. During the last revision of this work we heard of the death of Helmut Paul, an excellent friend and a better scientist, to whom we dedicate this work. We thank partial financial support by the Spanish Ministerio de Economía y Competitividad (Project FIS2014-58849-P) and the Murcia Regional Agency of Science and Technology (project 19907/GERM/15). PdV acknowledges financial support from the European Union's FP7-People Program (Marie Curie Actions) within the Initial Training Network No. 608163 "ARGENT".

References

1. Löeffler JS, Durante M (2013) Charged particle therapy-optimization, challenges and future directions. Nat Rev Clin Oncol 10:411–424
2. PTCOG: Particle Therapy Co-Operative Group webpage. http://www.ptcog.ch/
3. Sigmund P (2006) Particle penetration and radiation effects. Springer-Verlag, Berlin
4. Paul H (2007) The mean ionization potential of water, and its connection to the range of energetic carbon ions in water. Nucl Instr Meth B 255:435–437
5. Besemer A, Paganetti H, Bednarz B (2013) The clinical impact of uncertainties in the mean excitation energy of human tissues during proton therapy. Phys Med Biol 58:887–902
6. Surdutovich E, Solov'yov AV (2014) Multiscale approach to the physics of radiation damage with ions. Eur Phys J D 68:353
7. Uehara S, Nikjoo H, Goodhead DT (1992) Cross-sections for water vapour for the Monte Carlo electron track structure code from 10 eV to the MeV region. Phys Med Biol 37:1841–1858
8. Friedland W, Dingfelder M, Kundrát P, Jacob P (2011) Track structures, DNA targets and radiation effects in the biophysical Monte Carlo simulation code PARTRAC. Mutation Res 711:28–40
9. Pimblott SM, Mozumder A (1991) Structure of electron tracks in water. 2. Distribution of primary ionizations and excitations in water radiolysis. J Phys Chem 95:7291–7300
10. Nikjoo H, Uehara S, Emfietzoglou D (2012) Interaction of radiation with matter. CRC Press, Boca Raton
11. Champion C, Le Loirec C, Stosic B (2012) EPOTRAN: a full-differential Monte Carlo code for electron and positron transport in liquid and gaseous water. Int J Radiat Biol 88:54–61
12. Böhlen TT, Cerutti F, Chin MPW, Fassò A, Ferrari A, Ortega PG, Mairani A, Sala PR, Smirnov G, Vlachoudis V (2014) The FLUKA Code: developments and challenges for high energy and medical applications. Nucl Data Sheets 120:211–214
13. Waters LS (ed.) (2002) MCNPX User's manual Version 2.3.0. Report LA-UR-02-2607. Los Alamos Nat. Lab., Los Alamos, USA
14. Ziegler JF, Biersack JP, Ziegler MD (2008) SRIM—The stopping and range of ions in matter. SRIM Co., Chester, Maryland
15. Salvat F, Fernández-Varea JM, Sempau J (2011) PENELOPE-2011: a code system for monte carlo simulation of electron and photon transport. OECD NEA Data Bank/NSC DOC(2011)/5 (2011)
16. Agostinelli S, Allison J, Amako K, Apostolakis J, Araujo H, Arce P, Asai M, Axen D, Banerjee S, Barrand G, Behner F, Bellagamba L, Boudreau J, Broglia L, Brunengo A, Burkhardt H et al (2003) GEANT4-a simulation toolkit. Nucl Instrum Methods Phys Res A 506:250–303
17. Nikjoo H, Uehara S, Emfietzoglou D, Cucinotta F (2006) Track-structure codes in radiation research. Radiat Meas 41:1052–1074
18. Incerti S, Baldacchino G, Bernal M, Capra R, Champion C, Francis Z, Guatelli S, Gueye P, Mantero A, Mascialino B, Moretto P, Nieminen P, Rosenfeld A, Villagrasa C, Zacharatou C (2010) The GEANT4-DNA project. Int J Model Simul Sci Comput 1:157
19. Muñoz A, Perez J, Garcia G, Blanco F (2005) An approach to Monte Carlo simulation of low-energy electron and photon interactions in air. Nucl Instrum Methods Phys Res A 536:176–188
20. Muñoz A, Fuss MC, Cortés-Giraldo MA, Incerti S, Ivanchenko V, Ivanchenko A, Quesada JM, Salvat F, Champion C, García Gómez-Tejedor G (2012) In: García Gómez-Tejedor G (ed.) Radiation damage in biomolecular systems, Chap. 13. Springer, Dordrecht
21. Garcia-Molina R, Abril I, Heredia-Avalos S, Kyriakou I, Emfietzoglou D (2011) A combined molecular dynamics and Monte Carlo simulation of the spatial distribution of energy deposition by proton beams in liquid water. Phys Med Biol 56:6475–6493
22. Lindhard J (1954) On the properties of a gas of charged particles. K Dan Vidensk Selsk Mat.-Fys Medd 28(8):1
23. Garcia-Molina R, Abril I, Denton CD, Heredia-Avalos S, Kyriakou I, Emfietzoglou D (2009) Calculated depth-dose distributions for H^+ and He^+ beams in liquid water. Nucl Instrum Meth Phys Res B 267:2647–2652

24. Limandri S, de Vera P, Fadanelli RC, Nagamine LCCM, Mello A, Garcia-Molina R, Behar M, Abril I (2014) Energy deposition of H and He ion beams in hydroxyapatite films: a study with implications for ion-beam cancer therapy. Phys Rev E 89:022703
25. Abril I, Garcia-Molina R, Denton CD, Kyriakou I, Emfietzoglou D (2011) Energy loss of hydrogen- and helium-ion beams in DNA: calculations based on a realistic energy-loss function of the target. Rad Res 175:247–255
26. de Vera P, Abril I, Garcia-Molina R (2011) Inelastic scattering of electron and light ion beams in organic polymers. J Appl Phys 109:094901
27. Garcia-Molina R, Abril I, de Vera P, Kyriakou I, Emfietzoglou D (2014) A study of the energy deposition profile of proton beams in materials of hadron therapeutic interest. Appl Radiat Isotopes 83:109–114
28. Abril I, Garcia-Molina R, de Vera P, Kyriakou I, Emfietzoglou D (2013) Inelastic collisions of energetic protons in biological media. Adv Quantum Chem 65:129–164
29. Shimizu M, Kaneda M, Hayakawa T, Tsuchida H, Itoh A (2009) Stopping cross sections of liquid water for MeV energy protons. Nucl Instrum Meth Phys Res B 267:2667–2670
30. Shimizu M, Hayakawa T, Kaneda M, Tsuchida H, Itoh A (2010) Stopping cross-sections of liquid water for 0.3-2.0 MeV protons. Vacuum 84:1002–1004
31. Boudaïffa B, Cloutier P, Hunting D, Huels MA, Sanche L (2000) Resonant formation of DNA strand breaks by low-energy (3 to 20 eV) electrons. Science 287:1658–1660
32. Rezaee M, Cloutier P, Bass AD, Michaud M, Hunting DJ, Sanche L (2012) Absolute cross section for low-energy-electron damage to condensed macromolecules: a case study of DNA. Phys Rev E 86:031913
33. Garcia-Molina R, Denton CD, Abril I, Arista NR (2000) Energy-loss and exit-angle distributions of fragmented H_2^+ ions after traversing carbon foils. Phys Rev A 62(1):012901
34. Heredia-Avalos S, Garcia-Molina R, Abril I (2007) Energy-loss calculation of swift C_n^+ ($n =$ 2-60) clusters through thin foils. Phys Rev A 76:012901
35. Garcia-Molina R, Abril I, de Vera P, Kyriakou I, Emfietzoglou D (2012) In: Belkić D (ed) Fast ion-atom and ion-molecule collisions. World Scientific Publishing Company, Singapore
36. Allen MP, Tildesley DJ (1989) Computer simulation of liquids. Oxford University Press, Oxford
37. Box GEP, Müller ME (1958) A note on the generation of random normal deviates. Ann Math Stat 29:610–611
38. Press WH, Teukolsky SA, Vetterling WT, Flannery BP (1997) Numerical recipes in Fortran 77. The art of scientific computing, 2nd edn. Cambridge University Press, Cambridge
39. Abril I, Garcia-Molina R, Denton CD, Pérez-Pérez F, Arista NR (1998) Dielectric description of wakes and stopping powers in solids. Phys Rev A 58:357–366
40. Heredia-Avalos S, Garcia-Molina R, Fernández-Varea JM, Abril I (2005) Calculated energy loss of swift He, Li, B, and N ions in SiO_2, Al_2O_3, and ZrO_2. Phys Rev A 72:052902
41. Garcia-Molina R, Abril I, Kyriakou I, Emfietzoglou D (2012) Energy loss of swift protons in liquid water: role of optical data input and extension algorithms. In: García Gómez-Tejedor G, Fuss MC (eds) Radiation damage in biomolecular systems, Chap. 15. Springer, Dordrecht
42. Inokuti M (1971) Inelastic collisions of fast charged particles with atoms and molecules—The Bethe theory revisited. Rev Mod Phys 43:297–347
43. Fano U (1963) Penetration of protons, alpha particles, and mesons. Ann Rev Nucl Sci 13:1–66
44. de Vera P, Abril I, Garcia-Molina R (2014) Water equivalent properties of materials commonly used in proton dosimetry. Appl Radiat Isotopes 83:122–127
45. Fermi E (1940) The ionization loss of energy in gases and in condensed materials. Phys Rev 57:485
46. Ritchie RH, Howie A (1977) Electron excitation and the optical potential in electron microscopy. Phil Mag 36:463–481
47. Brandt W, Kitagawa M Effective stopping-power charges of swift ions in condensed matter. Phys Rev B 25:5631
48. Denton CD, Abril I, Garcia-Molina R, Moreno-Marín JC, Heredia-Avalos S (2008) Influence of the description of the target energy-loss function on the energy loss of swift projectiles. Surf Interface Anal 40:1481–1487

49. Mermin ND (1970) Lindhard dielectric function in the relaxation-time approximation. Phys Rev B 1:2362–2362
50. Egerton RF (2011) Electron energy-loss spectroscopy in the electron microscope, 3rd edn. Springer, New York
51. Bragg WH, Kleeman R (1905) On the alpha particles of radium, and their loss of range in passing through various atoms and molecules. Phil Magaz Series 10:318–340
52. Pines D (1999) Elementary excitations in solids. Perseus Books, Urbana, Illinois
53. Dressel M, Grüner G (2002) Electrodynamics of solids. Optical properties of electrons in matter. Cambridge University Press, Cambridge
54. Watanabe N, Hayashi H, Udagawa Y (1997) Bethe surface of liquid water determined by inelastic X-ray scattering spectroscopy and electron correlation effects. Bull Chem Soc Jpn 70:719–726
55. Hayashi H, Watanabe N, Udagawa Y, Kao C (2000) The complete optical spectrum of liquid water measured by inelastic x-ray scattering. Proc Nat Academy Sci USA 97:6264–6266
56. Garcia-Molina R, Abril I, Denton CD, Heredia-Avalos S (2006) Allotropic effects on the energy loss of swift H^+ and He^+ ion beams through thin foils. Nucl Instrum Meth Phys Res B 249:6–12
57. Abril I, Moreno-Marín JC, Fernández-Varea JM, Denton CD, Heredia-Avalos S, Garcia-Molina R (2007) Calculation of the energy loss of swift H and He ions in Ag using the dielectric formalism: the role of inner-shell ionization. Nucl Instrum Meth Phys Res B 256:172–176
58. Heredia-Avalos S, Abril I, Denton CD, Moreno-Marín JC, Garcia-Molina R (2007) Target inner-shells contributions to the stopping power and straggling for H and He ions in gold. J Phys Conden Matter 19:466205
59. Montanari C, Miraglia J, Heredia-Avalos S, Garcia-Molina R, Abril I (2007) Calculation of energy-loss straggling of C, Al, Si, and Cu for fast H, He, and Li ions. Phys Rev A 75:022903
60. Abril I, Behar M, Garcia-Molina R, Fadanelli RC, Nagamine LC, Grande PL, Schünemann L, Denton CD, Arista NR, Saitovitch EB (2009) Experimental and theoretical studies of the energy-loss straggling of H and He ion beams in HfO_2 films. Eur Phys J D 54:65–70
61. Behar M, Fadanelli RC, Abril I, Garcia-Molina R, Denton CD, Nagamine LCCM, Arista NR (2009) Energy loss of proton, α particle, and electron beams in hafnium dioxide films. Phys Rev A 80:062901
62. Behar M, Denton CD, Fadanelli RC, Abril I, Cantero ED, Garcia-Molina R, Nagamine LCC (2010) Experimental and theoretical determination of the stopping power of ZrO_2 films for protons and α-particles. Eur Phys J D 59:209–213
63. Limandri SP, Fadanelli RC, Behar M, Nagamine LCCM. Fernández-Varea JM, Abril I, Garcia-Molina R, Montanari CC, Aguiar JC, Mitnik D, Miraglia JE, Arista NR (2014) Stopping cross sections of TiO_2 for H and He ions. Eur Phys J D 68:194-1–8
64. Fadanelli RC, Behar M, Nagamine LCCM, Vos M, Arista NR, Nascimento CD, Garcia-Molina R, Abril I (2015) Energy loss function of solids assessed by ion beam energy-loss measurements: practical application to Ta_2O_5. J Phys Chem C 119:20561–20570
65. Garcia-Molina R, Abril I, Kyriakou I, Emfietzoglou D (2016) Inelastic scattering and energy loss of swift electron beams in biologically relevant materials. Surf Interf Anal. doi:10.1002/sia.5947
66. Schiwietz G, Grande PL (2001) Improved charge-state formulas. Nucl Instrum Meth Phys Res B 175–177:125–131
67. Siiskonen T, Kettunen H, Peräjärvi K, Javanainen A, Rossi M, Trzaska WH, Turunen J, Virtanen A (2011) Energy loss measurement of protons in liquid water. Phys Med Biol 56:2367–2374
68. Wenzel WA, Whaling W (1952) The stopping cross section of D_2O ice. Phys Rev 87:499–503
69. Andrews DA, Newton G (1977) The stopping power of heavy ice for low-energy (10–30 keV) deuterons. J Phys D 10:845–850
70. Bauer P, Kaferbock W, Necas V (1994) Investigation of the electronic energy loss of hydrogen ions in H_2O: influence of the state of aggregation. Nucl Instrum Meth Phys Res B 93:132–136

71. ICRU, Report 49 (1993) Stopping powers and ranges for protons and alpha particles. International Commission on Radiation Units and Measurements, Bethesda, Maryland
72. Möller W, Pospiech G, Schrieder G (1975) Multiple scattering calculations on ions passing through thin amorphous foils. Nucl Instrum Meth Phys Res B 130:265–270
73. Zajfman D, Both G, Kanter EP, Vager Z (1990) Multiple scattering of MeV atomic and molecular ions traversing ultrathin films. Phys Rev A 41:2482–2488
74. Turner JE, Wright HA, Hamm RN (1985) A Monte Carlo primer for health physicists. Health Phys 48:717–733
75. Brandt W, Sizmann R (1975) In: Datz S, Appleton BR, Moak CD (eds.) Atomic collisions in solids. Plenum Press, New York
76. ICRU (2000) Report 63. Nuclear data for neutron and proton radiotherapy and for radiation protection. International Commission on Radiation Units and Measurements, Bethesda, Maryland
77. Medin J, Andreo P (1997) Monte Carlo calculated stopping-power ratios, water/air, for clinical proton dosimetry (50–250 MeV). Phys Med Biol 42:89–106
78. Paul H (2010) Recent results in stopping power for positive ions, and some critical comments. Nucl Instrum Meth Phys Res B 268:3421–3425
79. Paul H (2013) On the accuracy of stopping power codes and ion ranges used for hadron therapy. Adv Quantum Chem 65:39–62
80. Garcia-Molina R, Abril I, de Vera P, Kyriakou I, Emfietzoglou D (2012) Role of the interaction processes in the depth-dose distribution of proton beams in liquid water. J Phys Confer Series 373:012015
81. Paganetti H (2012) Range uncertainties in proton therapy and the role of Monte Carlo simulations. Phys Med Biol 57:R99–R117
82. Zhang X, Liu W, Li Y, Li X, Quan M, Mohan R, Anand A, Sahoo N, Gillin M, Zhu XR (2011) Parameterization of multiple Bragg curves for scanning proton beams using simultaneous fitting of multiple curves. Phys Med Biol 56:7725–7735
83. Kundrát P (2007) A semi-analytical radiobiological model may assist treatment planning in light ion radiotherapy. Phys Med Biol 52:6813–6830
84. Abril I, de Vera P, Garcia-Molina R, Kyriakou I, Emfietzoglou D (2015) Lateral spread of dose distribution by therapeutic proton beams in liquid water. Nucl Instrum Meth Phys Res B 352:176–180
85. Kraft G (1990) The radiobiological and physical basis for radiotherapy with protons and heavier ions. Strahlenther Onkol 166:10–13
86. Hollmark M, Uhrdin J, Belkić D, Gudowska I, Brahme A (2004) Influence of multiple scattering and energy loss straggling on the absorbed dose distributions of therapeutic light ion beams: I. Analytical pencil beam model. Phys Med Biol 49:3247–3265
87. Solov'yov AV, Surdutovich E, Scifoni E, Mishustin I, Greiner W (2009) Physics of ion beam cancer therapy: a multiscale approach. Phys Rev E 79:011909
88. Scholz M, Kraft G (1996) Track structure and the calculation of biological effects of heavy charged particles. Adv Space Res 18:5–14
89. Schardt D, Elsässer T, Schulz-Ertner D (2010) Heavy-ion tumor therapy: physical and radiobiological benefits. Rev Modern Phys 82:383–425
90. Pan X, Cloutier P, Hunting D, Sanche L (2003) Dissociative electron attachment to DNA. Phys Rev Lett 90:208102
91. Rudd ME, Kim YK (1992) Electron production in proton collisions with atoms and molecules: energy distributions. Rev Modern Phys 64:441–490
92. Vriens L (1967) Binary-encounter proton-atom collision theory. Proc Phys Soc London 90:935–944
93. ICRU (1996) Report 55. Secondary electron spectra from charged particle interactions. International Commission on Radiation Units and Measurements, Bethesda, Maryland
94. de Vera P, Garcia-Molina R, Abril I, Solov'yov AV (2013) Semiempirical model for the ion impact ionization of complex biological media. Phys Rev Lett 110:148104

95. de Vera P, Garcia-Molina R, Abril I (2015) Angular and energy distributions of electrons produced in arbitrary biomaterials by proton impact. Phys Rev Lett 114:018101
96. Dingfelder M, Hantke D, Inokuti M, Paretzke HG (1998) Electron inelastic-scattering cross sections in liquid water. Radiat Phys Chem 53:1–18
97. Emfietzoglou D, Moscovitch M (2003) Secondary electron spectra for fast proton impact on gaseous and liquid water. Nucl Instrum Meth Phys Res B 209:239–245
98. de Vera P, Abril I, Garcia-Molina R, Solov'yov AV (2013) Ionization of biomolecular targets by ion impact: input data for radiobiological applications. J Phys Conf Series 438:012015
99. Bernhardt P, Paretzke HG (2003) Calculation of electron impact ionization cross sections of DNA using the Deutsch-Mark and Binary-Encounter-Bethe formalisms. Int J Mass Spectrom 224:599–611
100. de Vera P, Garcia-Molina R, Abril I: To be published
101. Wang M, Rudek B, Bennett D, de Vera P, Bug M, Buhr T, Baek WY, Hilgers G, Rabus H (2016) Cross sections for ionization of tetrahydrofuran by protons at energies between 300 and 3000 keV. Phys Rev A 93:052711-1–16
102. Inagaki T, Hamm RN, Arakawa ET, Painter LR (1974) Optical and dielectric properties of DNA in the extreme ultraviolet. J Chem Phys 61:4246–4250

Monte Carlo-Based Modeling of Secondary Particle Tracks Generated by Intermediate- and Low-Energy Protons in Water

Alexey Verkhovtsev, Pedro Arce, Antonio Muñoz, Francisco Blanco and Gustavo García

Abstract This chapter gives an overview of recent developments in the Monte Carlo-based modeling of the interaction of ionizing radiation with biologically relevant systems. Several track structure codes, such as Geant4 (GEometry ANd Tracking 4), Geant4-DNA, and LEPTS (Low-Energy Particle Track Simulation), are described. Main features, areas of application and current limitations of each tool are discussed. A special attention is focused on the energy range covered by primary and secondary charged particles and on the type of interactions included in the simulation. A recent development of LEPTS is presented, aimed at the simulation of full slowing-down of protons in water together with all molecular processes involving secondary particles. The utilized approach allows one to study radiation effects on the nanoscale in terms of the number and the type of induced molecular processes.

A. Verkhovtsev (✉) · G. García
Instituto de Física Fundamental, Consejo Superior de Investigaciones
Científicas (CSIC), Serrano 113-bis, 28006 Madrid, Spain
e-mail: verkhovtsev@iff.csic.es

G. García
e-mail: g.garcia@iff.csic.es

A. Verkhovtsev
MBN Research Center, 60438 Frankfurt am Main, Germany

P. Arce
Medical Applications Unit, Centro de Investigaciones Energéticas, Medioambientales y Tecnológicas (CIEMAT), Av. Complutense 40, 28040 Madrid, Spain
e-mail: pedro.arce@ciemat.es

A. Muñoz
Scientific Computing Unit, Centro de Investigaciones Energéticas, Medioambientales y Tecnológicas (CIEMAT), Av. Complutense 40, 28040 Madrid, Spain
e-mail: roldan@ciemat.es

F. Blanco
Departamento de Física Atómica, Molecular y Nuclear, Universidad Complutense de Madrid, Plaza de Ciencias 1, 28040 Madrid, Spain
e-mail: pacobr@fis.ucm.es

G. García
Centre for Medical Radiation Physics, University of Wollongong,
Wollongong, NSW 2522, Australia

A.V. Solov'yov (ed.), *Nanoscale Insights into Ion-Beam Cancer Therapy*,
DOI 10.1007/978-3-319-43030-0_3

Development of new tools for the simulation of biologically relevant materials opens the way for a more realistic, physically meaningful description of radiation damage in living tissue.

1 Introduction

Understanding radiation effects produced by charged projectiles traversing biological media is of great interest in radiation biology, radiation therapy, and environmental radiation protection. An important feature of the interaction of ionizing radiation with biological systems is the complexity of produced damage [1]. It is well-established nowadays that the great portion of biodamage resulting from ionizing radiation is related to secondary electrons, free radicals and other reactive species, which are produced by ionizing and exciting molecules of the medium [1–3]. All these secondary species have been found to be more efficient in producing damage than the primary radiation, because they can effectively trigger physicochemical processes leading to molecular structure alterations, for instance, to covalent bond breaking, ionization, or negative ion formation [4]. In this context, "event-by-event" Monte Carlo simulation codes [5–11] as well as the phenomenological multiscale approach to the assessment of radiation damage [3] have been developed in order to model the effects of radiation on the nanoscale and to explore their correlation with the observed damage.

The discovery of radiation damage in biomolecular systems by low-energy electrons [2, 12, 13] has led to the development of the concept of nanodosimetry. It aims at a detailed description of the interaction processes occurring in nanometer-size volumes of the medium and of implications of these processes in terms of radiation damage, such as the number of ionization or dissociative events, type of generated secondary species, etc. A thorough understanding of the mechanisms of biodamage done by ionizing radiation requires evaluation of molecular-level effects related to dose deposition on the nanoscale [5, 14]. For that purpose, deep knowledge of numerous interactions induced by charged particles traversing living matter is strongly essential. A comprehensive description of the mechanisms underlying these interactions may ultimately lead to the development of new strategies and protocols in modern treatment techniques with ionizing radiation [15].

One of the widely used methods to study these effects in detail is based on Monte Carlo simulations performed by the track structure codes [5–11]. By sampling a sufficiently large number of tracks and averaging over the ensemble obtained, Monte Carlo simulations can provide valuable information about the mechanisms of the interaction of radiation with matter [1].

A Monte Carlo approach aims at the detailed simulation of trajectories of single particles in a medium, i.e. the complete track structure of the projectile and all secondary particles generated in the medium [16]. Thus, a good quantification of interaction parameters in a broad energy range is required. A common way to precisely determine the physical and chemical events occurring on the nanoscale is to utilize models that can describe energy-loss processes in the medium in terms of interaction cross sections. Improving the accuracy of these models requires a considerable amount of interaction data that must be obtained from experiments and theoretical approaches. For modeling radiation damage in biological media, establishing an accurate and complete set of cross sections is thus of crucial importance. Being the primary input for track structure codes, such data should include appropriate integral and differential cross sections, energy loss spectra, and scattering cross sections for all kinds of inelastic events, in particular for those leading to molecular dissociations, chemical alterations and radical formation.

In recent years, substantial experimental and theoretical progress has been made to provide the essential data that describe how low-energy electrons, which are responsible for a significant non-repairable damage in biological systems, interact with the key molecular building blocks of living tissue, such as water and structural components of DNA and RNA molecules [17]. By means of the Low-Energy Particle Track Simulation (LEPTS) code (see the review paper [5] and references therein), it has become possible to model dynamics of secondary species down to the (sub-)electronvolt scale. This Monte Carlo-based tool has been developed to address the molecular level mechanisms of biodamage and to describe radiation effects in nanovolumes in terms of induced molecular dissociations [6]. LEPTS is based on reliable and self-consistent databases of interaction cross sections and energy-loss distributions compiled from experimental data and complemented with theoretical calculations. Up to now, these databases have been available for electrons and positrons [5].

The LEPTS methodology has been recently integrated [16] into the Geant4 (GEometry ANd Tracking 4) Monte Carlo toolkit [18, 19] as a new physics model for the simulation of low-energy electrons and positrons in relevant biological systems. As a result, it has become possible to select different sets of models for different energy intervals of the traced particles, for instance, using standard electromagnetic models, such as Livermore or Penelope, for high energies and LEPTS for low energies [16].

This chapter reports on the recent developments in the Monte Carlo-based modeling of the interaction of ionizing radiation with biologically relevant systems. We describe several widely utilized Monte Carlo track structure codes, such as Geant4 and Geant4-DNA [10]—an extension of Geant4 allowing for microdosimetric studies of biological damage induced by ionizing radiation. We also present an extension of the LEPTS methodology aiming at the explicit simulation of the slowing-down of heavy charged particles propagating through a biological medium, accounting for the production of secondary particles, including low-energy electrons, and a variety of induced molecular processes. Main features, areas of application and current limitations of each simulation tool are discussed. A special attention is paid to the energy range covered by primary and secondary charged particles and on the type of interactions included in the simulation.

As a case study, we present the results of the simulation of intermediate- and low-energy protons (starting from 1 MeV until their final thermalization down to the few-eV scale) traversing liquid water that is the main constituent of living tissue. Charged heavy particles of such energies contribute greatly to the maximum of energy deposition in the Bragg peak region [20]. The utilized approach allows one to study radiation effects on the nanoscale in terms of the number and the type of induced molecular processes. The analysis performed thus provides valuable information which may be used further to improve modern treatment techniques based on proton- or heavy ion therapy.

In the following sections, we briefly describe the main capabilities of the Geant4, Geant4-DNA and LEPTS packages. Then, we present the recent development of LEPTS aimed at the simulation of full slowing-down of protons in water. In order to include protons into the simulations, a comprehensive dataset of integral and differential cross sections of elastic and inelastic scattering of intermediate- and low-energy protons from water molecules has been compiled; this dataset is also described.

2 The Geant4 Monte Carlo Simulation Tooklit

The Geant4 toolkit [18, 19] provides a versatile and comprehensive software package for simulating the passage of particles through matter. It includes a complete range of functionality including geometry, tracking of particles through materials and external electromagnetic fields, physics models, and the visualization of geometry and particle trajectories. To manage particle interactions, a set of complementary or alternative physics models are offered, covering a comprehensive range of physics processes which include electromagnetic, hadronic and optical ones, over a wide energy range starting, in some cases, from eV energies and being extended up to the TeV energy range in other cases. To build these physics models, data and expertise have been drawn from many sources around the world and in this respect, Geant4 acts as a repository that incorporates a large part of all that is known about particle interactions. Moreover, it continues to be refined, expanded and developed.

The toolkit is the result of a worldwide collaboration of physicists and software engineers. It has been created exploiting software engineering and object-oriented technology and implemented in the C++ programming language. While its first release in 1998 was designed for High Energy Physics, thanks to its big flexibility, the utilities needed for other fields, like Nuclear Physics, Space Physics, and Medical Physics, were also added soon after. Concerning Medical Physics, several projects were also developed in the fields of radiation therapy (external beams and brachytherapy), hadron therapy, positron emission tomography, and later on, in microdosimetry and radiobiology.

In the field of proton therapy, it is still common to use for treatment planning commercial software based on analytical methods that employ different physics approximations. Nevertheless, Monte Carlo-based treatment planning systems are slowly entering the market being recognized as a precise tool for this type of calculations [21–29].

2.1 Geant4 Physics for Proton Therapy Simulation

The detailed simulation of a proton therapy treatment requires the selection of the appropriate physics models, including the electromagnetic interactions of protons and other particles [30]. Geant4 offers several models to handle the electromagnetic interactions of leptons, photons, hadrons, and ions. In most of these models, the interactions of charged particles are treated in a condensed approach to avoid excessive CPU time. This means that many ionization and bremsstrahlung interactions are not simulated but these interactions may result in the emission of secondary particles with the energy above a threshold, which is set by the user. The energy of the non-simulated interactions is then summed up and treated as a local energy deposit. In a similar way the elastic scattering of charged particles is treated in a "multiple scattering" approach, condensing all the interactions of a particle in the calculation of the global deviation in position and direction. Among the three main categories of models available to treat electromagnetic interactions of charged particles and energetic photons, i.e. standard, Livermore and Penelope, the standard model is the preferred one for proton therapy simulation [22–24] as it offers enough precision while keeping an optimized CPU time consumption. In the case of multiple scattering, the preferred model is the Urban model, which uses functions to determine the angular and spatial distributions after a step chosen in such a way as to give the same moments of the (angular and spatial) distributions as are given by the Lewis theory [31].

Concerning hadronic physics, Geant4 provides a vast number of possible models, so that the user is able to choose those best matching the particle types, energy ranges, and other characteristics particular to a given simulation. In the case of proton therapy, there is a vast literature discussing the most appropriate models, but it seems that the binary cascade model is the preferred choice [32–34]. The Geant4 binary cascade model is an intra-nuclear cascade model in which an incident hadron collides with a nucleon, forming two final-state particles, one or both of which may be resonances. The resonances then decay hadronically and the decay products are propagated through the nuclear potential along curved trajectories until they re-interact or leave the nucleus. The remaining fragment is treated by precompound and de-excitation models.

2.2 The Geant4 DNA Physics Package

As mentioned above, the electromagnetic physics in Geant4 treats the charged-particle interaction in a condensed way and does not allow one to simulate the interactions of low-energy electrons. Subsequently, it is not possible to assess with these models the biological damage induced by ionizing radiation at the cellular (micrometer and below) scale. To fill this gap, a preliminary set of physics processes adapted to microdosimetry in liquid water down to the electronvolt scale was delivered into the Geant4 toolkit in 2007 [35] and has been improved since then [10, 36–41]. The package named Geant4-DNA has been developed to introduce specific functionalities in Geant4, allowing for:

(i) The modeling of elementary physical interactions between ionizing particles of energies down to the electronvolt scale in biological media (liquid water and DNA), during the so-called "physical" stage.
(ii) The modeling of the "physico-chemical and chemical" stages corresponding to the production and diffusion of oxidative radical species, and the chemical reactions occurring between them. During the "physico-chemical" stage, the water molecules that were excited and ionized during the physics stage may de-excite and dissociate into molecular radical species. In the "chemical" stage, these radicals diffuse in the medium surrounding the DNA. They may eventually react among themselves or with the DNA molecule. Studies using radical scavengers have demonstrated that at low values of linear energy transfer (LET), these radical species are responsible for most of the damages caused to the DNA molecule, showing a selective behavior in the type and localization of the damage, and also play a significant contribution at high values of LET.
(iii) The modeling of a "geometrical" stage where the two above stages are combined with a geometrical description of biological targets, such as chromatin segments or the cell nucleus. In particular, it is possible to implement the geometry of biological targets with a high resolution at the sub-micron scale and to track particles within these geometries using the Geant4-DNA physics processes.

At present, the Geant4-DNA extension set covers the dominant interactions of light particles and ions, including electrons, protons, hydrogen atoms, neutral and charged helium atoms, down to the eV scale in liquid water which is the main component of biological matter. The physics models and their experimental validation are described and discussed in detail in Refs. [26–34]. Some of these models are purely analytical, others make use of interpolated cross section data tables for a faster computation. The list of available processes and models that are available in the public version 10.2 is the following (the kinetic energy range for each type of interaction is given in parentheses):

Electrons:

- Elastic scattering (7.4 eV–1 MeV)
  - screened Rutherford and Brenner-Zaider formula below 200 eV [42]
  - updated alternative version by Uehara et al. [43]
  - partial wave framework model by Champion et al. [37]
- Ionization (10 eV–1 MeV)
  - dielectric formalism and first Born approximation using the Heller optical data [44] up to 1 MeV and low-energy corrections, derived from the work of Emfietzoglou [45]
  - improved alternative version by Emfietzoglou and Kyriakou [40]

- Electronic excitation (8 eV–1 MeV)
  - dielectric formalism and first Born approximation using the Heller optical data [44] and semi-empirical low-energy corrections derived from the work of Emfietzoglou [45]
  - improved alternative version by Emfietzoglou and Kyriakou [40]
- Vibrational excitation (2–100 eV)
  - cross section measurements in amorphous ice by Michaud et al. [46]
  - a factor of 2 is included to account for phase effects
- Dissociative attachment (4–13 eV)
  - cross section measurements by Melton [47]

Protons and hydrogen atoms:

- Electronic excitation (10 eV–100 MeV)
  - Miller and Green [48] speed scaling of e^- excitation at low energies, and Born and Bethe theories above 500 keV, from Dingfelder et al. [49]
- Ionization (100 eV–100 MeV)
  - Rudd semi-empirical approach [50] by Dingfelder et al. [49], and Born and Bethe theories and dielectric formalism above 500 keV (relativistic + Fermi density)
- Charge transfer (100 eV–100 MeV)
  - analytical parameterizations by Dingfelder et al. [49]

He^0, He^+, He^{2+}:

- Electronic excitation and ionization (1 keV–400 MeV)
  - speed and effective charge scaling from protons by Dingfelder et al. [51]
- Charge transfer (1 keV–400 MeV)
  - semi-empirical models from Dingfelder et al. [49, 52]

Li, Be, B, C, N, O, Si, and Fe ions:

- Ionization (0.5 MeV/u–1 TeV/u)
  - speed scaling and global effective charge by Booth and Grant [53]

As follows from this list, Geant4-DNA is capable of simulating protons with the kinetic energy of up to 100 MeV. Although this energy covers almost the full range of proton therapy accelerators, it does not cover the highest accelerator energies, which need to be as high as 230 MeV in order to treat deeply-seated tumors in the human body.

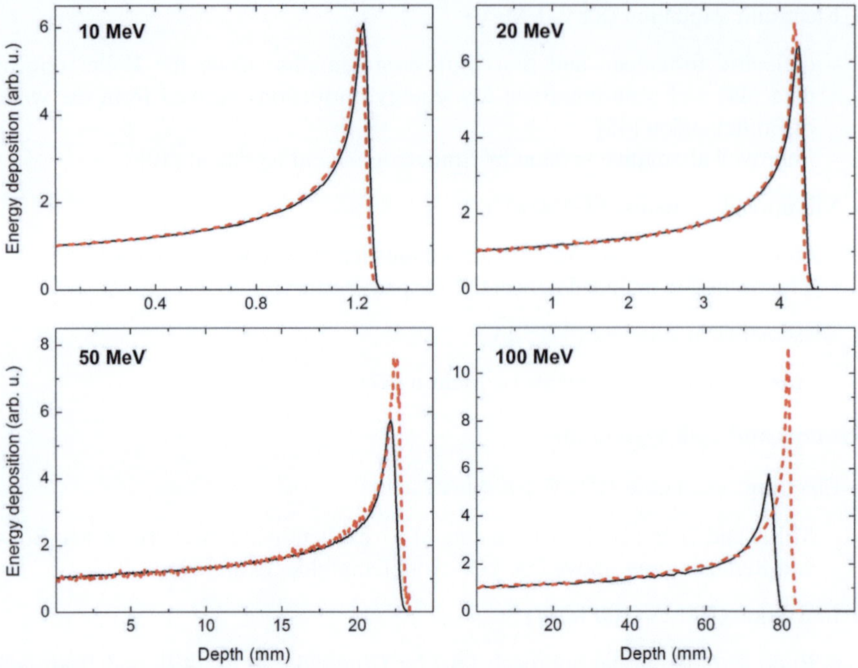

Fig. 1 Energy deposited in water medium along the path of 10, 20, 50, and 100 MeV protons. *Dashed curves* show the simulations performed by means Geant4-DNA physics, while *solid curves* shown the results obtained with binary cascade and standard electromagnetic physics

Figure 1 presents the distribution of dose calculated with the Geant4-DNA package (dashed curve) compared to that obtained with the standard physics plus binary cascade list (solid curve) for energies ranging from 10 to 100 MeV. In the case of a 10 MeV proton, the energy depositions obtained within the two approaches almost coincide with one another but the relative discrepancy in the position of the Bragg peak increases with increasing the projectile's energy. Although Geant4-DNA in some cases does not allow for an accurate quantitative description of the Bragg peak, this package is capable of determining the number of ionizations induced by protons as well as the electronic processes that may cause DNA damage. To illustrate this, we have simulated 1000 tracks of 10 MeV protons in liquid water. Table 1 shows the number of interactions of each kind which are produced by a single projectile. The distribution of these interactions along the proton's track for the three types of particles, namely, protons, hydrogen atoms, and electrons, is presented in Fig. 2. The Geant4-DNA simulations of the proton slow-down account for the ionization, excitation, and charge transfer interactions involving the projectile. However, they do not account for elastic scattering of the projectile from molecules of the medium. Thus, a projectile traversing the medium follows a straight line and the analysis of the total track does not give any additional information. As it will be discussed further

Table 1 Number of interactions of each type resulting from the propagation of a 10 MeV proton in water

Process name	No. of interactions
proton_G4DNAExcitation	68297.2
protons_G4DNAChargeDecrease	5315.6
hydrogen_G4DNAIonisation	6997.4
hydrogen_G4DNAExcitation	1117
hydrogen_G4DNAChargeIncrease	5312.25
e-_G4DNAElastic	1.18×10^8
e-_G4DNAIonisation	1.75×10^6
e-_G4DNAExcitation	296555
e-_G4DNAVibExcitation	1.33×10^7
e-_G4DNAAttachment	42529.9

in this chapter, elastic interactions play a significant role during the propagation of heavy particles in the medium, especially at the low kinetic energy, and this effect should be taken into consideration for a more accurate and complete picture. We note that the contribution of elastic scattering was included into the Geant4-DNA package very recently [41].

3 Extension of the Low-Energy Particle Simulation (LEPTS) Code

A general limitation of the most existing Monte Carlo track structure codes is that they do not describe very accurately the interaction of low-energy particles with molecular constituents of a medium [17]. Some codes actually stop modeling primary and secondary particles if their energies drop below 50–100 eV [54]. By means of the LEPTS code (see the review paper [5] and references therein), it has become possible to model dynamics of secondary species down to the (sub-)electronvolt scale. This Monte Carlo-based tool has been developed to address the molecular level mechanisms of biological damage and to describe radiation effects in nanovolumes in terms of induced molecular dissociations [6].

Recently, the LEPTS methodology has been extended to simulate explicitly the slowing-down of heavy charged particles propagating through a biological medium, accounting for the production of secondary particles, including low-energy electrons, and different molecular processes induced. As the first step, the attention has been focused on the simulation of intermediate- and low-energy protons traversing liquid water. In order to include protons into the simulation scheme, a comprehensive data set of integral and differential cross sections of elastic and inelastic scattering of protons from water molecules has been compiled. For that, experimental and

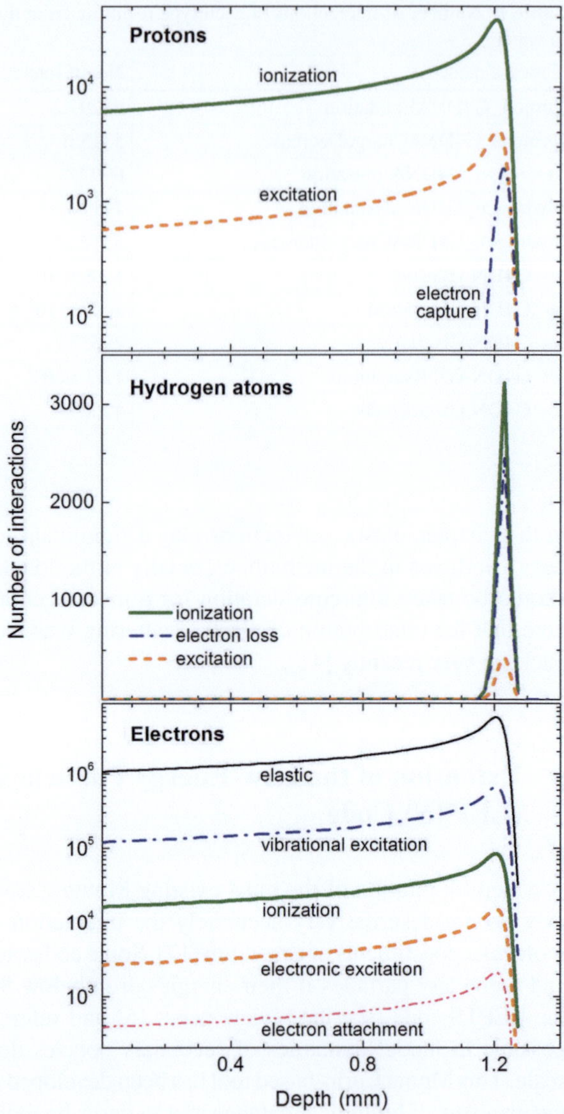

Fig. 2 Number of interactions of each type as a function of depth (in μm) for proton—(*upper panel*), neutral hydrogen—(*middle panel*) and electron—(*lower panel*) induced processes in water after the traverse of 1000 protons of 10 MeV energy

theoretical cross sections available in the literature have been carefully examined and verified. Development of a new database that includes adequate data for biologically relevant materials provides an opportunity for a more realistic, physically meaningful description of radiation damage in living tissue. Hence, the utilized approach allows one to study radiation effects on the nanoscale in terms of the number and the type of induced molecular processes. The next section provides an overview of the compiled data set used for the simulations performed with LEPTS.

3.1 Interaction Processes and Input Data

Generally, a track structure simulation of the charged particle propagation in a biological medium comprises a series of sampling steps that determine the distance between two successive interactions, as well as the type of interaction occurring at the selected point in space. These steps are routinely repeated for all primary and secondary particles until their kinetic energy becomes smaller than a pre-defined cutoff value. The interaction type is randomly selected according to the relative magnitude of the total cross section of all the processes. For the projectile-medium interaction, they are (i) ionization, capture, and excitation induced by a proton, and (ii) ionization, capture, excitation and electron loss induced by a hydrogen atom. The kinematics of the interaction is derived from single- and double-differential cross sections of the corresponding process. Secondary electrons are generated as a result of the ionization event; their energy is defined as the energy lost by the projectile minus the ionization potential of a target molecule. The formation and further evolution of all secondary species is simulated in full according to an explicit database of electron-induced molecular-level interactions. Up to now, the following processes involving electrons have been included in LEPTS: elastic scattering, ionization, electronic, vibrational and rotational excitations, dissociative electron attachment, and neutral dissociation (see Refs. [5, 17] and references therein).

One should note that several computer codes for proton transport in water have been reported in the literature so far (e.g., Refs. [7–9]). One of the most recently developed tools is the code called TILDA-V [11], which is based on quantum-mechanically calculated multiple differential and total cross sections for describing inelastic processes occurring during the slowing-down of protons in water and DNA. The advantage of the procedure implemented in LEPTS comes from much lower cut-off values for heavy charged projectiles and secondary electrons. In other words, all the particles are explicitly tracked in the simulation until they reach smaller energies. This allows one to get a more consistent picture of the radiation-induced processes occurring on the nanoscale. As noted above, this issue is crucial because low-energy secondary electrons, having the kinetic energy smaller than ionization or even excitation threshold of a water molecule, can produce significant biodamage as a result of dissociative electron attachment. In the TILDA-V code [11], the cutoff energy for protons and neutral hydrogen atoms is fixed to 10 keV, while the cutoff for secondary electrons corresponds to the electronic excitation threshold of a water molecule, that is 7.4 eV. In the simulation performed with LEPTS, the heavy projectiles are tracked down to approximately 1 eV as follows from the data set described below, and the electrons can be tracked until their final thermalization at the sub-eV scale [5, 16].

3.1.1 Integral Cross Sections

Integral cross sections for elastic and inelastic interactions of 1 eV–1 MeV protons with water molecules are summarized in Fig. 3. Ionization, excitation, and charge

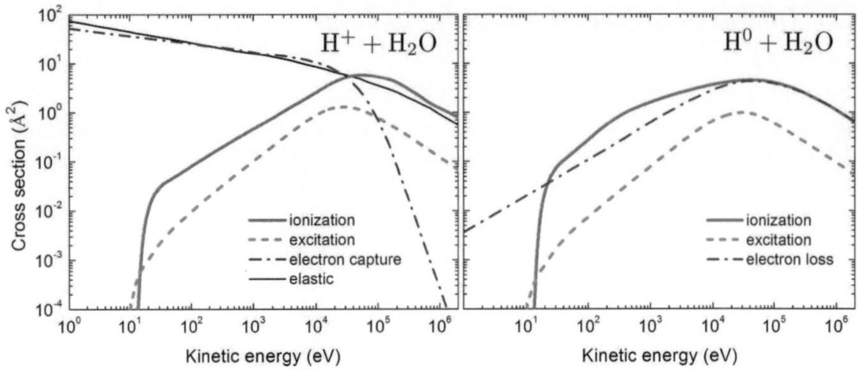

Fig. 3 Integral cross sections for collision of protons and neutral hydrogen atoms with water molecules that have been used as an input for the simulations. Details on data sources are provided in the text

transfer (electron capture) were considered as inelastic processes for H^+ projectiles (left panel). As a result of the charge transfer process, an electron from a water molecule is transferred to the moving slow proton to form a neutral hydrogen atom; the corresponding inelastic cross sections for the neutral projectile are shown in the right panel. We have included the processes of ionization and excitation of a water molecule by H^0 and also accounted for a probability of electron loss (stripping) by the neutral atom.

The ionization cross section by protons has been produced as a result of a thorough analysis of experimental and theoretical data, including recent measurements of the production of different charged fragments [55, 56], and the corresponding classical, semi-classical and *ab initio* calculations [57–59]. The excitation and charge transfer (both electron capture and loss) cross sections for both charged and neutral projectiles were taken from Refs. [11, 49] which are based on a semi-empirical model by Green et al. [48, 60]. As indicated in Ref. [49], parameters of the model were chosen to fit the calculated excitation cross sections to those obtained within the first Born approximation at higher projectile energies. We also accounted for elastic scattering of protons from water molecules (nuclear scattering) which becomes important at lower incident energies of about and below 10 keV. Integral elastic cross section data were taken from Refs. [41, 61].

3.1.2 Total Ionization Cross Section

Data on the total ionization cross section, which have been used in the simulations, are presented in Fig. 4. The data set includes the cross section taken from ICRU Report 49, as well as results of experimental measurements. Older experiments done by Rudd and co-workers [62, 63] were focused mainly on determining the total electron production cross section by the integration of their doubly differential electron

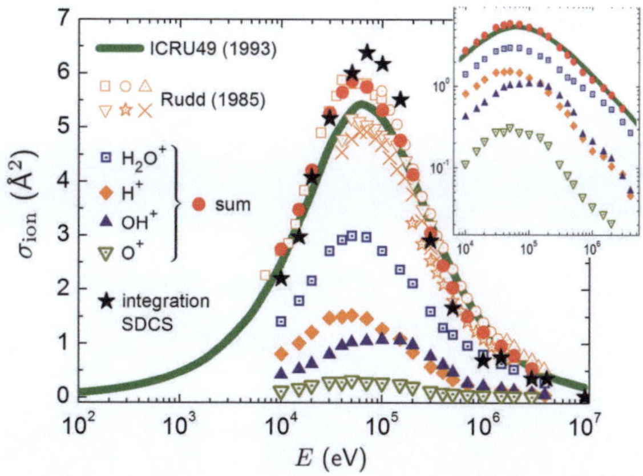

Fig. 4 Total ionization cross sections for collision of protons with water molecules that have been used as an input for the simulations. *All symbols* except for *filled stars* indicate experimental data from Ref. [62] and interpolated experimental data from Refs. [55, 56, 64] on the production of charged fragments. *Filled stars* correspond to the integrated values of single-differential cross sections described in Sect. 3.1.3

emission cross sections. More recent experiments [55, 56, 64] allowed one to get the information on production of different charged fragments, namely H_2O^+, H^+, OH^+, and O^+. In the compiled data set, we have used these data accompanying with the results of recent theoretical studies [57–59]. In order to get smooth cross sections (shown in Fig. 4 by symbols), we did spline interpolation of the experimental data from different measurements [55, 56, 64]. The figure illustrates that the results from the ICRU Report (solid curve) almost coincide with the recent experimental data (filled circles) at the energies about 10–20 keV and above 1 MeV. In the Bragg peak region, at about 50–100 keV, the new data exceed the already established ones by about 10 %. In the compiled data set, we used the new experimental data as a more preferred source. Thus, the resulting curve (shown by a thick solid line in Fig. 3) comprises the experimental data [55, 56, 64] in the range 10 keV–1 MeV and the data from the ICRU 49 Report at lower energies. Note that thus compiled data set is consistent, within the 10 % accuracy, with the integrated single-differential cross sections (filled stars in Fig. 4), described below.

3.1.3 Differential Ionization Cross Sections

Double-differential cross sections (in terms of the kinetic energy and angular distribution of secondary electrons) for 1.5-, 1.0-, 0.5-, 0.3-, 0.1-MeV, and for 15-keV protons were taken from the experimental data of Toburen and Wilson [65], Bolorizadeh and Rudd [63], and the calculations of Senger and Rechenmann [66]. The cited papers

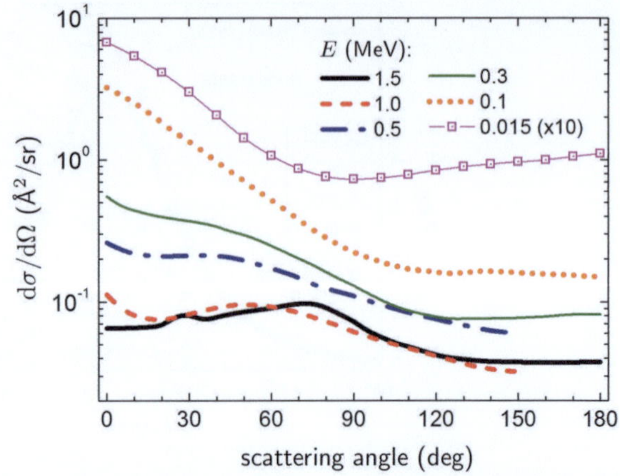

Fig. 5 Single-differential cross section $d\sigma/d\Omega$ describing angular distribution of secondary electrons ejected from a water molecule after the collision with protons. See the text for the details on data sources

presented the data for secondary electrons with kinetic energy ε from about 10 eV up to 2.2 keV. These data were interpolated, and the compiled dependencies were integrated over the kinetic energy of emitted electrons to get their angular distribution. Thus calculated single-differential (in terms of electron emission angle) cross sections, $d\sigma/d\Omega$, are shown in Fig. 5.

Single-differential (in terms of kinetic energy of secondary electrons) cross sections, $d\sigma/d\varepsilon$, were compiled based on the experimental data from Refs. [63, 67] and supplemented with the calculations from Refs. [9, 49, 68]. A thorough compilation of the data from different sources has allowed us to produce an explicit set of cross sections for 10-, 4.2-, 3.0-, 1.5-, 1.0-, 0.5-, 0.3-, 0.1-MeV and 70-, 50-, 30-, 20-, 15-, and 10-keV protons (see Figs. 6 and 7).

3.1.4 Self-consistency of the Data Set

An important issue of a database created from different experimental and theoretical sources is reliability of the input data. To elaborate on this issue, we have performed several self-consistency checks, namely we compared the integrated double-differential cross sections, $\int \frac{d^2\sigma}{d\Omega d\varepsilon} d\Omega$, with the single-differential cross section, $d\sigma/d\varepsilon$, taken from separate sources (see Fig. 7) and then also compared the integrated energy spectra $\int \frac{d\sigma}{d\varepsilon} d\varepsilon$ with the total ionization cross section σ_{ion} (see Fig. 4). The agreement between the differential cross sections is very good, while the relative discrepancy between the integrated $d\sigma/d\varepsilon$ and σ_{ion} does not exceed 10 % confirming the reasonable level of accuracy of the input data for simulations.

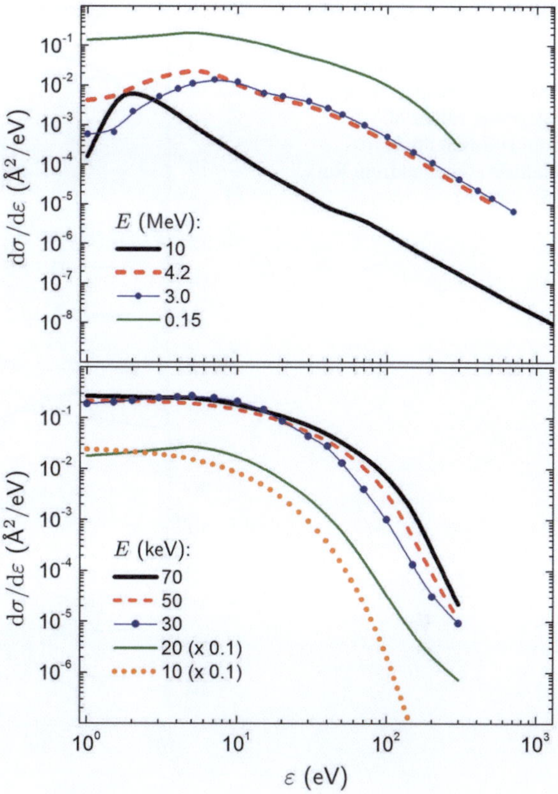

Fig. 6 Energy spectra of secondary electrons, $d\sigma/d\varepsilon$, emitted after irradiation with 10 MeV–10 keV protons. The data set is compiled from [63, 67] and theoretical calculations [9, 49, 68]

4 Results of the Simulations

As a case study, we present the results of the simulation of the slowing-down of 1 MeV protons in liquid water (1 g/cm^3 density) until their final thermalization at the few-eV scale. Charged heavy particles of such energies contribute greatly to the maximum of energy deposition in the Bragg peak region [20]. Hence, it is of significant interest to analyze the type and the number of molecular dissociations in the medium. In this study, we have simulated one million tracks to get good statistics. As noted above, data for single water molecules in the gas phase were used as input parameters but the tracks of all primary and secondary particles were modeled in the liquid phase by considering the liquid density and correcting the cross section values in order to introduce screening effects from the surrounding molecules [5]. Figure 8 illustrates the number of interactions as a function of the depth (in μm) for different scattering processes, including elastic collisions and different types of inelastic events. The maximum penetration of 1 MeV protons in water is about 25 μm, and the position of the Bragg peak corresponds to the kinetic energy of about 100 keV. The simulations performed by means of LEPTS provide a more detailed

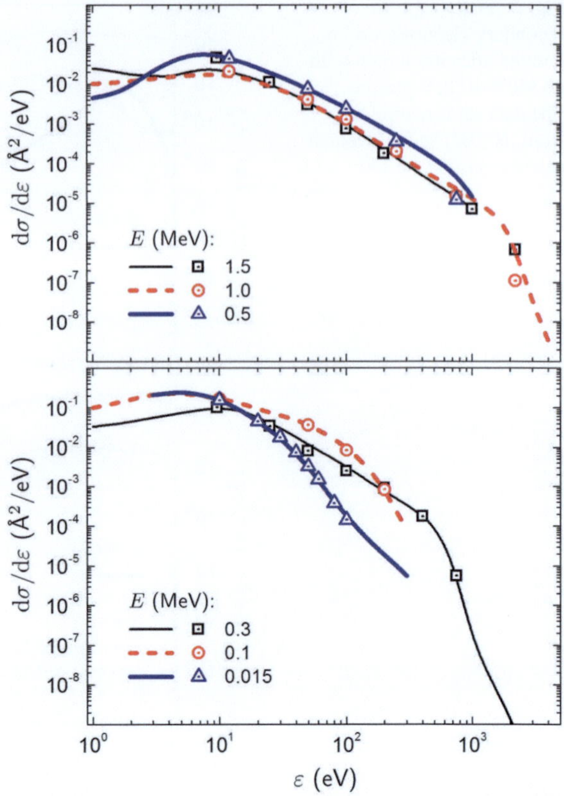

Fig. 7 Same as Fig. 6 but for different projectile energies (*solid* and *dashed curves*). *Symbols* correspond to the integrated values of double-differential cross sections compiled from Refs. [63, 65, 66]

description of different processes occurring on the nanoscale, as compared to the simulations performed with Geant4-DNA (see Fig. 2). As noted above, the LEPTS model explicitly accounts for elastic scattering of neutral and charged projectiles, which provides a substantial contribution to the total number of interaction events. Figure 8 demonstrates that all the interactions associated with protons and electrons stop at the depth of about 23 μm, while elastic collisions between neutral hydrogen atoms and the water molecules also contribute at further penetration distances up to 25 μm. Therefore, a detailed description of molecular dissociations induced by low-energy hydrogen atoms colliding with biologically relevant molecules is of significant importance. This information can be obtained, for instance, from advanced *ab initio* calculations.

The analysis of the interactions presented in Fig. 8 allows for a detailed evaluation of the energy deposition as a function of the penetration depth in the medium. This dependence is shown in Fig. 9. The figure illustrates that the dominating contribution to the energy deposition is related to the elastic and inelastic processes induced by secondary electrons which is a typical feature of irradiation with protons and heavier ions. The energy deposited by electrons has an interesting feature that results

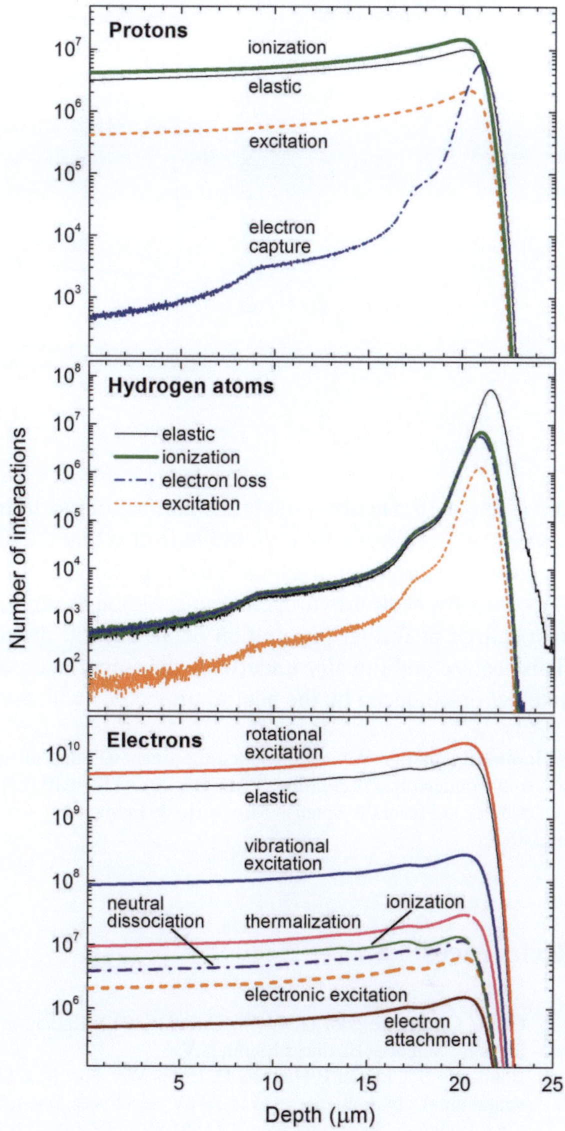

Fig. 8 Number of interactions of each type as a function of depth (in μm) for proton—(*upper panel*), neutral hydrogen—(*middle panel*) and electron—(*lower panel*) induced processes in water after the traverse of 10^6 protons of 1 MeV initial energy, as simulated by means of LEPTS

in a bump in the range between approximately 14 and 18 μm. This feature may be associated with an increased number of secondary electrons produced at these distances by the projectiles. As demonstrated in Fig. 8, this range of penetration corresponds to a gradual increase in the number of elastic and inelastic processes involving neutral hydrogen atoms. The electrons produced as a result of ionization of H atoms deposit their energy within several microns until they slow down to the energies below the ionization potential of a water molecule. This results in a small

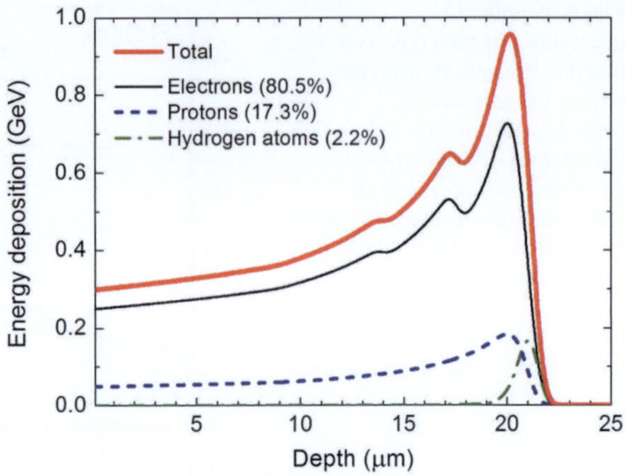

Fig. 9 Energy deposition of 10^6 incident protons of 1 MeV energy each in liquid water as a function of depth

dip at about 18 μm observed in the number of electron-induced ionizations (see the lower panel of Fig. 8). One should note also that the maximal energy deposition by electrons and protons corresponds to the position of the Bragg peak, while the energy deposited by neutral hydrogen atoms, although representing a minor contribution, is localized at further penetration depth beyond the maximum of the Bragg peak. This feature additionally underlines the importance of the accurate description of interactions induced by the neutral projectiles with very low kinetic energy.

Acknowledgements We acknowledge the financial support received from the European Union Seventh Framework Programme (PEOPLE-2013-ITN-ARGENT project) under grant agreement no. 608163 and from the Spanish Ministerio de Economia y Competitividad (Project no. FIS2012-31230).

References

1. Garcia Gomez-Tejedor G, Fuss MC (eds) (2012) Radiation damage in biomolecular systems. Springer Science+Business Media B.V
2. Boudaïffa B, Cloutier P, Hunting D, Huels MA, Sanche L (2000) Resonant formation of DNA strand breaks by low-energy (3 to 20 eV) electrons. Science 287:1658–1660
3. Surdutovich E, Solov'yov AV (2014) Multiscale approach to the physics of radiation damage with ions. Eur Phys J D 68:353
4. Oller JC, Ellis-Gibbings L, Ferreira da Silva F, Limão-Vieira P, García G (2015) Novel experimental setup for time-of-flight mass spectrometry ion detection in collisions of anionic species with neutral gas-phase molecular targets. EPJ Tech Instrum 2:13
5. Blanco F, Muñoz A, Almeida D, Ferreira da Silva F, Limão-Vieira P, Fuss MC, Sanz AG, García G (2013) Modelling low energy electron and positron tracks in biologically relevant media. Eur Phys J D 67:199
6. Muñoz A, Pérez JM, García G, Blanco F (2005) An approach to Monte Carlo simulation of low-energy electron and photon interactions in air. Nucl Instrum Meth A 536:176–188
7. Krämer M, Kraft G (1994) Calculations of heavy-ion track structure. Radiat Environ Biophys 33:91–109

8. Friedland W, Jacob P, Bernhardt P, Paretzke HG, Dingfelder M (2003) Simulation of DNA damage after proton irradiation. Radiat Res 159:401–410
9. Nikjoo H, Uehara S, Emfietzoglou D, Cucinotta FA (2006) Track-structure codes in radiation research. Radiat Meas 41:1052–1074
10. Incerti S et al (2010) Comparison of GEANT4 very low energy cross section models with experimental data in water. Med Phys 37:4692–4708
11. Quinto MA et al (2015) Proton track structure code in biological matter. J Phys: Conf Ser 583:012049
12. Huels MA, Boudaïffa B, Cloutier P, Hunting D, Sanche L (2003) Single, double, and multiple double strand breaks induced in DNA by 3–100 eV electrons. J Am Chem Soc 125:4467–4477
13. Pan X, Cloutier P, Hunting D, Sanche L (2003) Dissociative electron attachment to DNA. Phys Rev Lett 90:208102
14. Baccarelli I, Gianturco FA, Scifoni E, Solov'yov AV, Surdutovich E (2010) Molecular level assessments of radiation biodamage. Eur Phys J D 60:1–10
15. FP7 Initial Training Network Project "Advanced Radiotherapy, Generated by Exploiting Nanoprocesses and Technologies" (ARGENT). http://www.itn-argent.eu
16. Arce P, Muñoz A, Moraleda M, Ros JMG, Blanco F, Perez JM, García G (2015) Integration of the low-energy particle track simulation code in Geant4. Eur Phys J D 69:188
17. Fuss MC, Ellis-Gibbings L, Jones DB, Brunger MJ, Blanco F, Muñoz A, Limão-Vieira P, García G (2015) The role of pyrimidine and water as underlying molecular constituents for describing radiation damage in living tissue: a comparative study. J Appl Phys 117:214701
18. Agostinelli S et al (2003) Geant4—a simulation toolkit. Nucl Instr Meth A 506:250–303
19. Allison J et al (2006) Geant4 developments and applications. IEEE Trans Nucl Sci 53:270–278
20. Schardt D, Elsässer T, Schulz-Ertner D (2010) Heavy-ion tumor therapy: physical and radiobiological benefits. Rev Mod Phys 82:383–425
21. Paganetti H, Jiang H, Parodi K, Slopsema R, Engelsman M (2008) Clinical implementation of full Monte Carlo dose calculation in proton beam therapy. Phys Med Biol 53:4825–4853
22. Petti PL (1996) Evaluation of a pencil-beam dose calculation technique for charged particle radiotherapy. Int J Radiat Oncol Biol Phys 35:1049–1057
23. Schaffner B, Pedroni E, Lomax A (1999) Dose calculation models for proton treatment planning using a dynamic beam delivery system: an attempt to include density heterogeneity effects in the analytical dose calculation. Phys Med Biol 44:27–41
24. Soukup M, Alber M (2007) Influence of dose engine accuracy on the optimum dose distribution in intensity-modulated proton therapy treatment plans. Phys Med Biol 52:725–740
25. Soukup M, Fippel M, Alber M (2005) A pencil beam algorithm for intensity modulated proton therapy derived from Monte Carlo simulations. Phys Med Biol 50:5089–5104
26. Szymanowski H, Oelfke U (2002) Two-dimensional pencil beam scaling: an improved proton dose algorithm for heterogeneous media. Phys Med Biol 47:3313–3330
27. Tourovsky A, Lomax AJ, Schneider U, Pedroni E (2005) Monte Carlo dose calculations for spot scanned proton therapy. Phys Med Biol 50:971–981
28. Titt U et al (2008) Assessment of the accuracy of an MCNPX-based Monte Carlo simulation model for predicting three-dimensional absorbed dose distributions. Phys Med Biol 53:4455–4470
29. Paganetti H (2012) Range uncertainties in proton therapy and the role of Monte Carlo simulations. Phys Med Biol 57:R99–R117
30. Pia MG, Begalli M, Lechner A, Quintieri L, Saracco P (2010) Physics-related epistemic uncertainties in proton depth dose simulation. IEEE Trans Nucl Sci 57:2805–2830
31. Lewis HW (1950) Multiple scattering in an infinite medium. Phys Rev 78:526–529
32. Moteabbed M, España S, Paganetti H (2011) Monte Carlo patient study on the comparison of prompt gamma and PET imaging for range verification in proton therapy. Phys Med Biol 56:1063–1082
33. Robert C et al (2013) Distributions of secondary particles in proton and carbon-ion therapy: a comparison between GATE/Geant4 and FLUKA Monte Carlo codes. Phys Med Biol 58:2879–2899

34. Jarlskog CZ, Paganetti H (2008) Physics settings for using the Geant4 toolkit in proton therapy. IEEE Trans Nucl Sci 55:1018–1025
35. Chauvie S, Francis Z, Guatelli S, Incerti S, Mascialino B, Moretto P, Nieminen P, Pia MG (2007) Geant4 physics processes for microdosimetry simulation: design foundation and implementation of the first set of models. IEEE Trans Nucl Sci 54:2619–2628
36. Francis Z, Incerti S, Capra R, Mascialino B, Montarou G, Stepan V, Villagrasa C (2011) Molecular scale track structure simulations in liquid water using the Geant4-DNA Monte-Carlo processes. Appl Radiat Isot 69:220–226
37. Champion C, Incerti S, Aouchiche H, Oubaziz D (2009) A free-parameter theoretical model for describing the electron elastic scattering in water in the Geant4 toolkit. Radiat Phys Chem 78:745–750
38. Champion C, Incerti S, Tran HN, El Bitar Z (2012) Electron and proton elastic scattering in water vapour. Nucl Instrum Meth B 273:98–101
39. Champion C et al (2013) Proton transport in water and DNA components: a Geant4 Monte Carlo simulation. Nucl Instrum Meth B 306:165–168
40. Kyriakou I, Incerti S, Francis Z (2015) Improvements in Geant4 energy-loss model and the effect on low-energy electron transport in liquid water. Med Phys 42:3870–3876
41. Tran HN et al (2015) Modeling proton and alpha elastic scattering in liquid water in Geant4-DNA. Nucl Instrum Meth B 343:132–137
42. Brenner DJ, Zaider M (1984) A computationally convenient parameterisation of experimental angular distributions of low energy electrons elastically scattered off water vapour. Phys Med Biol 29:443–447
43. Uehara S, Nikjoo H, Goodhead DT (1993) Cross-sections for water vapour for the Monte Carlo electron track structure code from 10 eV to the MeV region. Phys Med Biol 38:1841–1858
44. Heller JM Jr, Hamm RN, Birkhoff RD, Painter LR (1974) Collective oscillation in liquid water. J Chem Phys 60:3483–3486
45. Emfietzoglou D (2003) Inelastic cross-sections for electron transport in liquid water: a comparison of dielectric models. Radiat Phys Chem 66:373–385
46. Michaud M, Wen A, Sanche L (2003) Cross sections for low-energy ($1 - 100$ eV) electron elastic and inelastic scattering in amorphous ice. Radiat Res 159:3–22
47. Melton CE (1972) Cross sections and interpretation of dissociative attachment reactions producing OH^-, O^-, and H^- in H_2O. J Chem Phys 57:4218–4225
48. Miller JH, Green AES (1973) Proton energy degradation in water vapor. Radiat Res 54:343–363
49. Dingfelder M, Inokuti M, Paretzke HG (2000) Inelastic-collision cross sections of liquid water for interactions of energetic protons. Radiat Phys Chem 59:255–275
50. Rudd ME, Kim Y-K, Madison DH, Gay TJ (1992) Electron production in proton collisions with atoms and molecules: energy distributions. Rev Mod Phys 64:441–490
51. Dingfelder M, Jorjishvili IG, Gersh JA, Toburen LH (2006) Heavy ion track structure simulations in liquid water at relativistic energies. Radiat Prot Dos 122:26–27
52. Dingfelder M, Toburen LH, Paretzke HG (2005) An effective charge scaling model for ionization of partially dressed helium ions with liquid water. In: Proceedings of the Monte Carlo 2005 Topical Meeting, Chattanooga, TN, 17–21 April 2005, American Nuclear Society, La Grange Park, IL, pp 1–12
53. Booth W, Grant IS (1965) The energy loss of oxygen and chlorine ions in solids. Nucl Phys 63:481–495
54. Muñoz A, Blanco F, Oller JC, Pérez JM, García G (2007) Advances in quantum chemistry. In: Sabin JR, Brändas E (eds). vol 52, pp 21–57. Academic Press
55. Gobet F et al (2004) Ionization of water by ($20 - 150$)-keV protons: separation of direct-ionization and electron-capture processes. Phys Rev A 70:062716
56. Luna H et al (2007) Water-molecule dissociation by proton and hydrogen impact. Phys Rev A 75:042711
57. Errea LF, Illescas C, Méndez L, Pons B, Rabadán I, Riera A (2007) Classical calculation of ionization and electron-capture total cross sections in $H^+ + H_2O$ collisions. Phys Rev A 76:040701(R)

58. Murakami M, Kirchner T, Horbatsch M, Lüdde HJ (2012) Single and multiple electron removal processes in proton-water-molecule collisions. Phys Rev A 85:052704
59. Errea LF, Illescas C, Méndez L, Rabadán I (2013) Ionization of water molecules by proton impact: two nonperturbative studies of the electron-emission spectra. Phys Rev A 87:032709
60. Green AES, McNeal RJ (1971) Analytic cross sections for inelastic collisions of protons and hydrogen atoms with atomic and molecular gases. J Geophys Res 76:133–144
61. Uehara S, Toburen LH, Nikjoo H (2001) Development of a Monte Carlo track structure code for low-energy protons in water. Int J Radiat Biol 2:139–154
62. Rudd ME, Goffe TV, DuBois RD, Toburen LH (1985) Cross sections for ionization of water vapor by 7–4000-keV protons. Phys Rev A 31:492–494
63. Bolorizadeh MA, Rudd ME (1986) Angular and energy dependence of cross sections for ejection of electrons from water vapor. II. 15–150-keV proton impact. Phys Rev A 33:888–892
64. Werner U, Beckord K, Becker J, Lutz HO (1995) 3D imaging of the collision-induced Coulomb fragmentation of water molecules. Phys Rev Lett 74:1962–1965
65. Toburen LH, Wilson WE (1977) Energy and angular distributions of electrons ejected from water vapor by 0.3–1.5 MeV protons. J Chem Phys 66:5202–5213
66. Senger B, Rechenmann RV (1984) Angular and energy distributions of δ-rays ejected from low-Z molecular targets by incident protons and α particles. Nucl Instrum Meth B 2:204–207
67. Wilson WE, Miller JH, Toburen LH, Manson ST (1984) Differential cross sections for ionization of methane, ammonia, and water vapor by high velocity ions. J Chem Phys 80:5631–5638
68. de Vera P, Garcia-Molina R, Abril I, Solov'yov AV (2013) Semiempirical model for the ion impact ionization of complex biological media. Phys Rev Lett 110:148104

Ion Collisions with Biomolecules and Biomolecular Clusters

Patrick Rousseau and Bernd A. Huber

Abstract In this chapter we describe the recent progress which has been made in experimental studies of ion collisions with biomolecular systems, either in form of isolated biomolecules in the gas phase or as clusters containing up to several tens of biomolecules. Most of the work has been performed with projectiles which play an important role in ion beam cancer therapy applications as protons or multiply charged ions of carbon and oxygen. The biomolecular targets are characterized by an increasing complexity and include water molecules, nucleobases, nucleosides and nucleotides, as well as amino acids and protein segments. Other complex targets are heterogeneous clusters containing biomolecular systems which are embedded in a water environment. After an introduction to ion-molecule collisions using C_{60} fullerene as a model system, we will review ionization and charge transfer processes as well as ion-induced fragmentation studies. Finally we will discuss the effect of the environment considering clusters of biomolecules including hydrated systems.

1 Introductory Remarks to Ion Collisions with Molecules: The Model System C_{60}

Ion collisions with molecular targets are facing a large variety of new dynamical phenomena and processes in comparison with ion-atom collisions. Thus, new classes of reactions have to be considered due to dissociation, fragmentation and the coupling between the light electron and the heavy particle systems. On the one hand, molecules, depending on their size, are characterized by a large number of degrees of

P. Rousseau (✉)
Normandie Univ, ENSICAEN, UNICAEN, CEA, CNRS, CIMAP,
14000 Caen, France
e-mail: patrick.rousseau@unicaen.fr

B.A. Huber
Centre de Recherche sur les Ions, les Matériaux et la Photonique (CIMAP),
Unité Mixte CEA-CNRS-EnsiCaen-Université de Caen Basse-Normandie,
UMR 6252, 6 Boulevard Maréchal Juin, 14050 Caen, France
e-mail: huber@ganil.fr

A.V. Solov'yov (ed.), *Nanoscale Insights into Ion-Beam Cancer Therapy*,
DOI 10.1007/978-3-319-43030-0_4

freedom (vibrational and rotational modes) in which a large amount of energy can be stored after collisions with ions. This amount may be much larger than the threshold energies for individual fragmentation channels and may lead to the observation of time-delayed processes, depending on the timescale of the experimental study. On the other hand, heavy particles of molecules are always embedded in an electronic environment, thus, the interaction and the energy exchange between the electronic system with the system of nuclei or ion cores becomes important. Via the electron-phonon coupling energy can be transferred from a hot electronic system to the heavy molecular atoms (vibrational and rotational excitation). This may lead to thermalization, to the damping of collective electron motion (plasmon decay) or inversely thermo-ionic electron emission from a vibrationally hot heavy particle system which may occur on very long timescales (μs).

Many of these processes have been studied in the last decades for ion collisions with the model system C_{60}. These fullerenes are characterized by covalent bonds between 60 carbon atoms, arranged in 20 hexagons and 12 pentagons forming the well-known spherical structure with a largely delocalized electron distribution. From the experimental point-of-view this system is easy to be studied, as C_{60} targets can be produced by simple evaporation of available C_{60} powder. Furthermore, the theoretical description is refined by many theoretical approaches. In this introductory part we will briefly summarize some characteristic results for C_{60} concerning ionization and fragmentation processes.

1.1 Electron Capture and Multi-ionization

In a pioneering study an experimental device has been developed which allowed to measure the produced recoil ions and the number of emitted electrons in coincidence with the charge state of the projectile after the collision [1, 2]. Thus, a total balance of the active electrons could be established (electrons captured and stabilized at the incoming ion + electrons emitted into the continuum = charge state of the intermediate molecular ion). In the collision system $Xe^{25+} + C_{60}$ it has been shown (see Fig. 1a) that in the reaction

$$Xe^{25+} + C_{60} \rightarrow Xe^{(25-s)+} + C_{60}^{r+} + (r - s)e^{-}$$

more than $r = 60$ electrons can be taken off the molecular target in a single collision, out of which s electron ($\langle s \rangle \sim 15$)) are stabilized at the projectile. The mechanism is based on multiple electron capture processes and fast electron emission due to Auger processes during the collision. This keeps the charge state of the projectile during the collision high which thus can play the role of an electron transmitter into the continuum.

The remaining highly charged fullerene will explode and for $s = 11$ only singly and multiply charged atomic carbon fragments are observed (see Fig. 1b) [3]. The C^{+} peak profile contains two different contributions: a low-energy central part stemming from C^{+} fragments which are pushed inside the C_{60} sphere by nuclear collisions

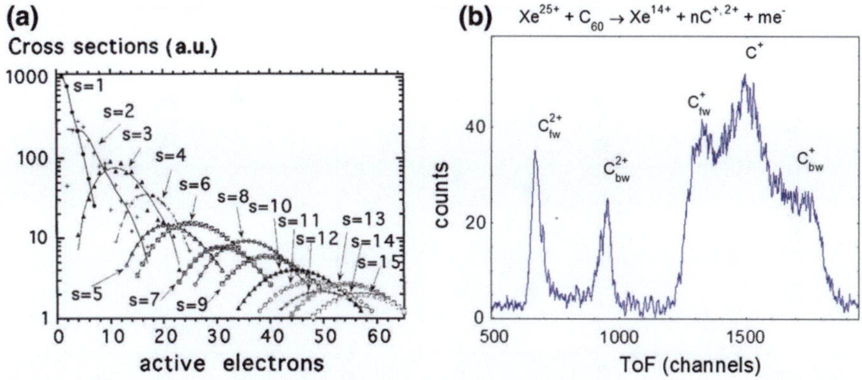

Fig. 1 **a** Cross sections for stabilizing s electrons for a given number of active electrons produced in 475 keV collisions of Xe^{25+} with C_{60} (Reproduced from [1] with the permission of AIP Publishing) **b** Fragmentation spectrum for energetic C^{+} and C^{2+} fragments ($s = 11$) in 500 keV Xe^{25+}/C_{60}. fw/bw indicate forward and backward emission angles with respect to the ToF spectrometer axis

and which therefore do not participate in the Coulomb explosion and the forward and backward parts—with respect to the extraction axis of the time-of-flight (ToF) spectrometer—which are emitted with kinetic energies of 100–200 eV. From these kinetic energies the total energy involved in the explosion process can be estimated. When comparing these values with the electrostatic energy of a corresponding highly charged sphere, the charge states of the highly charged fullerene ions can be estimated. The good agreement with the measured number of active electrons shows that for very high charge states the Coulomb explosion model yields a good description.

When the charge state of the incoming projectile is low, as in the case of proton collisions, electron capture becomes less prominent with respect to the ionization. Indeed, C_{60} molecules can be ionized by protons several times as shown by experiments performed for collision energies between 1 and 300 keV [4]. Whereas for the lowest kinetic energy single ionization (SI) clearly dominates, with increasing energy the contribution of multi-ionization (MI) increases up to the energy of the Bragg peak (~80 keV). For higher energies, the contribution of multi-ionization decreases again and single ionization becomes dominant (see Fig. 2). At the Bragg peak energy multi-ionization dominates over single ionization by a factor of ~2 (ratio: $f_{SI}/f_{MI} \sim 0.5$).

1.2 Fragmentation and Delayed Processes

Ion-induced fragmentation of C_{60} molecules has been studied in the past by several groups [5–8]. In particular, evaporation events (emission of neutral moieties) and fission processes (emission of charged particles) were studied for different experimental parameters (projectile velocity, mass, charge state, energy transfer…). An example is shown in Fig. 3. The upper part shows the mass spectrum obtained in

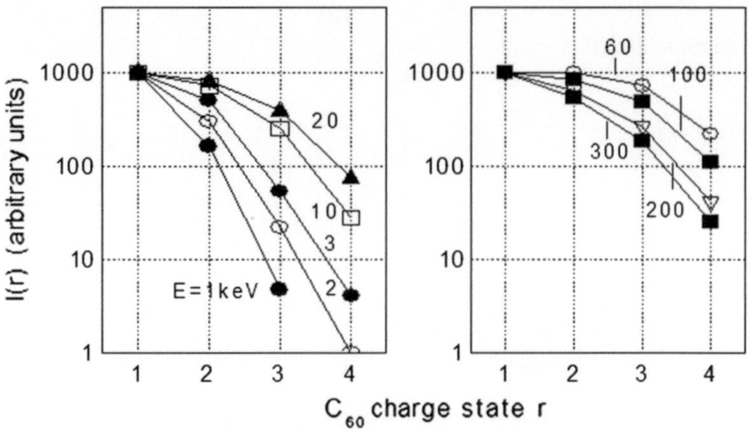

Fig. 2 Relative parent charge state distributions for proton/C_{60} collisions at energies *E* between 1 and 300 keV. For more details see Ref. [4]

proton collisions with C_{60} at 2 keV [9]. In addition to the dominant singly charged C_{60}^{+} fullerene ion, intact ions in charge states 2, 3 and 4 are observed. Furthermore, long evaporation chains due to the emission of up to about ten C_2 molecules (for C_{60}^{2+}) are measured. For proton collisions the interaction is dominated by electronic excitation processes leading to multi-ionization and after the transfer of the electronic excitation energy to the vibrational modes to the sequential emission of C_2 molecules.

At the same collision energy, for other projectiles like Ar a totally different spectrum is observed, as the interaction is dominated by elastic nuclear collisions (see middle part of Fig. 3). Now mainly single ionization occurs with the low signal for doubly charged fullerenes. The spectrum is dominated by a wide distribution of singly charged fragments centered at cluster sizes of about 15 carbon atoms. In the lower part of Fig. 3 the spectrum resulting from 500 keV collisions with Xe^{25+} projectiles is shown. Preferably at larger impact parameters electron capture occurs with low energy transfer yielding multiply charged intact fullerene ions. Evaporation processes are less important, however, in the size range between 10 and 20 atoms/charge, fission products due to C_2^{+} emission from $C_{60}^{3+,4+,5+}$ parent ions are observed. The small size fragment distribution (peaked at the monomer size) is produced in charge separating Coulomb processes.

Evaporation (emission of neutral particles) as well as charge-separating processes can be observed at different timescales depending on the internal energy and the experimental device. They become experimentally observable, if the decay times are of the same order of magnitude as the characteristic times required for the experimental mass analysis. Time-delayed evaporation processes show up in the time-of-flight spectrum as tails towards larger ion masses, for delayed charge-separating events, tails are observable in the coincidence map which shows the time-of-flight of both correlated fragments (see Fig. 4). The intense island, corresponding to C_2^{+}/C_{58}^{5+} coincidences, stemming from the decay of C_{60}^{6+} is widened by the kinetic energy release

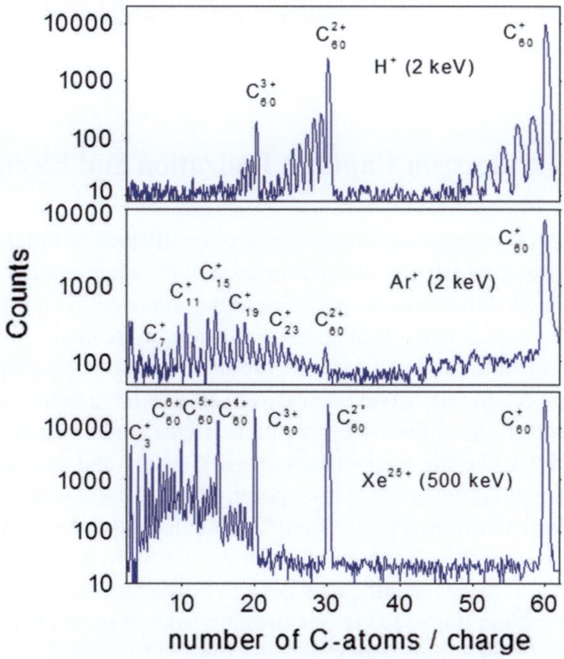

Fig. 3 Mass spectra obtained in collisions of fullerenes with: H^+ at 2 keV (*top* spectrum; electronic excitation); Ar^+ at 2 keV (*middle* spectrum; nuclear collisions); Xe^{25+} at 500 keV (*bottom*; multi-ionization and charge separating processes)

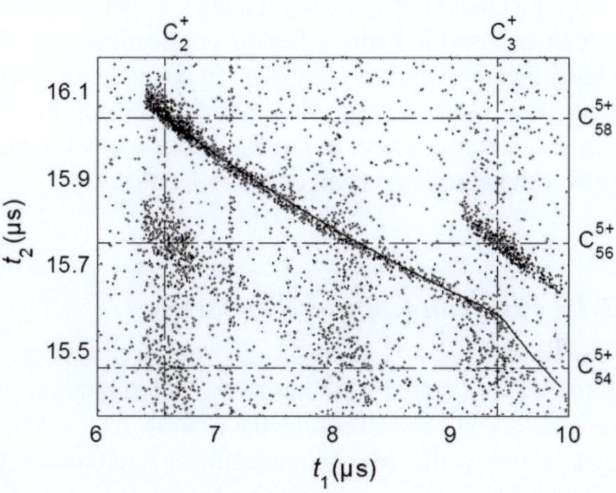

Fig. 4 Coincidence map showing the flight-times of two ions produced in one event (t_1 lighter fragment; t_2 heavier residue). The *thin line* corresponds to calculated flight-times for different decay positions in the TOF spectrometer

of the process. It is followed by a long tail, which is due to fragmentation occurring in the extraction system of the ToF mass spectrometer, i.e. in a time window of several µs. The full line in Fig. 4 represents the calculated time correlation.

In the following sections we will specify the described mechanisms for the case of ion collisions with molecules of biological relevance as well as with pure and mixed clusters in order to investigate the influence of the molecular environment on molecular processes. We will describe actual experimental results. As described before, we will concentrate on electron capture, ionization and fragmentation processes occurring in ion/biomolecule collisions.

2 Electron Capture, Ionization and Electron Loss Processes

Experimental investigations of ion collisions with biomolecular systems in the gas phase, which are relevant for modeling radiation damage at the molecular level, are more difficult to be performed than those with C_{60} molecules described above, in particular when biological systems in a "realistic" environment are to be studied. This difficulty is due to the problem of providing a sufficiently dense biomolecular target. In early experiments, oven devices have been used to produce targets of neutral intact molecules by evaporation. These studies concerned small building blocks of DNA like the nucleobases uracil [10, 11] and adenine [12], the sugar group [13], small amino acids [14] or porphyrins [15] and water [16]. However, this technique fails when larger "prepared" molecules are to be studied as during the evaporation process fragmentation of the molecules may occur, even at rather low temperatures. Thus, more recently, new devices based on electrospray ion sources in combination with ion traps [17] or electrostatic storage rings [18] have been developed to study ion interactions with molecular ions.

In the context of radiation damage, mostly ions have been chosen as projectiles which are used in hadron therapy applications, i.e. ions like protons, and multiply charged carbon and oxygen ions. In some cases, also heavy multiply charged ions have been used as projectiles in order to study the influence of the ion charge on ion-induced processes. In the following we will concentrate on collisions of light ions with neutral and charged biomolecules.

2.1 Electron Capture Processes

Experimental studies, aiming at the determination of absolute total electron capture cross sections, describing the process $A^{q+} + M \rightarrow A^{(q-r)+} + M^{r+}$ with A^{q+} = charged projectile, M = biomolecule, r = number of transferred charges, are rather limited. Some data are summarized in the review by Toburen concerning proton collisions with molecules like water [19, 20]. More recently, Farizon et al. have applied a coincidence technique to study collisions of protons with uracil and H_2O gas target [21–23]. By analyzing the charge state of the proton beam after the collision (H^+, H^0, H^-) in coincidence with the produced ions (intact or fragmented) the processes of direct ionization and single and double electron capture as well as for fragmentation into specific decay channels could be separated. Similar studies were performed by Afrosimov et al. yielding relative electron capture cross sections for multiply charged ions colliding with uracil [24, 25], adenine [24], methionine and norleucine [26].

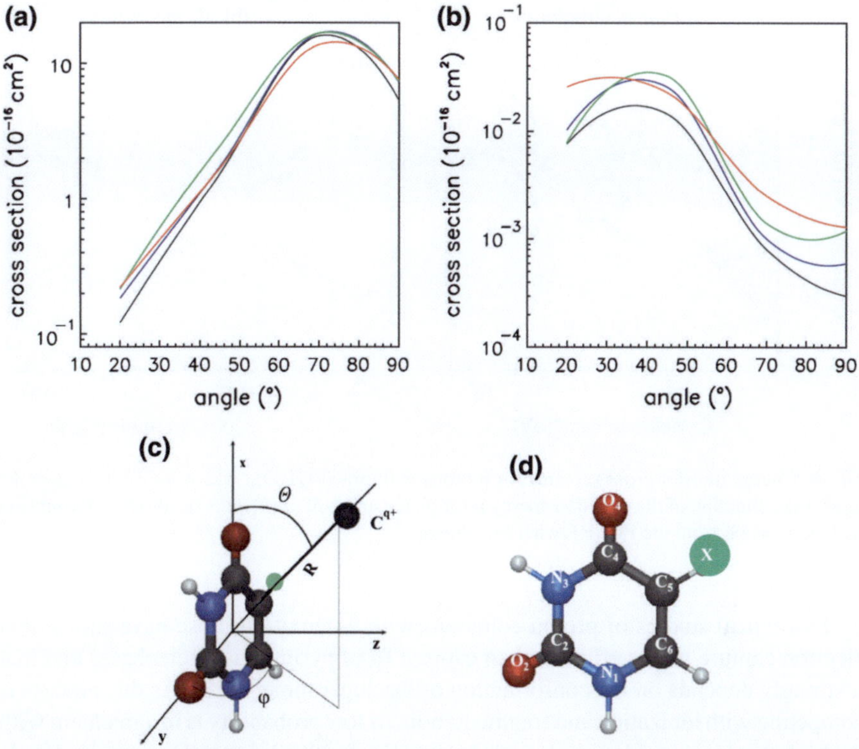

Fig. 5 Electron capture cross sections in units of 10^{-16} cm^2 as a function of the orientation angle θ. **a** Collision system: C^{4+} on uracil. **b** Collision system: C^{4+} on 5-bromouracil. The different curves correspond to projectile energies between 30 eV and 3 keV. **c** Internal coordinates of the C^{4+}-(halo)uracil system. **d** Geometry of the molecule, X=H in uracil and X=Br in 5-bromouracil (Reproduced from [27]. Copyright 2014, with permission from Elsevier)

Numerous theoretical studies have been performed to determine charge exchange cross sections for ions colliding with uracil, halo-uracil, thymidine [27, 28] and 2-deoxy-D-ribose [28, 29] and water molecules [30]. As an example we show in Fig. 5 the electron capture cross section for C^{4+} ions colliding with uracil (a) and 5-bromouracil (b) as a function of the molecular orientation angle θ. The calculations have been performed by means of ab initio quantum chemistry molecular methods followed by a semi-classical collision treatment. A strong anisotropy is observed which in the case of uracil favors the charge transfer around the perpendicular orientation with a maximum at about 70° (for more details see Fig. 5). Here the cross section is by two orders of magnitude larger than in the planar orientation indicating that the process is driven by π interactions. In the case of 5-bromouracil, the process is more efficient at ~45° certainly related to the steric effect by the large Br atom. Furthermore, the capture cross section is strongly reduced by a factor of 10^3–10^4. So far the observed anisotropy has not been confirmed experimentally.

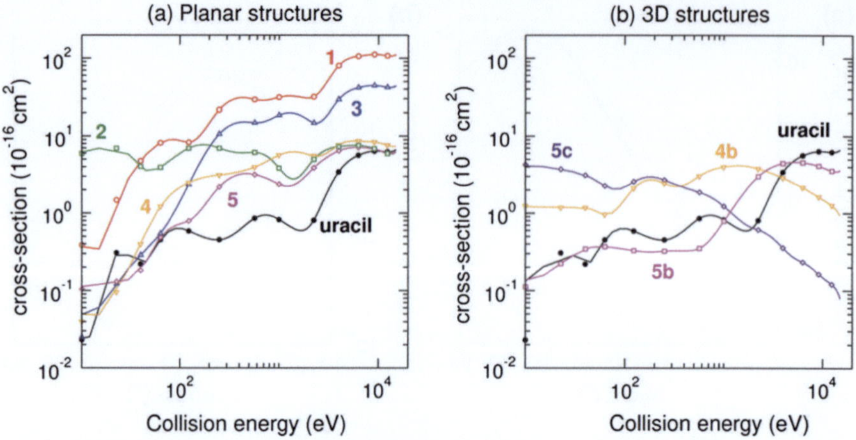

Fig. 6 Charge transfer cross sections for protons with uracil-$(H_2O)_n$ with $n = 1$–5 (given in the figure) as a function of the collision energy **a** for planar and **b** 3D structures (Reproduced from [31] with permission from the PCCP Owner Societies)

Theoretical studies of proton collisions with 2-deoxy-D-ribose have shown that electron capture is less efficient than capture from pyrimidine nucleobases and that it strongly depends on the conformation of the sugar moiety [28]. As this process is competing with ionization and fragmentation, its low probability is in agreement with the high fragment yield observed experimentally for the 2-deoxy-D-ribose molecule which is very fragile with respect to ion collisions [13].

Recently, proton-induced charge transfer from two- and three-dimensional structures of hydrated uracil molecules has been studied theoretically [31]. Structure calculations show that hydration starts in the uracil plane and that 3-dimensional structures become possible, when the number of water molecules becomes larger than 3. The calculated charge transfer cross sections are shown in Fig. 6 for different structures and numbers of attached water molecules varying from 0 to 5. It is found that hydration strongly changes the electron transfer probability. For planar geometries water addition increases the electron transfer cross section by up to two orders of magnitude, in particular when adding the first water molecule. However, with increasing hydration the cross sections become lower and approach those for pure uracil. In the case of 3-dimensional structures with peripheral hydration patterns this phenomenon is less pronounced.

A step forward in studying electron capture processes in larger biomolecular systems has been made by applying a new experimental technique [32]. Multiply protonated cytochrome-C ions (mass = 12 229 amu) in charge states of 15+ to 19+ were produced in an electrospray ion source, mass selected and stored in a Paul trap device [17] (see Fig. 7a). After irradiation with Xe^{8+} ions the reaction products (mainly intact molecular ions) are analyzed allowing to determine single and double electron capture cross sections. The structure and conformation of the protein depends strongly on its charge state. On the left hand of Fig. 7 two examples of conformations

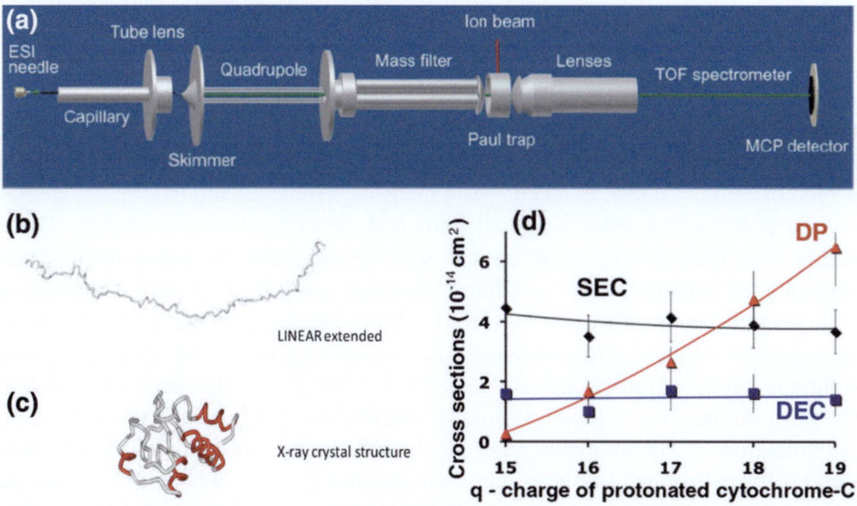

Fig. 7 **a** Experimental device of Schlathölter et al. for the study of ion-ion collisions. Model structures of cytochrome-C. **b** The extended structure expected for highly charged systems. **c** Native structure as obtained from X-ray diffraction, taken from the protein database pdb (1CYC). **d** Cross sections for the single (*diamonds*) and double (*squares*) capture and deprotonation (*triangle*) processes as a function of the charge of protonated cyt-C, $[M + qH]^{q+}$. The *lines* are to guide the eye (Reproduced from [32] with the permission of AIP Publishing)

are shown. For high charge states (15+ to 19+), the molecule is expected to exist in a linear extended configuration, with the length estimated to be 20.5 nm (see Fig. 7b). The native structure as obtained from X-ray diffraction is shown in Fig. 7c. Thus for highly charged proteins the protonation sites are expected to be widely distributed along the chain structure.

With the aid of the classical over-the-barrier model [33, 34] adapted to a long chain target, absolute cross sections for single (SEC) and double (DEC) electron capture are determined as shown in Fig. 7d. As to be expected for the peptide of this size, very large cross sections of the order of 4×10^{-14} cm^2 and 2×10^{-14} cm^2 are obtained, respectively. However, more surprising is the fact, that the values are nearly independent of the charge state of the protonated protein although the ionization potential increases with the charge state. This effect is compensated by the lowering of the potential barrier for the electron to be transferred due to the presence of the highly charged ion.

Figure 7d shows in addition the cross section for deprotonation (DP) of the protonated proteins which strongly increases with the charge state. This is in good accordance with the calculated proton affinities of cyt-C $[M + qH]^{q+}$ showing a monotonic decrease with increasing charge q. This deprotonation process is tentatively interpreted as due to the temporary barrier suppression for the binding of an H^+ ion in long-distance interactions between the multiply charged projectile and the binding electronic cloud.

2.2 Ionization Processes

Ionization is one of the important processes in atomic collisions physics, in particular when collisions with fast heavy ions in the MeV energy range are considered. Here, ionization dominates over other processes, such as electron capture. In the context of biomolecular systems, in recent years many experimental and theoretical studies have been performed concerning ionization and electron emission processes. These studies have been motivated by the fact that the produced secondary (slow) electrons play an important role in radiation damage and hadron therapy applications [35]. Accurate ionization cross sections for biomolecular systems are necessary as input data for Monte Carlo codes like GEANT4-DNA [36] or LEPTS [37] devoted to the modelling of radiation-induced cellular damages and ion track structures in biological tissues. The main experimental and theoretical effort has been put on the determination of double differential, single differential and total ionization cross sections for collisions of protons and light ions (He^{2+}, C^{4-6+}, O^{8+}, S^{13+}) with the molecules of water [38–45] and the RNA nucleobase uracil [11, 46–52] (in some cases also other nucleobases).

The theoretical description of ionization processes has to go beyond perturbative methods in order to account for details of the ionization mechanism. In particular, the distortion of the target wave function of both the initial and the final state have to be incorporated in the model descriptions. This still remains a challenging task. Most promising results are obtained with the CDW-EIS (continuum distorted wave-Eikonal initial state) approximation which is applied in different formulations: in the so-called prior-form the influence of the passive electrons on the dynamical evolution of the ejected one (dynamical screening) is implicitly included, whereas in the post-version it is only partially taken into account through the use of an effective charge in the final continuum representation [53]. We will compare in the following some of these results with recent experimental data for the molecules H_2O, uracil and adenine.

Water Molecules

A typical experimental device is shown in Fig. 8 (taken from [52]) which allows analyzing electrons which are produced in the overlap region of the primary ion beam and the angular acceptance of the electron spectrometer. Thus, the electron yield can be measured either at a given scattering angle and scanning the electron energy (yielding kinetic energy distributions) or for a fixed kinetic energy of the electrons and scanning the scattering angle (yielding the angular distributions). In both cases double differential cross sections (DDCS) are obtained which are defined as $d^2\sigma/dEd\Omega$. By integration over the energy or angular distribution the single differential cross sections $d\sigma/dE$ or $d\sigma/d\Omega$ (SDCS) are obtained and by further integration the total ionization cross section (TCS).

In Figs. 9 and 10, we compare DDCS values from experiment and theory for the collision system O^{8+}/H_2O at impact energy of 48 MeV. The general form of the electron energy distribution changes only weakly with the scattering angle of the

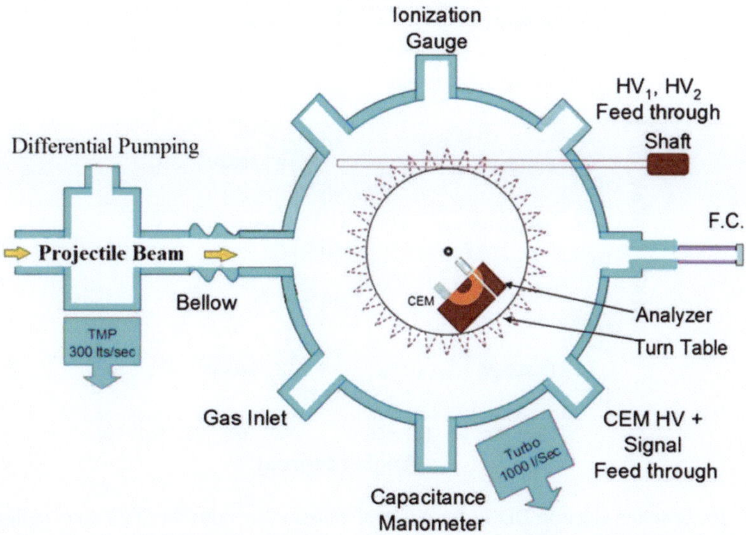

Fig. 8 Experimental device, consisting of an ion beam, water target, turnable electron spectrometer and channel electron multiplier. The magnetic rest field of the Earth is below a few mG (Reproduced from [52])

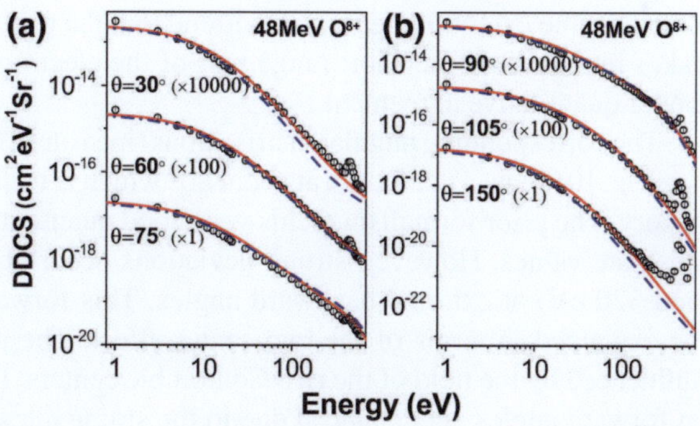

Fig. 9 The absolute electron DDCS for different emission angles for O^{8+}/water collisions at 48 MeV. In each plot, the *solid* (*dash-dotted*) *red* (*blue*) *line* corresponds to the CDW-EIS prior (post) version (Reproduced with permission from [43]. Copyright IOP Publishing. All rights reserved)

emitted electron (see Fig. 9). In all cases the DDCS $d^2\sigma/dEd\Omega$ reaches a maximum at the lowest kinetic energies which is due to the importance of collisions at large impact parameters. The values decrease by 4–5 orders of magnitudes when the energy is increased to ~500 eV. The structure observed in the experiment at higher electron energies corresponds to contributions from K-LL Auger lines from the oxygen atom. These are not included in the calculations based on the CDW-EIS approximation. The theoretical curves resulting from the prior (full red curve) and the post version (dashed-dotted blue curve) of the CDW-EIS approximation show a qualitative agreement with the experimental data, although at high energies (>100 eV) and larger angles (150°) the cross section is underestimated by a factor of up to 10. For $E < 10$ eV and small emission angles (30°) theory underestimates the experimental

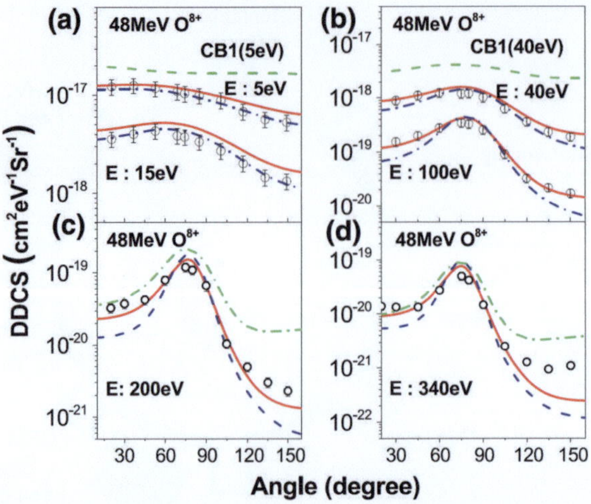

Fig. 10 The absolute electron DDCS for different emission energies for O^{8+}/water collisions at 48 MeV. The *solid* (*dash-dotted*) *red* (*blue*) *line* corresponds to the CDW-EIS prior (post) version. The *green dashed-dotted line* corresponds to calculations within the first-order Born approximation with initial and final wavefunctions satisfying correct boundary conditions (CB1) (Reproduced with permission from [43]. Copyright IOP Publishing. All rights reserved)

data to some extent. In general it is found that the prior version of CDW-EIS, which takes into account the short range part of the electron dynamic screening, yields a better quantitative agreement.

The corresponding angular distributions (measured for different electron energies, see Fig. 10) show a maximum at about 80° which is well reproduced by the CDW-EIS theory. The prior formalism yields very good quantitative agreement concerning the absolute values. However, strong deviations occur at high emission energies (240 and 320 eV) at extreme backward angles. This forward-backward asymmetry can be explained in terms of the two-center effect. The ejected electron gets strongly influenced by the field of the two Coulombic centers; in particular the cross sections at forward angles get enhanced due to the strong attraction of the emitted electrons by the receding projectile ions moving in the same direction. This effect is stronger for high energy electrons. As discussed before, the prior-formalism gives better agreement with the experiment than the post form.

Uracil, Adenine and Complex Systems

In a similar way, DDCS, SDCS and TCS values have been determined for ion collisions with nucleobases, in experiments and by calculations of several groups. An example is given in Fig. 11 showing the DDCS and SDCS for fast 2 MeV proton collisions with adenine [46]. Different emission angles are specified in the figure. In contrast to the above discussed water results, all the DDCSs spectra exhibit broad maxima at ejected electron energies of about 6 eV. Furthermore, low-energy electrons ($\sim$30 eV) are found to be ejected nearly isotropic. The two peaks located at about 250 and 400 eV are K-LL Auger electrons ejected from carbon and nitrogen atoms,

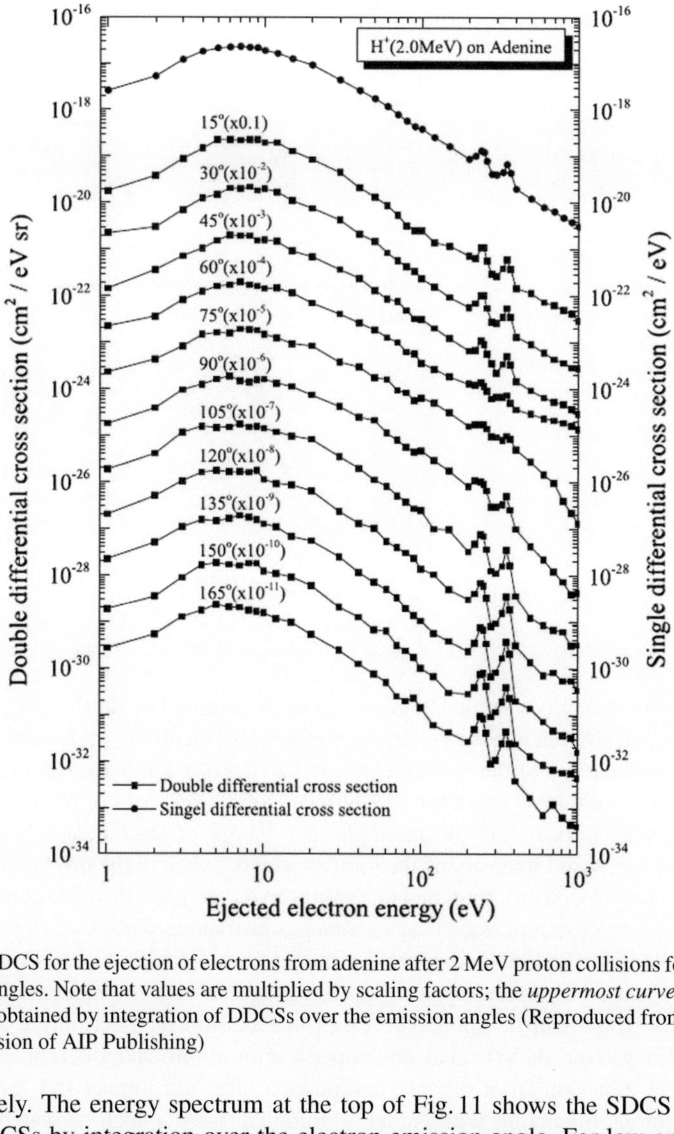

Fig. 11 DDCS for the ejection of electrons from adenine after 2 MeV proton collisions for different emission angles. Note that values are multiplied by scaling factors; the *uppermost curve* represents the SDCS obtained by integration of DDCSs over the emission angles (Reproduced from [46] with the permission of AIP Publishing)

respectively. The energy spectrum at the top of Fig. 11 shows the SDCS obtained from DDCSs by integration over the electron emission angle. For low energies of the ejected electrons the energy spectra should show optical properties of valence electrons, as the collision with fast protons is distant and the dipole interaction contributes predominantly. It is observed that the height and the peak position of the maxima vary as the proton impact energy changes and exhibit the characteristics of photoionization cross sections of adenine governed by dipole or multipole interactions. Similar findings are also reported for uracil. However, the mechanism of these results remains an open question as also contribution from plasmon excitations are discussed.

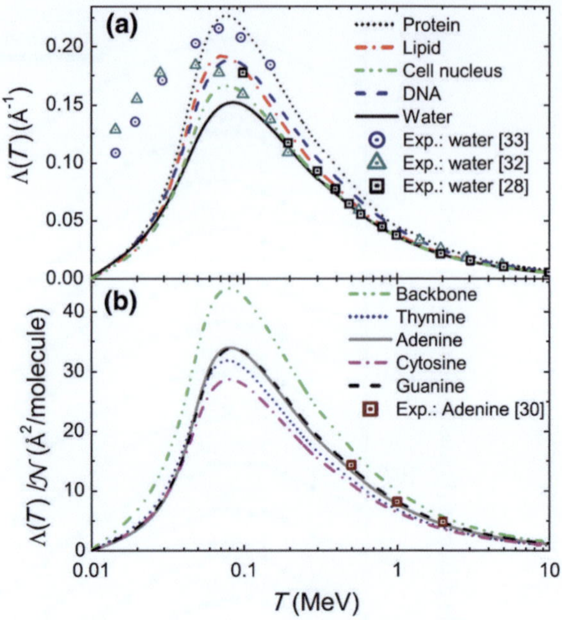

Fig. 12 **a** Calculated macroscopic total ionization cross sections Λ for proton impact in liquid water, with DNA, protein, lipid, and the cell nucleus. **b** Calculated microscopic TCS for proton impact with the DNA components adenine, cytosine, guanine, thymine and the sugar-phosphate backbone. Symbols represent experimental data. (Reproduced from [54] with the permission of AIP Publishing)

Total ionization cross sections (TCS) can be determined by double integration of DDCSs which have been measured by determining the electron yield at a given emission angle and energy of the ejected electron (as discussed above). Another method consists of the coincident detection of emitted electrons and ions or fragmented ions. In this case the information on the angle and energy of the ejected electron(s) is lost, but the measured intensity of the ionic products linked to the production of free electrons allows reconstructing total ionization cross sections. Possible contributions of transfer ionization processes may however contribute as well.

So far we described ionization cross sections for molecular targets such as water and nucleobases. More recently, a significant progress has been made in theoretical simulations which can be applied to electron emission in much larger and more complex biosystems. de Vera et al. developed a semi-empirical model for calculating the electron emission from organic compounds after ion impact [54, 55]. In the case of protons impacting with "small" biological targets such as water or DNA components a very good agreement is obtained with experimental data. Because of its simplicity and great predictive effectiveness, the method can be easily extended to any combination of large biological target like DNA, cell nuclei or proteins and charged particles of interest in ion beam cancer therapy (see Fig. 12).

Cross Section Scaling

As the number of studies is still limited to a restricted number of systems, many experimental groups made the effort to find scaling laws for the experimental data concerning their dependence on different parameters like projectile charge and energy and number of target valence electrons.

In the case of ion collisions with water molecules, Nandi et al. measured total ionization cross sections for the projectiles H^+, He^{2+}, C^{6+} and O^{8+} at MeV impact energies [52]. They assumed a quadratic projectile charge state dependence as to be expected from first-order theories employing an independent electron model [56]. Concerning the impact energy dependence, according to the Bethe theory [57], a scaling with the term $\frac{\ln(E/R)}{(E/R)}$ is proposed where E is the ion impact energy and R the Rydberg energy. Good agreement is obtained for H^+ and He^{2+} collisions whereas in the case of carbon and oxygen ions some deviations are observed [58].

In the case of the uracil target, Agnihotri et al. have measured TCS for collisions of several highly charged ions of carbon, oxygen and fluor at different collision energies [49, 50]. They analyzed the dependence of the total ionization cross section on the projectile charge state q in detail and found a deviation from the expected quadratic behavior. The fitting of the experimental data with a power law yielded exponents which varied between 1.1 and 2 with an average value of 1.5, i.e. TCS $\sim q^{1.5}$. The scaled data are shown in the left panel of Fig. 13. Concerning the impact energy dependence, a similar power law fitting yielded an exponent of −0.75, i.e. TCS $\sim E^{-0.75}$ (see right panel of Fig. 13). When introducing the ion velocity v, the scaling can be written as TCS $\sim (q/v)^{3/2}$. Such a scaling allows a representation of total ionization cross sections in uracil within 20 % and is therefore useful for the modelling of the radiation damage-induced by heavy ions.

Ionization cross sections have also been scaled with respect to molecular target properties. Toburen et al. investigated in the past high-energy proton-impact on var-

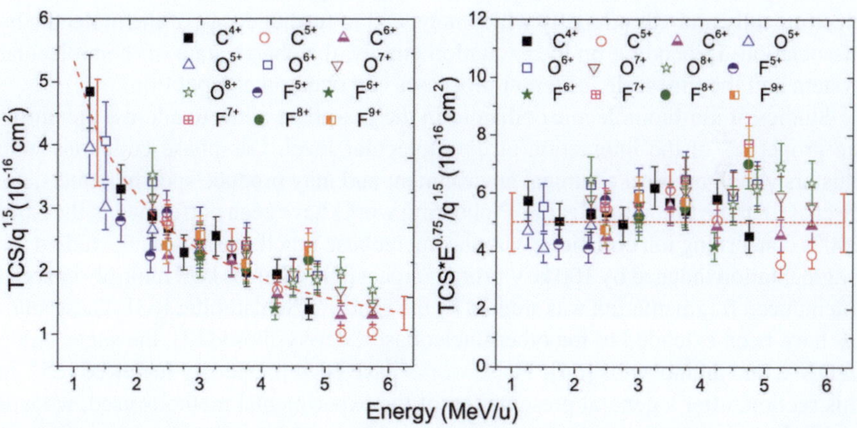

Fig. 13 Scaling of the total ionization cross sections for uracil obtained in collisions of C, O and F projectiles with **a** projectile charge state q and **b** projectile energy E. The *red dotted lines* are to guide the eyes and the *vertical lines* on both ends represent a tentative 20 % level of the values including most of the data points, i.e., roughly quantifying the scattering in the data (Reproduced with permission from [49]. Copyright IOP Publishing. All rights reserved)

ious polyatomic molecules [59, 60]. It was found that for a given proton energy, cross sections increase linearly with the number of weakly bound target electrons n_v showing a fitting accuracy of $\sim$15 % for molecular sizes up to benzene. Recently, Itoh et al. demonstrated that the scaling can be extended to larger molecules like uracil ($n_v = 42$) and adenine ($n_v = 50$) [51]. The validity of this scaling is explained by the fact that the velocities of weakly bound electrons of C, N, and O are much slower than that of the incident protons and that ionization cross sections for these electrons do not differ greatly. Also the scaling of the emission of electrons in a specific subshell of the target, (nl), has been reported recently by Montenegro et al. concerning total ionization cross sections of atomic and molecular targets by neutral to highly charged projectiles with energies ranging from a few keV/u to many MeV/u [61].

3 Fragmentation and Stability of Charged Biomolecules

Collisions with ions are always accompanied by an energy transfer to the biomolecular target. The energy loss of the ions occurs through the friction in the molecular electronic cloud (electronic stopping power) as well as in binary collisions with the nuclei of the molecule (nuclear stopping power) [62]. The importance of these two processes depends on the nature of the projectile (mass, charge state) and its velocity. As a rule of thumb, heavier and slower projectiles show higher nuclear stopping power. Faster projectiles are associated with higher electronic stopping power with a maximum in the Bragg peak velocity range. Thus, after the collision with the incoming ion, the biomolecular target is frequently ionized but also excited both electronically and vibrationally which may lead to further decay of the molecule by dissociation. Depending on the excitation energy, the charge state of the molecular system and the timescale, different processes can occur in competition.

Studies of ion-biomolecule collisions in the gas-phase allow to address the intrinsic properties of the interaction at the molecular level. Gas-phase collisions with clusters which mimic a chemical environment and may produce specific effects, are discussed in the following Sect. 4. Pioneering works have been performed in the early 2000s concerning ion collisions with the nucleobase uracil. Farizon et al. studied the fragmentation induced by 100 keV proton impact [21] whereas keV multiply charged ion induced fragmentation was studied by the group of Schlathölter [63]. These studies have been extended to the other nucleobases, deoxyribose [13], the sugar group in DNA, and amino acids [64]. These works have been previously reviewed [65]. In this section, after a general presentation of the experimental methods used, we will highlight some results on the fragmentation of isolated biomolecules.

Methods

A typical experimental arrangement allowing to study the fragmentation of biomolecules under ion impact is a crossed beam device where the ion beam interacts with a molecular beam. The charged products of the interaction are analyzed by time-of-flight mass spectrometry [66]. Additional insight into the fragmentation dynamics is obtained by applying coincidence measurements [67]. These allow to distinguish collision events with respect to the charge state of the molecule, the electron emission, the number of captured electrons etc.

The molecular target can be produced by different types of sources. To study the ion interaction with isolated molecules, a resistively heated oven allows to produce in the gas phase dense effusive beams of different nucleobases and amino acids which are the building blocks of the nucleic acids and proteins, respectively. Larger molecular systems are less volatile and too fragile to be produced with such a technique. Alternatively, the use of electrospray (ESI) ion sources allows to produce in the gas phase beams of charged (macro) biomolecules from a solution [68]. However, due to the limited target beam density obtained with such sources, the experimental device needs the addition of an ion trap in order to bunch the target biomolecular ions [17].

With the dense target typical fragmentation timescales ranging from ns to microseconds are accessible. With the introduction of ion traps coupled with electrospray ion sources longer irradiation times are required to compensate for the low target density. Thus, the timescales studied are in the order of several hundreds of microseconds to ms. In any case, the electronic excitation following the collisions is converted on the ps timescale into vibrational excitation by efficient electron-phonon coupling.

From the theoretical point of view, the complete treatment of the collision and competitive fragmentation processes is a real challenge due to the different timescales involved. Thus, a possible method consists of treating independently the collision process and the fragmentation pathways as timescales for the collision are typically of the order of (sub) fs and that for fragmentation range from ps to microseconds. Such two-steps approaches have been applied to the prototypical systems namely water and C_{60} fullerene [69]. The group of Tom Kirchner treated the collisions resolving the time-dependent Schrödinger equation restricted to an independent-electron model and they used empirical branching ratios for the fragmentation processes [45, 70]. A good agreement is obtained with the experimental results in the case of single and multiple electron removals in collision of protons and He^+ ions with H_2O molecules, thus, validating such a two-steps approach. Ismanuel Rabadan et al. also considered the collision of singly charged ions with water using the classical trajectory Monte-Carlo method [71]. They further treated the fragmentation of excited water molecules by wave-packet propagation on ab initio potential energy surfaces [72]. So far, the ion collision with a more complex system, such as a nucleobase or an amino acid, has not been entirely treated theoretically. On the one hand, cross sections of ionization and electron capture have been calculated mainly for nucleobases by different methods as discussed in the Sect. 2.1. On the other hand, the fragmentation channels are theoretically studied through the exploration of the potential energy surface [67, 73] or using molecular dynamics in the ground state [74] or in excited states [75].

Mass Spectrometry of Biomolecular Fragmentation Under Ion Impact

Most of the studies performed before 2012 have been previously reviewed by Thomas Schlathölter [65]. They mainly focus on the fragmentation of nucleobases and amino acids and report on the mass spectra obtained after electron capture or ionization. Whereas a rather large fraction of the nucleobase molecules survives the interaction, mass spectra of amino acids following collisions with ions are largely dominated by the fragmentation along the carbon backbone. In DNA, 2-deoxy-D-ribose (the sugar part) appears as a weak link leading to an extensive fragmentation after ion collisions [13, 29]. Analyzing the projectile charge state after the collision in coincidence with the mass spectra of the interaction products, Farizon and coworkers were able to discriminate electron capture and ionization processes and to determine the associated fragmentation branching ratios in collisions of 20–150 keV protons with nucleobases [23, 76]. With a similar coincidence method, Afrosimov and coworkers have carried out measurements of the fragmentation branching ratios of singly and doubly charged nucleobases [24, 25] and amino acids [26, 77] produced in collisions with keV protons or doubly charged ions (He^{2+}, C^{2+}, O^{2+}). Recently, Rabus and coworkers also measured for 0.3–16 MeV proton collisions the relative fragmentation cross sections of pyrimidine, tetrahydrofuran, and trimethylphosphate [78], respectively, constituents of the pyrimidinic nucleobases (cytosine, thymidine, and uracil), and of the sugar and the phosphate group. In the case of pyrimidine, this work gives an extension to higher energies of the data obtained by Montenegro and coworkers [79] concerning fragmentation branching ratios and cross sections.

Recent studies also focus on halogenated uracil [80, 81] (5-fluorouracil is commonly used as radiosensitizer [82]) as well as on nucleosides [83]. For the latter ones, the cleavage of the glycosidic bond, binding the nucleobase to the sugar moiety, has been found to be the dominating fragmentation process. Depending on the nature of the nucleobase and of the sugar forming the nucleoside molecule, different charge localizations have been observed: the charge is more located on the base in the case of guanine (guanosine) whereas it is centered on the sugar part in the case of deoxyribose (deoxythymidine) [83]. In the case of 5-halouracils, Champeaux and coworkers studied the collisions with 100 keV protons (i.e. the proton energy near the Bragg peak). They found that an important process is the loss of the halogen species both in the neutral and cationic form [80]. The emission of halogen cations is increasing with the atomic number and is accompanied by a large amount of kinetic energy with a mean value between 1 eV for I^+ and 3 eV for F^+. The kinetic energy distributions show a tail extending to values above 10 eV. The fragmentation dynamics of 5-bromouracil (5BrU) induced in collisions with 100 keV multiply charged ions (O^{6+}, Ar^{11+} and Xe^{19+}) was studied with a special emphasis on delayed fragmentation processes [81]. In the case of the singly charged molecule, a metastable population of $5BrU^+$ was observed with a lifetime of several microseconds, decaying by HNCO loss according to a power law. This shows a rather large excitation energy distribution after the collision. Moreover, a transient 5-membered ring dication, formed by CO loss and surviving also on the microsecond timescale, was inferred from coincidence measurements.

Influence of the Excitation Energy

The opening of different decay channels as well as their relative branching ratios strongly depend on the excitation energy and the charge state of the biomolecule after the ion collision. Depending on the impact parameter, collisions of ions with biomolecules lead to a distribution of excitation energy and charge removal. For example, in the case of nucleobases, calculations of the electronic energy loss of 80 keV protons show a maximum at around 70 eV and a tail towards energies higher than 150 eV [84].

To disentangle the contributions of the different molecular charge states and excitation energies to the fragmentation mass spectrum, the group of S. Martin developed a method based on the triple coincidence measurements of the projectile (charge state and energy loss), of the emitted electrons (number) and the cationic products (fragmentation mass spectrum) [85]. Thus, with the so-called CIDEC method (Collision Induced Dissociation under Energy Control), it is possible to consider the fragmentation as a function of the internal energy of the molecule in the case of collisions where the initial and final state of the projectile is well known, typically in the case of collisions $H^+ + M \rightarrow H^- + M^{++}$. This technique has been successfully applied and monitors as a function of the internal energy the fragmentation channels of the dication of 2-deoxy-D-ribose [86], the sugar group in DNA, and of adenine both singly [87] and doubly charged [12]. Figure 14 shows the excitation energy determined for different fragments of the adenine dication in the case of collisions with 3 keV Cl^+ ions. In the case of the adenine molecule ($C_5H_5N_5$) where the distinction between different channels giving a similar fragment $C_xH_yN_z$ is difficult, the CIDEC method allowed to propose a detailed scheme of the fragmentation pathways of the adenine dication [12].

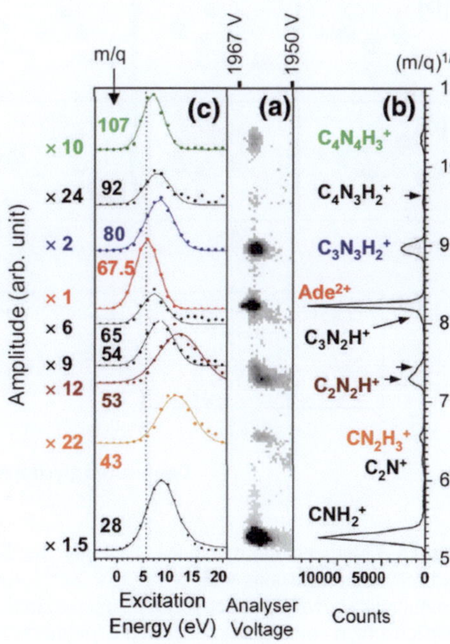

Fig. 14 **c** Excitation energy distributions for different fragments produced in the dissociation of the adenine dication formed in the collision with 3 keV Cl^+ ions, **a** map between the projectile energy loss (a voltage of 1967 V correspond to zero loss) and the recoil ion time-of-flight, **b** time-of-flight spectrum integrated along the energy loss (Reproduced from [12] with the permission of AIP Publishing)

Molecular Rearrangement Versus "Coulomb Explosion"

At the Bragg peak energy, the ionization of biomolecules occurs through removal of inner-shell electrons. This is followed by Auger decay leading to the formation of doubly positively charged species and the emission of secondary low-energy electrons. Even in the case of the rather stable nucleobase molecules, dications of biomolecules are unstable due to the charge localization on a limited size system and the subsequent rather large Coulomb repulsion. Thus, charge separation processes are expected to produce secondary (radical) ions with energies of a few eV. These species can in turn trigger additional reactions.

In addition to the expected charge repulsion dissociation channels, recently, a competition with ultrafast (few tens of fs) molecular rearrangements has been observed allowing to transiently stabilize the two charges in amino acids [74, 88]. An experimental proof of such intramolecular transfer is the observation of a series of dicationic fragments in the fragmentation mass spectra obtained after collisions with keV multiply charged ions. Ab-initio molecular dynamics calculations of the dication showed that a hydrogen transfer occurs within 40 fs (Fig. 15c), i.e. on the same timescale

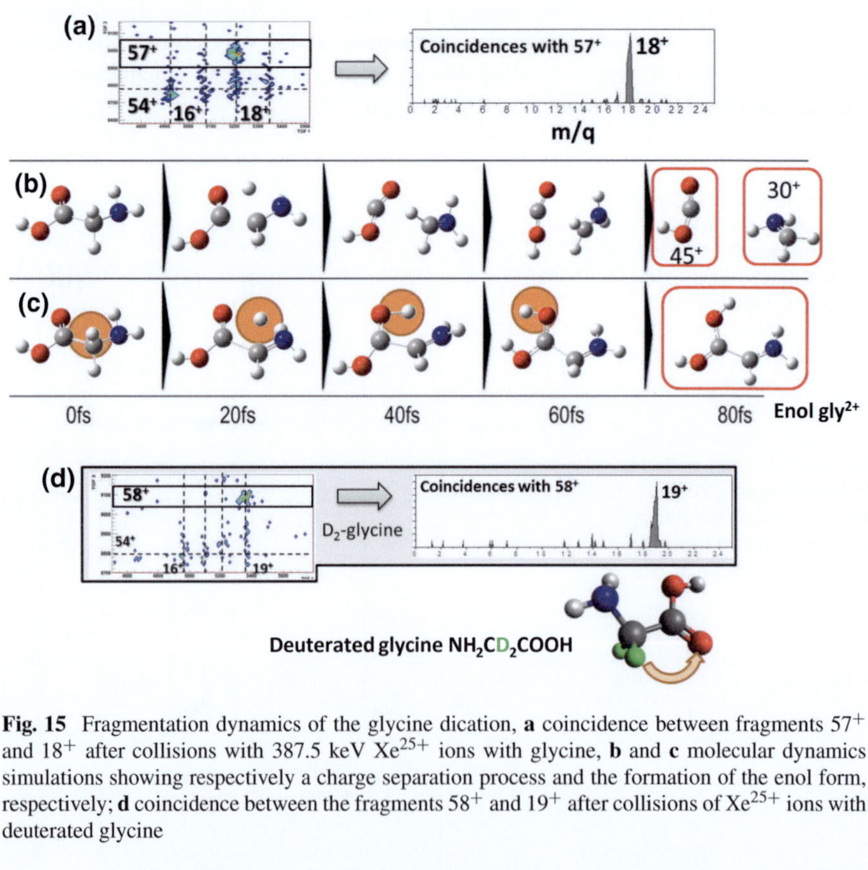

Fig. 15 Fragmentation dynamics of the glycine dication, **a** coincidence between fragments 57^+ and 18^+ after collisions with 387.5 keV Xe^{25+} ions with glycine, **b** and **c** molecular dynamics simulations showing respectively a charge separation process and the formation of the enol form, respectively; **d** coincidence between the fragments 58^+ and 19^+ after collisions of Xe^{25+} ions with deuterated glycine

as the charge separation (Fig. 15b), leading to the formation of the more stable enol form of the amino acids. The enol dication further dissociates by loss of neutral moieties, giving the doubly charged fragments observed, or by charge separation channels which could be specific of the isomerization, e.g. in the case of the glycine the loss of H_2O^+ giving the coincidence $57^+/18^+$ (Fig. 15a) or in the case of deuterated glycine $58^+/19^+$ (Fig. 15d) [74]. Apart from hydrogen transfer, the isomerization can also involve more unexpected mechanisms, like OH migration which has been observed in the case of the fragmentation of the β-alanine dication ($NH_2CH_2CH_2COOH^{++}$) [88]. Further loss of CO from this isomer is the only mechanism which can explain the observation of coincidences between the fragment $NH_2CH_2^+$ and $CH_2OH_2^+$. Statistical analyses of the simulations show that the "Coulomb explosion" is the dominating process but isomerization processes represent about 5 % depending on the excitation energy.

Towards Macrobiomolecules

As mentioned before, most of the studies on ion-biomolecules collisions concern nucleobases and amino acids which are the building blocks of nucleic acids and proteins, respectively. However, the presence of a single peptide bond changes the charge localization and therefore the fragmentation dynamics is modified as shown in the model molecule, N-acetylglycine [89]. In the case of larger biomolecules, one can expect different dissociation channels from the basic building blocks.

Following the pioneering work by Bari et al. on keV ion-induced peptide dissociation [17, 90] the fragmentation of larger molecules such as proteins [32] and oligonucleotides [91] after collisions with slow multiply charged ions has been studied. These works used the home-build apparatus of the group of Schlathölter which coupled an ESI source to a 3D Paul trap where biomolecular ions are colliding with an ion beam, the interaction products are then extracted into a linear time-of-flight mass spectrometer (see Fig. 7). In the case of the oligonucleotide dGCAT, the fragmentation pattern obtained after collisions with 40 keV C^{4+} (Fig. 16) is dominated by nucleobase ions and fragments arising from the deoxyribose molecule [91]. Only a weak signal of intact nucleosides is detected accordingly with the strong cleavage of the glycosidic bond observed in proton collisions with isolated nucleosides [83]. Nucleobase fragments reflect both the initial protonation sites of the target and the charge localization of the electron removal associated with the ion collision. Protonated and non-protonated species are formed which in any case require at least one H transfer from the sugar group to the nucleobase. Finally, the fragmentation of the sugar moiety contributes strongly to the mass spectrum as shown by the intense peak at $m/z = 81$ amu. It has to be noted that this fragment was not observed in the fragmentation of isolated 2-deoxy-D-ribose as in the gas phase it has a different structure (pyranose, a six-membered ring) than in nucleosides, nucleotides or DNA (furanose, a five-membered ring) giving different fragmentation pathways [92, 93].

Collision experiments with singly charged keV ions have been performed recently which might be considered as an alternative activation method in mass spectrometry coupling an ion gun with a commercial mass spectrometer. The group of Zubarev used an orbitrap based analyzer and "air cations" (a mixed ion beam) as projectile

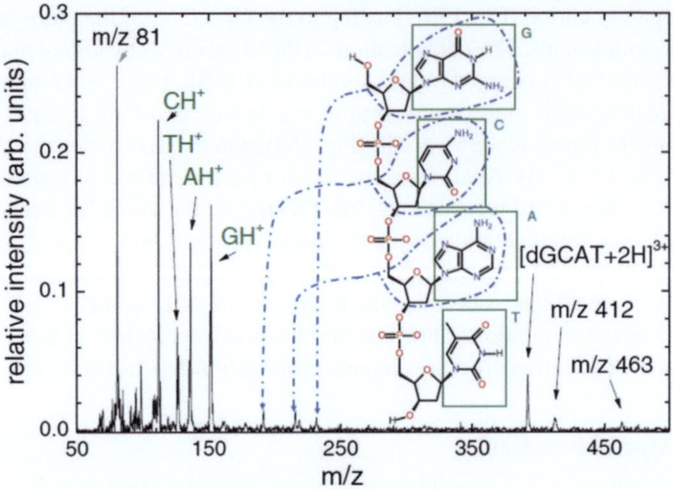

Fig. 16 Fragmentation mass spectrum of doubly protonated oligonucleotides [dGCAT + 2H]$^{2+}$ after collisions with 40 keV C^{4+} ions (Reproduced from [91] with the permission of AIP Publishing)

[94], whereas Hoffmann and Jackson modified a linear quadrupole mass spectrometer to inject a He^+ ion beam inside [95]. In order to compare with well-established mass spectrometry methods, both groups studied reference peptides (angiotensin I, ubiquitin and substance P). An interesting method is the preparation of higher charge state species due to the electron removal by an incoming projectile, especially for those species which cannot be produced by electrospraying. Thus this opens the possibility to perform further mass spectrometry studies on "supercharged" systems [94]. Additionally, the electron removal by the projectile also triggers radical mechanisms driving the fragmentation [95]. Such mechanisms are sensitive to the tertiary structure of the proteins. Moreover, in the case of multiply protonated targets, a deprotonation can occur as shown by a charge reduction [32, 94] or by typical fragmentation channels linked to the cleavage of N–C_α bonds [94]. Thus, it is possible to observe in a single ion-induced activation mass spectrum fragments resulting from different interaction processes (see Fig. 17). This can increase the sensitivity with respect to the molecular composition. This is shown by a higher scoring in the proteomics sequence analysis based on the mass spectrum [94]. Both groups conclude that the activation by ion collisions is a promising method to study low-charge state gas phase ions.

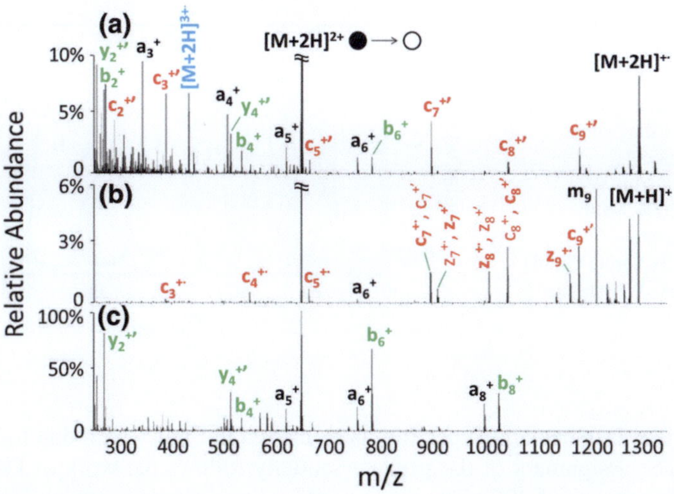

Fig. 17 Fragmentation mass spectrum of doubly protonated angiotensin I **a** after collisions with air cations, **b** after electron transfer (ETD electron transfer dissociation) and **c** after collision with gas (higher energy collision induced dissociation HCID) (Reproduced from [94]. Copyright 2014 American Chemical Society.)

4 Biomolecules in a Cluster Environment

Ion collisions with biomolecules which are embedded in a cluster environment are a first approach to study the fragmentation of biomolecules in a more natural environment where they can interact with the surrounding molecules before, during and after the ion collision. Thus, new phenomena have been observed which are linked to the opening of new fragmentation channels, the redistribution of energy and charge transferred in the collision and protonation and proton transfer processes occurring in the cluster.

Clusters can be produced in ESI sources by optimizing the solvation process. In this case the target density is rather low, but the formation of charged species allows for the size selection of the target. Alternatively a distribution of neutral clusters can be obtained using a gas aggregation cluster source [96] or a supersonic expansion which can produce heterogeneous clusters as well as hydrated ones, using a mixture of different molecular vapors.

4.1 Pure Clusters

New Fragmentation Pathways

In early experiments the fragmentation behavior of isolated biomolecules has been compared with that observed in a cluster environment [97]. For the thymine molecule

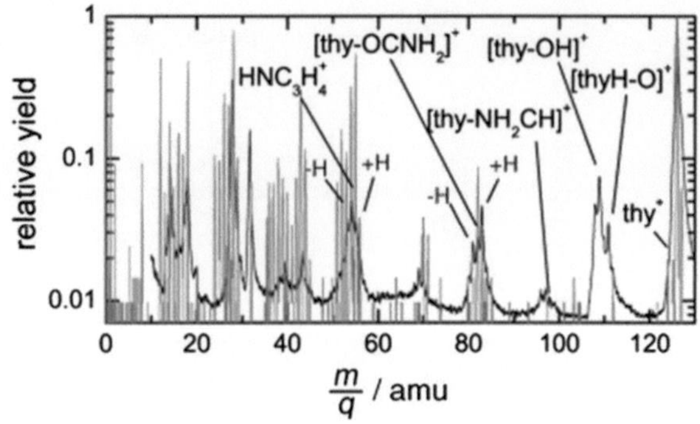

Fig. 18 Mass spectra for collisions of O^{5+} ions with thymine (*thin grey lines*) and thymine clusters (*full curve*). thy indicates the parent thymine molecule (Reproduced from [97]. Copyright 2006, with permission from WILEY-VCH Verlag GmbH & Co. KGaA, Weinheim)

the molecular fragmentation spectrum is compared in Fig. 18 with that for thymine clusters. The assignment of the peaks essentially follows the work of Deng et al. [98] and Jochims et al. [99]. The fragmentation mass spectrum which is measured in the cluster case with lower resolution, is obviously quite different from that of the isolated molecule. This concerns on the one hand the relative intensities of individual peaks and on the other hand the opening of new fragmentation channels which are not present in the isolated molecule case. The new channels which are characterized by the m/q values of 97 and 109 are due to the loss of NH_2CH and OH, respectively. These channels have neither been observed in photoionization nor in electron impact ionization experiments. The OH loss process has been measured in ion collisions with the condensed phase thymine [98], which indicates that this process can only be due to an exocyclic bond cleavage. To further analyze the mechanism the authors performed coincidence measurements showing that the OH-loss occurs already for the dimer case and gains importance for larger clusters. The most stable dimer case is planar and bound by two hydrogen bonds between the H donor and the O acceptor of each molecule [100]. Quantum-chemical calculations indicate that also for larger clusters (up to the pentamer) the planar geometry prevails and that the systems are based on hydrogen bonds between O and H atoms. It is concluded that the characteristic OH-loss fragmentation channel in the cluster and in the condensed phase is a fingerprint for hydrogen bonding in these systems.

Redistribution of Energy and Charge

As mentioned in the introductory part of this chapter, early studies with C_{60} molecules and clusters of C_{60} have shown that in keV collisions many evaporation processes (loss of C_2) occur even in the case of single ionization. In the case of the cluster target under identical collision conditions a much lower yield of evaporation and of fragmentation of individual molecules is observed [101]. This was attributed to the fact that the energy transferred in the collision is rapidly redistributed over all degrees of freedom of the cluster system, thus lowering the locally deposited energy and hence the fragmentation yield. In similar experiments it was concluded that also the charge is highly mobile and can be transferred during the ion collision to

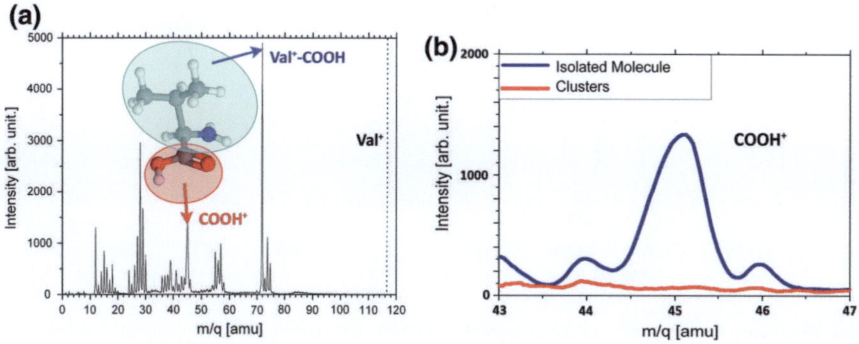

Fig. 19 **a** Inclusive fragmentation mass spectrum of valine irradiated with 300 keV Xe^{20+} ions, specifying the two fragments produced by the cleavage of the C–C_α bond; **b** comparison of the $COOH^+$ fragment intensity for the isolated valine molecule (*blue line*) and the cluster case (*red line*) (Reproduced from [14]. Copyright 2011, with permission from WILEY-VCH Verlag GmbH & Co. KGaA, Weinheim)

neighbored molecules [101]. Thus in the fullerene case, the redistribution of energy and charge occurs in very fast processes (sub-fs to ps).

Similar studies have been performed more recently in the case of amino acid clusters [14]. The fragmentation of small amino acids like glycine or valine is governed by the cleavage of the C–C_α bond, yielding as fragments the $COOH^+$ carboxyl group and the residual ion $(M–COOH)^+$ [102, 103]. Both ions are shown in the mass spectrum of valine irradiated with 300 keV Xe^{20+} ions (see left part of Fig. 19). When analyzing separately single ionization events, dominantly the $(M–COOH)^+$ residue is observed, indicating that the charge is most likely located at the amino group. In the case of two charges both fragments ions are observed in coincidence [74]. In that case one charge is located at the carboxyl group; the other one at the amino group, i.e. the valine molecule has to be charged at least twice. When one compares the results for the isolated valine molecule with that obtained for the valine cluster, a total suppression of the $COOH^+$ signal occurs (see right panel of Fig. 19), proving that the fragmenting molecule within the cluster is only singly charged. Furthermore, one observes that the signal of the intact molecule, which is hardly observable in the isolated molecule case shows up in the cluster environment proving a "protective" effect due to the redistribution of the transferred energy. Thus, fast redistribution of energy and charge is also observed in clusters of biomolecular systems [14].

Protonation

Several experiments have been performed recently concerning ion collisions with neutral clusters of amino acids like alanine, valine and glycine as well as with clusters of lactic acid. It was found in the fragmentation mass spectra, obtained in 30 keV collisions with O^{3+} projectiles, that monomers and dimers stemming from cluster fragmentation are to a large extend protonated. This is due to the proton transfer within amino acid clusters after ionization as shown by ab-initio quantum-chemical

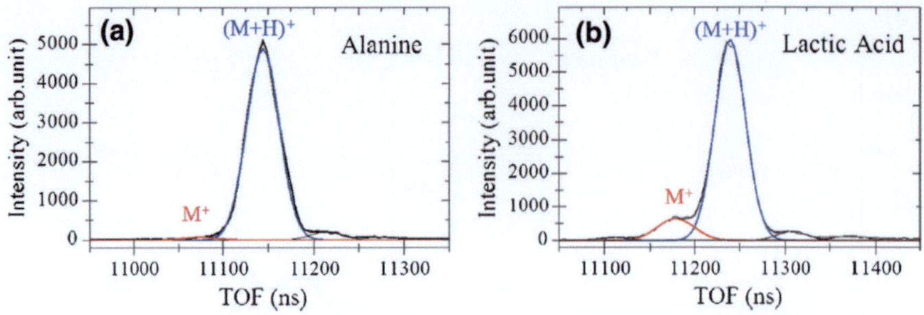

Fig. 20 Gaussian functions are fitted to radical cations M^+ (*red line*) and protonated monomers $(M+H)^+$ (*blue line*) of **a** alanine and **b** lactic acid. Peaks on the *right side* of the $(M+H)^+$ peak are assigned to isotopic contributions of the molecules (Reproduced from [104]. Copyright 2015, with permission from WILEY-VCH Verlag GmbH & Co. KGaA, Weinheim)

calculations [104]. For lactic acid, which has a lower proton affinity than amino acids, a non-negligible amount of the radical cation monomer was observed (see Fig. 20) while it almost disappears in the case of alanine clusters. The chemical structures of L(+)-lactic acid and α-alanine are very similar: they differ only by one group attached to the asymmetric carbon, namely hydroxyl group for lactic acid and amino group for alanine. In the inclusive mass-over-charge spectra Gaussian functions have been fitted to the radical cation and the protonated monomer peaks (see Fig. 20). In the alanine case, the radical cation is barely noticeable and corresponds to only (1.6 ± 0.2) % of the protonated molecule. In the case of lactic acid this value increases to (12.3 ± 0.7) %. This finding is consistent with the gas-phase proton affinities showing that lactic acid has a much lower proton affinity (817.4 ± 4.3 kJ mol^{-1}) than alanine (902 kJ mol^{-1}) [105]. Furthermore, typical fragments of the protonated monomer are observed in all collision systems which are interpreted as a result from statistical fragmentation and may occur delayed on a microsecond timescale.

4.2 Water Clusters

At the molecular level, the most "realistic" target in the gas phase consists of nanohydrated molecules, i.e. a cluster formed by a biomolecule (partially) solvated in water molecules. The effect of the hydration on the ionization and fragmentation processes will be discussed in the Sect. 4.3. Here, we focus on the fragmentation of water clusters induced by ion collisions. Beside the fact that a water cluster is a brick of the nanohydrated biomolecular target, it is important to understand its fragmentation behavior in order to consider the balance between the so-called direct and indirect effects of radiation damage [106]. Indeed as water is the main constituent of the body, its ionization/fragmentation will be a huge reservoir of secondary particles and will play an important role for indirect effects. For example, it is well known that the radiolysis of water produces several radical species [107] and that the ionization of water clusters $(H_2O)_n$ leads to the formation of protonated water clusters $H^+(H_2O)_{n-1}$ and the release of one OH radical [108]. Moreover, recently it has been shown that ionized water clusters are the source of low-energy secondary electrons through intermolecular coulombic decay processes [109, 110].

Both, low energy ion and swift ion collisions with water clusters have been reported. The experiments were performed at GANIL (Caen, France). The ionization of water clusters by swift Ni^{25+} ions (12 MeV/u) have been studied by Adoui et al. [114]. In this work water clusters were produced by a supersonic expansion, while a gas aggregation source was used by Maisonny et al. when studying the interaction of 300 keV Xe^{20+} ions with water clusters [112]. In both velocity regimes, where ionization and electron capture respectively dominate, it has been observed that the ionization of water clusters leads to the formation of protonated species $H^+(H_2O)_n$ [112, 114]. Only a weak contribution of radical cation $(H_2O)_n^+$ is observed for the monomer and the dimer (one to two orders of magnitude lower in intensity) [112]. TD-DFT molecular dynamics calculations demonstrate a high proton mobility, showing that the proton transfer occurs very fast on the femtosecond timescale [114]. Coincidence measurements of the charged fragments allow to consider the fragmentation dynamics of doubly charged water clusters. The dominant dissociation channels involve the emission of at least one H_3O^+ ion. Interestingly the kinetic energy distribution of this fragment for different channels (see panels a and b of Fig. 21) shows a tail towards larger energies of a few eV and a peak at around 900 meV [111, 114]. In the case of low-energy Xe^{20+} ion collisions [112], coincidence measurements give an average kinetic energy of about 1 eV for the H_3O^+ ion. Moreover, a typical peak shape is observed for clusters containing about 10 water molecules (see panels c and d of Fig. 21). This shape reflects the rather high kinetic energy (about 2 eV) of clusters in this intermediate size range. Also protons emitted are associated with an average kinetic energy of about 15 eV. Thus the fragmentation of water clusters induced by ion collisions is associated with the emission of rather energetic molecular fragments.

Abdoul-Carime et al. studied recently the fragmentation dynamics of protonated water clusters $H^+(H_2O)_n$ following collisions at 8 keV with Ar atoms [113]. In these so-called kinematically inverse experiments, it is possible to analyze the emitted neutrals which have sufficient kinetic energy to be detected, in particular the experimental set-up allows for mass and velocity selection of the neutral fragments using a time and position sensitive detector. The velocity distributions of the evaporated water molecule, measured in the case of collisions of small protonated water clusters ($n = 2$–8, see Fig. 21e), show that most of the fragments are emitted with a velocity according to a Maxwell-Boltzmann statistics, i.e. that the energy transferred during the collision is efficiently redistributed inside the cluster. Experimental distributions are well represented by statistical molecular dynamics calculations showing a total conversion of the transferred excitation energy into the vibrational system [113]. However, an interesting feature of the experimental velocity distributions is an additional distribution of fragments with a higher velocity corresponding to a kinetic energy of about 1 eV. This feature could be reproduced in molecular dynamics sim-

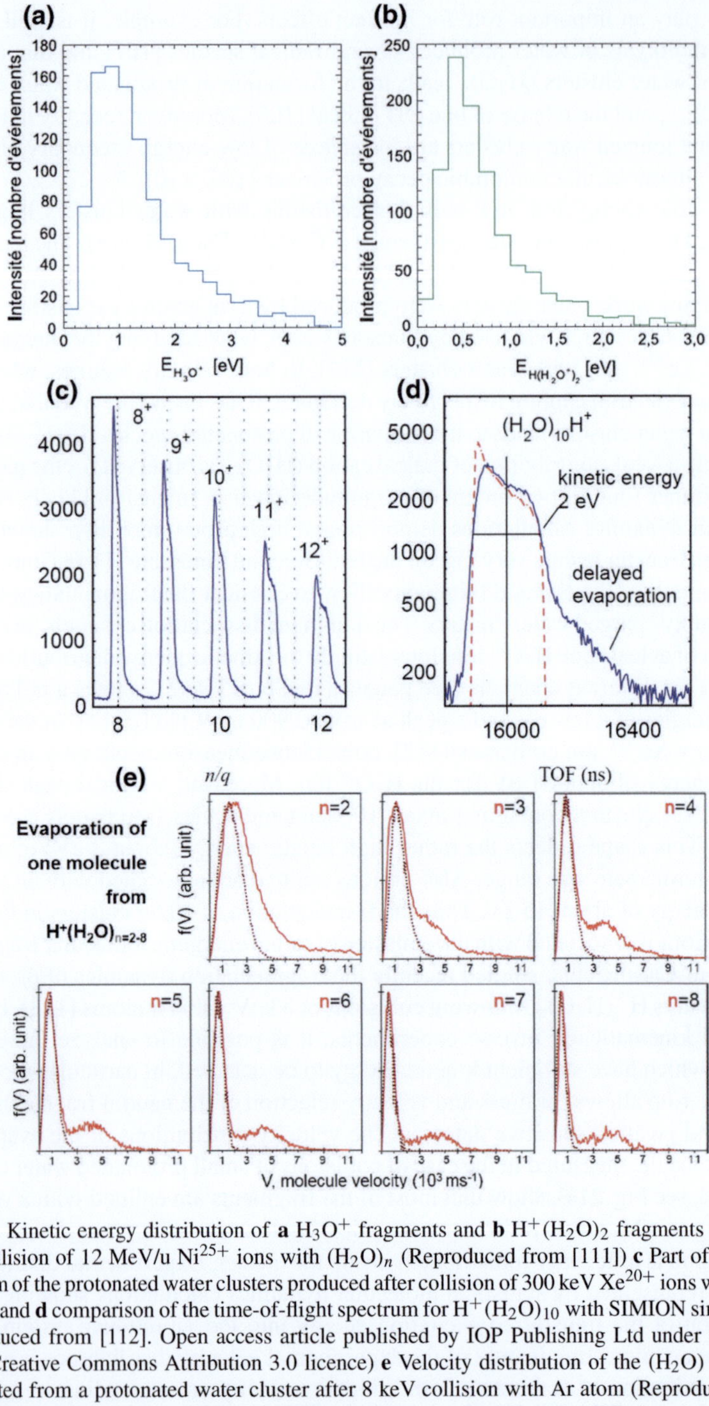

Fig. 21 Kinetic energy distribution of **a** H_3O^+ fragments and **b** $H^+(H_2O)_2$ fragments produced after collision of 12 MeV/u Ni^{25+} ions with $(H_2O)_n$ (Reproduced from [111]) **c** Part of the mass spectrum of the protonated water clusters produced after collision of 300 keV Xe^{20+} ions with water clusters and **d** comparison of the time-of-flight spectrum for $H^+(H_2O)_{10}$ with SIMION simulations (Reproduced from [112]. Open access article published by IOP Publishing Ltd under the terms of the Creative Commons Attribution 3.0 licence) **e** Velocity distribution of the (H_2O) molecule evaporated from a protonated water cluster after 8 keV collision with Ar atom (Reproduced from [113]. Open access article published by Wiley-VCH Verlag GmbH & Co. KGaA under the terms of the Creative Commons Attribution Non-Commercial NoDerivs License)

ulations assuming a local excitation. Thus, it has been shown that in competition to the dominant energy redistribution process producing statistical fragmentation, a localized energy transfer can lead to the emission of rather high energy neutral fragments before the energy is redistributed over the cluster. Such localized excitations are typical for low energy atomic collisions, where the energy deposit occurs also through the nuclear stopping power. Thus, specific non-statistical fragmentation channels are opened as shown in the case of C_{60} fullerenes [115, 116] or polycyclic aromatic hydrocarbon molecules [117, 118].

4.3 Nanosolvated Mixed Clusters

In contrast to homogeneous molecular clusters which have been described before, we will now review the work which has been performed in relation with the fragmentation of mixed clusters. In particular, biomolecular systems which are nanosolvated in water will be considered which mimic in a realistic way the conditions in biological tissues. Here protective effects, proton transfer reactions as well as the formation of hydrated molecular fragments will be discussed.

First gas-phase experiments have been performed by the Aarhus group [119, 120] where fragmentation of singly charged anions of adenosine 5-monophosphate (AMP) was induced by collisions with neutral atoms (Ne, Na). The anions were produced with the aid of an electrospray ion source and the corresponding beams were accelerated to a kinetic energy of 50 keV. In inelastic collisions CID (collision-induced dissociation) or ECID (electron capture induced dissociation) may occur. The ESI-source allowed to prepare anions of AMP with a certain number of water molecules attached ($m = 0$–20), thus allowing to compare fragmentation patterns of the isolated and nanosolvated molecular anion. In Fig. 22 a possible geometric structure of the $AMP^-(H_2O)_{20}$ anion is shown, demonstrating that the water cluster is mainly attached to the phosphate group, not covering the whole biomolecule. In the right part of Fig. 22 CID fragmentation spectra are monitored as a function of the number of the attached water molecules. The top spectrum represents the case of the isolated AMP^- anion without any water attached. It shows that in addition to the primary anion AMP^- (346 amu) several small-size fragments are observed which correspond to the loss of O or adenine, and in particular to the formation of negative phosphate groups PO_3^- and $H_2PO_4^-$. When the number of water molecules is increased, the loss of water molecules which are loosely bound, becomes the dominating process. For $m > 13$, the water evaporation chain arrives at the naked AMP^- anion but the internal energy of the system is no longer sufficient to fragment the AMP^- molecule. This is true also for larger water coverage. The observed fragment distributions are well described with the model of an evaporative ensemble and a fitting procedure allows to estimate the internal energy transferred in the collision which is of the order of several eV.

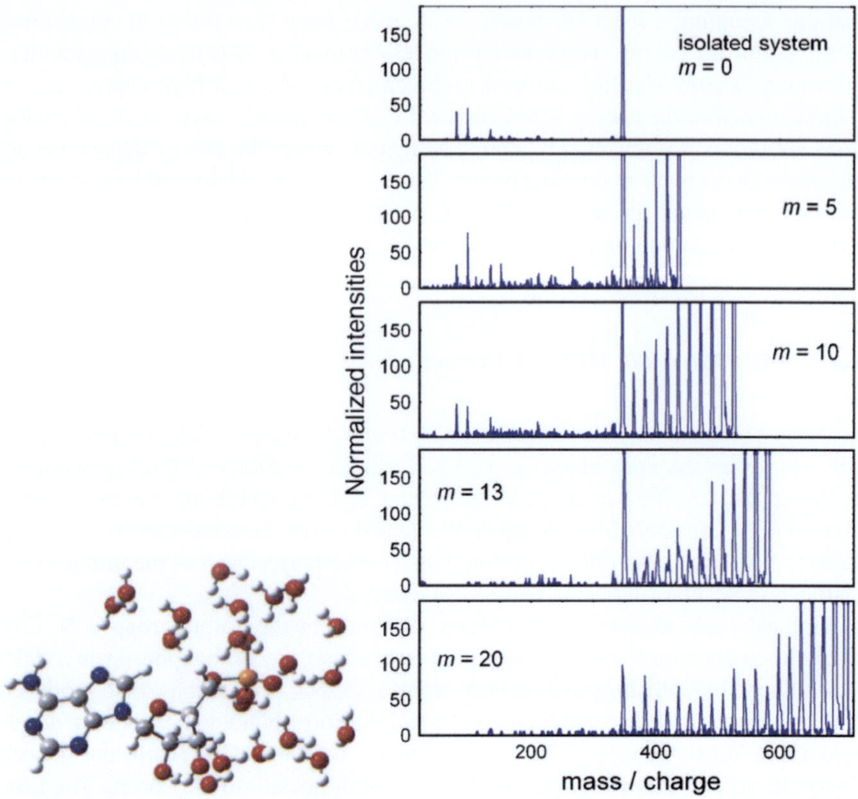

Fig. 22 Fragmentation spectra of $AMP^{-}(H_2O)_m$ anions for different numbers *m* of attached water molecules (obtained in collisions with sodium atoms at 50 keV). The spectra have been normalized to the total ion count rate. One possible structure of the system $AMP^{-}(H_2O)_{20}$. Color code: *gray* (C atoms), *blue* (N atoms), *red* (O atoms), *orange* (P atom), *small white spheres* (H atoms) (Reproduced from [119] with the permission of AIP Publishing)

In contrast to the CID results, ECID experiments evidenced an inverse effect [120]. After electron capture from a sodium target the AMP^{-} anion always loses an H atom. This fragmentation channel increases in intensity when the number of attached water molecules is increased.

In another recent experiment [121], nano-hydrated protonated pyridine (Py) clusters $PyH^{+}(H_2O)_n$ ($n = 0$–5) have been produced and the fragmentation spectra have been measured after collisions with Ar atoms (CID). The observed fragment ions show that a proton transfer occurs from the protonated pyridine to water molecules. This is surprising as in the case of $PyH^{+}(H_2O)$, the energy required for dissociation into PyH^{+} and H_2O is substantially lower than that for the dissociation into Py and $H^{+}H_2O$, namely by a factor of approximately 7 (see Fig. 23a). However, in the case of $PyH^{+}(H_2O)_2$, the solvation by the second water molecule leads to a lower value of about 3. This is in agreement with the proton affinity in water clusters of different

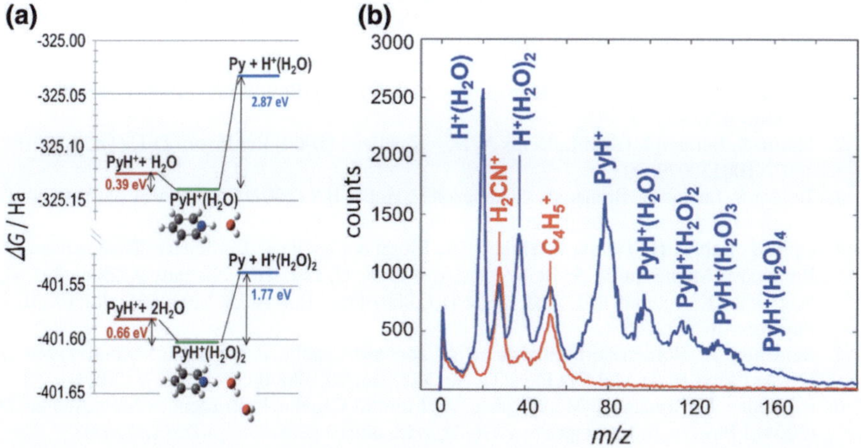

Fig. 23 **a** Schematic representation of the Gibbs free energy of the fundamental state of the different species involved in the fragmentation of $PyH^+(H_2O)$ and $PyH^+(H_2O)_2$ ions derived from DFT calculations. **b** The charged-fragment mass spectrum (*blue*) obtained through CID of $PyH^+(H_2O)_5$ ions on Ar, compared to the spectrum obtained through CID of PyH^+ ions (*red*) for the same experimental conditions (Reproduced from [121]. Copyright 2015, with permission from WILEY-VCH Verlag GmbH & Co. KGaA, Weinheim)

size. The mass spectrum which is displayed in Fig. 23b shows, that in addition to water loss channels, a competition between the production of the PyH^+ ions and of H^+H_2O or $H^+(H_2O)_2$ ions occurs. The intensity of protonated water fragments increases as a function of the number of attached H_2O molecules in the parent cluster ion. It reaches a fraction of 50 % for the largest cluster studied, $PyH^+(H_2O)_5$. The results underline the importance of the proton mobility and proton transfer reactions in the relaxation of excited nano-solvated biomolecular systems.

A third example concerns the fragmentation of isolated and nano-hydrated uracil clusters induced by C^{4+} ions at 36 keV energy. The most intriguing feature observed in the nano-hydrated cluster case is the presence of hydrated fragments [122]. The series of fragments starts with the protonated HCNH (or CO) molecule, which are observed as intense fragments in the non-hydrated cluster case. Such observation of hydrated fragments was not reported in other hydrated nucleobases like thymine and adenine [123, 124]. Four hydrophilic sites are present in the uracil molecule, where the water molecules participate in the formation of two H-bonds with two adjacent H/O atoms of uracil. Moreover, when the number of water molecules increases, calculations shows that small water clusters are situated in the hydration sites [31, 125]. This is consistent with the observation of a series of hydrated HCNH (CO) fragment, where the H bonds between the water cluster and the uracil molecule survive the molecular fragmentation.

References

1. Martin S, Chen L, Denis A, Désesquelles J (1999) Phys Rev A 59(3):R1734. doi:10.1103/PhysRevA.59.R1734
2. Martin S, Bernard J, Chen L, Denis A, Désesquelles J (2000) Eur Phys J D 12(1):27. doi:10.1007/s100530070039
3. Tomita S, Lebius H, Brenac A, Chandezon F, Huber BA (2002) Phys Rev A 65(5):053201. doi:10.1103/PhysRevA.65.053201
4. Opitz J, Lebius H, Tomita S, Huber BA, Moretto Capelle P, Bordenave Montesquieu D, Bordenave Montesquieu A, Reinköster A, Werner U, Lutz HO, Niehaus A, Benndorf M, Haghighat K, Schmidt HT, Cederquist H (2000) Phys Rev A 62(2):022705. doi:10.1103/PhysRevA.62.022705
5. Rentenier A, Bordenave-Montesquieu A, Moretto-Capelle P, Bordenave-Montesquieu D (2004) J Phys B: At Mol Opt Phys 37(12):2429. doi:10.1088/0953-4075/37/12/001
6. Rentenier A, Bordenave-Montesquieu A, Moretto-Capelle P, Bordenave-Montesquieu D (2004) J Phys B: At Mol Opt Phys 37(12):2455. doi:10.1088/0953-4075/37/12/002
7. Rentenier A, Ruiz LF, Díaz-Tendero S, Zarour B, Moretto-Capelle P, Bordenave-Montesquieu D, Bordenave-Montesquieu A, Hervieux PA, Alcamí M, Politis MF, Hanssen J, Martín F (2008) Phys Rev Lett 100(18):183401. doi:10.1103/PhysRevLett.100.183401
8. Chen L, Wei B, Bernard J, Brédy R, Martin S (2005) Phys Rev A 71(4):043201. doi:10.1103/PhysRevA.71.043201
9. Reinköster A, Siegmann B, Werner U, Huber BA, Lutz HO (2002) J Phys B: At Mol Opt Phys 35(24):4989. doi:10.1088/0953-4075/35/24/304
10. de Vries J, Hoekstra R, Morgenstern R, Schlathölter T (2003) Phys Rev Lett 91(5):053401. doi:10.1103/PhysRevLett.91.053401
11. Moretto-Capelle P, Le Padellec A (2006) Phys Rev A 74(6):062705. doi:10.1103/PhysRevA.74.062705
12. Brédy R, Bernard J, Chen L, Montagne G, Li B, Martin S (2009) J Chem Phys 130(11):114305. doi:10.1063/1.3080162
13. Alvarado F, Bari S, Hoekstra R, Schlathölter T (2006) Phys Chem Chem Phys 8(16):1922. doi:10.1039/B517109A
14. Maclot S, Capron M, Maisonny R, Ławicki A, Méry A, Rangama J, Chesnel JY, Bari S, Hoekstra R, Schlathölter T, Manil B, Adoui L, Rousseau P, Huber BA (2011) ChemPhysChem 12(5):930. doi:10.1002/cphc.201000823
15. Bernigaud V, Manil B, Maunoury L, Rangama J, Huber BA (2009) Eur Phys J D 51(1):125. doi:10.1140/epjd/e2008-00153-8
16. Werner U, Beckord K, Becker J, Lutz HO (1995) Phys Rev Lett 74(11):1962. doi:10.1103/PhysRevLett.74.1962
17. Bari S, Hoekstra R, Schlatholter T (2010) Phys Chem Chem Phys 12(14):3376. doi:10.1039/B924145K
18. Schmidt HT (2015) Phys Scr 2015(T166):014063. doi:10.1088/0031-8949/2015/T166/014063
19. Toburen LH (1995) Atomic and molecular data for radiotherapy and radiation research, no. 799 in IAEA-TECDOC (International Atomic energy Agency, Vienna, 1995). In: Ionization by fast charged particles, pp 47–162. http://www-pub.iaea.org/books/IAEABooks/5444/Atomic-and-Molecular-Data-for-Radiotherapy-and-Radiation-Research
20. Toburen LH (1998) Radiat Environ Biophys 37(4):221. doi:10.1007/s004110050122
21. Coupier B, Farizon B, Farizon M, Gaillard M, Gobet F, de Castro N, Jalbert FG, Ouaskit S, Carré M, Gstir B, Hanel G, Denifl S, Feketeova L, Scheier P, Märk T (2002) Eur Phys J D 20(3):459. doi:10.1140/epjd/e2002-00166-3
22. Gobet F, Eden S, Coupier B, Tabet J, Farizon B, Farizon M, Gaillard MJ, Carré M, Ouaskit S, Märk TD, Scheier P (2004) Phys Rev A 70(6):062716. doi:10.1103/PhysRevA.70.062716
23. Tabet J, Eden S, Feil S, Abdoul-Carime H, Farizon B, Farizon M, Ouaskit S, Märk TD (2010) Phys Rev A 82(2):022703. doi:10.1103/PhysRevA.82.022703

24. Afrosimov VV, Basalaev AA, Morozov YG, Panov MN, Smirnov OV, Tropp EA (2012) Tech Phys 57(5):594. doi:10.1134/S1063784212050027
25. Afrosimov VV, Basalaev AA, Vasyutinskii OS, Panov MN, Smirnov OV (2015) Eur Phys J D 69(1):3. doi:10.1140/epjd/e2014-50435-5
26. Afrosimov VV, Basalaev AA, Morozov YG, Panov MN, Smirnov OV, Tropp EA (2013) Tech Phys 58(9):1243. doi:10.1134/S1063784213090028
27. Bacchus-Montabonel MC (2014) Appl Radiat Isotopes 83(Part B):95. doi:10.1016/j.apradiso.2012.12.017
28. Bacchus-Montabonel MC (2015) Eur Phys J D 69(4):107. doi:10.1140/epjd/e2015-60049-0
29. Hervé Du Penhoat MA, López-Tarifa P, Ghose KK, Jeanvoine Y, Gaigeot MP, Vuilleumier R, Politis MF, Bacchus-Montabonel MC (2014) J Mol Model 20(6):2221. doi:10.1007/s00894-014-2221-9
30. Errea L, Illescas C, Méndez L, Rabadán I (2014) Appl Radiat Isotopes 83(Part B):86. doi:10.1016/j.apradiso.2013.01.021
31. Bacchus-Montabonel MC, Calvo F (2015) Phys Chem Chem Phys 17(15):9629. doi:10.1039/C5CP00611B
32. Martin S, Ortega C, Chen L, Brédy R, Vernier A, Dugourd P, Antoine R, Bernard J, Reitsma G, Gonzalez-Magaña O, Hoekstra R, Schlathölter T (2014) Phys Rev A 89(1):012707. doi:10.1103/PhysRevA.89.012707
33. Niehaus A (1986) J Phys B: At Mol Phys 19(18):2925. doi:10.1088/0022-3700/19/18/021
34. Forsberg BO, Alexander JD, Chen T, Pettersson AT, Gatchell M, Cederquist H, Zettergren H (2013) J Chem Phys 138(5):054306. doi:10.1063/1.4790164
35. Boudaiffa B, Cloutier P, Hunting D, Huels MA, Sanche L (2000) Science 287(5458):1658. doi:10.1126/science.287.5458.1658
36. Incerti S, Baldacchino G, Bernal M, Capra R, Champion C, Francis Z, Guèye P, Mantero A, Mascialino B, Moretto P, Nieminen P, Villagrasa C, Zacharatou C (2010) Int J Model Simul Sci Comput 1(2):157. doi:10.1142/S1793962310000122
37. Fuss M, Sanz AG, Muñoz A, Blanco F, Téllez M, Huerga C, Garcia G (2011) Biomedical engineering, trends in electronics, communications and software. In: Laskovski AN (ed) LEPTS—a radiation-matter interaction model at the molecular level and its use in biomedical applications. InTech, pp 277–294. doi:10.5772/13061
38. Errea L, Illescas C, Méndez L, Rabadán I, Suárez J (2015) Chem Phys 462(1):17. doi:10.1016/j.chemphys.2015.08.009
39. Nandi S, Biswas S, Khan A, Monti JM, Tachino CA, Rivarola RD, Misra D, Tribedi LC (2013) Phys Rev A 87(5):052710. doi:10.1103/PhysRevA.87.052710
40. Rivarola RD, Galassi ME, Fainstein PD, Champion C (2013) Theory of heavy ion collision physics in hadron therapy. Computation of distorted wave cross sections for high-energy inelastic collisions of heavy ions with water molecules. In: Belkić D (ed) Advances in quantum chemistry, vol 65. Academic Press, Amsterdam, pp 231–267. doi:10.1016/B978-0-12-396455-7.00009-1
41. Champion C, Hanssen J, Rivarola RD (2013) Theory of heavy ion collision physics in hadron therapy. In: Belkić D (ed) Advances in quantum chemistry, vol 65. Academic Press, Amsterdam. The first born approximation for ionization and charge transfer in energetic collisions of multiply charged ions with water, pp 269–313. doi:10.1016/B978-0-12-396455-7.00010-8
42. Tachino CA, Monti JM, Fojón OA, Champion C, Rivarola RD (2014) J Phys B: At Mol Opt Phys 47(3):035203. doi:10.1088/0953-4075/47/3/035203
43. Bhattacharjee S, Biswas S, Bagdia C, Roychowdhury M, Nandi S, Misra D, Monti JM, Tachino CA, Rivarola RD, Champion C, Tribedi LC (2016) J Phys B: At Mol Opt Phys 49(6):065202. doi:10.1088/0953-4075/49/6/065202
44. Ohsawa D, Tawara H, Soga F, Galassi ME, Rivarola RD (2014) J Phys: Conf Ser 488(10):102030. doi:10.1088/1742-6596/488/10/102030
45. Murakami M, Kirchner T, Horbatsch M, Lüdde HJ (2012) Phys Rev A 85(5):052713. doi:10.1103/PhysRevA.85.052713

46. Iriki Y, Kikuchi Y, Imai M, Itoh A (2011) Phys Rev A 84(3):032704. doi:10.1103/PhysRevA.84.032704
47. Champion C, Galassi ME, Fojón O, Lekadir H, Hanssen J, Rivarola RD, Weck PF, Agnihotri AN, Nandi SN, Tribedi LC (2012) J Phys: Conf Ser 373(1):012004. doi:10.1088/1742-6596/373/1/012004
48. Tribedi LC, Agnihotri AN, Galassi ME, Rivarola RD, Champion C (2012) Eur Phys J D 66(11):1. doi:10.1140/epjd/e2012-30330-y
49. Agnihotri AN, Kasthurirangan S, Nandi S, Kumar A, Champion C, Lekadir H, Hanssen J, Weck PF, Galassi ME, Rivarola RD, Fojón O, Tribedi LC (2013) J Phys B: At Mol Opt Phys 46(18):185201. doi:10.1088/0953-4075/46/18/185201
50. Agnihotri AN, Nandi S, Kasthurirangan S, Kumar A, Galassi ME, Rivarola RD, Champion C, Tribedi LC (2013) Phys Rev A 87(3):032716. doi:10.1103/PhysRevA.87.032716
51. Itoh A, Iriki Y, Imai M, Champion C, Rivarola RD (2013) Phys Rev A 88(5):052711. doi:10.1103/PhysRevA.88.052711
52. Nandi S (2014) Collisions of highly charged ions with small molecules: some new facets. Ph.D. thesis, Tata Institute of Fundamental Research, Mumbai
53. Monti JM, Fojón OA, Hanssen J, Rivarola RD (2010) J Phys B: At Mol Opt Phys 43(20):205203. doi:10.1088/0953-4075/43/20/205203
54. de Vera P, Garcia-Molina R, Abril I, Solov'yov AV (2013) Phys Rev Lett 110(14):148104. doi:10.1103/PhysRevLett.110.148104
55. de Vera P, Garcia-Molina R, Abril I (2015) Phys Rev Lett 114(1):018101. doi:10.1103/PhysRevLett.114.018101
56. Olson RE, Berkner KH, Graham WG, Pyle RV, Schlachter AS, Stearns JW (1978) Phys Rev Lett 41(3):163. doi:10.1103/PhysRevLett.41.163
57. Bethe H (1930) Ann Phys 397(3):325. doi:10.1002/andp.19303970303
58. Tribedi LC (2016) Private communications
59. Wilson WE, Toburen LH (1975) Phys Rev A 11(4):1303. doi:10.1103/PhysRevA.11.1303
60. Toburen LH, Wilson WE, Porter LE (1977) J Chem Phys 67(9):4212. doi:10.1063/1.435401
61. Montenegro EC, Sigaud GM, DuBois RD (2013) Phys Rev A 87(1):012706. doi:10.1103/PhysRevA.87.012706
62. Wilhelm RA, El-Said AS, Krok F, Heller R, Gruber E, Aumayr F, Facsko S (2015) Prog Surf Sci 90(3):377. doi:10.1016/j.progsurf.2015.06.001
63. de Vries J, Hoekstra R, Morgenstern R, Schlathölter T (2002) J Phys B: At Mol Opt Phys 35(21):4373. doi:10.1088/0953-4075/35/21/304
64. Bari S, Sobocinski P, Postma J, Alvarado F, Hoekstra R, Bernigaud V, Manil B, Rangama J, Huber B, Schlathölter T (2008) J Chem Phys 128(7):074306. doi:10.1063/1.2830032
65. Schlathölter T (2012) Radiation damage in biomolecular systems. In: Gómez-Tejedor GG, Fuss M (eds) Biological and medical physics, biomedical engineering. Springer Netherlands, Dordrecht, Heidelberg, London, New York. Ion-induced radiation damage in biomolecular systems, pp 177–190. doi:10.1007/978-94-007-2564-5
66. Guilhaus M (1995) J Mass Spectrom 30(11):1519. doi:10.1002/jms.1190301102
67. Capron M, Díaz-Tendero S, Maclot S, Domaracka A, Lattouf E, Ławicki A, Maisonny R, Chesnel JY, Méry A, Poully JC, Rangama J, Adoui L, Martín F, Alcamí M, Rousseau P, Huber BA (2012) Chem-Eur J 18(30):9321. doi:10.1002/chem.201103922
68. Fenn J, Mann M, Meng C, Wong S, Whitehouse C (1989) Science 246(4926):64. doi:10.1126/science.2675315
69. da Silva H, Oller J, Gatchell M, Stockett MH, Hervieux PA, Adoui L, Alcamí M, Huber BA, Martín F, Cederquist H, Zettergren H, Rousseau P, Díaz-Tendero S (2014) Phys Rev A 90(3):032701. doi:10.1103/PhysRevA.90.032701
70. Murakami M, Kirchner T, Horbatsch M, Lüdde HJ (2012) Phys Rev A 86(2):022719. doi:10.1103/PhysRevA.86.022719
71. Illescas C, Errea LF, Méndez L, Pons B, Rabadán I, Riera A (2011) Phys Rev A 83(5):052704. doi:10.1103/PhysRevA.83.052704
72. Suárez J, Méndez L, Rabadán I (2015) J Phys Chem Lett 6(1):72. doi:10.1021/jz5022894

73. Sadr Arani L, Mignon P, Abdoul-Carime H, Farizon B, Farizon M, Chermette H (2012) Phys Chem Chem Phys 14(28):9855. doi:10.1039/C2CP40384F
74. Maclot S, Piekarski DG, Domaracka A, Méry A, Vizcaino V, Adoui L, Martín F, Alcamí M, Huber BA, Rousseau P, Díaz-Tendero S (2013) J Phys Chem Lett 4(22):3903. doi:10.1021/jz4020234
75. López-Tarifa P, Hervé du Penhoat MA, Vuilleumier R, Gaigeot MP, Tavernelli I, Le Padellec A, Champeaux JP, Alcamí M, Moretto-Capelle P, Martín F, Politis MF (2011) Phys Rev Lett 107(2):023202. doi:10.1103/PhysRevLett.107.023202
76. Tabet J, Eden S, Feil S, Abdoul-Carime H, Farizon B, Farizon M, Ouaskit S, Märk T (2010) Int J Mass Spectrom 292(1–3):53. doi:10.1016/j.ijms.2010.03.002
77. Smirnov OV, Basalaev AA, Boitsov VM, Vyaz'min SY, Orbeli AL, Dubina MV (2014) Tech Phys 59(11):1698. doi:10.1134/S1063784214110231
78. Rudek B, Arndt A, Bennett D, Wang M, Rabus H (2015) Eur Phys J D 69(10):237. doi:10.1140/epjd/e2015-60204-7
79. Wolff W, Luna H, Sigaud L, Tavares AC, Montenegro EC (2014) J Chem Phys 140(6):064309. doi:10.1063/1.4864322
80. Champeaux JP, Carcabal P, Rabier J, Cafarelli P, Sence M, Moretto-Capelle P (2010) Phys Chem Chem Phys 12(20):5454. doi:10.1039/B926803K
81. Delaunay R, Champeaux JP, Maclot S, Capron M, Domaracka A, Méry A, Manil B, Adoui L, Rousseau P, Moretto-Capelle P, Huber BA (2014) Eur Phys J D 68(6):162. doi:10.1140/epjd/e2014-50073-y
82. Wardman P (2007) Clin Oncol 19(6):397. doi:10.1016/j.clon.2007.03.010
83. Poully JC, Miles J, De Camillis S, Cassimi A, Greenwood JB (2015) Phys Chem Chem Phys 17(11):7172. doi:10.1039/C4CP05303F
84. Mishra P, Kadhane U (2014) Nucl Instrum Meth B 336(1):12. doi:10.1016/j.nimb.2014.06.007
85. Chen L, Martin S, Bernard J, Brédy R (2007) Phys Rev Lett 98(19):193401. doi:10.1103/PhysRevLett.98.193401
86. Alvarado F, Bernard J, Li B, Brédy R, Chen L, Hoekstra R, Martin S, Schlathölter T (2008) ChemPhysChem 9(9):1254. doi:10.1002/cphc.200800110
87. Chen L, Brédy R, Bernard J, Montagne G, Allouche AR, Martin S (2011) J Chem Phys 135(11):114309. doi:10.1063/1.3621713
88. Piekarski DG, Delaunay R, Maclot S, Adoui L, Martin F, Alcami M, Huber BA, Rousseau P, Domaracka A, Diaz-Tendero S (2015) Phys Chem Chem Phys 17(26):16767. doi:10.1039/C5CP01628B
89. Kocisek J, Piekarski DG, Delaunay R, Huber BA, Adoui L, Martín F, Alcamí M, Rousseau P, Domaracka A, Kopyra J, Díaz-Tendero S (2015) J Phys Chem A 119(37):9581. doi:10.1021/acs.jpca.5b06009
90. Bari S, Hoekstra R, Schlatholter T (2011) Int J Mass Spectrom 299(1):64. doi:10.1016/j.ijms.2010.09.019
91. González-Magaña O, Tiemens M, Reitsma G, Boschman L, Door M, Bari S, Lahaie PO, Wagner JR, Huels MA, Hoekstra R, Schlathölter T (2013) Phys Rev A 87(3):032702. doi:10.1103/PhysRevA.87.032702
92. Ha DT, Huels MA, Huttula M, Urpelainen S, Kukk E (2011) Phys Rev A 84(3):033419. doi:10.1103/PhysRevA.84.033419
93. Herve du Penhoat MA, Ghose KK, Gaigeot MP, Vuilleumier R, Fujii K, Yokoya A, Politis MF (2015) Phys Chem Chem Phys 17(48):32375. doi:10.1039/C5CP05196G
94. Chingin K, Makarov A, Denisov E, Rebrov O, Zubarev RA (2014) Anal Chem 86(1):372. doi:10.1021/ac403193k
95. Hoffmann WD, Jackson GP (2014) J Am Soc Mass Spectrom 25(11):1939. doi:10.1007/s13361-014-0989-6
96. Zimmermann U, Malinowski N, Näher U, Frank S, Martin TP (1994) Phys Z D 31(1):85. doi:10.1007/BF01426583

97. Schlathölter T, Alvarado F, Bari S, Lecointre A, Hoekstra R, Bernigaud V, Manil B, Rangama J, Huber B (2006) ChemPhysChem 7(11):2339. doi:10.1002/cphc.200600361
98. Deng Z, Bald I, Illenberger E, Huels M (2005) Phys Rev Lett 95(15):153201. doi:10.1103/PhysRevLett.95.153201
99. Jochims HW, Schwell M, Baumgärtel H, Leach S (2005) Chem Phys 314(1–3):263. doi:10.1016/j.chemphys.2005.03.008
100. Kelly REA, Kantorovich LN (2006) J Phys Chem B 110(5):2249. doi:10.1021/jp055552o
101. Manil B, Maunoury L, Huber BA, Jensen J, Schmidt HT, Zettergren H, Cederquist H, Tomita S, Hvelplund P (2003) Phys Rev Lett 91(21):215504. doi:10.1103/PhysRevLett.91.215504
102. Bari S, Alvarado F, Postma J, Sobocinski P, Hoekstra R, Schlathölter T (2008) Eur Phys J D 51(1):81. doi:10.1140/epjd/e2008-00054-x
103. Maclot S, Piekarski DG, Delaunay R, Domaracka A, Méry A, Vizcaino V, Chesnel JY, Martín F, Alcamí M, Huber BA, Adoui L, Rousseau P, Díaz-Tendero S (2014) Eur Phys J D 68(6):149. doi:10.1140/epjd/e2014-40819-x
104. Poully JC, Vizcaino V, Schwob L, Delaunay R, Kocisek J, Eden S, Chesnel JY, Mry A, Rangama J, Adoui L, Huber B (2015) ChemPhysChem 16(11):2389. doi:10.1002/cphc.201500275
105. Bouchoux G, Huang S, Inda BS (2011) Phys Chem Chem Phys 13:651. doi:10.1039/C0CP00775G
106. von Sonntag C (1987) The chemical basis of radiation biology. Taylor & Francis, New York
107. Farhataziz MAJ (1987) Rodgers, radiation chemistry: principles and applications. VCH Publishers, New York
108. Tachikawa H (2004) J Phys Chem A 108(39):7853. doi:10.1021/jp049269l
109. Jahnke T, Sann H, Havermeier T, Kreidi K, Stuck C, Meckel M, Schoffler M, Neumann N, Wallauer R, Voss S, Czasch A, Jagutzki O, Malakzadeh A, Afaneh F, Weber T, Schmidt-Bocking H, Dorner R (2010) Nat Phys 6(2):139. doi:10.1038/nphys1498
110. Mucke M, Braune M, Barth S, Forstel M, Lischke T, Ulrich V, Arion T, Becker U, Bradshaw A, Hergenhahn U (2010) Nat Phys 6(2):143. doi:10.1038/nphys1500
111. Legendre S (2006) Étude de l'ionisation et de la dissociation d'H_2O induites par collision avec des ions multichargés rapides. Ph.D. thesis, université de Caen Basse-Normandie. https://tel.archives-ouvertes.fr/tel-00090898
112. Maisonny R, Capron M, Maclot S, Poully JC, Domaracka A, Méry A, Adoui L, Rousseau P, Huber BA (2013) J Phys: Conf Ser 438(1):012007. doi:10.1088/1742-6596/438/1/012007
113. Abdoul-Carime H, Berthias F, Feketeová L, Marciante M, Calvo F, Forquet V, Chermette H, Farizon B, Farizon M, Mrk TD (2015) Angew Chem Int Ed 54(49):14685. doi:10.1002/anie.201505890
114. Adoui L, Cassimi A, Gervais B, Grandin JP, Guillaume L, Maisonny R, Legendre S, Tarisien M, López-Tarifa P, Politis MF, du Penhoat MAH, Vuilleumier R, Gaigeot MP, Tavernelli I, Alcamí M, Martín F (2009) J Phys B: At Mol Opt Phys 42(7):075101. doi:10.1088/0953-4075/42/7/075101
115. Larsen MC, Hvelplund P, Larsson MO, Shen H (1999) Eur Phys J D 5(2):283. doi:10.1007/s100530050257
116. Zettergren H, Rousseau P, Wang Y, Seitz F, Chen T, Gatchell M, Alexander JD, Stockett MH, Rangama J, Chesnel JY, Capron M, Poully JC, Domaracka A, Méry A, Maclot S, Schmidt HT, Adoui L, Alcamí M, Tielens AGGM, Martín F, Huber BA, Cederquist H (2013) Phys Rev Lett 110(18):185501. doi:10.1103/PhysRevLett.110.185501
117. Stockett MH, Zettergren H, Adoui L, Alexander JD, Bērziņš U, Chen T, Gatchell M, Haag N, Huber BA, Hvelplund P, Johansson A, Johansson HAB, Kulyk K, Rosén S, Rousseau P, Støchkel K, Schmidt HT, Cederquist H (2014) Phys Rev A 89(3):032701. doi:10.1103/PhysRevA.89.032701
118. Delaunay R, Gatchell M, Rousseau P, Domaracka A, Maclot S, Wang Y, Stockett MH, Chen T, Adoui L, Alcamí M, Martín F, Zettergren H, Cederquist H, Huber BA (2015) J Phys Chem Lett 6(9):1536. doi:10.1021/acs.jpclett.5b00405

119. Liu B, Nielsen SB, Hvelplund P, Zettergren H, Cederquist H, Manil B, Huber BA (2006) Phys Rev Lett 97(13):133401. doi:10.1103/PhysRevLett.97.133401
120. Liu B, Haag N, Johansson H, Schmidt HT, Cederquist H, Brøndsted Nielsen S, Zettergren H, Hvelplund P, Manil B, Huber BA (2008) J Chem Phys 128(7):075102. doi:10.1063/1.2839597
121. Berthias F, Feketeová L, Chermette H, Forquet V, Morell C, Abdoul-Carime H, Farizon B, Farizon M, Märk TD (2015) ChemPhysChem 16(15):3151. doi:10.1002/cphc.201500465
122. Markush P, Bolognesi P, Cartoni A, Rousseau P, Maclot S, Delaunay R, Domaracka A, Kocisek J, Castrovilli MC, Huber BA, Avaldi L (2016) Phys Chem Chem Phys 18(25):16721. doi:10.1039/C6CP01940D
123. Domaracka A, Capron M, Maclot S, Chesnel JY, Méry A, Poully JC, Rangama J, Adoui L, Rousseau P, Huber BA (2012) J Phys: Conf Ser 373(1):012005. doi:10.1088/1742-6596/373/1/012005
124. Kim NJ, Kim YS, Jeong G, Ahn TK, Kim SK (2002) Int J Mass Spectrom 219(1):11. doi:10.1016/S1387-3806(02)00547-X
125. Wang FF, Zhao DX, Yang ZZ (2009) Chem Phys 360(1–3):141. doi:10.1016/j.chemphys.2009.04.022

Dissociative Electron Attachment to Biomolecules

Ilko Bald, Roman Čurík, Janina Kopyra and Michal Tarana

Abstract During the last two decades it became clear that a significant fraction of the biological cellular damage caused by high-energy radiation is actually due to reactions induced by low-energy electrons (<20 eV). In this energy regime electrons can efficiently decompose molecules such as DNA or DNA building blocks by dissociative electron attachment (DEA). Experiments on single DNA building blocks have been performed in the gas phase revealing that DEA can proceed with remarkable site selectivity. Low-energy electron-induced DNA strand breakage is typically investigated using plasmid DNA in the condensed phase. Very recently, a pronounced dependence of electron induced DNA strand breakage on the nucleotide sequence was found using different experimental approaches suggesting that at least part of the observed strand breaks are due to initial electron attachment to the nucleobases. Currently, a strong research focus is on the fundamental understanding of DEA to therapeutically administered radiosensitizers. In the near future DEA to novel potential radiosensitizers will be explored, and the electron induced damage of biomolecules within complex environments has to be investigated. Considerable attention has been paid to the theoretical research of the DEA in the context of the DNA damage. With respect to this, the theoretical part of the chapter reviews all the computational approaches that have been used to study DEA to biomolecules over the last decade. These approaches are divided into two classes. The first class

I. Bald (✉)
Institute of Chemistry—Physical Chemistry, University of Potsdam, Karl-Liebknecht-Str. 24-25, D-14476 Potsdam, Germany
e-mail: bald@uni-potsdam.de

I. Bald
BAM Federal Institute for Materials Research and Testing, Richard-Willstätter Str. 11, D-12489 Berlin, Germany

R. Čurík · M. Tarana
J. Heyrovský Institute of Physical Chemistry of the ASCR, v.v.i., Dolejš kova 2155/3, 182 23 Prague 8, Czech Republic

J. Kopyra
Faculty of Sciences, Siedlce University, 3 Maja 54, 08-110 Siedlce, Poland

A.V. Solov'yov (ed.), *Nanoscale Insights into Ion-Beam Cancer Therapy*, DOI 10.1007/978-3-319-43030-0_5 consists of electronic structure methods studying the transient negative ions formed by electrons captured by the neutral building blocks of the DNA. Approaches dealing with the complicated nuclear dynamics of the DEA to biomolecules form the second class explored in this chapter.

1 Introduction

Low-energy electrons (LEEs) are produced in copious amounts in biological media along the ionization track of high-energy radiation, which is for instance used in tumor radiation therapy [1]. It was shown in Monte Carlo simulations of proton radiation in the MeV range impinging on water, that more than 80 % of the energy of the primary radiation is transferred to electrons mainly by ionization of water molecules [2]. When considering only a single ionization event per primary particle (protons in the MeV range) the most probable energy of LEEs is around 10 eV, and further secondary ionization processes can shift the energy distribution further to lower energies. In the traditional notion of radiation chemistry and biology these LEEs were considered inactive, but in 2000 it was shown by the group of Leon Sanche that LEEs can induce single strand breaks (SSBs) and double strand breaks (DSBs) in plasmid DNA [3]. The strand break (SB) yields are displayed in Fig. 1 and show very characteristic maxima at specific electron energies below 20 eV with the most pronounced maximum for SSBs and DSBs around 8 and 10 eV, respectively. Later on, the experiments have been extended to the lower energy regime of 0–4 eV, where again a pronounced maximum of SSBs was observed just below 1 eV [4]. This is a remarkable result, since it demonstrates that LEE induced SBs are even possible below the threshold for electronic excitation. The clear resonant structures in the SB yield curves indicate that the basic mechanism leading to strand breakage is dissociative electron attachment (DEA).

The DEA process can be illustrated in a simplified two-dimensional potential energy diagram as shown in Fig. 2 [5]. The black curve represents the total energy E of the neutral ground state of a complex multi-atom molecule denoted as ABCD. Electron attachment is fast compared to the movement of the nuclei, and thus the transition to the anionic potential energy curve (blue) is vertical (Franck-Condon transition) resulting in the formation of a transient negative ion (TNI). Since the potential energy curve of the anionic state is typically repulsive within the Franck-Condon region a transient negative ion is formed, which is unstable towards autodetachment (loss of the electron) and dissociation (DEA):

$$
\begin{array}{lll}
e^- + ABCD \rightarrow ABCD^{*-} & & \text{(TNI)} \\
 & \rightarrow ABCD + e^- & \text{(autodetachment)} \\
 & \rightarrow AB + CD^- & \text{(dissociation)}
\end{array}
$$

Autodetachment and dissociation are competitive processes. Autodetachment can take place until the crossing point of the anionic and neutral potential energy curves

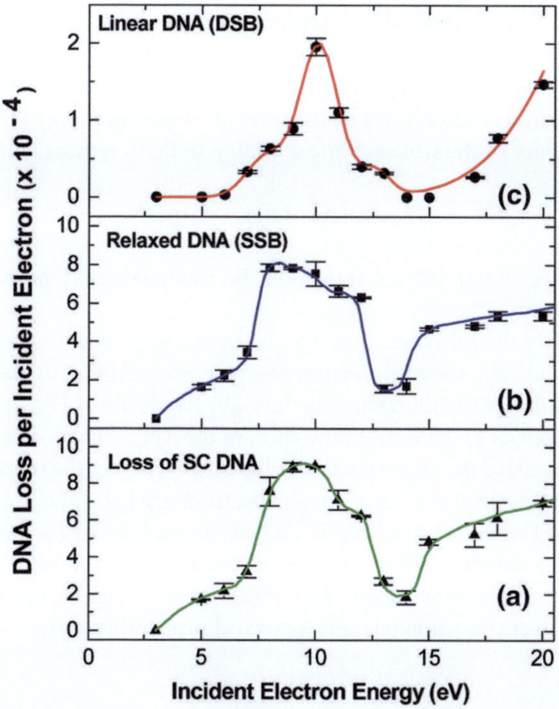

Fig. 1 First experimental evidence of low-energy electron induced DNA single- and double strand breaks (SSBs and DSBs) via dissociative electron attachment (DEA). The experiments have been performed using plasmid DNA and analysis of strand breaks by gel electrophoresis. Adapted from Ref. [3], with permission of AAAS

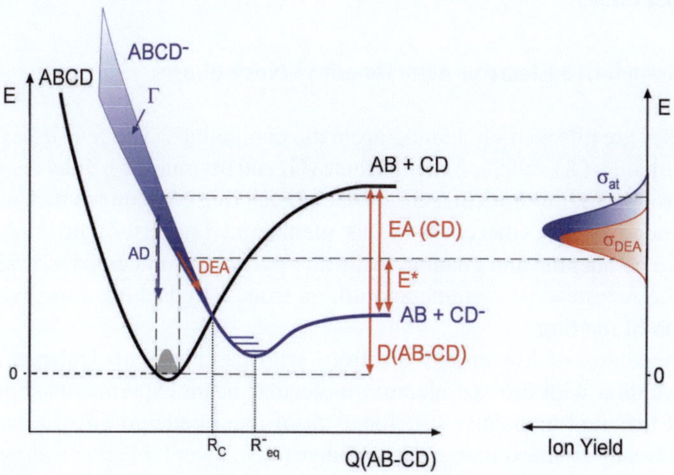

Fig. 2 Schematic two-dimensional potential energy diagram illustrating the formation of the transient negative ion and the competition between autodetachment (AD) and dissociation (DEA). Adapted from Ref. [5], with permission of Elsevier

is reached (R_C). After passing R_C the return to the neutral state is no longer possible, and the anion dissociates for instance along the B–C bond as is shown in Fig. 2 to form the stable anion CD^- and the neutral radical AB. The thermodynamic threshold (ΔH) of the reaction is determined by the bond dissociation energy (D(AB–CD)) and the electron affinity of the neutral corresponding to the fragment anion (EA(CD)):

$$\Delta H = D(\text{AB–CD}) - EA(\text{CD}).$$

Depending on the initial vertical transition the fragments may possess some additional excess energy E^*.

Experimentally the DEA reaction may be observed as a strand break in the case of DNA that occurs at a specific electron energy corresponding to the initial Franck-Condon transition. This is illustrated on the right-hand side of Fig. 2. The resonance structure corresponds to the initial formation of the TNI. The most straight-forward experiment to describe the DEA reaction is the observation of the fragment anion as a function of the electron energy using mass spectrometry [5]. Such experiment results in an ion yield curve as shown in Fig. 2. The resonance profiles for the fragmentation reactions may be slightly shifted to lower electron energies when compared to the resonance for the formation of the TNI since the dissociation occurs with higher probability at lower transition energies (survival probability shift).

2 Experimental Research

2.1 Dissociative Electron Attachment to Gas-Phase Molecules

2.1.1 Dissociative Electron Attachment to Nucleobases

Nucleobases are nitrogen-containing aromatic compounds. The four bases found in DNA are adenine (A), cytosine (C), guanine (G) and thymine (T). They are classified into two groups; pyrimidines that are six-membered ring compounds such as thymine and cytosine and purines that consist of six-membered rings fused with imidazol rings represented by adenine and guanine. A further pyrimidine nucleobase, uracil (U), is found in RNA instead of thymine and differs from T by lacking a methyl group at C5 position of the ring.

The interaction of low energy electrons with gas phase nucleobases A, C, G, T and U studied with crossed electron-molecular beam experiments showed that, in spite of different molecular structures, all of them undergo DEA at two energy domains, already at subexcitation energy range ($<$3 eV) and at higher energy between 6–12 eV exhibiting quite similar fragmentation patterns [7–9]. At this low energy range the molecules exclusively generate a closed shell dehydrogenated parent anion $(\text{Nb–H})^-$ (where Nb = nucleobase) which is most likely formed with preservation of the cyclic structure. On the contrary, at higher energy the formation of the fragment anions is mainly associated with deterioration of the ring structure, which results in the formation of a variety of light fragments.

Later on, more detailed studies in thymine revealed an unexpected bond and site selectivity for the loss of neutral hydrogen atom. In general, the H losses can occur either from the N site (position 1 and 3) or from the C site (position C6 and methyl group at C5 position). From the studies on partially deuterated thymine at the C positions it appears that the loss of H atom occurs exclusively from the N sites [9, 10]. Further studies on thymine methylated at the N1 position and uracil methylated at the N3 position revealed that hydrogen loss is also subjected to remarkable site selectivity as electrons of 1 eV energy exclusively induce N1–H bond cleavage whilst electrons of 1.8 eV energy preferentially induce N3–H bond cleavage (Fig. 3) [6].

Besides the prominent selectivity in the low energy range associated with the loss of the neutral hydrogen atom mentioned above it has been shown that bond selectivity is also observed in the high energy domain (>5 eV) and visible via formation of the complementary H^- ion. The H^- ion yield in the case of non-labeled thymine comprises four overlapping peaks appearing in the energy range between 5 and 12 eV [11]. In the case of deuterated thymine (T_D) at the C positions the loss of D^- is almost completely quenched within the first two peaks (at 5.5 and 7 eV) and is mainly generated above 7.5 eV. On the other hand the loss of the H^- ion from the N positions is already observed at 5.5 eV and extends until 12 eV, however with a very low intensity at the energy range where the C–H bond cleavage occurs (i.e., above 7.5 eV).

More recently remarkable site selectivity has been reported for a much more complex reaction resulting in the formation of NCO^- from thymine, uracil and their N-methylated derivatives [12]. Utilizing a variety of methods it has been shown that

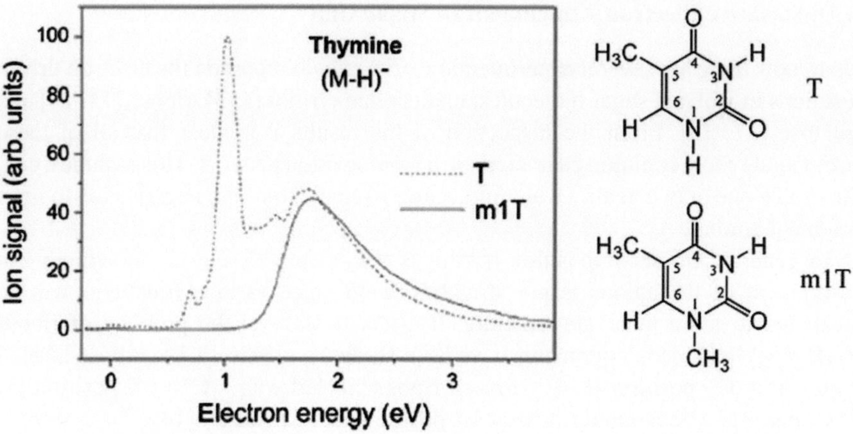

Fig. 3 Selective abstraction of neutral hydrogen from thymine (T) and N1-methylthymine (m1T). Adapted from Ref. [6]

this pseudohalogen anion is generated via both a prompt dissociation of transient negative ion and metastable decay of a dehydrogenated parent anion. Remarkably, it has been shown that site selectivity observed for the formation of $(Nb–H)^-$ is preserved for its subsequent decay reaction into NCO^-. It has been postulated that this effect is related to different electronic states of the TNI subjected to metastable decay.

From the material presented above it is obvious that nucleobases are sensitive towards electron attachment and the decomposition process is both bond and site selective. Apart from electron attachment to Nbs, electrons can attach directly to the DNA backbone causing single and double strand breaks. In the following we will present the results obtained for the DNA backbone components namely sugar and phosphate moieties.

2.1.2 Electron Attachment to Model Compounds for the DNA Backbone

The DNA backbone is composed of a monosaccharide sugar called 2′-deoxy-D-ribose and a phosphate unit, which are covalently bound via the C5′–O–P and C3′–O–P ester bonds to form an alternating sugar-phosphate backbone. The backbone further couples to the nucleobase via N-glycosidic bond existing between the sugar and the base. Naturally, the cleavage of any of the C–O–P ester bonds and the C4′–C5′ bond within the sugar-phosphate unit would represent a single strand break. Additionally, degradation of the sugar unit within DNA would also result in base release causing destabilization of the DNA backbone. The sensitivity of the backbone components with respect to LEEs is therefore of particular interest as decomposition reactions within this part of DNA could represent one of the most direct mechanism for SSBs.

A Dissociative Electron Attachment to Sugar Unit

Up to now there are several experimental works, which report on the electron driven reactions in isolated sugar molecules, namely deoxyribose [13], ribose [14–16] and fructose [17, 18]. From the inspection of the results it is clear that all of these compounds show common characteristic decomposition patterns. This includes, e.g., the loss of one or two neutral water molecules from the transient negative ion to form fragment anions.

In general, the decomposition pattern is very complex and the stoichiometric assignment of the anions is not straightforward since various fragments can be attributed to more than one structure. In order to unravel the problem of ribose ($C_5H_{10}O_5$) its three isotopomeres have been studied extensively, i.e., ribose labeled with ^{13}C at C1 position (1-^{13}C-ribose), ribose labeled with ^{13}C at C5 position (5-^{13}C-ribose) and deuterated ribose at C1 position (C,1-D-ribose) [14].

In non-labeled ribose the following dissociation reactions at around 0 eV have been observed:

$$
\begin{aligned}
e^- + C_5H_{10}O_5 \rightarrow C_5H_{10}O_5^{*-} &\rightarrow C_5H_8O_4^- + H_2O && (m/z\ 132)\\
&\rightarrow C_5H_6O_3^- + 2H_2O && (m/z\ 114)\\
&\rightarrow C_4H_6O_3^- + CH_2O + H_2O && (m/z\ 102)\\
&\rightarrow C_4H_5O_3^- + CH_2O + H_2O + H && (m/z\ 101)\\
&\rightarrow C_3H_4O_2^- + 2CH_2O + H_2O && (m/z\ 72)\\
&\rightarrow C_3H_3O_2^- + 2CH_2O + H_2O + H && (m/z\ 71)\\
&\rightarrow C_2H_3O_2^- + 3CH_2O + H && (m/z\ 59)\\
&\rightarrow HCOO^- + \text{neutral fragments} && (m/z\ 45),
\end{aligned}
$$

where $C_5H_{10}O_5^{*-}$ corresponds to the TNI. As already mentioned the predominant reactions are due to the loss of one or two water molecules. Besides, there are several other fragment anions, which are produced with significant intensities. The loss of water molecules most likely leads to the formation of anions with cyclic structure whilst the other fragments arise from the degradation of the sugar ring.

As for the loss of carbon-containing neutral counterparts it was found that the reactions are highly selective and they almost exclusively proceed by the cleavage of the C5 containing units. As shown for the exemplary anionic fragments detected at m/z 101 and 102 the same masses are observed for non-labeled ribose and 5-^{13}C-ribose. For the other isotopomers, 1-^{13}C-ribose and C,1-D-ribose, the signals are shifted by one mass unit respectively to m/z 102 and to m/z 103 indicating that the C1 atom remains on the anionic fragment.

Apart from these low energy resonances ribose decomposes via high energy features in the energy range of 5–9.5 eV [15]. These features attributed to shape resonances are visible via formation of the anionic fragments at m/z 72 ($C_3H_4O_2^-$) and 59 ($C_2H_3O_2^-$) and formation of the OH^- ion. It appears that the decomposition of the high energy transient anions is no longer as selective as that observed for the TNIs generated at energies below 1 eV. To exemplify, for non-labeled ribose and 5-^{13}C-ribose the $C_2H_3O_2^-$ ion appears exclusively at m/z 59. For 1-^{13}C-ribose and deuterated ribose at C1 position $C_2H_3O_2^-$ appears at m/z 59 and 60 and both signals exhibit the same intensity. It follows from this that there is no longer a strong tendency for the localization of the excess charge on C1 at energies above 5 eV.

B Electron Attachment to a More Complex Sugar Surrogate. The Case of Tetra-Acetylribose

The results obtained for ribose clearly show that the fragmentation is site selective and leads preferentially to the excision of C5 containing units. However, ribose exists as a six-membered heterocyclic ring. Thus, in order to better mimic the behavior of the sugar unit in DNA/RNA 1,2,3,5-tetra-O-acetyl-β-D-ribofuranose (tetra-acetylribose, TAR) has been investigated [19]. Like the sugar unit in DNA/RNA, tetra-acetylribose is a five-membered ring in which acetyl groups mimic both the coupling to the neighbouring phosphate groups and to the nucleobase.

The predominant signal from TAR was observed at m/z 59 and identified as the closed shell acetate anion CH_3COO^-. It is generated within three energy domains via a sharp peak close to 0 eV, a broader peak between 1 and 3 eV and via a series of overlapping structures in the energy range 7–10 eV. The most direct pathway for its formation is the rupture of one of the C–O bonds at C1, C2, C3 or C5 position.

The stoichiometric and structural assignment of the fragments recorded at higher masses is not straightforward since such fragments are likely the result of structural and electronic rearrangement in the precursor ion. One group of these ions has been observed at energies close to 0 eV with an additional weaker and broader structure in the energy range of 7–11 eV. This group includes ions detected at m/z 100 ($C_4H_4O_3^-/C_5H_8O_2^-$), 84 ($C_4H_4O_2^-/C_5H_8O^-$) and 70 ($C_4H_6O^-$). All of these fragment anions could represent five-membered ring radical anions for which the shape and the position of the resonances resemble those obtained from ribose. The remaining fragments were observed at m/z 215 ($C_9H_{11}O_6^-$), 161 ($C_6H_9O_5^-$), 154 ($C_7H_6O_4^-$), 119 ($C_4H_7O_4^-$) and 113 ($C_5H_5O_3^-$). They appear within the energy range of 1–2.5 eV and have been attributed to π^* shape resonances localized outside the ribose ring [19].

The results indicate that processes triggered by electron attachment to the sugar moiety may play a significant role in the formation of single strand breaks. Especially the group of anionic fragments, which are formed at energies near 0 eV may be associated with the degradation of the sugar ring. However, the strongest signal is due to the formation of the acetate anion, which is generated from the cleavage of a single C–O bond. The cleavage of a C5'–O or C3'–O bond within DNA would correspond to a SSB. On the other hand, cleavage of a C1'–O bond can mimic an excision of a DNA base. Interestingly, the molecule exhibits further core excited resonances and thus may also be active in a more complex route to form DSBs.

C Dissociative Electron Attachment to the Phosphate Unit: Dibutylphosphate and Triethylphosphate

The first experimental gas phase studies on low energy electron attachment to phosphoric acid derivatives concern dibutylphosphate (DBP) and triethylphosphate (TEP). Both molecules are subjected to various DEA reactions at different energies, from 0 eV up to 10 eV. The predominant reaction in DBP is loss of a neutral hydrogen atom resulting in the formation of $(DBP–H)^-$ that is visible via an intense resonance at 1 eV [20]. The corresponding reaction was not observed for TEP indicating that hydrogen loss in DBP occurs from the hydroxyl group. However, the most significant reaction within the context of DNA damage is the loss of an entire neutral butyl group leading to an ion at m/z 153 $(DBP–C_4H_9)^-$. This DEA reaction takes place within the two energy ranges 2–4 eV and 7–10 eV. A similar reaction was observed for TEP, i.e., the loss of C_2H_5O radical due to P–O bond rupture. In this case, however, only a high energy resonance between 7–9 eV was generated. Both reactions, the loss of the butyl group or the ethoxy radical from DBP and TEP, respectively, would correspond to a strand break in DNA.

A further set of reactions was observed from both DBP and TEP and associated with the excision of P containing fragments already at very low energy < 1 eV. For DBP this results in the formation of $((C_3H_7O)P(O)(OH)_2)^-$, $H_2PO_3^-$, PO_3^- and PO^-, whilst for TEP only $H_2PO_3^-$ and PO_3^- were observed. These anions are due to complex reactions that require multiple bond breaking, which would lead to a strand break in DNA.

2.1.3 Dissociative Electron Attachment to Larger DNA Units

The gas phase experiments on nucleobases, sugar and phosphate surrogates show that all of them are fragile with respect to low energy electron attachment. An important question remains whether the components covalently coupled within DNA are still sensitive towards LEEs and if so what is the mechanism behind, which is responsible for the formation of single strand breaks.

In order to track that problem, larger units of DNA consisting of two (thymidine (Td) [9, 22] and ribose-5′-phosphate (RP) [23]) or three (2′-deoxycytidine-5′-monophosphate (dCMP) [21]) building blocks of DNA have been studied recently in the gas phase by means of a crossed electron-molecular beam technique or the LIAD technique (Laser Induced Acoustic Desorption).

From the experiments on thymidine it becomes obvious that electrons with very low energy (<3 eV) can effectively decompose Td into the sugar and thyminyl moieties [9, 22]. The reaction proceeds via a cleavage of the N1–C glycosidic bond with the excess charge localized on either of the fragments. It should be stressed that the results for Td do not give a direct evidence for an electron transfer mechanism (a transfer of the electron from the nucleobase to the sugar moiety), theoretically predicted for a nucleotide dCMP [24], that would be responsible for the formation of SSBs. Instead it has been assumed that the fragmentation of the molecule associated with N1–C bond rupture occurs through electron localization on either of the two moieties [22].

The results on ribose-5′-phosphate (RP) obtained utilizing a soft technique for transfer of the molecules into the gas phase (LIAD) demonstrate the propensity of RP towards decomposition triggered by the capture of near 0 eV electrons [23]. The fragmentation of RP occurs either via a cleavage of the P–O bond resulting in the formation of $H_2PO_4^-$ or via C–O bond leading to $C_5H_9O_5^-$. These observations support the hypothesis that the formation of single stand breaks in DNA can be initiated by the direct attachment of low energy electrons to the DNA backbone. This has been recently confirmed by the measurements performed for an entire gas phase nucleotide dCMP [21]. It has been shown that dCMP subjected to low energy electrons decomposes at two energy regions below 2 eV and above 5 eV. At the latter energy range two fragments at m/z 16 and 1 have been observed. They have been attributed to NH_2^-/O^- and H^-, respectively. In the low energy range the formation of rather heavy fragments has been reported, i.e., m/z 45, 95 and 110. The fragment at m/z 45 has been assigned to the $HCOO^-$ ion most likely generated from the decomposition of the sugar unit. In fact, such fragment has been reported previously

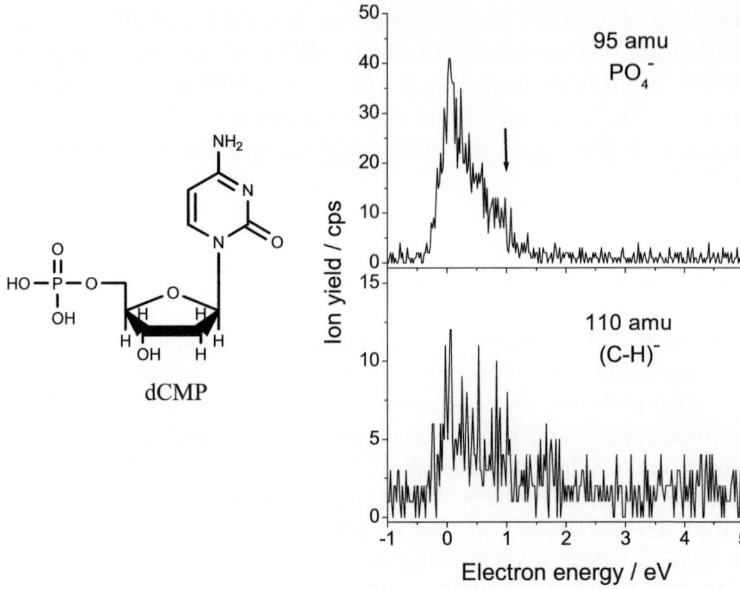

Fig. 4 Yield of the PO_4^- (95 amu) and the (C–H)$^-$ (110 amu) anions resulting from electron attachment to dCMP. The *arrow* indicates the contribution from initial electron localisation at cytosine. Adapted from Ref. [21], with permission of Royal Society of Chemistry

from DEA to sugar compound [14] and observed via a narrow peak already at the threshold energy. Two further anionic fragments m/z 95 and m/z 110 have been assigned to PO_4^- and (C–H)$^-$ (where C = cytosine) (Fig. 4).

The formation of the fragment at m/z 95 (PO_4^-) requires a cleavage of the C–O bond between the phosphate and the sugar unit with concomitant loss of the neutral hydrogen molecule. PO_4^- appears at low energy range with a dominant structure appearing near 0 eV and a shoulder like structure at around 1 eV. The fragment at m/z 110 is generated from the N1–C glycosidic bond cleavage and has been attributed by stoichiometry to the closed shell dehydrogenated cytosine anion (C–H)$^-$. It appears at energies below 2 eV via a broad and rather low-intensity structure.

Comparing the electron attachment profiles of the fragments generated from dCMP with those observed from phosphate surrogates and sugars it has been assumed that both mechanisms, i.e., direct electron attachment to the DNA backbone and electron transfer from the base to the backbone, contribute to the SSBs. Furthermore, based on the results from dCMP, it has been estimated that approximately 60 % of the SSBs is due to a direct attachment of LEEs to the phosphate group, around 25 % is due to attachment to the sugar unit, and around 15 % from initial localization of the excess charge at the base with subsequent electron transfer to the backbone [21]. It has been also suggested that these numbers may change with changing the nucleobase within a nucleotide.

Indeed, recent results obtained for gas phase adenosine monophosphate (AMP) show that the intensities of the PO_4^- and the $(A\text{–}H)^-$ anions are almost the same. Such observation points out towards the role of different nucleobases in electron attachment processes [25].

2.1.4 Sensitization of DNA Towards Radiation

Radiation therapy, in spite of being one of the most frequently used to treat cancer, is a non-selective method of treating different types of solid tumors. Therefore during its administration not only the cancerous cells are affected but also the healthy ones. A susceptibility of tumor cells to ionizing radiation may be enhanced by a synchronous application of a chemotherapeutic drug that can specifically and selectively sensitize cancer cells [26] (see also references therein). The most important groups of radio-sensitizers are the so-called alkylating agents and antimetabolites. Reactive drugs belong to the first group of agents that bind to the chemical groups found in nucleic acids causing chemical changes in DNA. The most significant molecular changes induced by these compounds are the formation of cross-links between strands of DNA and the loss of its basic components. The result is that the nucleic acid loses its ability to replicate and hence to carry out the functions of the cell. This eventually leads to either cell death or mutagenic changes. This group can be represented by platinum based compounds, e.g., *cis*-diamminedichloroplatinum(II). On the other hand, antimetabolites are characterized by high structural similarity to the naturally occurring metabolites. Therefore they readily become incorporated into DNA or RNA and interfere with the cells normal metabolic functioning. In particular they exhibit the ability of disrupting specific metabolic pathways inside the cell through replacement of metabolites. This group of compounds includes, inter alia, halogenated nucleobases and nucleosides.

A Dissociative Electron Attachment to Platinum Containing Compounds (Alkylating-Like Agents)

As mentioned above, *cis*-diamminedichloroplatinum(II), commonly assigned as cisplatin, is a coordination complex ($Pt(NH_3)_2Cl_2$) that belongs to the group of alkylating agents. Although cisplatin does not alkylate the DNA directly, the function is similar and, therefore, it is actually classified as alkylating-like drug. Its cytotoxicity is based on its ability to bind to DNA hence inhibiting replication. It can form both inter- and intrastrand cisplatin-DNA adducts. However, the predominant and most relevant for DNA damages are intrastrand adducts, which are formed at adjacent guanine bases of the same DNA strand [27]. It has been shown recently that the combination of the drug with radiation increases the survival of cancer patients compared to those who were treated with radiotherapy [28]. The largest effect has been achieved in a concomitant application of both chemo- and radiotherapy. This observation has been attributed to unidentified synergetic interaction between the drug and radiation. As already mentioned, for the application of radiation the major amount of energy is channeled into the formation of secondary electrons. Therefore, it is very likely that these electrons are involved in the activation of the drug in chemoradiotherapy. Indeed, the experiments by Sanche and co-workers on solid films with and without cisplatin bound to guanine and irradiated with electrons of variable energies clearly show that the number of strand breaks is enhanced in the synchronous action of drug and low energy electrons (see Sect. 2.2.3) [29]. It was suggested that the transient negative ions that are formed from electron attachment play a key role in the enhancement of DNA damage.

The role of dissociative processes in degradation of cisplatin prior to the formation of cisplatin-DNA adducts has been also imposed from ultrafast transient absorption spectroscopy experiments [30, 31]. In these experiments the presolvated electrons (e_p^-), which are produced by the laser pulse, attach to cisplatin, thereby causing a rupture of the Pt–Cl bond, via:

$$e_p^- + Pt(NH_3)_2Cl_2 \rightarrow Pt(NH_3)_2Cl + Cl^-.$$

The generated neutral $Pt(NH_3)_2Cl$ radical may be subjected to further release of the Cl^- ion resulting in the formation of the $Pt(NH_3)_2$ radical, which can bind to DNA.

Likewise the release of the Cl^- ion as well as its complementary ion $Pt(NH_3)_2Cl^-$ has been recently observed from dissociative electron attachment to gas phase cisplatin [32]. However, more interesting was the observation that a single slow electron triggers the loss of both Cl atoms resulting in the formation of two further complementary anions, namely $(Pt(NH_3)_2)^-$ and Cl_2^- [32], the former one being more than 25 times more intense:

$$e^- + Pt(NH_3)_2Cl_2 \rightarrow Pt(NH_3)_2^- + Cl_2$$
$$e^- + Pt(NH_3)_2Cl_2 \rightarrow Pt(NH_3)_2 + Cl_2^-.$$

These findings have been confirmed by the theoretical work published by Gianturco's group [33]. It was hence suggested that the $(Pt(NH_3)_2)^-$ ion can be regarded as the reactive intermediate to form cisplatin-DNA adducts, which inhibit replication (Fig. 5).

Fig. 5 Structure of cisplatin and intrastrand cross-link as the dominant binding mode of cisplatin and DNA. Adapted from Ref. [32], with permission of John Wiley and Sons

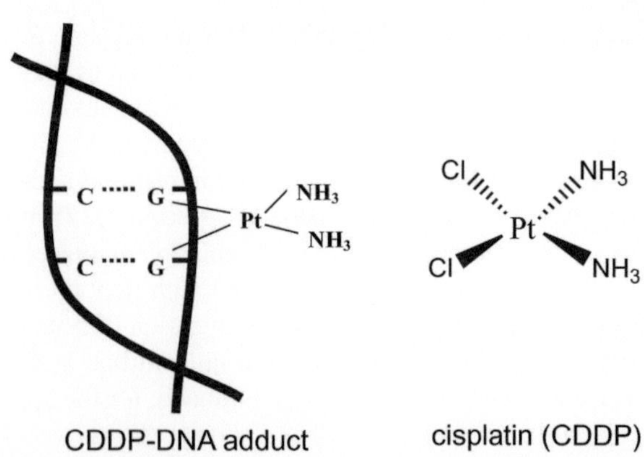

B Dissociative Electron Attachment to Antimetabolites

The importance of halo-containing antimetabolites in treatment of tumors became obvious more than half a century ago. It has been observed that thymine deficiency in growing bacteria, which is caused by the incorporation of halouracil into DNA results in inhibiting their growth and survival [34]. Unusual antitumor activity has been observed for 5-fluorouracil (5-F-U) [35]. Almost simultaneously an increase of radiosensitivity related to replacement of thymine by 5-bromouracil (5-Br-U) in bacterial DNA has been reported [36]. Later on, impressive results of combined chemo- and radiotherapy have been reported from in vivo experiments for brain tumors treated with either 5-bromo-deoxyuridine or 5-iodo-deoxyuridine and X-rays [37]. These results laid the foundation for development of a new cancer treatment methodology combining halogenated compounds with radiation. Since then, much efforts have been undertaken to unravel the mechanism that is responsible for the radiosensitization. One of the proposed mechanisms based on various studies of the radiation processes in aqueous biologically relevant systems involves reactions with hydrated electrons (e^-_{aq}) [38]. It was suggested that e^-_{aq} reacts with halouracils to form anion radicals ((halouracils)$^-$), which then undergo decomposition via release of a bromide ion and a formation of uracilyl radical with the unpaired electron localized at C_5 in a σ orbital [39, 40]. This uracilyl radical is believed to be the reactive precursor responsible for the enhancements in DNA damage.

Later on, as mentioned in the introduction, it has been shown that not only hydrated electrons but also ballistic free electrons are able to damage plasmid DNA [3]. The damage of the nucleic acids by low energy electrons occurs at the molecular scale, where the fragmentation is localized at particular molecular components of the DNA and thus may be the initial step towards the observed strand breaks. The question was whether the selective radiosensitivity can be achieved via irradiation of the cells containing halouracil-containing DNA with nonhydrated nonthermal secondary electrons. From the studies of isolated 5-halouracils bombarded with LEEs it became obvious that they induce several reactions [41, 42]. It has been reported that besides the formation of parent anions, also diverse fragment anions are generated with a very high intensity (Fig. 6). One of the most abundant among them is a halogen anion X^- (where X = Cl, Br, I) (and an uracilyl radical as a neutral counterpart) also observed in aqueous systems. In addition, a further reactive dissociation pathway is reported to lead to the formation of an uracilyl anion (U–yl)$^-$ and a free halogen radical. Both of these radicals (U-yl)$^\bullet$ and $X^\bullet$ are believed to contribute to the DNA damage. However, taking into account the yield of the radicals it has been concluded that for 5-I-U the uracilyl radical is likely responsible for the radiosensitizing effect whilst for 5-Br-U and for 5-Cl-U these are the halogen radicals. Hence the DNA damage in the case of the two latter compounds has been attributed to the reactions of the $X^\bullet$ within their vicinity [41]. On the other hand, in the case of 5-I-U the strand break can be induced via hydrogen abstraction from the neighboring sugar unit. More recently it has been shown that also 2-chloroadenine efficiently dissociates into reactive species upon interaction with low energy electrons, corroborating its potential as a radiosensitizer [43].

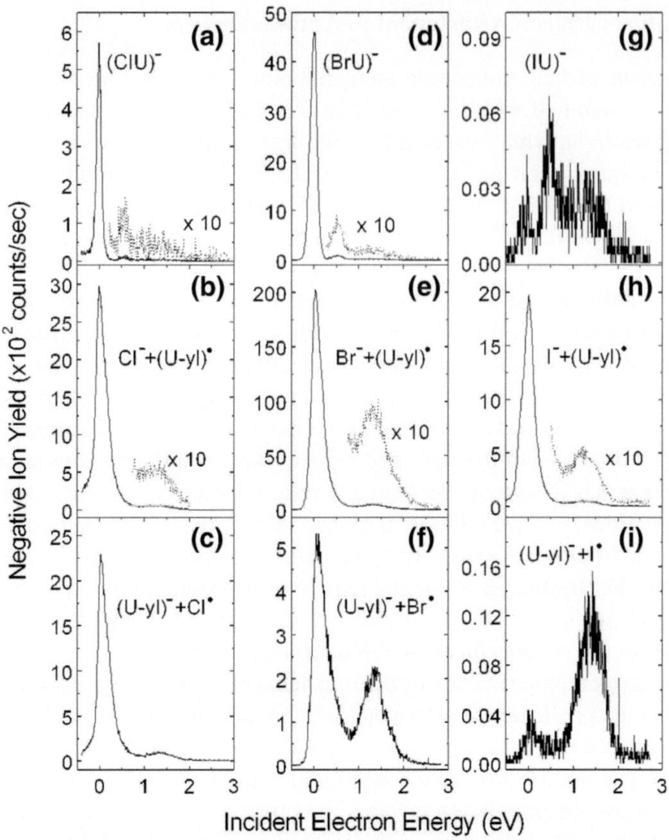

Fig. 6 Yield of the fragment anions produced from electron attachment to 5-X-uracil (where X = Cl, Br, I) as a function of electron energy. Adapted from Ref. [41], with permission of American Chemical Society

The fact that the sugar molecules are effectively decomposed by low energy electrons raises a question of whether the sugar group may also be selectively sensitized to radiation, e.g., by introducing a halogen atom at the sugar unit within a nucleoside or nucleotide. The results obtained for 2′,2′-difluorocytidine (gemcitabine, a drug commonly used in chemotherapy) confirm this possibility [44]. It appears that the presence of fluorine atoms substituting the 2′ hydrogen atoms at the sugar site substantially increases the electron attachment cross section (Fig. 7). This in turn is reflected in higher intensity of the fragment anions produced already at the threshold energy ≈ 0 eV.

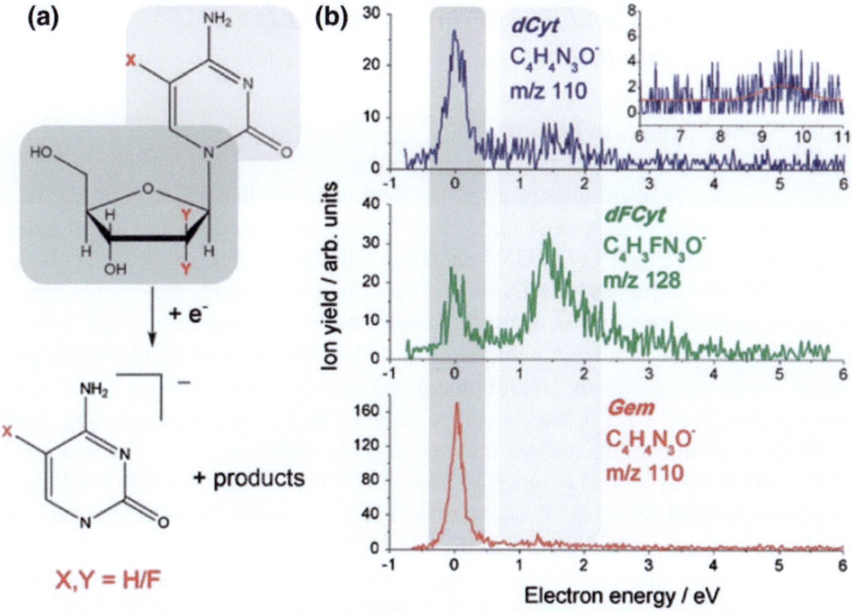

Fig. 7 Gemcitabine (X = H, Y = F) is a drug commonly used in chemotherapy. The presence of fluorine atoms at the sugar unit substantially increases the yield of a threshold peak as appears from the comparison with 2′-deoxycytidine (dCyt) and 2′-deoxy-5-fluorocytidine (dFCyt). Adapted from Ref. [44], with permission of Royal Society of Chemistry

Besides the halogenated nucleobases and nucleosides thiolated analogues of canonical nucleobases are well known as anti-cancer drugs [45]. Recently, it has been shown that the combined use of such analogues and light develop sensitizing properties [46]. However, until recently there was a lack of knowledge about the degradation of these molecules by ionizing radiation and in particular by secondary electrons. Results obtained from dissociative electron attachment to sulphur nucleobase analogues, i.e., 2-thiothymine and 2-thiouracil clearly show that the fragmentation arises mainly at the sulphur site of the molecules [47, 48]. This means that the presence of the S atom within the molecular structure facilitates the decomposition of the molecules initiated by the capture of low energy electrons. In a recent paper Chomicz et al. [49] were looking for potential radiosensitizers and suggested that another sulphur containing analogue, 5-thiocyanatouracil (5-NCS-U), can be a good candidate. They suggested that efficient radiosensitizers should contain an electron withdrawing group and the bond between substituent and nucleobase should be relatively weak in order to make the dissociation feasible.

2.2 DEA in the Condensed Phase: Plasmid DNA and Oligonucleotides

2.2.1 Strand Breaks in Plasmid DNA

Early experiments by Boudaïffa et al. [3] demonstrated that DEA is operative in rather complex systems such as plasmid DNA (see introduction) and could indeed be a major damage pathway in DNA radiation damage [3, 50].

Plasmid DNA is typically used for the determination of strand break yields, since a single and double strand break results in a conformational change from supercoiled to relaxed circular or linear DNA, which is easy to detect by agarose gel electrophoresis (AGE). In this way, already small amounts of damaged materials can be detected, which is crucial for LEE induced damage since LEEs have a low penetration depth.

The SSB yield exhibits distinct maxima at energies below 12 eV. The first report by Boudaïffa et al. [3] already reported SSB and DSB yield maxima at 8-12 eV using the plasmid pGEM 3Zf(-) (3199 base pairs). A very interesting observation was made by Martin et al. [4], who reported SSB yield maxima at 0.8 eV and around 2 eV. At these energies electronic excitation is not possible and the strand breakage must occur through shape resonances. Although the work has been reproduced many times by the Sanche group [52, 53], other researchers found SB yields with different resonance profiles [51, 54, 55] indicating that the experimental results depend strongly on the type of plasmid used and the specific preparation conditions (Fig. 8) [56–58].

Orlando et al. [55] probed the 5-25 eV range and observed a local maximum of SSBs and DSBs around 10 eV and a global maximum at 20 eV [54]. Additionally, they coupled the determination of SBs by AGE to yields of neutral fragments determined by single photon ionization mass spectrometry [54]. In another study by the same group non-dried p14 plasmid DNA (6360 base pairs) was used and different resonance positions have been found (a global maximum of SSBs at 7 eV, and a second local maximum at 15 eV) [55]. The differences in resonance positions have been ascribed to different water and salt content of the investigated plasmid DNA samples [55].

Kumar et al. [51] used the plasmid pQE30 (3481 base pairs) and showed that the resonance position depends critically on experimental parameters such as the electron dose, and possibly also on the type of plasmid used and the specific preparation conditions (Fig. 8). The SSB yield shown at the top (Fig. 8a) is from the original work by Boudaïffa et al. [3], whereas the graph shown in Fig. 8b is from the Orlando group [55]. The SSB yields in Fig. 8c and d have been obtained by Kumar et al. [51].

The exact resonance positions yield information about the involved electronic states. Apart from recording the energy dependence of strand breakage there was a constant effort to determine the exact magnitude of damage through effective SB cross sections within a broad electron energy range. These cross sections are effective since they depend on the experimental conditions. The determination of absolute strand break cross sections is more difficult. For multilayer DNA films a molecular survival model is applied to determine the absolute damage cross sections [59].

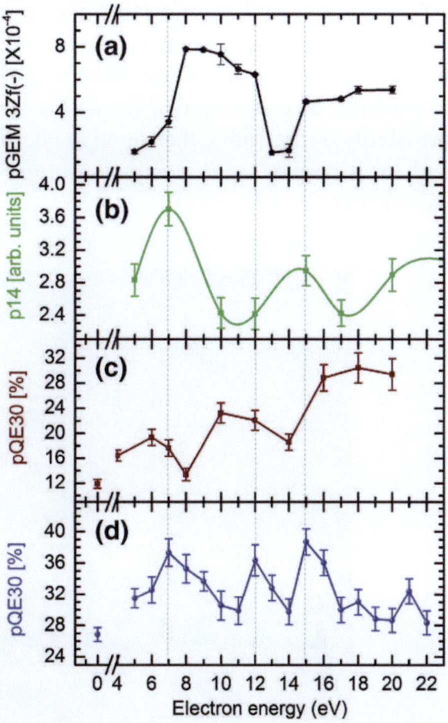

Fig. 8 Comparison of single strand yields as a function of electron energy obtained by electron irradiation of plasmid DNA in different experimental setups. Adapted from Ref. [51], with permission of AIP Publishing

Very recently, the absolute cross sections for DNA strand breaks in plasmid DNA (pUC21, 3151bp) induced by 10 eV electrons has been determined to be 3.0×10^{-14} cm^2 (weighted mean value from different film thicknesses) [60]. Homogeneous films of DNA have been prepared on graphite (HOPG) using doubly protonated 1,3-diaminopropane as a linker between the negatively charged DNA layers to control the DNA thickness. The cross section for the loss of supercoiled DNA was found to be strongly dependent on the film thickness, which allowed to determine the penetration depth (or information depth) to be 14.1 nm [60].

More recently, a simplified approach to determine absolute strand break cross sections was introduced that is based on the use of DNA nanoarrays, which can be formed from DNA origami nanostructures [61]. Selected target DNA sequences are arranged on DNA origami platforms, which are adsorbed with sub-monolayer coverage on a conducting solid support, such as Si [62]. A strand breakage in the target DNA is detected by directly imaging the DNA nanoarrays with atomic force microscopy (AFM). In this way the damage is recorded on a single-molecule level making the absolute cross section for strand breakage easily accessible by determination of the relative number of strand breaks as a function of the electron fluence [61, 63]. To be able to directly image the intact target DNA on the DNA nanoarrays a biotin label is introduced and after the electron irradiation, the remaining intact strands are visualized by binding the protein streptavidin to the biotin markers. In

AFM images the intact strand then appears as a bright dot (Fig. 9). For determination of the absolute strand break cross section also the electron induced damage to the biotin label must be taken into account [64]. This approach is not only attractive due to the direct determination of absolute strand break cross sections, but also due to the possibility to determine the strand break cross section for specific DNA sequences, and for its capability to compare multiple DNA sequences in a single irradiation experiment.

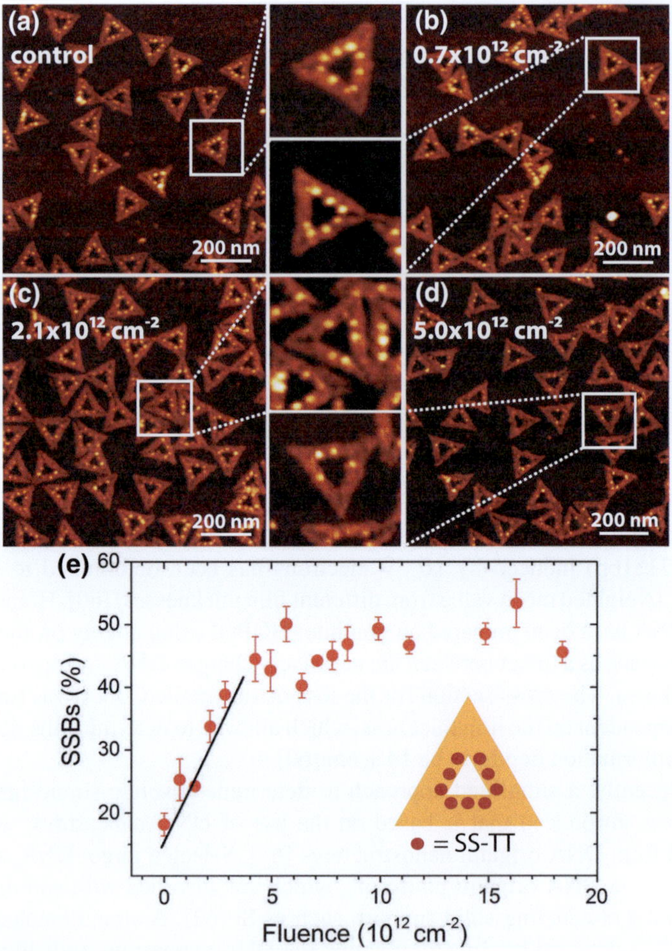

Fig. 9 Irradiation of DNA origami templates with nine protruding target strands. **a** AFM image of a nonirradiated control sample. **b–d** Samples irradiated with different fluences of 18 eV electrons. The number of specifically bound streptavidin decreases with electron fluence, indicating low-energy electron-induced DNA strand breaks. **e** Number of single strand breaks (in %) versus electron fluence. Adapted from Ref. [62], with permission of American Chemical Society

2.2.2 Oligonucleotides: Sequence Dependence of DNA Damage

Depending on the electron energy the initial electron attachment (eventually resulting in a DNA strand break) can occur either directly to the phosphate-sugar backbone, or to the DNA bases with subsequent charge transfer to the DNA backbone and bond cleavage within the sugar-phosphate moieties (see discussion on dCMP above). For a direct electron attachment to the phosphate-sugar backbone it is not expected that it matters significantly, which nucleobase is present at the respective site. However, if the nucleobases act as antenna for LEEs then there should be a pronounced dependence of the strand break cross section on the type of base. Additionally, in DNA the nucleobases are involved in π-π stacking interactions between the adjacent nucleobases and in intermolecular hydrogen bonding to the complementary DNA strand. These non-covalent interactions can significantly change the electronic properties of the nucleobases giving rise to a sequence-dependence of strand break cross sections. To elucidate such effects, different experimental approaches have been pursued to study LEE interactions with oligonucleotides of defined sequence. In early experiments thin films of oligonucleotides have been irradiated with LEEs to determine the electron stimulated desorption (ESD) of small anions such as CN^- using mass spectrometry. Although a clear damage of the oligonucleotides could be observed, a clear information about strand breakage could not be obtained [65, 66].

A detailed information about the bond cleavage sites is available from analyses of damaged material by high performance liquid chromatography (HPLC). Using HPLC analyses of the oligonucleotide GCAT and its abasic analogue GCXT upon irradiation with 4-15 eV electrons it was concluded that most of the LEE induced damage in this energy regime proceeds through initial electron attachment to the nucleobases [67]. In further studies using oligonucleotide trimers a clear sequence-dependence of strand breakage and base release was found with TTT showing the highest total damage [68, 69]. However, the most serious constraint of HPLC analysis is that due to the detection limit for specific fragments the analysis of longer sequences than tetramers is not feasible. For longer sequences the number of different fragmentation products increases and at the same time the amount of a specific fragment decreases below the detection threshold [68]. A complementary technique makes use of self-assembled monolayers (SAMs) of DNA arranged within DNA microarrays [70]. After irradiation with LEEs the DNA SAMs are incubated with the complementary sequence carrying a fluorescent dye. The hybridization is only successful for intact strands and the DNA damage by strand breakage can be analysed by fluorescence spectroscopy. In this way a pronounced dependence of the LEE (1 eV) induced DNA damage on the G content was found [71].

By using the DNA origami based DNA nanoarrays described above the absolute cross sections for strand breakage for different sequences after electron irradiation (18 eV) were determined. It was found that the strand break cross section decreases in the descending order [61]

$$\mathrm{TT(ATA)_3TT} > \mathrm{TT(CTC)_3TT} > \mathrm{TT(GTG)_3TT}.$$

Furthermore, the replacement of the three central T bases by the radiosensitizer 5-Br-U increases the strand breakage by a factor of up to 1.66 (depending also on the sequence). Very recently, it was demonstrated that this technique can be also applied to the study of UV photon induced strand breakage [72]. Further studies covering a wide variety of DNA sequences and electron energies is currently in progress to elucidate sequence dependent effects on the strand breakage and different damage mechanisms at different electron energies as was suggested also by gas phase and condensed phase experiments using simpler model compounds [21, 67].

2.2.3 Modification of DNA Strand Breakage by Radiosensitizers

In order to improve the efficiency of tumor chemoradiotherapy a fundamental understanding of the mode of action of therapeutically used radiosensitizers is very beneficial. As described above halogenated nucleosides and cisplatin derivatives are the most important therapeutics. They show interesting effects in their interaction with LEEs indicating that there is a physicochemical basis of DNA radiosensitization [44].

The ability of halogenated nucleobases to substantially modify the interaction with LEEs was clear rather early from gas phase experiments using single halogenated nucleobases (see above), but also from the condensed phase using modified oligonucleotides [65]. An increased DEA cross section was observed especially for 5-Br-U, whereas 5-F-U shows a very similar reactivity towards LEEs as non-modified U or T [65]. In electron-stimulated desorption (ESD) experiments only small anions such as H^-, O^-, CN^- and OCN^- could be detected and thus a correlation with strand breaks could not clearly be deduced. In later experiments using HPLC analysis and the DNA nanoarray technique a higher total damage and strand break cross section for 5-Br-U modified oligonucleotides were found [61, 69]. Recent studies show that the presence of Br also modifies the reactivity of the other nucleobases, but 5-Br-U remains the most active one in ESD (desorption of Br^-) [73], whereas HPLC analysis of LEE (10 eV) irradiated films indicates that the enhancement of damage decreases in the following order: BrG $\approx$ BrA $>$ BrC $\approx$ BrU [74]. Future studies in the condensed phase and using oligonucleotides modified with relevant radiosensitizers such as gemcitabine [44] need to be conducted.

Cisplatin and its derivatives belong to the most widely applied therapeutics in cancer radiotherapy. Their interaction with LEEs has been studied in detail also in the condensed phase [26]. Initial experiments using mixed films of plasmid DNA and cisplatin report enhancement of SSBs up to a factor of 4.4 at 10 eV electron energy and a cisplatin–DNA ratio of 8:1 [29]. Very recently it was shown that the DSB yield is increased by a factor of 3.1 upon 10 eV electron irradiation in the presence of two carboplatin molecules per plasmid [75]. For Oxaliplatin and Cisplatin slightly lower values have been found [75]. In the presence of Pt compounds it was demonstrated that a single electron with close to zero eV energy (0.5 eV) is able to induce a DSB in plasmid DNA, which is not possible without the Pt drug, or with other types of radiation such as X rays [76].

Other potential radiosensitizers are gold nanoparticles (AuNPs), since they release LEEs upon irradiation with high energy radiation, which may increase the local LEE dose when administered to cell tissue [77–79]. Irradiation of mixtures of plasmid DNA and AuNPs with 60 keV electrons showed an enhancement of SSBs and DSBs by a factor of 2.5 in the presence of AuNPs [80]. A combination of AuNPs and Cisplatin can enhance the yield of DSBs even by a factor of 7.5 [81], which opens up interesting novel strategies for the application of combined nanoparticle-cisplatin drugs [82].

2.2.4 The Role of the Environment

A very critical issue is how the LEE induced reactions observed in the gas phase with single molecules or in dry DNA in the ultrahigh vacuum is modified in an aqueous solution. In initial attempts to understand the influence of water on electron-induced DNA damage the anion desorption from thin films of oligonucleotides covered with three monolayers of water was studied. It was found that the anion desorption increased by a factor of 1.6 indicating more effective DNA damage in the presence of water [83]. This is confirmed by a recent study indicating a significant damage enhancement as soon as a bulk-like hydration is reached [84]. There are several theoretical studies, which indicate that the adiabatic electron affinities of dissolved nucleotides increase and the activation barrier for phosphodiester bond cleavage increases [85, 86]. Nevertheless, the barriers at room temperature are still small enough to be overcome by small thermal fluctuations [87]. To arrive at a complete understanding of LEE induced DNA damage, DNA in aqueous buffer solution needs to be investigated. This is experimentally extremely challenging and new experimental approaches need to be developed. Very recently, the LEE induced damage to T, U and BrU in aqueous solution was monitored using AuNPs as an electron source and surface-enhanced Raman spectroscopy as a detection method for the LEE induced nucleobase damage [88].

3 Computational Treatment

The physical picture of the DEA nuclear dynamics driven by a resonant scattering process (proposed half a century ago [89]) indicates that the theoretical study of electron impact induced nuclear motion can be performed in two steps. The first step involves the calculation of electronic properties (energies) for a fixed molecular geometry (fixed-nuclei approximation). The resulting energy surfaces are then utilized in the second step to describe the evolution of the nuclear wave function. In contrast to routine quantum chemistry applications, the electronic potential energy surfaces are often complex, i.e. the resonances have both an energy and a lifetime (or width) associated to them, allowing for the autodetachment process.

The calculation of the real (i.e. bound) segment of the potential energy surfaces is, at present, a reasonably straightforward task. These calculations describe a stable negative ion or a neutral molecule and at least a dozen commercially available quantum chemistry packages are up to the task. On the other hand, the situation is less straightforward for the complex (resonant) parts of the surfaces. The electronic resonant states are not directly linked to bound states that can be obtained by variational or non-variational methods used in quantum chemistry computations. The molecular Hamiltonian often requires a modification in the form of a complex absorbing potential, rotation of electronic coordinates into complex plane (complex scaling methods), or an additional external field that binds the resonant state (analytical continuation in coupling constant methods [90]). Another option for determination of the resonance parameters, avoiding altering the Hamiltonian, is to carry out scattering calculations. In the scattering calculations the resonant state is searched directly among the continuum states.

Among all theoretical methods developed to treat the nuclear dynamics of the DEA, only the resonance *R*-matrix theory developed by Fabrikant [91] has been successfully employed to study this process in molecules of biological relevance [92, 93]. Other studies with an ambition to elucidate the complicated mechanisms of the nuclear dynamics in the DEA of biologically relevant molecules [94] are based on the ab initio calculations of the electron-molecule collisions in the fixed-nuclei approximation performed for a set of nuclear geometries. The authors analyzed the parameters of the electronic resonance for various nuclear geometries to identify the dissociation pathways as well as the symmetries of the electronic states involved in the DEA. However, this approach did not yield any cross section of the DEA.

The separation of the computational treatment to the electronic and nuclear is followed by the organization of this section. Furthermore, we attempt to outline computational tools that have been applied to studies of biomolecules over the last decade. Our interest is focused on explaining the basic principles of the methods, we also try to emphasize strong points of the approaches, note possible limitations and point a reader to its applications relevant for this monography. The text represents our limited view of the subject and by no means can be considered as an objective and exhaustive review of the field.

3.1 Electronic Structure Calculations

3.1.1 Methods Based on Scattering Calculations

A Extracting the Resonance Parameters

Essential output of the scattering methods discussed in this section is provided by a phase shift in the scattering wave function that was gained by the electron-molecule interaction. This phase shift determines (and is determined by) the typical outputs of the scattering codes, so called S-matrices, K-matrices, and T-matrices. These matrices and the phase shifts are computed at real collision energies. On the other hand the resonance, in the original meaning of Siegert [95], is characterized by a pole of the S-matrix in the complex energy plane

$$E_c = E_r - i\frac{\Gamma}{2} , \tag{1}$$

with the position of the resonance E_r and the resonance width Γ. Proximity of such a pole to real-energy axis will strongly influence scattering quantities evaluated along the real energy. It was shown by Hazi [96] that an isolated resonance in electron-molecule collisions manifests itself as a characteristic jump by π in the calculated eigenphase sum $\delta_s(E)$ via the Breit-Wigner formula

$$\delta_s(E) = \arctan\left[\frac{2(E - E_r)}{\Gamma}\right] + \delta_{bg}(E) , \tag{2}$$

where $\delta_{bg}(E)$ is a background (non-resonant) phase and it is assumed to possess a weak energy dependence in the resonant region.

Larger polyatomic molecules of a biological relevance, however, rarely exhibit a single isolated resonance in the low-energy region (under 10 eV). On contrary, series of overlapping broad resonances are very common and their separation and identification becomes a challenging task. A powerful tool to overcome such difficulty is provided by the time delay matrix $Q(E)$ presented by F.T. Smith in 1960 [98] (originally introduced as the lifetime matrix)

$$Q(E) = iS\frac{\mathrm{d}S^+}{\mathrm{d}E}, \tag{3}$$

where S^+ describes hermitian conjugate of the S-matrix. Later it has been shown [99, 100] that the Tr $Q(E)$ can be linked to the change of the eigenphases sum. Generalization of the Breit-Wigner formula (2) to the case of N overlapping resonances takes the following form:

$$\mathrm{Tr}\, Q(E) = 2\frac{\mathrm{d}\delta_s(E)}{\mathrm{d}E} = \sum_{i=1}^{N} \frac{\Gamma_i}{(E - E_i)^2 + (\Gamma_i/2)^2} + 2\frac{\mathrm{d}\delta_{bg}(E)}{\mathrm{d}E} . \tag{4}$$

This formula was applied (see for example Refs. [97, 99, 101]) to fit numerically the calculated eigenphase sum, $\delta_s(E)$, and to determine the unknown resonance positions E_i and resonance widths Γ_i together with the background contribution which is often approximated by a constant or by a low-order polynomial.

The efficiency of the time-delay approach is demonstrated in Fig. 10 obtained from a recent computational study [97]. While the integral cross section displayed on the left panel of Fig. 10 shows some hints of possible overlapping broad resonances, the eigenphase sums (upper right panel) raise so steeply that it becomes impossible to identify any π jump required by Breit-Wigner fit (2). However, the positions and

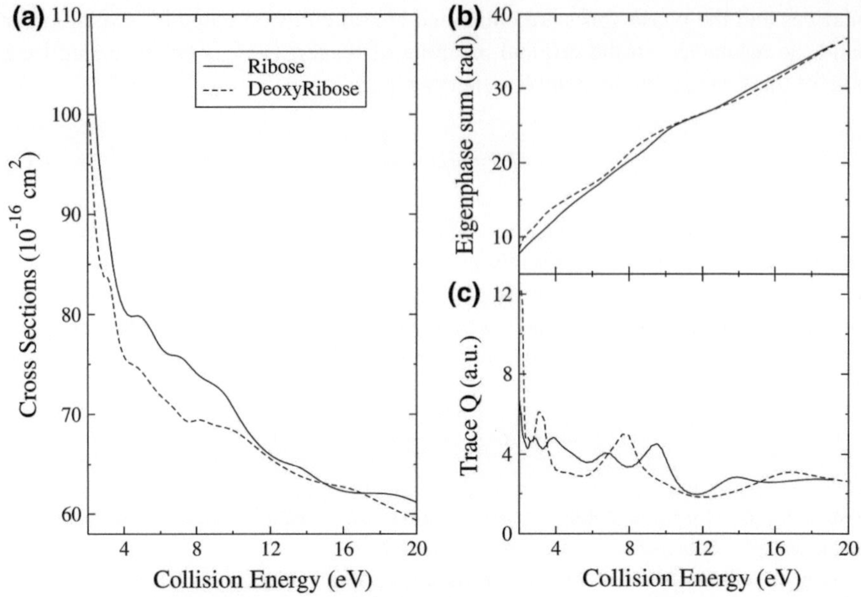

Fig. 10 *Left panel* computed cross sections for electron scattering by β-D-ribose and its $2'$-deoxy analogue. *Right panels* The corresponding eigenphases (*upper right panel*) and the time delay matrix traces (*bottom right panel*). Adapted from Ref. [97]

widths of the resonances are more clearly visible in the trace of the time-delay matrix $Q(E)$ shown in bottom right panel of Fig. 10.

B Single Centre Expansion Approach

The core idea of the single-center expansion (SCE) method lies in an angular expansion of the scattering-electron degrees of freedom around a chosen center (typically the center of mass of the molecule). SCE was also employed in the pioneering times of ab initio bound-state calculations of diatomic systems with basis sets formed by Slater-type orbitals [102]. Later, for calculations of polyatomic molecules, the SCE approach was abandoned, when merits of Gaussian basis sets were recognized and thus the evaluation of three- and four-center integrals ceased to be a problem.

The SCE approach belongs to a class of one-electron methods employing an optical potential. The optical potential is a one-electron, generally non-local interaction obtained by projecting out degrees of freedom of the bound electrons. The resulting interaction is usually described by three different components [103]:

$$V = V_s + V_x + V_{cp} \ , \qquad (5)$$

where V_s denotes the electrostatic interaction of the impinging electron with the molecular charge density (formed by nuclei and the bound electrons), V_x stands for generally non-local exchange interaction, and the correlation-polarization potential V_{cp} describes a response of the bound molecular electrons to the presence of the scattered electron. Such response is typically modeled by Density Functional Theory (DFT) [1]. Moreover, calculations for larger biomolecular targets often require further simplification for the exchange interaction V_x. Computationally intensive non-local interaction V_x is often replaced by a local exchange models V_{xl} of Hara [104] or Riley and Truhlar [105].

As mentioned above, the SCE method is based on the partial-wave expansion of the continuum wave function $\psi(\mathbf{r})$

$$\psi(\mathbf{r}) = \frac{1}{r} \sum_{lm} u_{lm}(r) Y_{lm}(\hat{\mathbf{r}}) \,. \tag{6}$$

The three-dimensional Schrödinger equation is then projected onto the orthonormal angular basis of the spherical harmonics $Y_{lm}(\hat{\mathbf{r}})$. Resulting set of radial Schrödinger equations

$$\left[\frac{\mathrm{d}^2}{\mathrm{d}r^2} - \frac{l(l+1)}{r^2} + k^2 \right] u_{lm}(r) = 2 \sum_{l'm'} V_{lm,l'm'}(r)\, u_{l'm'}(r) \,, \tag{7}$$

is coupled by the anisotropy of the local interaction potential V_{loc} expressed in the coupling elements

$$V_{lm,l'm'}(r) = \langle Y_{lm}(\hat{\mathbf{r}}) | V_{\mathrm{loc}}(\mathbf{r}) | Y_{l'm'}(\hat{\mathbf{r}}) \rangle \,, \tag{8}$$

where the scalar products are carried out solely on the angular space $\hat{\mathbf{r}}$ leaving the dependence on the radial coordinate r.

The angular functions chosen here are the spherical harmonics Y_{lm} typical in implementation of Itikawa and collaborators [106, 107]. In a different implementation of Gianturco and collaborators [103] the spherical harmonics are combined in a subspace of the fixed l-value in such a way that they belong to irreducible representations of the molecular point group [103, 108]. Different symmetries do not mix and the set of Eq. (7) decouples into independent groups according to the corresponding molecular symmetries.

An advantage of the SCE method is that it works in a physical picture in which the scattered electron interacts with the target molecule by a physically motivated interaction potential V. Its tractable implementation permits calculations for larger polyatomic targets. In fact, the SCE approach belongs to very few ab initio methods applied to the DNA building blocks. Unfortunately, this merit constitutes also the weak point of the method. Very little space for the improvement of the electron-molecule interaction may become a limitation of the SCE approach in cases of more complicated target electronic structures (open shells, radicals, etc.).

Despite of the possible limitations, the SCE method was successfully applied to about a dozen of biomolecules over the last decade. Radiation damage mechanisms were studied [110] for several aminoacids (glycine, alanine, proline, and valine) by investigation of the resonant fragmentation patterns. In the case of uracil the authors show [109, 111] that the colliding electrons which get temporarily attached to uracil in the gas phase are able to efficiently couple with the nuclear degrees of freedom during the attachment process in a very selective, resonance-specific way. Therefore, it could be construed as dissipating part of their excess initial energy via different mechanisms into the molecular vibrational network:

(i) By direct electron-nuclear coupling via a dissociative potential energy surface (PES)
(ii) By indirect nonadiabatic conical intersections between different PESs

An example of the less effective π^* resonances that undergo conical intersections with σ^* resonances of A' symmetry is show in Fig. 11 for the uracil molecule.

Similar conclusions were also found for the simplest molecule with a peptide bond, formamide. Computational investigation [112] revealed that its dissociation from a TNI state has to occur via a symmetry-breaking and non-adiabatic curve crossing. The SCE approach was also employed for a computational and experimental study [15] in which the authors demonstrate that the furanosic blocks of the RNA backbone are sensitive to electron attachment processes and can undergo irreversible damage with electrons up to about 10 eV.

More applications of the SCE method to molecules of a biological relevance can be found in the excellent review [1] that summarizes available experimental and theoretical models (up to year of 2010) for studying the electron-induced DNA damage.

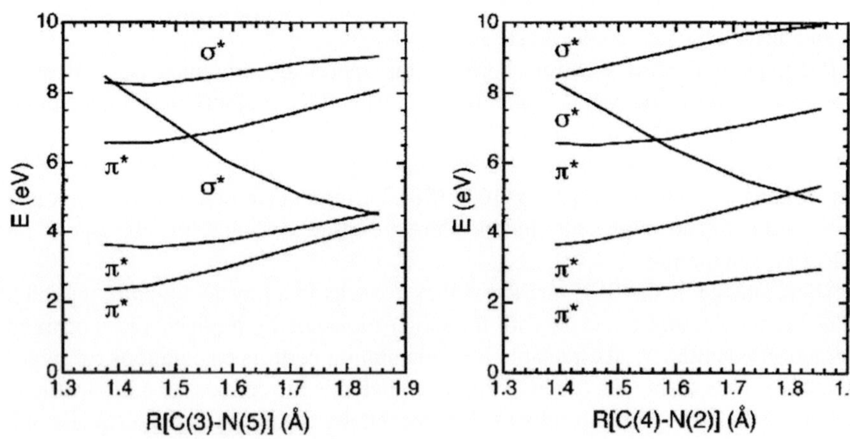

Fig. 11 Computed real parts of the A' and A'' resonance energies as a function of ring-breaking deformations in the uracil ring. The *left panel* deals with the C(3)-N(5) stretch while the *right panel* reports the C(4)-N(2) stretch. Adapted from Ref. [109], with permission of AIP Publishing

C *R*-matrix method

The UK-molecular *R*-matrix codes, which are the subject of the present article, have been applied to polyatomic molecular targets since the original implementation of a Gaussian type orbitals (GTOs) within the polyatomic code [113] in 1997. Since then the codes have undergone a comprehensive rewriting and modernization. A number of important algorithms have been changed to increase their functionality. The suite of codes is now called UKRmol. The *R*-matrix method and its application to the electron-molecule collisions have been described in detail elsewhere [114], thus only a brief summary is presented here. In a manner similar to SCE method above, the *R*-matrix approach works within the fixed-nuclei approximation in which the nuclei of the molecule are held fixed during the collision.

The core idea of the *R*-matrix is the partitioning of configuration space into an inner and an outer region, separated by a sphere of radius r_0 as shown in Fig. 12. The electronic density of the target states must be contained inside the inner region. The eigenfunctions of the total Hamiltonian are expanded in two major contributions:

$$\Psi_k(\mathbf{r}_1 \ldots \mathbf{r}_{N+1}) = \mathcal{A} \sum_{ij} a_{ijk} \phi_i(\mathbf{r}_1 \ldots \mathbf{r}_N)\, \psi_{ij}(\mathbf{r}_{N+1}) + \sum_l b_{lk}\, \chi_l(\mathbf{r}_1 \ldots \mathbf{r}_{N+1}) \,, \quad (9)$$

where $\mathcal{A}$ is the antisymmetrization operator, continuum wave functions $\psi_{ij}(\mathbf{r}_{N+1})$ describe the scattering electron, and $\mathbf{r}_1 \ldots \mathbf{r}_N$ are coordinates of the bound electrons. If the first sum over the target state functions $\phi_i(\mathbf{r}_1 \ldots \mathbf{r}_N)$ was taken as complete the second term in Eq. (9) would not be necessary. However, an introduction of the (*N*+1)-electron functions $\chi_l(\mathbf{r}_1 \ldots \mathbf{r}_{N+1})$ fully contained inside the sphere, can

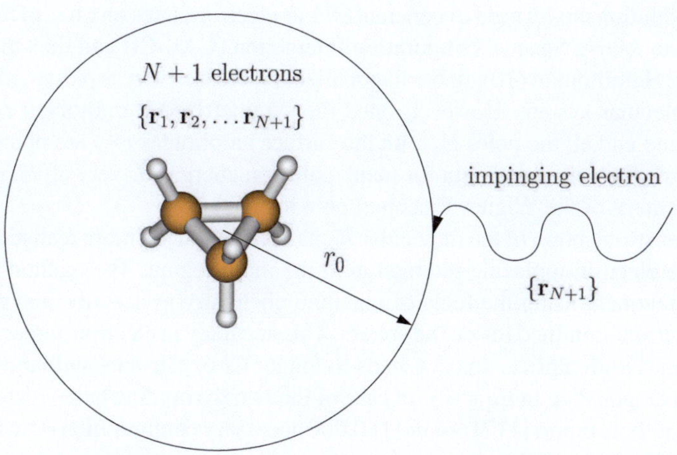

Fig. 12 The *R*-matrix method divides space into an inner region which contains the wave function of all the $N + 1$ electrons and into the outer region. The boundary is given by a sphere of radius r_0 centered on the target center-of-mass. Only the scattering electron can be present in the outer region

drastically decrease the number of necessary terms in the expansion over the target states functions $\phi_i(\mathbf{r}_1 \ldots \mathbf{r}_N)$.

The use of GTOs to describe the continuum wave functions $\psi_{ijk}(\mathbf{r}_{N+1})$ simplifies calculations of the repulsion integrals, however, it also limits the size of the R-matrix sphere r_0 (to around 15–18 bohrs [115]). The coefficients a_{ijk} and b_{lk} are determined variationally by diagonalization of the symmetrized (N+1)-electron Hamiltonian $\bar{H} = H - L$ (L is the Bloch operator [116]) in the inner region

$$\langle \Psi_k | \bar{H}(\mathbf{r}_1 \ldots \mathbf{r}_{N+1}) | \Psi_{k'} \rangle = E_k \, \delta_{kk'} \; . \tag{10}$$

The inner-region wave functions and their associated eigenvalues allow to build the R-matrix on the spherical surface:

$$R_{\alpha\beta} = \frac{1}{2r_0} \sum_k \frac{w_{\alpha k}(r_0)\, w_{\beta k}(r_0)}{E_k - E} \; , \tag{11}$$

where E_k, known as R-matrix poles, are the eigenvalues of the symmetrized Hamiltonian (10) and $w_{\alpha k}(r_0)$, known as surface amplitudes, are the surface values of the eigenfunctions Ψ_k in the channel denoted here by symbol α.

The R-matrix carries all the necessary information about the scattering wave function on the surface specified by radius r_0. This wave function is often propagated in the outer region by a local propagator not much different from the one used in the Sect. B. It is assumed that the one-electron wave function in the outer region can be described by a local interaction with the target molecule. Such a local interaction is formed by the lowest electrostatic multipoles and the static molecular polarizability of the target molecule [114].

The excitation model used to generate (N+1)-electron states on r.h.s. of the Eq. (9) is Complete Active Space Configuration Interaction (CAS-CI) and thus the dimension of the Hamiltonian (10), to be diagonalized, increases very rapidly with the size of the molecular system. However, once the symmetrized Hamiltonian $\bar{H}$ is fully diagonalized and all the poles E_k with the surface amplitudes $w_{\alpha k}$ are obtained, one can generate the collisional data for many collision energies E very efficiently. This is a major merit of the Wigner-Eisenbud pole formula (11).

Another strong point of the molecular R-matrix method is a more realistic description of the electron-molecule interaction in the inner region. The method is an all-electron treatment using the tools of quantum chemistry to describe a system with (N+1) electrons confined inside the sphere. This accuracy in electron-molecule interaction comes with a price, since it leads to longer CI expansions and larger Hamiltonians to diagonalize in Eq. (10). In case of the two pyrimidine bases (cytosine and thymine) authors report [117] around 180.000 necessary configuration state functions for the Hamiltonian (10).

The R-matrix method was applied to study positions and widths of low-lying π^* resonances of pyrazine [118]. Authors also report presence of numerous core-excited resonances above 5 eV of the collision energy. The presence of the three low-

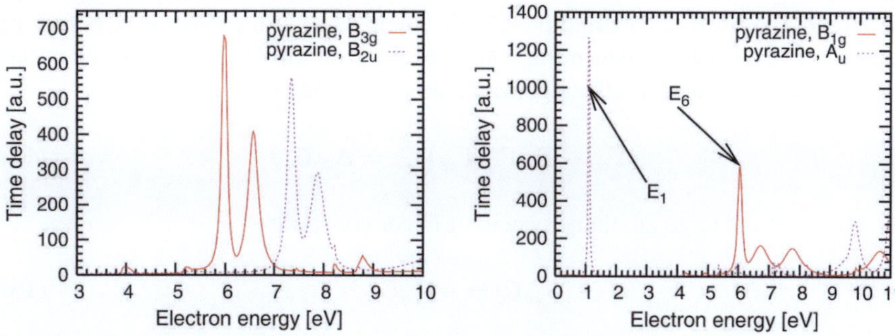

Fig. 13 The largest eigenvalue of the time-delay matrix as a function of energy for pyrazine. Adapted from Ref. [101], with permission of AIP Publishing

energy shape resonances of the π^* symmetry was also confirmed in case of the uracil molecule [119, 120]. Very detailed analysis [101] of the resonances and comparison among several diazines (pyrazine, pyrimidine, and pyridazine) reveals that many, but not all, of the resonances characterized in uracil can be traced back to equivalent resonances in diazines. An example of the time-delay analysis of resonances in several symmetries is shown, for pyrazine, in Fig. 13.

In comparison to the uracil calculations, where three π^* shape resonances were found, the computational studies of the pyrimidine [117] (cytosine and thymine) and the purine [121] bases (adenine and guanine) revealed, in all the four cases, four shape resonances in the energy range below 10 eV. Besides the analysis of the resonances necessary for understanding of the DEA nuclear dynamics, the *R*-matrix method was also used to study non-resonance collisions of the electrons with several biomolecules. Studied processes were elastic collisions [122, 123] and electronic excitation processes [124, 125].

D Finite Element *R*-Matrix Approach

In contrast to previous section, this formulation [126] attempts to avoid full diagonalization of large matrices. It starts from the variational principle for the logarithmic derivative b of the wave function Ψ, constant throughout the surface specified by r_0 in Fig. 12

$$-b = \frac{\partial \ln(r\Psi)}{\partial r} \, . \tag{12}$$

It can be shown [127] that such a variational principle results in a generalized eigenvalue problem for an eigenvalue of the logarithmic derivative

$$\Gamma C = \left(E - \bar{H}\right) C = \Lambda C b \, , \tag{13}$$

where the matrix Λ is the surface overlap matrix. However, the finite element basis set employed in this method allows to partition the basis functions into two subspaces, the closed and open, depending on whether their value at the surface r_0 is zero or nonzero. This allows to reduce the burden of the solution of Eq. (13) to the easier task of solving a much smaller eigenvalue problem in the open space

$$\left(\Gamma_{oo} - \Gamma_{oc}\Gamma_{cc}^{-1}\Gamma_{co}\right) C_o = \Lambda_{oo} C_o b \tag{14}$$

in addition to the large auxiliary system of linear equations

$$\Gamma_{cc} C_c = -\Gamma_{co} C_o \,, \tag{15}$$

where the subscripts "c" and "o" indicate the closed and open blocks, respectively. While the system of linear equations (15) is generally less demanding than a full-spectrum diagonalization (10), this set of equations must be now solved for every collision energy E, since the matrix $\Gamma = E - \bar{H}$.

Finite element R-matrix approach is a one-electron method. Thus again, the electron-molecule interaction is described by a model very similar to the one outlined in Sect. B. The method was successfully applied [128] to uracil and each of the DNA bases (thymine, cytosine, adenine, and guanine) taken as isolated molecules in their equilibrium geometry. Computed total time-delay, shown in Fig. 14, reveals resonances in a fairly good agreement with findings described in the previous section. In case of tetrahydrofuran and phosphoric acid, models of the subunits in the DNA backbone, authors observed [129] two shape resonances. They also employed the spatial shapes of the resonant structures relevant for the scattering wave functions in

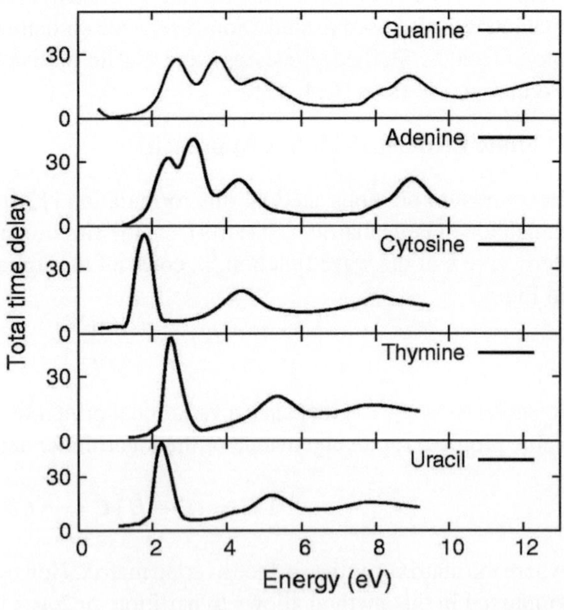

Fig. 14 Total time delay for the uracil and the DNA bases. Adapted from Ref. [128], with permission of AIP Publishing

order to gain a further insight in the possible systematic connections with the DEA in these subunits.

All the gas-phase data calculated for the DNA building blocks were later combined by use of the multiple scattering approach to explore an interaction of the free electron with the DNA double strand in the condensed phase [130, 131]. The targets considered were the idealized A-form and B-forms of GCGAATTGGC decamer together with its regularly sequenced poly(A)·poly(B) base pairs decamer.

E Schwinger Multichannel Approach

The heart of this method, the Schwinger variational principle (SVP), was introduced by Julian Schwinger in his lectures at Harvard University in 1947. It is based on the integral form of the Schrödinger equation—the Lippmann-Schwinger equation [132]. Since there exist several excellent reviews on the topic (for reviews related to atomic and molecular physics see, for example, Refs. [133–135]), only a brief description of the variational method will be given here. The essential quantity for calculation of cross section for transition from the channel β to the channel α is the scattering T-matrix defined as

$$T_{\alpha\beta} = \langle S_\alpha | V | \Psi_\beta^{(+)} \rangle = \langle \Psi_\alpha^{(-)} | V | S_\beta \rangle = \langle \Psi_\alpha^{(-)} | V - V G^{(+)} V | \Psi_\beta^{(+)} \rangle \,, \qquad (16)$$

where S_α is the asymptotic solution in the channel α, i.e. product of one-electron free solution with the target state. Interaction potential V describes interaction of the continuum electron with the target electrons and nuclei. The retarded Green's function $G^{(+)}$ is a resolvent of the free Hamiltonian, i.e. resolvent of sum of the target Hamiltonian and the kinetic energy of the scattered electron. The Lippmann-Schwinger equations have the following form:

$$\begin{aligned} |\Psi_\alpha^{(+)}\rangle &= |S_\alpha\rangle + G^{(+)} V |\Psi_\alpha^{(+)}\rangle \,, \\ |\Psi_\alpha^{(-)}\rangle &= |S_\alpha\rangle + G^{(-)} V |\Psi_\alpha^{(-)}\rangle \,. \end{aligned} \qquad (17)$$

By combining all the three expressions for the T-matrix (16) one can write a simple identity (of the type of $T + T - T$):

$$T_{\alpha\beta} = \langle S_\alpha | V | \Psi_\beta^{(+)} \rangle + \langle \Psi_\alpha^{(-)} | V | S_\beta \rangle - \langle \Psi_\alpha^{(-)} | V - V G^{(+)} V | \Psi_\beta^{(+)} \rangle \,. \qquad (18)$$

If the scattering wave functions $\Psi_\beta^{(+)}$ and $\Psi_\alpha^{(-)}$ were exact, the above equation would reduce to one of the three identities in Eq. (16). However, numerical solutions are never exact and small variations of the scattering wave functions $\delta\Psi_\beta^{(+)}$ and $\delta\Psi_\alpha^{(-)}$ in Eq. (16) lead to a non-zero variation of $T_{\alpha\beta}$ on l.h.s. of Eq. (16). The merit of the combined expression (18) is that the first-order variations on r.h.s. cancel out and the resulting T-matrix is stationary, i.e. $\delta T_{\alpha\beta} = 0$. The expression (18) is thus variationally stable and represents a variational principle.

An important feature of the SVP is that the trial functions $\langle \Psi_\alpha^{(-)}|$ and $|\Psi_\beta^{(+)}\rangle$ are always multiplied by the electron-molecule interaction operator V. Thus a trial basis for these functions needs only to cover the region of space where V does not vanish. Another advantage of the Schwinger variational method comes from the use of the Green's function $G^{(+)}$ that incorporates correct asymptotic boundary conditions and therefore allows more flexibility for the choice of the basis functions. On the other hand, a major drawback of the method lies in the difficulty of calculating the Green's function matrix elements $\langle \Psi_{\alpha}^{(-)} | V G^{(+)} V | \Psi_{\beta}^{(+)} \rangle$ that result in nine-dimensional integrals. The multichannel implementation of the SVP, the Schwinger multichannel (SMC) method, belongs to the class of the all-electron methods, i.e., the wave functions $\Psi_{\beta}^{(+)}$, $\Psi_{\alpha}^{(-)}$, and the interaction potential V are functions of the (N+1) electrons coordinates.

There exist a large number of applications of the SMC method to electron-molecule scattering and photoionization. Impressive progress was made possible by availability of higher computational power and massively parallel computational techniques [136]. This development made the SMC method feasible for computations of the elastic cross sections for polyatomic biomolecules. About two dozens of such studies from a few groups were published over the last decade. Among them we select a non-exhaustive set of samples in order to represent the capabilities of the method. The presence of π^* ring resonances for several conjugated ring molecules was confirmed in computed elastic cross sections [137–141]. Unfortunately, in these studies authors did not attempt to extract information about the resonance lifetimes. As an example, in Fig. 15 we show the computed spatial shapes of the three lowest π^* resonances of pyrazine.

Further development in the SMC method focused on application of non-local norm-conserving pseudopotentials [142]. Implementation of such pseudopotentials helps to reduce the number of the electrons in the target molecule by replacing the core electrons (typically the full shells) by a nonlocal interaction. The SMC method enhanced by pseudopotentials is denoted in the literature as SMCPP. The SMCPP

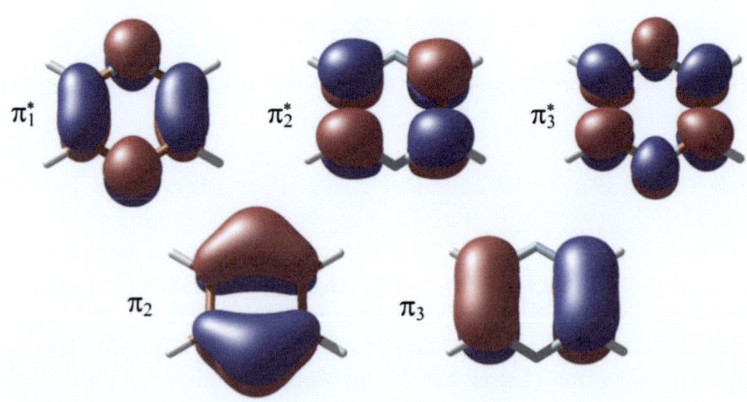

Fig. 15 Occupied and vacant π molecular orbitals of pyrazine involved in formation of low-energy electron-scattering resonances. Adapted from Ref. [138], with permission of American Physical Society

implementation was used intensively, over last years, to study electron driven reactions of molecular compounds used in cancer treatment. For example, incorporation of halogenated pyrimidines in the DNA to replace thymidine sensitizes the cell in the radiotherapy of tumors. Therefore, the radiosensitizers explored with the SMCPP methods are conjugated one-ring molecules with a hydrogen atom replaced by a halogen atom [143–145], the two-ring case of 2-chloroadenine [43], or the sulphur containing analogue of uracil, the 2-thiouracil that is also used as an anti-cancer drug [48].

3.1.2 Methods Based on Quantum Chemistry Calculation

While all the scattering methods presented in this chapter were efficient enough to be applied to calculations of the resonant parameters at the equilibrium nuclear geometry, attempts to compute a larger piece of the complex resonant surface of biomolecules are very rare [109, 110]. On the other hand, calculation of ground and excited real-energy potential surfaces describing bound states of polyatomic molecules is a fairly routine task in the field of quantum chemistry. Hence, there are several approaches available that attempt to exploit the tremendous development of quantum chemistry computational methods during the last few decades and to calculate complex energies of the metastable states that are relevant in the DEA description of the nuclear dynamics. These approaches tend to transfer the continuum state into the bound space and, therefore, they do not provide any dynamical information about the scattering wave function. However, they can often provide a complex resonant surface with a very good accuracy.

A Complex Absorbing Potentials

Particularly promising for large scale calculations of molecular resonances is the complex absorbing potential (CAP) method. The idea underlying the CAP method is to introduce an absorbing boundary condition in the exterior region of the molecular system [146]. In this way the wave function of the scattered electron becomes bound. Electron absorption is accomplished by replacing the Hamiltonian H by

$$H(\eta) = H - \mathrm{i}\eta W . \tag{19}$$

The soft-box absorbing potential W raises outside the molecule quadratically

$$W(\mathbf{r}, \mathbf{d}) = \sum_{i=x,y,z} W_i(x_i, d_i) , \tag{20}$$

$$W_i(x_i, d_i) = \begin{cases} 0, & |x_i| \le d_i \\ (|x_i| - d_i)^2, & |x_i| > d_i \end{cases} \quad (i = x, y, z) , \tag{21}$$

where the positive distances d_i must be chosen such that the resulting rectangular box, inside of which the CAP vanishes, fully encloses the molecule. For medium and large-sized polyatomic molecules a different type of the CAP was proposed [147, 148]. This smoothed Voronoi-CAP possesses an ability to follow the general shape and the symmetry of the polyatomic molecule.

Complex eigenvalues of the Hamiltonian $H(\eta)$ form a discrete set and they generally depend on the linear strength parameter η. The absorbing potential is too steep for large η parameters and causes an unwanted modification of the resonant wave function at relevant distances. If the parameter η is too small, the compact L^2 basis used in the calculations is not able to support the continuum resonant function which is damped too slowly. Hence, there exists an interval of η for which the Hamiltonian (19) and the basis set provide stable complex energies. This interval of stability is very clearly visible as characteristic cusp in the η-trajectory in the complex plane. This cusp is the resonant pole in the complex plane. An example of such a trajectory is displayed in Fig. 16.

The method was successfully applied to identification of the lowest π^* resonances of several heteroaromatic compounds (pyridine, pyrazine, and furan) [147]. The general agreement of the resonance positions with the available experiment is better when compared to results of the scattering methods described in the previous section. In case of chlorouracil authors computed complex potential energy curves for abstraction of the chlorine atom [150]. They observed several π^* resonances that appear ineffective for the dissociation of the C–Cl bond. However, in a manner similar to Fig. 11, these π^* resonances undergo a crossing with dissociative σ^* resonance and thus providing a pathway for an indirect dissociation mechanism.

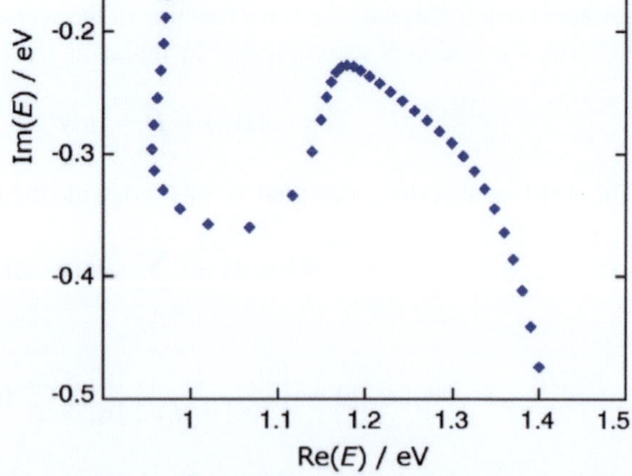

Fig. 16 Typical η-trajectory of the $^2B_1$ resonance of H_2CO at the ground equilibrium geometry. The CAP strength parameter η is varied from 10^{-8} to 10^{-2} in exponentially increasing steps. Adapted from Ref. [149], with permission of Elsevier

B Stabilization Methods

The so-called stabilization method [151] implements another tool for identifying the resonance or metastable states of negative ions. The underlying idea of the approach is based on an additional confinement of the studied system. Role of the confinement, as in the case of the CAP method, is to make the continuum resonant wave function bound. There are multiple variants of the stabilization method, they differ in ways how the confinement is implemented. It can be realized by an additional potential of a spherical-box shape [152], by a rotation of the coordinate into the complex plane (complex scaling methods) [153, 154], or even by scaling of the exponents of the Gaussian basis [155] thereby scaling the space available for the resonant wave function.

In a stabilization study of the anion metastable states, one carries out a series of calculations for different "box-size" parameters (they can be implemented as angles of the rotation in the complex scaling method or the exponent scaling factors in the orbital exponent stabilization approach). One then analyzes plots of the lowest energies (or electron affinities) of such system as functions of the box-size parameter. An example of such a plot is shown in Fig. 17. The eigenvalues associated with discretization of the continuum, so called box states, strongly depend on the box size and their energies fall steeply down as the box size increases. However, if the metastable state exists, one also finds a series of avoided crossings as seen in Fig. 17. This behavior results from the coupling of one state, whose energy is relatively insensitive to the box size, to the box states. Since the lifetime of the resonance state describes its coupling to the continuum, discretized to box states in this scenario, the splitting of the avoided crossing can be directly tied to the resonance width [151, 154, 156].

Simons and coworkers applied the method of the exponent scaling to several analyses of the neutral and lowest adiabatic anion curves for systems consisting

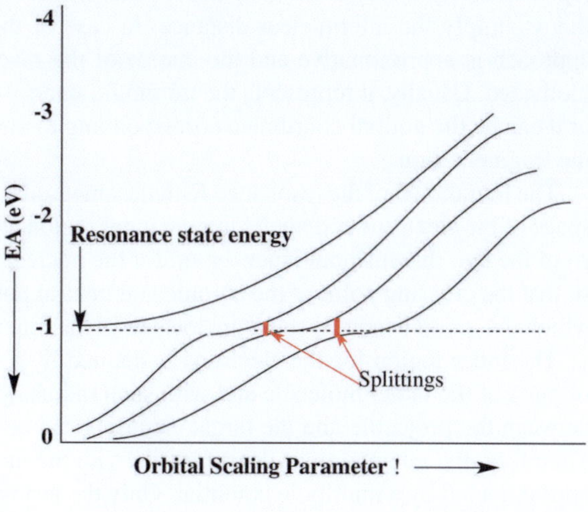

Fig. 17 Example of a stabilization plot showing four eigenvalues as a function of the orbital scaling parameter, illustrating the avoided crossings due to the presence of a metastable state. Adapted from Ref. [155]

of a base and a sugar-phosphate group [157–160]. Their interest focused on the indirect DEA processes in which the low energy electron ($\sim$1 eV) attaches to a resonant π^* orbital of a DNA base. Subsequently, this π^* anion then undergoes a σ^* sugar-phosphate C–O bond rupture and thus produces a SSB. The study was done for cytosine [157, 158] and thymine [159]. Their quantum ab initio calculations show that the lower adiabatic anion state is dominantly described by the excess electron occupying a π^* orbital of the base at stretched C–O bond lengths. Once the C–O bond is elongated, this π^* resonance crosses diabatically with the ribose-phosphate σ^* state that possesses a dissociative character. Authors also propose a theoretical model that helps to eliminate the necessity of large number of calculations on metastable anion states [160]. They introduce a simple 2×2 matrix model for the diabatic crossing point together with a few physically reasonable and computationally efficient approximations to the diabatic states. Such method then allows to compute the rate of transition from the capturing π^* state to the dissociating σ^* state. Different DNA damage mechanism is explored in a study [161] which suggests a possibility of cleaving a thymine N3–H bond to generate a nitrogen-centered anion and a hydrogen radical which might proceed to induce further bond cleavages.

An overview of these mechanisms by which low-energy free electrons attach to DNA and cause strong ($\sim$4 eV) covalent bonds to break, can be found in an interesting review by Simons [162].

3.2 *Nuclear Dynamics in the DEA*

3.2.1 Resonance *R*-Matrix Theory

The resonance *R*-matrix theory [163] has been applied to study the DEA and resonant VE of molecules where the motion of the nuclei during the dissociation can be represented by a single reaction coordinate ρ. In case of the diatomic molecules, this is simply the internuclear distance. In case of the polyatomic molecules, this approach is approximative and the choice of the reaction coordinate is physically motivated. Usually, it represents the minimum-energy dissociation path of the target or it can be the normal coordinate corresponding to the separation of the products of the fragmentation.

The foundation of the resonance *R*-matrix method is a separation of the coordinate space of the electrons and nuclei into an inner region and outer region. The boundary ρ_0 of the one-dimensional inner region for the nuclear reaction coordinate is chosen so that the crossing point of the anionic and neutral potential-energy curves, beyond which the anion becomes stable, is located in the inner region.

The inner region for the electrons is defined by a sphere centered in the center of mass of the target molecule and with such radius r_0 that the exchange interaction between the projectile and the target electrons can be neglected outside the sphere. Therefore, the interaction of the projectile with the target in the outer region can be approximated by a multipole potential. Only the terms representing the static dipole moment $\mathbf{D}(\rho)$ and isotropic part of the polarizability $\alpha(\rho)$ of the target molecule are considered in the practical calculations [163]:

$$W(\mathbf{r}_{N+1}, \rho) = -\frac{\mathbf{D}(\rho) \cdot \mathbf{r}_{N+1}}{r_{N+1}^3} - \frac{\alpha(\rho)}{2r_{N+1}^4}. \tag{22}$$

The total wave function in the outer region is in the resonance R-matrix theory expressed as follows:

- In the attachment region, where $r_{N+1} > r_0$ and $\rho < \rho_0$,

$$\psi(\mathbf{r}_1, \ldots, \mathbf{r}_N, \mathbf{r}_{N+1}, \rho) = \mathcal{A}\psi_0(\mathbf{r}_1, \ldots, \mathbf{r}_N, \rho) \sum\!\!\!\!\!\!\int_{\nu} \varphi_\nu(\rho) F_\nu(\mathbf{r}_{N+1}), \tag{23}$$

  where $\mathcal{A}$ is the antisymmetrization operator of the electrons, $\psi_0(\mathbf{r}_1, \ldots, \mathbf{r}_N, \rho)$ is the electronic wave function of the neutral molecule in the Born-Oppenheimer approximation, $\varphi_\nu(\rho)$, $\nu = 0, 1, \ldots$ are the vibrational wave functions of the target. Substituting Eq. (23) into the Schrödinger equation, projecting onto $\psi_0(\mathbf{r}_1, \ldots, \mathbf{r}_N, \rho)\varphi_{\nu'}(\rho)$ and employing Eq. (22) yields a set of coupled differential equations for the wave functions of the scattered electron $F_\nu(\mathbf{r}_{N+1})$. However, the off-fdiagonal dipole and polarization coupling terms $\langle\varphi_\nu(\rho)|\mathbf{D}(\rho)|\varphi_{\nu'}(\rho)\rangle$ and $\langle\varphi_\nu(\rho)|\alpha(\rho)|\varphi_{\nu'}(\rho)\rangle$, respectively, are neglected for $\nu \neq \nu'$ since the resonance R-matrix method assumes that the direct excitation of the target molecule by an electron impact is weak.
- In the dissociation region, where $r_{N+1} < r_0$ and $\rho > \rho_0$,

$$\begin{aligned}\psi(\mathbf{r}_1, \ldots, \mathbf{r}_N, \mathbf{r}_{N+1}, \rho) =& \mathcal{A}\psi_A(\mathbf{r}_1, \ldots, \mathbf{r}_M) \left[\psi_{B^-}(\mathbf{r}_{M+1}, \ldots, \mathbf{r}_N, \mathbf{r}_{N+1})\phi(\rho)\right. \\ &\left. + \psi_B(\mathbf{r}_{M+1}, \ldots, \mathbf{r}_N) \int \psi_{\mathbf{p}}(\mathbf{r}_{N+1})\phi_{\mathbf{P}}(\rho) d\mathbf{p}\right],\end{aligned} \tag{24}$$

  where $\psi_A(\mathbf{r}_1, \ldots, \mathbf{r}_M)$ and $\psi_{B^-}(\mathbf{r}_{M+1}, \ldots, \mathbf{r}_N, \mathbf{r}_{N+1})$ are the electronic wave functions of the neutral and anionic products of the DEA, respectively. The wave function of the nuclei in the dissociation-attachment (DA) channel is denoted as $\phi(\rho)$. The wave function $\psi_B(\mathbf{r}_{M+1}, \ldots, \mathbf{r}_N)$ represents the electronic state of the neutral fragment B, $\psi_{\mathbf{p}}(\mathbf{r}_{N+1})$ is the continuum wave function of the electron interacting with this neutral fragment in the channel where the target molecule dissociates into neutral fragments. Corresponding wave function of the nuclei in this channel is $\phi_{\mathbf{P}}(\rho)$. The relation between the relative momentum of the fragments in the neutral dissociation channel $\mathbf{P}$ and the momentum of the scattering electron in the continuum is established by the conservation of the total energy

$$E_e = E_D + \frac{p^2}{2} + \frac{P^2}{2M}. \tag{25}$$

The energy interval of the interest in the studies of the DEA is well below the threshold of the electronic excitation. Therefore, the expansions of the total wave function in Eqs. (23) and (24) include only the ground electronic state of the target molecule as well as of the products and the electronically excited states are neglected.

The smooth matching between the wave function in the outer region discussed above and the solutions of the Schrödinger equation in the inner region is formulated using the R-matrix in the fixed-nuclei approximation. It can be expressed in terms of the pole expansion [114] as

$$R(E_e, \rho) = \frac{1}{2r_0} \sum_i \frac{\gamma_i^2(\rho)}{E_i(\rho) - E_e}, \tag{26}$$

where $E_i(\rho)$ and $\gamma_i^2(\rho)$ are the poles and amplitudes of the electronic R-matrix, respectively, evaluated for the reaction coordinate ρ. In order to represent the electronic resonance, the R-matrix (26) can be approximated by a one-pole function [164, 165]

$$R_a(\rho) = R^{(0)}(\rho) + R_r = \frac{\gamma^2(\rho)}{E_1(\rho) - E_e} + R_r(E_e), \tag{27}$$

where the amplitude $\gamma(\rho)$, the pole $E_1(\rho)$ and the smooth function $R_r(E_e)$ are the parameters of the theory and they are fitted to the results of the ab initio fixed-nuclei scattering calculations as discussed later in this section.

The denominator in Eq. (27) can be rewritten as

$$E_1(\rho) - E_e = U(\rho) - V(\rho) + \mathcal{E} - E, \tag{28}$$

where $V(\rho)$ and $\mathcal{E}$ are the potential and total energy of the neutral molecule, respectively, $U(\rho) = V(\rho) + E_1(\rho)$ and E is the total energy of the whole anionic system. Following Schneider et al. [166], the coupling of the electronic and nuclear motion is taken into account by replacing the kinetic energy in Eq. (28) by the corresponding operator:

$$\mathcal{E} - V(\rho) \rightarrow T \equiv -\frac{1}{2M}\frac{\partial^2}{\partial \rho^2}. \tag{29}$$

This replacement turns R_a defined by Eq. (27) into an operator. Its matrix elements are obtained by its projection onto the asymptotic states defined by Eq. (23) and Eq. (24). Smooth matching of the solutions of the Schrödinger equation with the asymptotic boundary conditions of the incoming and outgoing wave performed on the boundary of the inner region then yields the S-matrix of the DEA and resonant VE [163].

Application of the resonance R-matrix method as it was formulated by Fabrikant [163] to study the VE and DEA of HCl showed that the calculated cross sections of the VE near the channel thresholds are sensitive to the value of ρ_0 [91]. This issue was solved by a reformulation of the theory in terms of the non-local complex potential [167, 168] where ρ_0 does not explicitly appear.

The parameters of the theory are the fixed-nuclei functions that describe the neutral target molecule and the anionic complex. The potential-energy curve of the neutral target $V(\rho)$ is usually parametrized by the Morse potential with the parameters chosen to reproduce the experimental dissociation energy and the vibrational frequency of the normal mode corresponding to the reaction coordinate.

The potential-energy curve of the anionic complex $U(\rho)$ is also parametrized by the Morse potential. The parameters that determine the asymptotic energy are chosen to reproduce the experimental electron affinity of the anionic fragment. Remaining parameters are optimized to correctly reproduce the energy of the negative ion at those values of ρ where it is stable and the fixed-nuclei resonance energy for those values of ρ where the anionic complex is metastable. These energies are usually obtained from the ab initio scattering calculations performed in the fixed-nuclei approximation.

The fixed-nuclei surface amplitude $\gamma(\rho)$ and the background term $R_r(E_e)$ are optimized using the fixed-nuclei phase shifts obtained from the ab initio calculations of the electron collisions with the target. The functions $\gamma(\rho)$ and $R_r(E_e)$ are optimized in such way that the electronic R-matrix (27) yields the fixed-nuclei phase shifts in agreement with those obtained from the ab initio scattering calculations for all relevant values of ρ.

3.2.2 Application of Resonance *R*-matrix Theory to Biological Molecules

In order to study the DEA process in amino-acids and closely related molecules, Gallup et al. [92] carried out the calculation of the DEA to formic acid and glycine using the resonance R-matrix method [91]. The main goal of this work was to find a mechanism of breaking the O–H bond. Previously published ab initio fixed-nuclei calculations of the anionic complex potential-energy surfaces [94] suggested that the out-of-plane distortions can couple the initially formed π^* resonance with a σ^* anion state and this makes the fragmentation possible. However, Scheer et al. [169] in their later experimental study pointed out the arguments for invoking a single σ^* resonance; the peak of the DEA cross sections in this fragmentation channel lies 0.5 eV below the energy of the COOH π^* resonance. Gallup et al. [92] in their work addressed the question, whether the planar geometry of the system can be preserved during the fragmentation process or whether the coupling of the σ^* and π^* state plays a significant role.

The reaction coordinate used in this study corresponds to stretching the O–H bond along the minimum-energy dissociation path. The scattering phase shifts in the fixed-nuclei approximation necessary to provide the parameters of the resonance R-matrix method were obtained using the finite-element discrete model (FEDM) method [170] combined with the variational technique for quasi-bound states [171]. As it is illustrated in Fig. 18 for case of HCOOH, calculated width of the σ^* resonance is relatively large at the equilibrium geometry R_e and decreases slowly as the reaction

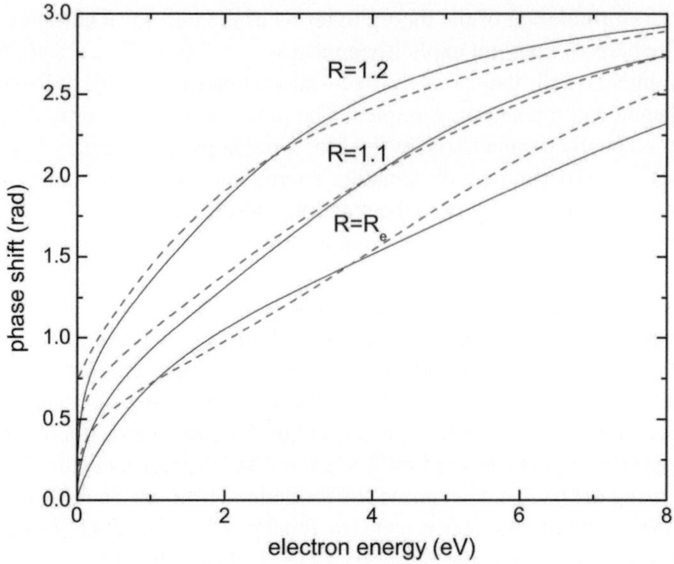

Fig. 18 Resonant contribution to the scattering phase shift for formic acid for the internuclear distances $R = R_e = 0.9695$ (equilibrium geometry of the neutral molecule), $R = 1.1$, and $R = 1.2$ Å. Corresponding values of the reaction coordinate ρ are 0, 0.247, and 0.436 a.u. The *solid lines* represent the results of the FEDM calculation, fit of the one-pole *R*-matrix [Eq. (27)] is represented by the *red dashed lines*. Adapted from Ref. [92], with permission of American Physical Society

coordinate approaches the point where the negative ion becomes stable (see Fig. 19). As can be seen in Figs. 20 and 21, even this broad σ^* resonance yields cross sections of the DEA that are in good qualitative agreement with the experimental results [172–174]. The DEA cross section calculated for HCOOH is also quantitatively consistent with the experimental data for energies above 1.1 eV. Although the theoretical peak cross section for glycine is much higher than the normalized results of the experiment, the qualitative agreement of the calculated and measured structures is encouraging.

Since the model of the nuclear dynamics by Gallup et al. [92] based only on the broad σ^* resonance and single reaction coordinate reproduced the experimental results, the authors concluded that the σ^*/π^* coupling does not play any significant role in breaking the O–H bond by the electron impact, while the non-local effects of the broad σ^* resonance are essential. Similar mechanism was found responsible for the DEA to the hydrogen halides [91, 163].

This conclusion stimulated a discussion between Gallup et al. [92] and the authors of the previous theoretical study [94] proposing the DEA mechanism via σ^*/π^* coupling [175–177]. Rescigno et al. [175] in their comment admitted that the non-local effects can play an important role in the DEA to the formic acid. However, they claimed that their detailed ab initio study of the electron collisions with HCOOH in the fixed-nuclei approximation [94] does not show any sign of a σ^* resonance at equilibrium nuclear geometry within the energy range below 6 eV. Another issue

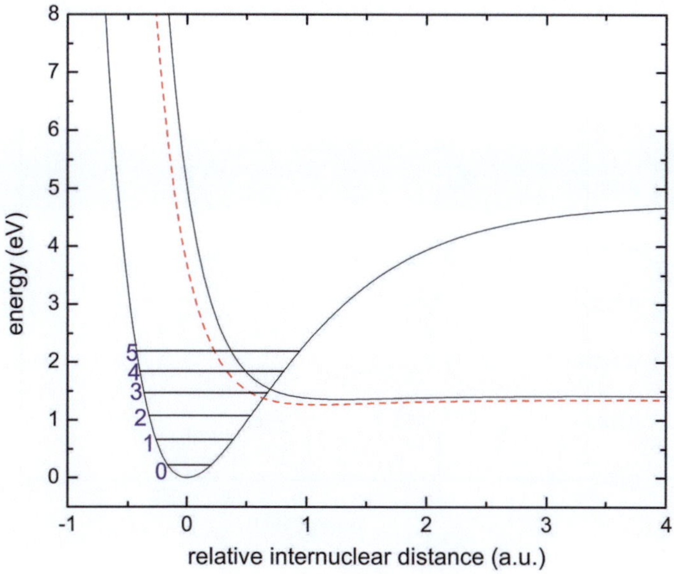

Fig. 19 Potential-energy curves as functions of the effective reaction coordinate ρ (representing the O–H bond). Vibrational energy levels are shown by horizontal lines. The *solid anion curve* corresponds to formic acid and the *dashed line* represents the anionic curve for glycine. Adapted from Ref. [92], with permission of American Physical Society

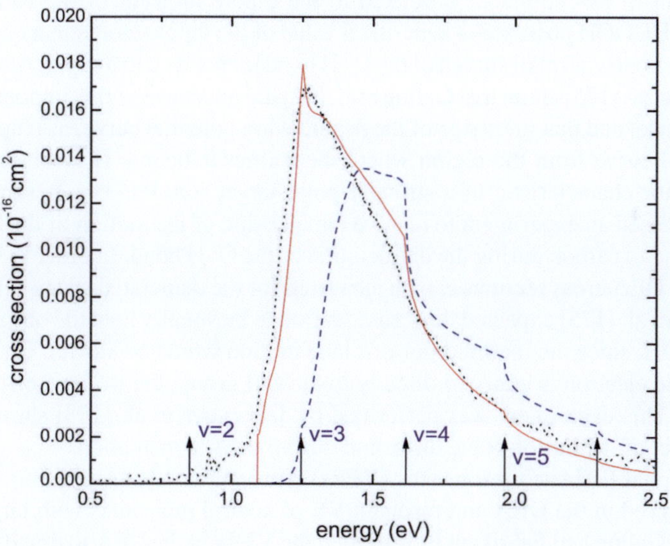

Fig. 20 DEA cross sections for formic acid. *Solid* and *dashed curves* represent the results of the calculation for two different fits of the fixed-nuclei ab initio results. *Dotted curve* corresponds to the experimental data [172, 173]. Adapted from Ref. [92], with permission of American Physical Society

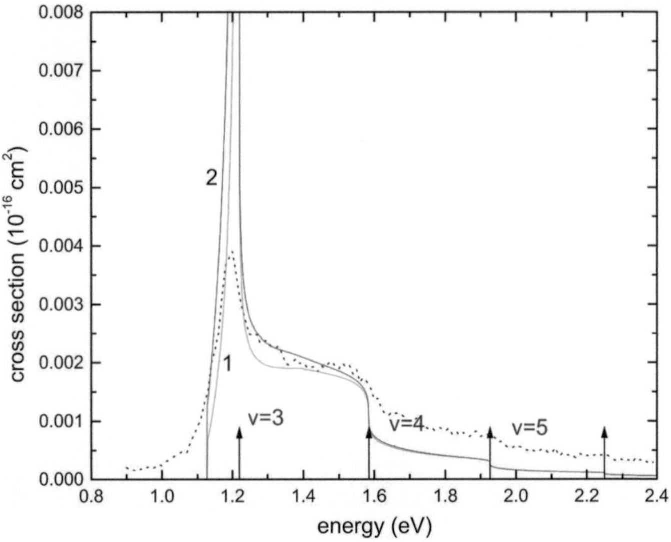

Fig. 21 DEA cross sections for glycine. *Solid* and *dashed curves* represent the results of the calculation for two different fits of the fixed-nuclei ab initio results. *Dotted curve* corresponds to the experimental data [174]. Adapted from Ref. [92], with permission of American Physical Society

pointed out in this comment is related to the dipole moment of the formic acid. Although HCOOH possesses a subcritical value of the dipole moment at equilibrium nuclear geometry, a small stretch of the O–H bond increases it to a supercritical value. Rescigno et al. [175] claim that Gallup et al. [92] did not consider this important effect in their model and that the shape of the negative ion potential curve, as it approaches the neutral curve from the region where the former is bound, is inconsistent with the behavior characteristic of a strongly polar target (see Fig. 19). Rescigno et al. [175] proposed an experiment to resolve the question of the motion of the hydrogen atom bound to carbon during the dissociation of the O–H bond. In order to study this effect, the DEA cross sections can be measured for the deuterated species, DCOOH. Rescigno et al. [175] expected this cross section to be smaller than the one measured for HCOOH, since the required out-of-plane motion would be slower. On the other hand, if the electron is attached directly to the OH group, the isotopic effect would be small. This experiment was performed by Janečková et al. [178] showing only small effect of the out-of-plane distortion during the fragmentation.

Vibrational Feshbach resonances (VFRs) represent another interesting phenomenon observed in the DEA and resonant VE of neutral molecules with large dipole moments. Gallup and Fabrikant [93] studied the VFRs in the DEA to uracil using the resonant *R*-matrix method. The authors performed an one-dimensional calculation of the nuclear dynamics for the DEA loss of the hydrogen atom from the N_1 site (see Fig. 22). This fragmentation involves a broad σ^* shape resonance and VFRs, and, therefore, requires the non-local approach. The dipole moment of uracil is $D = 4.7$

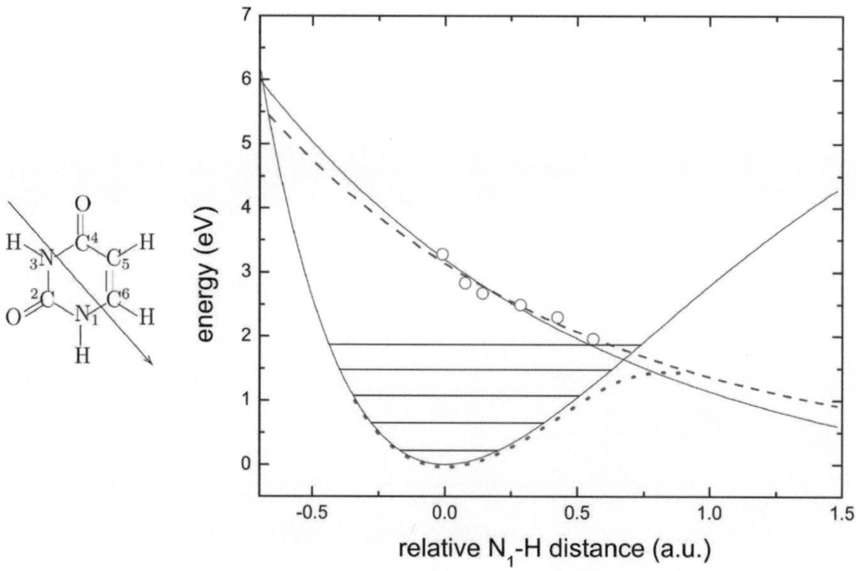

Fig. 22 *Left panel* Molecular structure of uracil. The *arrow* indicates the direction of the dipole moment. *Right panel* Potential-energy curves as functions of the N_1–H distance relative to equilibrium ($R_e = 1.890$ a.u.). The two anion curves (*solid black* and *dashed blue*) are the *R*-matrix poles for two different parametrizations of the model. *Circles* are corresponding energies from the FEDM calculations. The *red dotted line* is the dipole anion curve corresponding to the same model as the *blue dashed line*. The *horizontal lines* indicate the positions of vibrational energy levels of the neutral molecule. Adapted from Ref. [93], with permission of American Physical Society

D. Although no evidence of a broad σ^* resonance at low energies was found in previously published ab initio fixed-nuclei scattering calculations, Winstead and McKoy [137] found a broad peak in the partial A' cross sections centered around 8.5 eV and the authors claim that it is due to a σ^* shape resonance. Gallup and Fabrikant [93] conclude that this situation is similar to formic acid and glycine [92] and that the resonance is very broad and "hidden" in the background scattering predominated by the long-range dipole and polarization interaction.

Calculated DEA cross section is plotted in Fig. 23. Very pronounced VFRs are seen, particularly two resonances below the $\nu = 2$ and $\nu = 3$ excitation thresholds. This is the first case when two VFRs in DEA are produced near the same vibrational threshold. This is a consequence of a high dipole moment of uracil.

DEA to biologically relevant molecules in water environment represents another difficult challenge for the theoretical research. Smyth et al. [179] studied the DEA to uracil in the water cluster environment using a multiple-scattering method developed by Caron and Sanche [180] and later employed to study the DEA to molecules in a cluster environment by Fabrikant et al. [181]. Smyth et al. [179] performed a calculation for five water molecules surrounding uracil where their mutual orientation

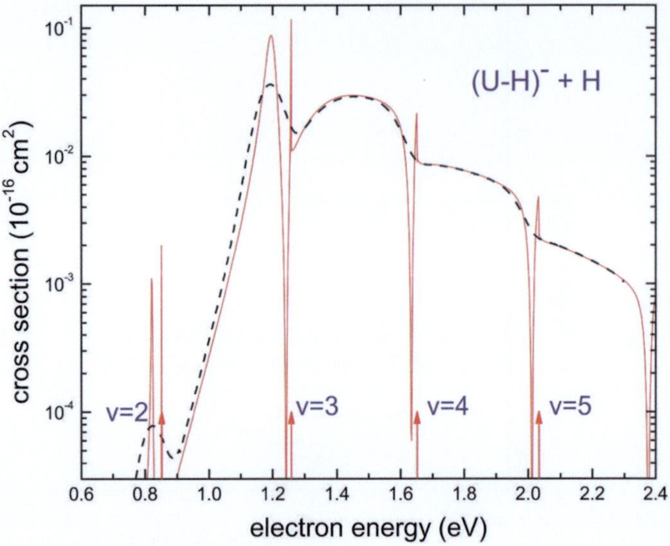

Fig. 23 DEA cross section for uracil. The *solid curve* is the non-averaged cross section, the *dashed curve* is the cross section folded with a Gaussian distribution of width 0.07 eV. The vibrational excitation thresholds are indicated by arrows. Adapted from Ref. [93], with permission of American Physical Society

was optimized using a density-functional theory (DFT). The DEA cross section calculated for the case of the water environment is considerably enhanced compared to the cross section calculated for an isolated molecule (see Fig. 24).

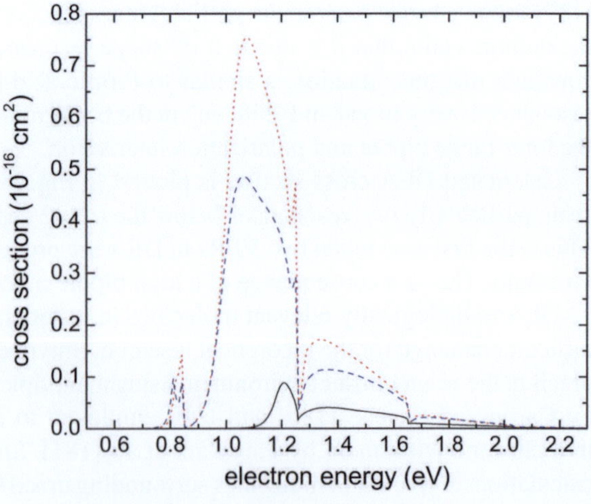

Fig. 24 Dissociative electron attachment cross section for isolated uracil (*solid curve*) and uracil embedded in the $(H_2O)_5$ water cluster. *Dotted red curve* is the cross section calculated with inclusion of scattering by all molecules in the cluster. *Dashed blue curve* is the cross section calculated with inclusion of scattering by only water molecules. Adapted from Ref. [179], with permission of AIP Publishing

The authors attributed this enhancement to the reduction of the resonance width due to electron trapping in the water environment [181]. In addition, the water clusters cause a negative shift of the potential-energy curves that leads to further increase of the DEA cross section.

References

1. Baccarelli I, Bald I, Gianturco FA, Illenberger E, Kopyra J (2011) Phys Rep 508:1
2. Pimblott SM, LaVerne JA (2007) Radiat Phys Chem 76:1244
3. Boudaïffa B, Cloutier P, Hunting D, Huels MA, Sanche L (2000) Science 287:1658
4. Martin F, Burrow PD, Cai Z, Cloutier P, Hunting D, Sanche L (2004) Phys Rev Lett 93:068101
5. Bald I, Langer J, Tegeder P, Ingólfsson O (2008a) Int J Mass Spectrom 277:4
6. Ptasińska S, Denifl S, Scheier P, Illenberger E, Märk TD (2005a) Angew Chem, Int Ed 44:6941
7. Huels MA, Hahndorf I, Illenberger E, Sanche L (1998) J Chem Phys 108:1309
8. Hanel G, Gstir B, Denifl S, Scheier P, Probst M, Farizon B, Farizon M, Illenberger E, Märk TD (2003) Phys Rev Lett 90:188104
9. Abdoul-Carime H, Gohlke S, Fischbach E, Scheike J, Illenberger E (2004) Chem Phys Lett 387:267
10. Ptasińska S, Denifl S, Mróz B, Probst M, Grill V, Illenberger E, Scheier P, Märk TD (2005b) J Chem Phys 123:124302
11. Ptasińska S, Denifl S, Grill V, Märk TD, Scheier P, Gohlke S, Huels MA, Illenberger E (2005c) Angew Chem, Int Ed 44:1647
12. Silva FF, Matias C, Almeida D, García G, Ingólfsson O, Flosadóttir HD, Ómarsson B, Ptasińska S, Puschnigg B, Scheier P, Limão-Vieira P, Denifl S (2013) J Am Soc Mass Spectrom 24:1787
13. Ptasińska S, Denifl S, Scheier P, Märk TD (2004) J Chem Phys 120:8505
14. Bald I, Kopyra J, Illenberger E (2006) Angew Chem, Int Ed 5:4851
15. Baccarelli I, Gianturco FA, Grandi A, Sanna N, Lucchese RR, Bald I, Kopyra J, Illenberger E (2007) J Am Chem Soc 129:6269
16. Bald I, Flosadóttir HD, Kopyra J, Illenberger E, Ingólfsson O (2009) Int J Mass Spectrom 280:190
17. Sulzer P, Ptasińska S, Zappa F, Mielewska B, Milosavljević AR, Scheier P, Märk TD, Bald I, Gohlke S, Huels MA, Illenberger E (2006) J Chem Phys 125:044304
18. Flosadóttir HD, Bald I, Ingólfsson O (2011) Int J Mass Spectrom 305:50
19. Bald I, Kopyra J, Dabkowska I, Antonsson E, Illenberger E (2007) J Chem Phys 126:074308
20. König C, Kopyra J, Bald I, Illenberger E (2006) Phys Rev Lett 97:018105
21. Kopyra J (2012) Phys Chem Chem Phys 14:8287
22. Ptasińska S, Denifl S, Gohlke S, Scheier P, Illenberger E, Märk TD (2006) Angew Chem, Int Ed 45:1893
23. Bald I, Dabkowska I, Illenberger E (2008b) Angew Chem, Int Ed 47:8518
24. Berdys J, Anusiewicz I, Skurski P, Simons J (2004a) J Am Chem Soc 126:6441
25. Kopyra J in preparation
26. Sanche L (2009) Chem Phys Lett 474:1
27. Fichtinger-Schepman AMJ, van Oosterom AT, Lohman PHM, Berends F (1987) Cancer Res 47:3000
28. Brabek V (2001) Platinum-based drugs in cancer therapy. In: Kelland LR, Farrell NP. Humana Press Inc., Totowa, NJ
29. Zheng Y, Hunting DJ, Ayotte P, Sanche L (2008a) Phys Rev Lett 100:198101
30. Lu Q-B, Kalantari S, Wang C-R (2007) Mol Pharm 4:624
31. Lu Q-B (2010) Mutat Res Rev Mutat 704:190

32. Kopyra J, Koenig-Lehmann C, Bald I, Illenberger E (2009) Angew Chem, Int Ed 48:7904
33. Carey R, Lucchese RR, Gianturco FA (2013) J Chem Phys 138:204308
34. Cohen SS, Flaks JG, Barner HD, Loeb MR, Lichtenstein J (1958) Proc Natl Acad Sci USA 44:1004
35. Duschinsky R, Pleven E, Heidelberger C (1957) J Am Chem Soc 79:4559
36. Zamenhof S, de Giovanni R, Greer S (1958) Nature 181:827
37. Szybalski W (1974) Cancer Chemother Rep 58:539
38. Zimbrick JD, Ward JF, Myers LS (1969) Int J Radiat Biol 16:505
39. Bhatia K, Schuler RH (1973) J Phys Chem 77:1888
40. Rivera E, Schuler RH (1983) J Phys Chem 87:3966
41. Abdoul-Carime H, Huels MA, Illenberger E, Sanche L (2001) J Am Chem Soc 123:5354
42. Abdoul-Carime H, Huels MA, Illenberger E, Sanche L (2003) Int J Mass Spectrom 228:703
43. Kossoski F, Kopyra J, Varella MTDN (2015) Phys Chem Chem Phys 17:28958
44. Kopyra J, Keller A, Bald I (2014a) RSC Adv 4:6825
45. Elion G (1989) Science 244:41
46. Cui G, Fang W (2013) J Chem Phys 138:044315
47. Kopyra J, Freza S, Abdoul-Carime H, Marchaj M, Skurski P (2014b) Phys Chem Chem Phys 16:5342
48. Kopyra J, Abdoul-Carime H, Kossoski F, Varella MTDN (2014) Phys Chem Chem Phys 16:25054
49. Chomicz L, Zdrowowicz M, Kasprzykowski F, Rak J, Buonaugurio A, Wang Y, Bowen KH (2013) J Phys Chem Lett 4:2853
50. Huels MA, Boudaïffa B, Cloutier P, Hunting D, Sanche L (2003) J Am Chem Soc 125:4467
51. Kumar SVK, Pota T, Peri D, Dongre AD, Rao BJ (2012) J Chem Phys 137:045101
52. Alizadeh E, Orlando TM, Sanche L (2015) Annu Rev Phys Chem 66:379
53. Alizadeh E, Sanche L (2012) Chem Rev 112:5578
54. Chen Y, Aleksandrov A, Orlando TM (2008) Int J Mass Spectrom 277:314
55. Orlando TM, Oh D, Chen Y, Aleksandrov AB (2008) J Chem Phys 128:195102
56. Śmiałek MA, Jones NC, Balog R, Mason NJ, Field D (2011) Eur Phys J D 62:197
57. Śmiałek MA, Balog R, Jones NC, Field D, Mason NJ (2010) Eur Phys J D 60:31
58. Śmiałek MA, Moore SA, Mason NJ, Shuker DEG (2009) Radiat Res 172:529
59. Rezaee M, Cloutier P, Bass AD, Michaud M, Hunting DJ, Sanche L (2012) Phys Rev E 86:031913
60. Boulanouar O, Fromm M, Bass AD, Cloutier P, Sanche L (2013) J Chem Phys 139:055104
61. Keller A, Rackwitz J, Cauet E, Lievin J, Körzdörfer T, Rotaru A, Gothelf KV, Besenbacher F, Bald I (2014) Sci Rep 4:7391
62. Keller A, Bald I, Rotaru A, Cauet E, Gothelf KV, Besenbacher F (2012) ACS Nano 6:4392
63. Panajotovic R, Martin F, Cloutier P, Hunting D, Sanche L (2006) Radiat Res 165:452
64. Keller A, Kopyra J, Gothelf KV, Bald I (2013) New J Phys 15:14
65. Dugal P-C, Abdoul-Carime H, Sanche L (2000) J Phys Chem B 104:5610
66. Ptasińska S, Sanche L (2006) J Chem Phys 125:144713
67. Zheng Y, Wagner JR, Sanche L (2006) Phys Rev Lett 96:208101
68. Li ZJ, Cloutier P, Sanche L, Wagner JR (2010) J Am Chem Soc 132:5422
69. Li ZJ, Cloutier P, Sanche L, Wagner JR (2011) J Phys Chem B 115:13668
70. Solomun T, Sturm H (2007) J Phys Chem B 111:10636
71. Solomun T, Seitz H, Sturm H (2009) J Phys Chem B 113:11557
72. Vogel S, Rackwitz J, Schürman R, Prinz J, Milosavljević AR, Réfrégiers M, Giuliani A, Bald I (2015) J Phys Chem Lett 6:4589
73. Polska K, Rak J, Bass AD, Cloutier P, Sanche L (2012) J Chem Phys 136:075101
74. Park Y, Polska K, Rak J, Wagner JR, Sanche L (2012) J Phys Chem B 116:9676
75. Rezaee M, Hunting DJ, Sanche L (2013) Int J Radiat Oncol Biol Phys 87:847
76. Rezaee M, Alizadeh E, Cloutier P, Hunting DJ, Sanche L (2014) ChemMedChem 9:1145
77. McMahon SJ, Hyland WB, Muir MF, Coulter JA, Jain S, Butterworth KT, Schettino G, Dickson GR, Hounsell AR, O'Sullivan JM, Prise KM, Hirst DG, Currell FJ (2011) Sci Rep 1:18

78. Jain S, Hirst DG, O'Sullivan JM (2012) Br J Radiol 85:101
79. Jeremic B, Aguerri AR, Filipovic N (2013) Clin Transl Oncol 15:593
80. Zheng Y, Hunting DJ, Ayotte P, Sanche L (2008b) Radiat Res 169:19
81. Zheng Y, Sanche L (2009) Radiat Res 172:114
82. Setua S, Ouberai M, Piccirillo SG, Watts C, Welland M (2014) Nanoscale 6:10865
83. Ptasińska S, Sanche L (2007) Phys Rev E 75:031915
84. Alizadeh E, Sanz AG, García G, Sanche L (2013) J Phys Chem Lett 4:820
85. Kumar A, Sevilla MD (2009) Chem Phys Chem 10:1426
86. Schyman P, Laaksonen A (2008) J Am Chem Soc 130:12254
87. Smyth M, Kohanoff J (2012) J Am Chem Soc 134:9122
88. Schürmann R, Bald I (2016) J Phys Chem C
89. O'Malley T (1966) Phys Rev 150:14
90. Horáček J, Paidarová I, Čurík R (2015) J Chem Phys 143:184102
91. Fabrikant II (1991) Phys Rev A 43:3478
92. Gallup GA, Burrow PD, Fabrikant II (2009a) Phys Rev A 79:042701
93. Gallup GA, Fabrikant II (2011) Phys Rev A 83:012706
94. Rescigno TN, Trevisan CS, Orel AE (2006) Phys Rev Lett 96:213201
95. Siegert AJF (1939) Phys Rev 56:750
96. Hazi AU (1979) Phys Rev A 19:920
97. Baccarelli I, Sebastianelli F, Gianturco FA, Sanna N (2009) Eur Phys J D 51:131
98. Smith FT (1960) Phys Rev 118:349
99. Igarashi A, Shimamura I (2004a) Phys Rev A 70:012706
100. Igarashi A, Shimamura I (2004b) J Phys B: At, Mol Opt Phys 37:4221
101. Mašín Z, Gorfinkiel JD (2012) J Chem Phys 137:204312
102. Allen LC, Karo AM (1960) Rev Mod Phys 32:275
103. Gianturco FA, Thompson DG, Jain A (1995) Computational methods for electron-molecule collisions. In: Huo WM, Gianturco FA, Computational Methods for Electron-Molecule Collisions. Plenum Press, New York, pp 75–118
104. Hara S (1969) J Phys Soc Jap 27:1009
105. Riley M, Truhlar D (1975) J Comp Phys 63:2182
106. Nishimura T, Itikawa Y (1996) J Phys B: At, Mol Opt Phys 29:4213
107. Takekawa M, Itikawa Y (1999) J Phys B: At, Mol Opt Phys 32:4209
108. Altmann SL, Cracknell AP (1965) Rev Mod Phys 37:19
109. Gianturco FA, Sebastianelli F, Lucchese RR, Baccarelli I, Sanna N (2008) J Chem Phys 128:174302
110. Panosetti C, Baccarelli I, Sebastianelli F, Gianturco FA (2010) Eur Phys J D 60:21
111. Gianturco FA, Sebastianelli F, Lucchese RR, Baccarelli I, Sanna N (2009) J Chem Phys 131:249901
112. Goumans TPM, Gianturco FA, Sebastianelli F, Baccarelli I, Rivail JL (2009) J Chem Theory Comput 5:217
113. Morgan L, Gillan C, Tennyson J, Chen X (1997) J Phys B: At, Mol Opt Phys 30:4087
114. Tennyson J (2010) Phys Rep 491:29
115. Carr JM, Galiatsatos PG, Gorfinkiel JD, Harvey AG, Lysaght MA, Madden D, Mašín Z, Plummer M, Tennyson J, Varambhia HN (2012) Eur Phys J D 66:58
116. Burke P (2011) R-Matrix Theory of Atomic Collisions: Application to Atomic, Molecular and Optical Processes, Springer series on atomic, optical, and plasma physics. Springer, Berlin Heidelberg
117. Dora A, Bryjko L, van Mourik T, Tennyson J (2012a) J Phys B: At, Mol Opt Phys 45:175203
118. Mašín Z, Gorfinkiel JD (2011) J Chem Phys 135:144308
119. Dora A, Tennyson J, Bryjko L, van Mourik T (2009) J Chem Phys 130:164307
120. Mašín Z, Gorfinkiel JD (2014) Eur Phys J D 68:112
121. Dora A, Bryjko L, van Mourik T, Tennyson J (2012b) J Chem Phys 136:024324
122. Palihawadana P, Sullivan JP, Buckman SJ, Mašín Z, Gorfinkiel JD, Blanco F, Garcia G, Brunger MJ (2013) J Chem Phys 139:014308

123. Fujimoto MM, Tennyson J, Michelin SE (2014) Eur Phys J D 68:67
124. Mašín Z, Gorfinkiel JD, Jones DB, Bellm SM, Brunger MJ (2012) J Chem Phys 136:144310
125. Sieradzka A, Blanco F, Fuss MC, Mašín Z, Gorfinkiel JD, Garcia G (2014) J Phys Chem A 118:6657
126. Tonzani S, Greene C (2005) J Chem Phys 122:014111
127. Aymar M, Greene C, LucKoenig E (1996) Rev Mod Phys 68:1015
128. Tonzani S, Greene C (2006a) J Chem Phys 124:054312
129. Tonzani S, Greene CH (2006b) J Chem Phys 125:094504
130. Caron L, Sanche L, Tonzani S, Greene CH (2008) Phys Rev A 78:042710
131. Caron L, Sanche L, Tonzani S, Greene CH (2009) Phys Rev A 80:012705
132. Lippmann BA, Schwinger J (1950) Phys Rev 79:469
133. Lucchese RR, Takatsuka K, McKoy V (1986) Phys Rep 131:147
134. Watson DK (1989) Schwinger Variational Methods. In: Bates D, Bederson B, Advances in Atomic and Molecular Physics. Academic Press, Vol 25, pp 221–250
135. Winstead C, McKoy V (1996) Electron scattering by small molecules. In: Prigogine I, Rice S, Advances in chemical physics. Wiley Interscience, vol 96, pp 103–190
136. Winstead C, Mckoy V (1996) Highly Parallel Computational Techniques for Electron-Molecule Collisions. In: Berman PR, Gavrila M, Inokuti M, Philips WD. Advances in Atomic, Molecular, and Optical Physics. Academic Press, vol 36, pp 183–219
137. Winstead C, Mckoy V (2006) J Chem Phys 125:174304
138. Winstead C, McKoy V (2007) Phys Rev Lett 98:113201
139. Winstead C, Mckoy V (2007) Phys Rev A 76:012712
140. Winstead C, Mckoy V, Sanchez SD (2007) J Chem Phys 127:085105
141. Gauf A, Hargreaves LR, Jo A, Tanner J, Khakoo MA, Walls T, Winstead C, McKoy V (2012) Phys Rev A 85:052717
142. Bettega MHF, Ferreira LG, Lima MAP (1993) Phys Rev A 47:1111
143. Barbosa AS, Bettega MHF (2013) J Chem Phys 139:214301
144. Kossoski F, Bettega MHF, Varella MTDN (2014) J Chem Phys 140:024317
145. Kossoski F, Varella MTDN (2015) Phys Chem Chem Phys 17:17271
146. Santra R, Cederbaum L, Meyer H (1999) Chem Phys Lett 303:413
147. Ehara M, Sommerfeld T (2012a) Chem Phys Lett 537:107
148. Sommerfeld T, Ehara M (2015) J Chem Theory Comput 11:4627
149. Ehara M, Sommerfeld T (2012b) Chem Phys Lett 537:107
150. Sommerfeld T (2001) Chem Phys Chem 2:677
151. Hazi AU, Taylor HS (1970) Phys Rev A 1:1109
152. Zhou S-G, Meng J, Zhao E-G (2009) J Phys B: At, Mol Opt Phys 42:245001
153. Mandelshtam VA, Taylor HS, Ryaboy V, Moiseyev N (1994) Phys Rev A 50:2764
154. Simons J (1981) J Chem Phys 75:2465
155. Jordan KD, Voora VK, Simons J (2014) Theor Chem Acc 133:1445
156. Ryaboy V, Moiseyev N, Mandelshtam VA, Taylor HS (1994) J Chem Phys 101:5677
157. Barrios R, Skurski P, Simons J (2002) J Phys Chem B 106:7991
158. Berdys J, Anusiewicz I, Skurski P, Simons J (2004b) J Phys Chem A 108:2999
159. Berdys J, Skurski P, Simons J (2004c) J Phys Chem B 108:5800
160. Anusiewicz I, Sobczyk M, Berdys-Kochanska J, Skurski P, Simons J (2005) J Phys Chem A 109:484
161. Theodore M, Sobczyk M, Simons J (2006) Chem Phys 329:139
162. Simons J (2007) How very low-energy (0.1-2 eV) electrons cause DNA strand breaks. In: Sabin JR, Brändas E, Advances in Quantum Chemistry. Elsevier, vol 52, pp 171–188
163. Fabrikant I (1986) Phys Z D 3:401
164. Lane AM, Thomas RG (1958) Rev Mod Phys 30:257
165. Fabrikant II (1985) J Phys B: At Mol Phys 18:1873
166. Schneider BI, LeDourneuf M, Burke PG (1979) J Phys B: At Mol Phys 12:L365
167. Fabrikant II (1990) Comments At Mol Phys 24:37
168. Domcke W (1991) Phys Rep 208:97

169. Scheer AM, Mozejko P, Gallup GA, Burrow PD (2007) J Chem Phys 126:174301
170. Nesbet RK (1981) Phys Rev A 24:1184
171. Froelich P, Brändas E (1975) Phys Rev A 12:1
172. Pelc A, Sailer W, Scheier P, Mason N, Illenberger E, Märk T (2003) Vacuum 70:429
173. Pelc A, Sailer W, Scheier P, Märk T (2005) Vacuum 78:631
174. Abouaf R (2008) Chem Phys Lett 451:25
175. Rescigno TN, Trevisan CS, Orel AE (2009) Phys Rev A 80:046701
176. Gallup GA, Burrow PD, Fabrikant II (2009b) Phys Rev A 80:046702
177. Gallup GA (2013) Phys Rev A 88:052705
178. Janečková R, Kubala D, May O, Fedor J, Allan M (2013) Phys Rev Lett 111:213201
179. Smyth M, Kohanoff J, Fabrikant II (2014) J Chem Phys 140:184313
180. Caron LG, Sanche L (2003) Phys Rev Lett 91:113201
181. Fabrikant II, Caprasecca S, Gallup GA, Gorfinkiel JD (2012) J Chem Phys 136:184301

Photoprocesses with Biomolecules in the Gas Phase

Paola Bolognesi and Lorenzo Avaldi

Abstract The basic processes in molecules of biological interest induced by the absorption of VUV and soft X-rays are reviewed. The study of excitation, ionisation and dissociation in the gas phase on the one hand provides detailed information on the electronic structure and geometry that determine the functioning of these molecules in macroscopic systems and, on the other hand, sheds light on the microscopic effects of radiation damage in living cells.

1 Introduction

The electronic structure and geometrical arrangement (conformation, isomerisation, tautomerisation) of atoms and molecules, the basic constituents of matter, determine the functioning of systems at the macroscale. This is particularly true, in the case of biomatter, where for example the functionality of complex molecules, like enzymes and proteins built up by 20 different amino acids up to a size in the nanometer to micrometre range, is in close relation to the details of their conformation. Another example is the radiation damage where the macroscopic effects induced in living cells by the absorption of ionizing radiation are closely related to the structural and chemical properties of their molecular constituents, with processes initiated at the atomic and molecular level. It is well known, for example that substantial damage to DNA/RNA can be produced by slow electrons with energy of a few eV [1], produced either directly or via ionisation of the medium they are immersed in. In this field, atomic and molecular physics can provide a valuable contribution both experimentally and theoretically. Gas phase studies enable to disentangle the intrinsic properties of the molecules from those due to the interaction with the environment. Indeed the approach which begins with the characterisation of the building blocks of complex biological systems (bases, nucleosides and nucleotides for DNA, amino acids and peptides for proteins) and the understanding of the very basic physical chemical

P. Bolognesi (✉) · L. Avaldi
CNR-Istituto di Struttura della Materia, Area della Ricerca di Roma 1,
via Salaria km 29.300, CP10, 00015 Monterotondo, Italy
e-mail: paola.bolognesi@cnr.it

A.V. Solov'yov (ed.), *Nanoscale Insights into Ion-Beam Cancer Therapy*,
DOI 10.1007/978-3-319-43030-0_6

processes due to the interaction with ionising sources to systems of increasing complexity (clusters, hydrated clusters, nanoaggregates) provides important benchmark data for studies in liquid solutions or in the solid state. Moreover, studies of isolated biomolecules can benefit from the armoury of all theoretical techniques developed for polyatomic molecules, such as ab initio calculations and DFT methods.

The understanding of the physics and chemistry of isolated molecules of biological interest can also provide relevant contribution to biotechnological applications, such as sensors and molecular electronics, astrochemistry and astrobiology, where for example key information on the origin of life in the universe is provided by the understanding of the chemistry of relatively simple molecules in an environment subject to ionizing radiation.

This chapter is devoted to the description of photon induced processes in molecules of biological interest in gas phase. Several radiation sources have been used to investigate the interaction of "light" with biomolecules, from IR and UV lasers, to VUV and X-ray sources at fixed wavelength as well as synchrotron radiation. Synchrotron radiation with its tunability over a broad energy spectrum and synchrotron based spectroscopic techniques represent a unique combination to investigate the absorption of a defined amount of energy by the molecule, sometimes even at a specific bond or molecular site, and then to probe the effects of this excitation on the electronic structure and stability against fragmentation of the molecule. Thus most of this chapter will be devoted to the study of photoprocesses excited by synchrotron radiation. It can be easily predicted that the advent of Free Electron Lasers, FELs, and High Harmonic Generation, HHG, sources with intense pulses, whose duration is only a few *fs* or even hundreds of *as*, will produce a step forward in the understanding of the dynamics and energy flow in biomolecules. Indeed these sources allow a time-resolved study of the processes occurring between the energy absorption and the manifestation, for example of the damage or de-excitation and 'repair'. This will lead to the control and proper handling of the process itself.

The chapter is organised as follows. Section 2 is devoted to the experimental methods used in the studies of molecules of biological interest. In Sect. 3 the applications of the different techniques to nitrogenous bases, amino acids and peptides are reported briefly. The subsections address the different wavelength regimes and one of them is devoted to time dependent studies based on pump-probe experiments, which are benefitting of the advent of FEL and HHG sources. Finally Sect. 4 is devoted to some perspectives and conclusions.

2 Methods

Most biomolecules, including even the smallest such as nucleobases and amino acids, are solid at room temperature and they have to be brought into the gas phase to be studied at the single molecule level. This poses new challenges compared with the investigation of atoms or smaller and more volatile molecules. Thermal evaporation has the advantage of producing a beam of neutral molecules. In the case of

DNA/RNA bases a solid ring structure makes these molecules (except for guanine) quite resistant to thermal decomposition, so that they can be evaporated from ovens to produce effusive or supersonic beams of neutral molecules. In other cases, e.g. some nucleosides and amino acids, great care has to be taken to characterise the working conditions that guarantee the evaporation of intact molecules. For most of the more complex systems, especially if they contain reactive side groups such as carboxylic acid, hydroxyl, and sulfhydryl thermal evaporation cannot be employed, due to the fragility of the targets. In such cases, alternative and more elaborated approaches like Electrospray Ionisation (ESI) [1], Matrix-Assisted Laser Desorption Ionisation (MALDI) [2] and Laser-Induced Acoustic Desorption (LIAD) [3] have been successfully used instead of thermal evaporation. These techniques allow to bring large species intact in the gas phase. However, they also present some drawbacks, as the formation of protonated/de-protonated or multiply charged biomolecules in the ESI, possible contaminations from the matrix molecules and solvents in the MALDI, a pulsed source in the LIAD and, in all cases, the low density of the sample.

The approaches to produce the target beam are briefly described in Sect. 2.1, while the typical spectroscopic techniques regarding the detection of electrons and ions produced in the photo-induced processes, detected both separately and in time coincidence, are briefly reviewed in Sect. 2.2.

2.1 The Targets

The typical experimental set-up used for the evaporation of small biomolecules is composed of a resistively heated oven that heats a crucible containing powder of the target molecule. The heating is often provided by ceramic insulated wires or commercial Thermocoax heaters [4], which provide up to several tens of Watts of power. The winding of the oven should be non-inductive to avoid spurious magnetic fields in the set-up. Furthermore, the top and the bottom of the crucible are normally held at slightly different temperatures, with the top being hotter than the bottom in order to prevent condensation and blockage of the narrow orifice at the exit of the crucible. The oven is contained in a high or, even better, ultrahigh vacuum chamber, bakeable and equipped with a cold finger facing the oven, to trap the vapour. The cold trap limits the contamination of the set-up and helps maintaining a low background pressure, which is particularly important in mass spectrometric studies of biomolecules, where the relatively low density of the vapour beam requires low background pressure in order to produce a good signal-to-noise ratio. Efficient trapping is also necessary to prevent deposition of an insulating layer on the electrodes of the electron and ion analysers, which can seriously affect their efficiency and stability. Last but not the least, baking the entire vacuum system is an efficient way to clean the set-up of residual contaminations; this is a standard procedure for Ultra High Vacuum systems and it is best to provide this capability even for high vacuum systems. Design considerations for ovens have been published for evaporation of metal vapours [5] and similar ideas have been recently adapted and optimised for the evaporation of fragile biomolecules [6–11]. These more specific approaches are based on the use of non-metallic and more inert materials for the crucible, strict control of the working conditions and a careful characterisation of the thermal decomposition of the target molecules. The biomolecular sample can also be applied as saturated solution onto pre-cleaned fibreglass wool, which can be then packed tightly inside, for example an externally heated glass tube. This method serves to increase the surface area, thus enhancing the thermal desorption rate over the thermal decomposition rate [12].

2.2 Analytical Methods

Despite laboratory sources, like rare gases discharge lamps, have been used in the early photoionisation experiments on valence shell, synchrotron radiation has become the most 'popular' and effective radiation source for these studies due to the broad range of tunability of the photon energy, continuous from the UV to hard X-rays. This allows for photoionisation and photoexcitation experiments, able to map out both occupied and empty states respectively, of the valence as well as inner shell orbitals of the target. *Photo Emission and X-ray Photoemission Spectroscopies* (*PES* and *XPS*), where the kinetic energy (KE) of the photoionised electron is measured at a fixed photon energy ($h\nu$), allows reconstruction of the electronic distribution of the molecular orbitals of binding energy BE = $h\nu$ – KE. The typical PES/XPS set-up is based on electron optics principles and uses well defined geometries and electric fields to guide and select the electrons according to their kinetic energy and angle of emission. One of the most common, for example is the hemispherical deflector analyser, typically composed of an electrostatic lens, the energy selector and the detector. The electrostatic lens is responsible for the acceptance, transport and focusing of the electrons from the interaction region to the entrance of the hemispherical analyser, where the electron trajectories are 'deflected' according to their kinetic energy passing through a radial field. One or more detectors placed at the exit of the hemispheres detect the 'selected' electron, and their kinetic energy is reconstructed according to well-known equations.

In the valence shell, see Fig. 1, especially from comparison with theoretical predictions, important information about the electronic charge distribution of the orbitals in the molecule can be derived. In the inner shell the localised nature of the core electrons implies that each atom is affected by its surrounding chemical environment and site-selective information can be obtained. The use of the binding energy shifts to extract chemical information is also known as Electron Spectroscopy for Chemical Analysis (ESCA) [13]. Subtle differences among families of similar molecules (e.g. isomers or analogues), or the effect of functionalisation can be identified, assessed and discussed in terms of the measured and calculated inner shell chemical shifts.

Complementary information to that provided by electron emission, which probes the occupied states, can be obtained by probing the empty states. *Near Edge X-ray Absorption Fine Structure (NEXAFS)* spectroscopy provides electronic structure data, promoting core electrons to empty (excited) states [14]. The experiments, which

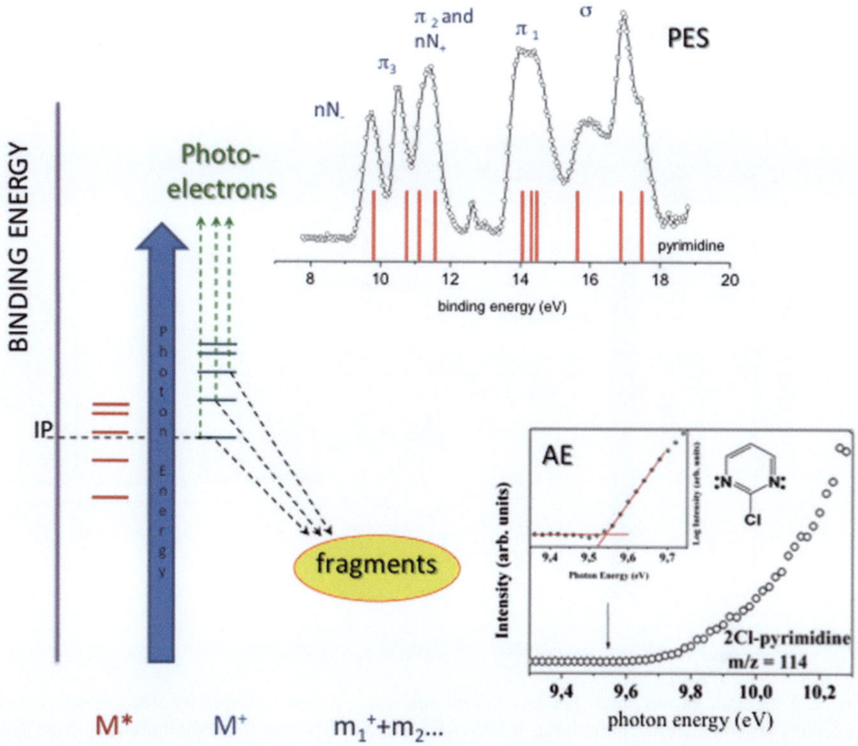

Fig. 1 Schematic of the valence shell photoionisation processes leading to photoemission and molecular fragmentation, with the corresponding measurements of the PES spectrum and the AE of an ionic fragment (see text)

necessarily require tunable photon energy, can be performed by scanning the photon energy across an inner shell threshold and measuring a signal that is proportional to the absorption cross section, for example the total ion yield, see Fig. 2. In biomolecules, C, N, O and S are typically the atoms that can be probed in the soft X-ray region, near their respective K-edges. A simple method of measuring the NEXAFS spectrum is to use an ion detector, placed at the magic angle, consisting of a channel electron multiplier or channel plate, and to measure the current, or number of ions (pulse counting).

Electron energy distributions measured over a broad range of kinetic energies from zero up to $h\nu$ can be used to characterise the complete electron emission spectrum due to the photoemission as well as to the Auger decay of core ionic states or autoionisation of neutral excited states. If the electron spectrum is measured at a photon energy corresponding to a resonant core excited state, previously determined using NEXAFS, the technique is known as *Resonant Auger* spectroscopy. Similarly, *autoionisation* photoelectron spectra are due to excitation to a resonant valence excited state, which subsequently decays by emitting an electron. They differ from normal

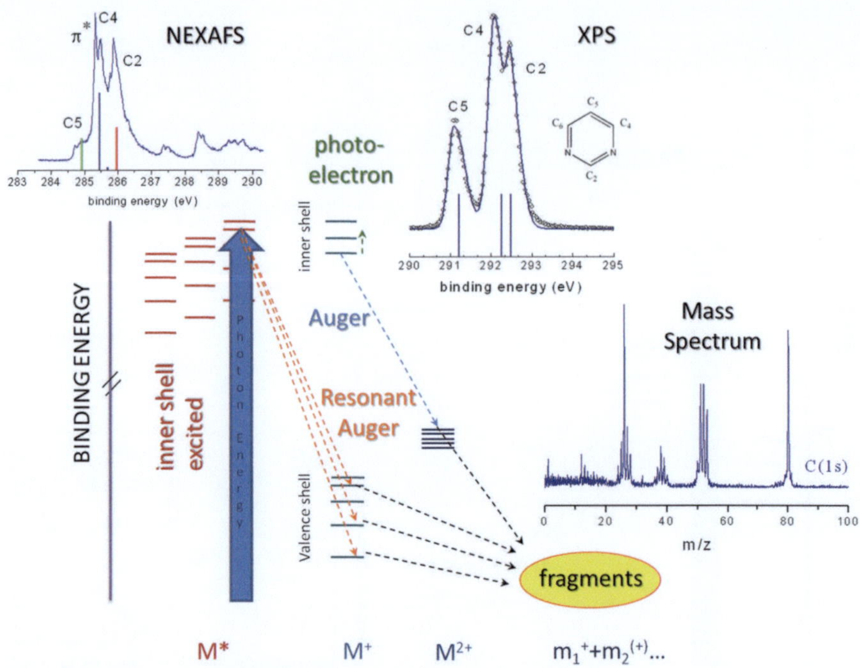

Fig. 2 Schematic of the inner shell excitation/ionisation process followed by electron decay and molecular fragmentation. In the insets a NEXAFS and XPS spectra of the pyrimidine K-edge and a mass spectrum taken at few eV above C(1s) threshold are shown

Auger spectroscopy because they involve the decay of a neutral excited state to a singly charged ionic state, whereas in the *normal Auger* process a singly charged ion decays to a doubly charged ion. Specially on biomolecules, the full electron emission spectra, including photoelectron, autoionisation, Auger and Resonant Auger electron characterised at different photon energies are useful benchmark data for Monte Carlo and Ion Tracking simulation codes, used to evaluate the direct and indirect radiation damage due to the interaction of ionising radiation with a biological medium. The valence and core electron spectra are also useful as benchmark data for calculations of molecular properties based on the electronic structure, particularly in the valence. If the calculation can reproduce accurately the measured valence and core spectra, and also the geometric structure (usually the calculated structure is compared with crystallographic data), then it is reasonable to expect that other calculated properties are accurately predicted, for example dipole and multipole momenta, appearance energies, electron and proton affinities, reaction rates, etc.

If the ion detection is performed using a technique that disperses the fragments according to their mass, then even more information is obtained, and the technique is known as *photoionisation mass spectrometry (PIMS)*. The most common detectors are time of flight (TOF) [15] and quadrupole mass spectrometers (QMS). In the TOF spectrometer the ions are dispersed according to their mass over charge ratio (m/z), which determines their flight time to the detector. In the QMS, a radiofrequency field provides the m/z selection, where only a specific m/z, within the instrumental resolution, will travel to the detector while the other ones are lost through the QMS bars. With these detectors, the molecular fragmentation pattern is measured at fixed photon energy, and it is possible to relate the absorbed photon energy to the products of dissociation. The NEXAFS spectrum can be reconstructed by scanning the photon energy and reporting the total ion signal at each photon energy. Valence band ionisation can leave the molecule in its ground ionic state, which may or may not be stable, or in other excited ionic states up to a binding energy equal to the exciting photon energy, $h\nu$. Since excited electronic states contain more internal energy than the ground ionic state, they have a greater tendency to fragment. As the photon energy increases, more of these excited ionic states are produced, leading to stronger dissociation and to the opening of new fragmentation channels, with the possibility for molecular rearrangements and multiple bond breakings. These processes can be investigated and quantified in the study of the appearance energy (AE), which is the minimum energy required to produce a particular fragment ion in a specific molecule. Experimentally, the AE is determined by selecting a fragment and measuring its ion yield versus photon energy, see inset of Fig. 1. By fitting the spectrum with two straight lines, the first onset of the ion signal is determined. Interestingly, while the experimental determination of the AE is unique, the theoretical value depends not only on the geometry of the charged fragment, but also on the structure of the neutral fragments and on the fragmentation mechanism, providing essential information in the dynamics of molecular fragmentations.

In the soft X-ray region, excitation of core level resonances most likely gives rise to fast electron decay via Resonant Auger emission on the fs timescale, followed by molecular fragmentation on the ps or longer timescale. The situation is complex and the molecular fragmentation hard to predict, because even if the excitation was initially localised on a specific molecular site, after the electronic decay a delocalised valence excited state is produced in which the energy flows through the molecule before fragmentation. For ionisation, rather than excitation of a core level, the core hole decays via an Auger process, again on the timescale of a few fs, and most likely produces a doubly charged ionic state with both holes in the valence band (KVV decay). The fragmentation is then very rapid (if the molecule is not too large) as there is a strong Coulomb repulsion between the two holes. However, this situation is also rather complicated, as there are very many two-hole states, and it may be difficult to identify them.

The simultaneous detection of electrons and ions in time coincidence, i.e. from the same ionisation event, allows for a better control over the many variables in the physical process, adding further insights into molecular fragmentation. In *photoelectron photoion* [17] and *photoelectron photoion photoion coincidence* (*PEPICO and PEPIPICO*) [18] experiments, electrons and ions are detected in time coincidence in order to identify the ones which are generated in the same ionisation event. Figure 3 shows the example of a set-up used in electro-ion coincidence experiments. The basic idea of a coincidence experiment is that correlated particles are generated at the

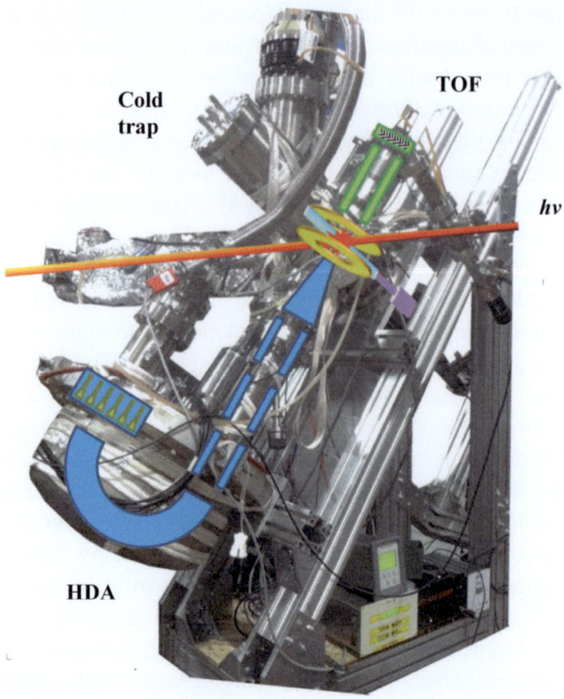

Fig. 3 The PEPICO set-up used at the Gas Phase photoemission beamline of Elettra is composed of a hemispherical electron analyser and a TOF mass spectrometers mounted opposite to each other at the magic angle for synchrotron radiation experiments. The superimposed schematics show the electron (*blue*) and ion (*green*) analysers as well as the photon beam path (*violet arrow*) [16, courtesy of Maclot]

same time and they have a fixed time delay in their arrival time to the detectors, while uncorrelated particles are generated in different ionisation events and therefore will have a random distribution in their arrival times. Therefore the measurement of the time distribution of the difference in the arrival time of these particles will display correlated events, called 'true coincidences', as a peak on a flat background of uncorrelated events, called 'random coincidences'. In practice, in these measurements the ionisation rate is reduced to a sufficiently low value that in a given time interval (the time resolution of the detection system), there is likely to be only one ionisation event. In these experiments the electron kinetic energy determines which valence or inner shell orbital has been ionised, providing state- and site-selectivity (for valence and core ionisation respectively), while the measurement of the mass spectrum in coincidence gives information about the molecular fragmentation of that specific state. In the PEPIPICO experiments, also the detailed description of the dissociation process of doubly/multiply charged ions and the determination of kinetic energy released in the process (available from TOF spectrometers) can be achieved providing a very comprehensive picture of the dynamics of fragmentation.

3 Applications to Nitrogenous Bases, Amino Acids and Peptides

In this section some applications of the above-mentioned techniques to a few classes of biomolecules like nucleobases, their parent compounds, derivatives and analogues, amino acids and their derivatives and oligomers, pharmaceuticals and other molecules with biochemical roles like antibiotics, sugar or lipids are collected. Among the very many excellent results that can be found in the literature concerning the study of biomolecules in the gas phase, the reported literature is far from being a comprehensive review, and we will display in some more details just a few cases in order to provide very brief examples of the implementation of the spectroscopic techniques previously illustrated.

3.1 Excitation and Ionisation in the VUV Range (laboratory Source and Synchrotron)

In the VUV range the electronic charge distribution of the empty orbitals below the first ionisation potential and of the first few molecular orbitals in the valence and inner-valence shells can be accessed. The comparison with theoretical predictions helps to understand the electronic properties and to model the chemical behaviour of the investigated molecules. PES and PEPICO experiments in the valence region can be performed by both rare gas lamps laboratory source (see, for example [17, 19–22]) as well as synchrotron radiation in DNA bases [23–26], amino acids [27–31], pharmaceuticals and other related biomolecules [32–39], while the tunability of the synchrotron radiation is compulsory for the study of empty orbitals [40, 41].

The valence photoelectron spectra allow the understanding of the nature of the outer orbitals and how the molecule is bound. Due to the complexity of the target these spectra generally exhibit broad overlapping bands, often resulting from several unresolved tautomers and conformers, with little evidence of resolved vibrational structure so that their interpretation much relies on ab initio and sophisticated theoretical calculations, like for example ADC(3) [25, 26] and TDDFT B-spline LCAO methods [24, 40], for both the outer and inner valence regions. Figure 4 shows an example of the joint experimental and theoretical study of the photoelectron spectra of halogenated pyrimidines, where the role of different halogen substitutions to the pyrimidine ring is investigated. These molecules are prototype radiosensitisers for selectively enhanced radioterapic effect in cancer treatment. Even though belonging to the same family and therefore very similar, the spectra clearly display some differences due to the presence of the halogen substitution. The substituent effects on the orbitals of the pyrimidine ring can be discussed as a function of the identity and position of the halogen atom. The shifts of the binding energies of these orbitals can be accounted for by a combination of the inductive and resonance effects of the halogen atoms of the ring orbitals [21].

An experimental approach to support the theoretical assignment and overcome the limited resolution in PES spectra is based on the measurement of the photoelectron angular distribution [25, 42] that, based on the different behaviour of the asymmetry parameter of π- and σ-type orbitals, is able to distinguish between orbital types and is particularly useful in binding energy regions containing overlapping photoelectron bands. The photoelectron angular distribution displays also a chiroptical effect, observed as a strong forward/backward asymmetry, with respect to the photon propagation axis, when circularly polarised radiation is used to photoionise gas phase enantiomers. It has been shown that the new observable in the photoelectron angular distribution with circularly polarised radiation, the chiral parameter, is extremely sensitive to static molecular structures, chemical substitution, conformers, dimerisation and clustering [43]. The application to the amino acid alanine [44] showed that the technique provides a plausible conformer population in a genuine biological floppy system and it can be a precious tool for the study the electronic structure as well as molecular structures of biopolymer building blocks in a bottom/up approach of biomolecular complexity.

An alternative method to traditional PES measurements consists of high-resolution Threshold PhotoElectron (TPES) and Threshold PhotoElectron Photoion Coincidence (TPEPICO) spectroscopy. These techniques have been recently used to measure vibrationally resolved photoelectron spectra of DNA/RNA bases and amino acids [45–49], for both the ground and lowest electronic states, and allowed precise determination of vertical ionisation energies.

From the simple molecules of DNA/RNA and amino acids, the natural evolution in the study of elementary biomolecules leads to the investigation of their more complex structures, the nucleosides and peptides for DNA chain and proteins, respectively.

In the two cases, the glycosidic and the peptide bonds play a fundamental role in the formation of long and complex chains of biomolecules that, beginning from relatively few and simple building blocks provide the essential mean for the development of life, with storage and replication of the genetic information as well as for the most differentiated functions performed by proteins. Therefore, it is extremely interesting to investigate the electronic structure of these more complex compounds, also to understand how the properties of the single, isolated molecule is affected by the presence of neighbouring molecules present in their natural biological environment. Unfortunately, due to the fragility of these relatively complex molecules against thermal decomposition, only some selected targets have been investigated in their neutral form, mostly by mass spectrometry, which requires less density than photoemission experiments. However, several photoemission studies are reported in the case of dipeptides containing glycyl [29, 50] and cyclic dipeptides like Glycyl-Glycyl (cGG), Leucyl-Prolyl (cLP), Phenylalanyl-Prolyl (cPP) [30], Histidyl-Glycyl (cHisGly), Tyrosyl-Prolyl (cTyrPro) and Phenylalanyl-Phenylalanyl (cPhePhe) [31]. These last compounds have a solid ring structure, which makes them resistant in the evaporation. These joined experimental and theoretical studies via valence and inner shell photoemission spectroscopy compared with similar results obtained for

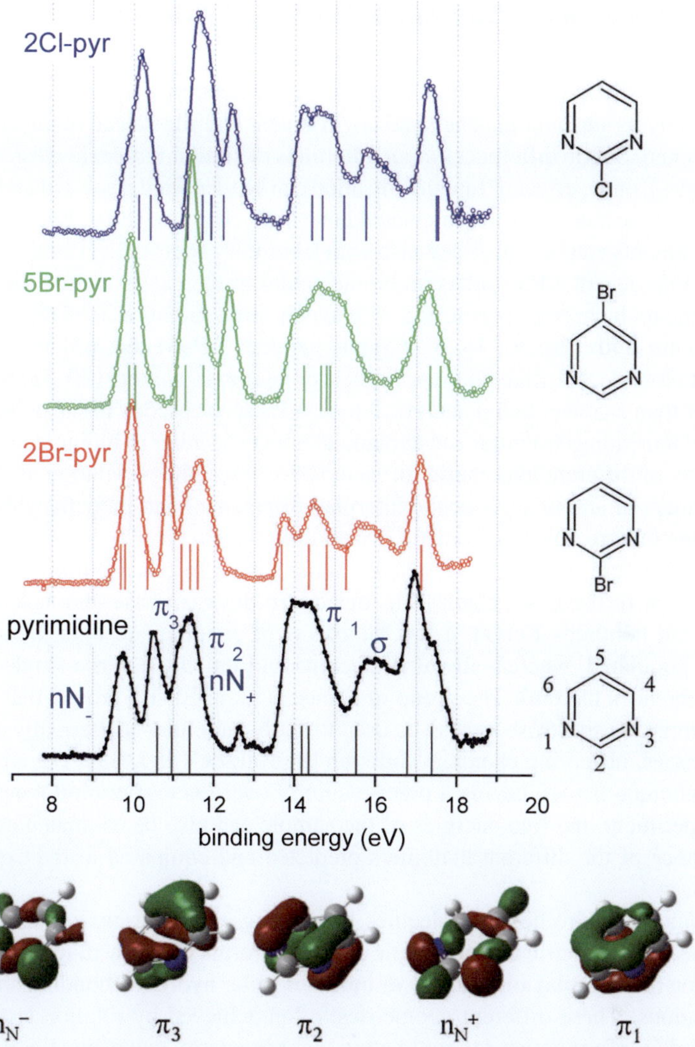

Fig. 4 He I photoelectron spectra of pyrimidine and some halopyrimidines. The sticks indicate the ab initio ionisation energies calculated using the B3LYP [21] while the Hartree–Fock molecular orbitals refer to the case of pyrimidine

their constituent isolated molecules, concluded that in most cases the side-chains interact weakly with the central moiety. Thus a building block approach, in which the chemical properties can be considered to be the sum of those of the functional groups making up the cyclic dipeptide, can be substantially justified.

3.2 Excitation and Ionisation in the Soft X-Ray

In the inner shell, the localised nature of the core electrons implies that each atom is affected by its surrounding chemical environment and site-selective information can be obtained. Subtle differences among families of similar molecules (e.g. isomers or analogues), or the effect of functionalisation can be identified, assessed and discussed in terms of the measured and calculated inner shell chemical shifts. Even though XPS measurements can be performed also with laboratory sources [51] as analytical tool, most of the recent XPS studies on biomolecules in the gas as well as the NEXAFS experiments have been performed with synchrotron radiation for DNA bases [52–60], amino acids [28, 30, 46, 61–63] and peptides [29–31, 64, 65] as well as other related biomolecules like pharmaceuticals or neurotransmitters [35–38, 59, 66–69]. Further than a conventional analytical tool, XPS spectroscopy has also been used to unravel intriguing structural and dynamical effects peculiar of biomolecules, like the interplay of different tautomeric forms in DNA bases, the population of a variety of conformers in amino acids or the interplay of resonant and inductive effects in the shielding of core holes in aromatic pyrimidinic rings.

The structure of DNA bases plays a fundamental role in the proper base paring mechanism in the DNA chain [70]. Indeed, in the gas phase and at temperatures of several hundreds Kelvin, a few tautomers of guanine and cytosine are significantly populated, whereas thymine, adenine and uracil exist in a single form [55, and references therein]. The group of Prince et al. at Elettra proved that core level photoemission spectroscopy can be used to study tautomers in thermally evaporated DNA bases, observing chemical shifts up to several eV [54, 55]. The advantage of this technique is that, having a precise control over the evaporation temperature of the experiment, the free energies of the sample can also be estimated and relative abundance of the different tautomers predicted and compared to the experimental results.

Amino acids are floppy molecules, presenting a rich variety and complexity in the possible geometric and electronic structural forms that they may adopt, depending upon the interplay of a variety of intramolecular hydrogen bonding and electron correlations. These different geometries, often achieved by rotation around a single bond (conformers) are sometimes so close in energy that even slightly different evaporation temperatures can populate several of them, so that they normally coexist during the experimental measurements. The different conformers can be difficult to be clearly identified in PES spectra, due to the delocalised nature of the valence orbitals. However, it has been proved that in XPS spectra, where the core electrons are very sensitive to the chemical environment, the binding energies of different conformers can be as distant as 1 eV [62] so that XPS can be used to identify and determine the relative populations of conformers of amino acids in the gas phase and reveal important information on their 'shape'. Conformer effects, clearly observable in photoemission, appear to be more difficult to resolve in photoabsorption, probably due to partial cancellation of the energy shifts of opposite signs from the core hole states and unoccupied orbitals [52]. Nevertheless, the NEXAFS spectroscopy

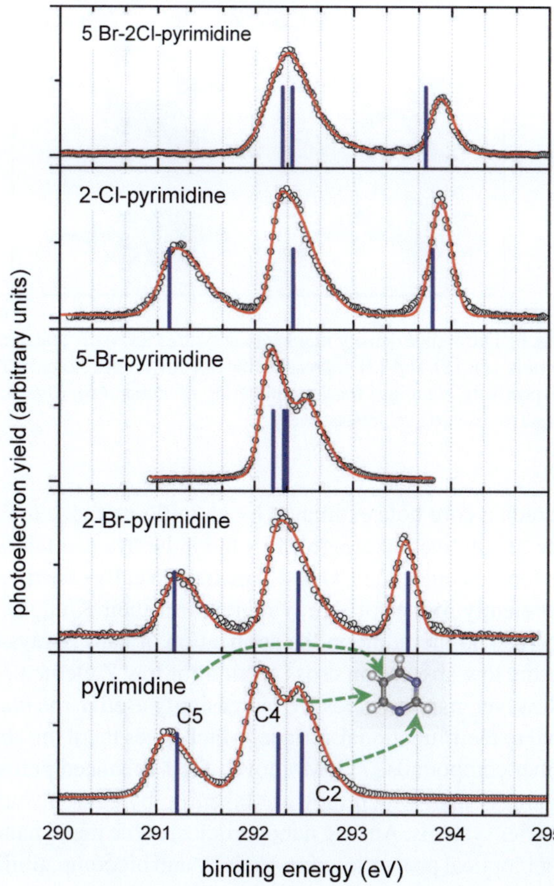

Fig. 5 C(1s) XPS spectra of pyrimidine and halogenated pyrimidines: experimental results (*open dots*), fits with asymmetric Gaussian lineshapes (*full line*) and DTF theoretical calculations [66]

remain a useful tool in the study of the molecular electronic structure, providing insights about the nature of empty orbitals as well as some geometrical information on molecular bond lengths [67].

As a last example, Fig. 5 reports the results of the XPS study of a series of the halopyrimidine molecules [66] showing that, due to the higher electronegativity of the halogen atom compared to the substituted hydrogen, all the carbon atoms but in particular the one where the substitution has taken place are affected by halogenation and their binding energies shift towards higher energies. However, deeper insights regarding the resonant and inductive effects ruling the charge distribution in these aromatic molecules can be gained with the support of DFT calculations. By analysing the displacement of the electron charge density in molecules with a C(1s) core hole (see Fig. 6) it can be clearly observed that, regardless of the location of the carbon site, inductive effect drugs electron charge from neighbouring atoms while resonance contribution (π) drugs electron charge from the atom in *para*-position. Therefore, rearrangements of the electronic charge occurring in the pyrimidine derivatives to

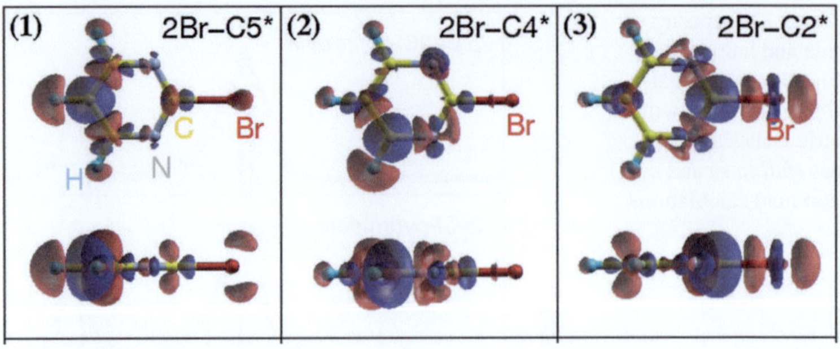

Fig. 6 Difference density maps (*top* and *side* views) of "standard" versus "core hole" calculations. *Panels* (*1*), (*2*) and (*3*) represent the case of a core hole localised on the C5, C4 and C2 site, respectively, see Fig. 4 for the numbering of atoms [66]. *Blue* and *red* zones represent positive and negative isosurfaces, respectively

screen a core hole estimated by ab initio methods are in very good agreement with the qualitative picture given by the inductive resonance model.

In biomolecules C, N, O and S are the typically constituent atoms. They can be probed efficiently in the soft X-ray region near their K-edges. Thus it is not surprising that there is no literature on the application of hard X-rays to isolated biomolecules, due to the low absorption cross section for low Z elements.

However, nanoparticles are increasingly used in the biological field for drug delivery due to their functional surface, which gives them the ability to bind, adsorb and carry other compounds [71]. Moreover, their enhanced permeability and retention and the easiness with which they are taken up by cells favours them with respect to other carrier systems. Among nanoparticles, noble metal nanoparticles are preferred due to their optical properties, non-toxicity and biocompatibility compared to the other metals. Wang et al. [72] have presented a comprehensive review of synchrotron radiation based techniques using radiation up to several keV for the study of nanomaterials at cellular and subcellular interfaces and their transformation it in vitro as well as of the molecular mechanism of the reaction of nanomaterials with biomolecules, but to our knowledge no studies of isolated nanoparticles have been yet reported.

3.3 Photofragmentation Studies

When the ionisation involves orbitals deeper than the ground state the charged molecule begins to fragment. Thus mass spectrometry, AE as well as the PEPICO measurements can characterise the molecular decomposition and link the production of certain fragments to specific photon energies and ionic states. In the most simple experiments, the fragmentation mass spectrum is measured at a fixed photon energy displaying all the fragmentation channels already opened. These experiments can be performed either at the fixed wavelength provided by discharge lamp laboratory

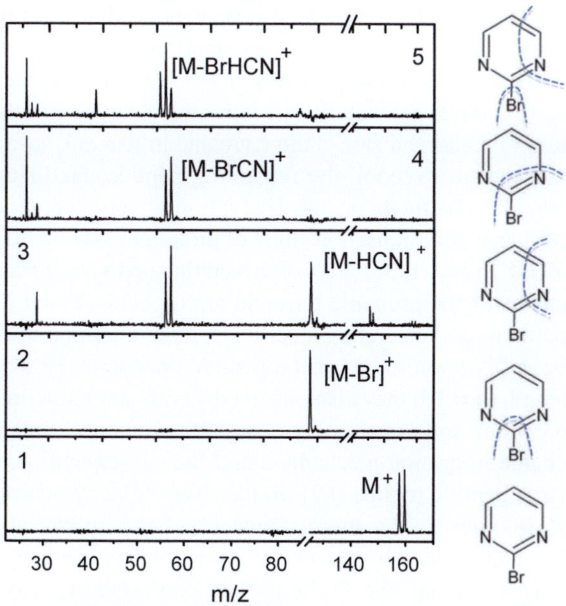

Fig. 7 The PEPICO mass spectra of 2-Bromopyrimidine measured in the BE range between 9.8 and 20 eV (*panels* from *1* to *5*). On the *right*-hand side, the main fragmentation patterns are also indicated

sources [27, 34] or by tunable synchrotron radiation [73–76]. When a TOF spectrometer is used in conjunction with a continuous ionisation source, then the trigger for the measurement of the fragment flight time is typically given by the detection of an unresolved photoelectron in a PEPICO experiment. When the detected photoelectron is energy dispersed, then the PEPICO measurement also carry information of the state-selected fragmentation of the different molecular orbitals [17], which display marked selectivity, see Fig. 7.

A different approach to photofragmentation is represented by multi-photon ionisation, MPI, or Resonant Enhanced Multiphoton Ionisation, REMPI. In these cases a laser in the wavelength range 220–270 nm, a region typical of the π^* excitation in most of the molecules of biological interest, is used and the absorption of two or more photons leads to the ionisation and fragmentation of the sample. This method has been used by Barc et al. [77, 78] to investigate fragmentation of uracil, uracil–water and adenine–water clusters.

The mass spectra of DNA bases and amino acids measured at different photon energies have shown how the DNA bases, thanks to their cyclic structure, have a higher survival probability to radiation exposition with respect to amino acids [74], even though molecular fragmentation increases dramatically as the photon energy increases [73]. Furthermore, it has been observed that photons seem to be more harmful than electrons inducing molecular decomposition. Interestingly, peptides have been observed to be more radiation resistant that their constituent amino acids [79, 80] and synchrotron radiation VUV mass spectrometric studies of free polypeptides and proteins in the gas phase [81–91] were able to shine some light in this observation.

The study of the protonated pentapeptide (leu–enk) measured in the VUV photon energy range 8–20 eV [84] has identified a possible mechanism to explain such increased resistance of peptides. Specially in the higher photon energy range, Bari et al. [84] observed that the fast intramolecular flow of the electron charge leads to the loss of a charged side chain (tyrosine in leu–enk case), which could be an efficient mechanism to 'cool' the remaining peptide, facilitating the survival of functional peptide substructures after absorption of very energetic photons. González-Magaña [81], in a photoionsation study of protonated synthetic peptides of increasing length, YGnF (n = 0, 1, 3, 5, 10), observed that up to n = 5 fragment ions related to the side-chains of the aromatic terminal amino acids Y and F dominate the fragmentation patterns, demonstrating an efficient hole migration towards the terminal amino acids upon photoionisation of the peptide backbone. However, beyond a certain peptide length (n = 10) they also observed significant reduction in fragmentation, with large dications and large singly charged ions that was attributed to a quenching of the charge migration mechanism and the subsequent increased stability typical of the large peptide regime [82]. In the case of the cytochrome c protein Milosavljevic et al. [82] observed a strong connection between ionisation potential and protein conformation, ruled by the charged state, on one hand suggesting that the conformation plays a crucial role in the protein photostability and, the other hand, proposing a novel experimental approach to investigate protein structure in the gas phase [65].

A key feature of synchrotron radiation is the tunability of the photon beam over a large energy range, which opens up the possibility of continuously monitoring the consequences of photon irradiation on the mass spectra as a function of the wavelength, that is often called 'action spectroscopy' [86, 91]. In the VUV range, this allows for the measurement of the AE of biomolecules, which are very important in order to explore their degradation pathways. In the small molecules, this information permits assessment, on thermochemical grounds, of the possible ionic and neutral products, as well as to propose the likely dissociation processes. Joint experimental and theoretical investigations of DNA/RNA bases [92, 93] and their functionalised analogues as halosubstituted DNA bases [32] and amino acids [49, 94, 95] for example, have allowed to discuss the role that these biomolecules could have played in the origin and development of life on earth as well as in applications like radiosensitisers in radiotherapy for cancer treatment, providing deep insights in the very basic mechanism of their decomposition. In the inner shell regions, where basically all fragmentation channels are opened, the resonant excitation of core electrons can be used to selectively deposit the energy on specific atoms in the molecules, performing a site-selective study of the molecular fragmentation. The mass spectra measured at several photon energies across inner shell thresholds display significant differences depending on the localisation of the core hole [58, 96, 97]. The more selective PEPICO and Resonant Auger-Ion Coincidence experiments allow for both site- and state-selective mass spectroscopy. The experiments performed in pyrimidine [96] and 2-Bromopyrimidine [97] have clearly demonstrated that fragmentation following inner shell excitation and valence photoionisation is exactly the same as long as the final ion state reached by the decay of the core excited state or electron ionisation are the same, see Fig. 8. This can be explained [97] by the fact that the fragmentation

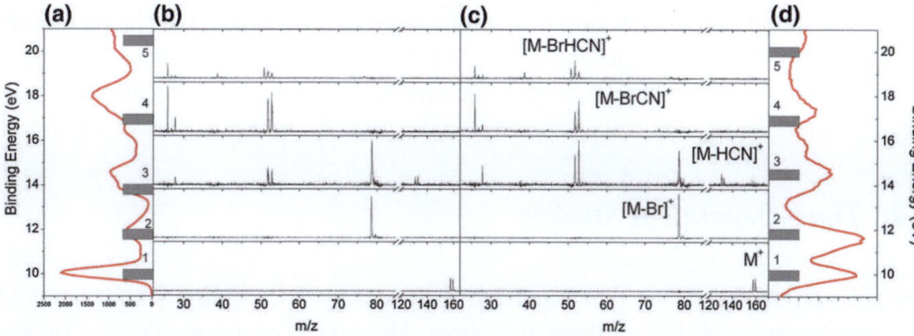

Fig. 8 The energy selected mass spectra of 2-Bromopyrimidine at a few BEs for the case of the C_2 (1s → π*) excitation (*b*) and valence photoionisation at 100 eV photon energy (*c*). In *panels* (*a*) and (*d*) the C_2 (1s → π*) RAE and PES spectra, respectively, are shown. The bars labelled 1 to 5 in *panels* (*a* and *d*) represent the selected energies, with corresponding energy resolution, for the PEPICO measurements [97]

occurs on a timescale longer than the non-radiative relaxation of the inner shell vacancy. Thus, it is the charge distribution of the final singly charged ion that controls the fragmentation, and the enhancements observed in the intensity of some fragments of the mass spectra measurement across inner shell thresholds of pyrimidine and 2-Bromopyrimidine is due to a combination of the state-selective fragmentation of the molecular orbitals and an altered branching ration in the population of the valance orbitals due to Resonant Auger decay. For the large biomolecules [63, 64, 98], the mass resolved NEXAFS spectra of protonated peptides and proteins in the gas phase qualitatively resemble the ones in the condensed phase, carrying information on the inner shell excitations localised at different sites within the peptide.

At photon energies above the core ionisation, the most likely de-excitation process is the Auger decay that efficiently populates fast dissociating doubly and multiply charged ion state. In this case, in order to 'localise' the energy deposition into a specific molecular site, and have a complete control over the decay and fragmentation, a multiple coincidence experiment, where the photoelectron, the Auger electron ad the two charged fragments are detected in coincidence, should be performed. However, due to the extremely differential information to be measured, such experiments are at the limit of feasibility even for the most efficient set-ups. In alternative, several experiments have been performed by detecting only one of the two ejected electrons, either the photoelectron in nucleobases [57, 99], nucleosides [59, 100], ribose [69] and amino acids [101], or the Auger electron [102] in electron energy resolved photoion photoion coincidence (PEPIPICO) measurements. These experiments provide detailed information of the dynamics of fragmentation and bond cleavage patterns in doubly ionised molecules. These PEPIPICO experiments on nucleobases have shown that the final fragments are produced directly by simple ring bond fractures with some possibility of hydrogen migration without involving complex geometrical rearrangements of the parent molecule, while in ribose a very strong damage is observed following core ionisation, with the residual of very small fragments. In nucleosides, due to redistribution of the final valence shell holes, the fragmentation does not exhibit any significant site-specific fragmentation and indeed cannot be described by the fragmentation of separated nucleobase and ribose dications.

3.4 *Time-Dependent Studies*

The absorption of a photon from the IR to X-ray can initiate processes and chemical reactions in molecules of biological interest. How these processes depend on the time scales, energetics, and molecular distances has been the subject of considerable research effort. The advent of radiation sources with tunable energy and duration, which is comparable with the time scale of the nuclei, and, more recently, electronic dynamics has disclosed the possibility to perform time-resolved studies with the aim to understand as well as to control these processes. The technique adopted in these studies, named pump-probe spectroscopy, involves the excitation of the sample by one pulse train (pump) and the monitoring of the changes induced in the sample by the second pulse train (probe), which is suitably delayed with respect to the pump.

The two examples reported below have been chosen considering the new challenges and opportunities disclosed by the advent of new excitation sources like the soft X-ray Free Electron Lasers (FEL) and the ultra fast lasers.

UV(λ < 400 nm)-induced damage to DNA has profound biological consequences, including photocarcinogenesis [103–105], thus it is not surprising that the excited states of the nucleic acid are highly stable to photochemical decay, perhaps as a result of natural selection during a long period of molecular evolution. This photostability due to the rapid decay for electronic energy, has attracted and is attracting a lot of interest. These studies benefit a lot by the development of fs lasers, VUV and soft X-ray FEL's and molecular beam techniques that allow the study of isolated DNA bases despite their low vapour pressure and easy thermal decomposition, as well as advances in quantum chemistry that have made possible the modelling of excited states. Crespo-Hernandez et al. [106] reviewed in 2005 the impressive armoury of experimental and theoretical techniques that have been used to study excited states in nucleic acids and their constituents. All that work has shown convincingly that the fluorescence lifetimes of single DNA bases in solution at room temperature are in the subpicosecond range. For base monomers, the lack of solvent effects in the condensed phase and the ultrashort lifetimes in supersonic jet experiments suggest [103] that internal conversion is not the result of strong solute–solvent interactions, but the inevitable outcome of nonadiabatic dynamics on the complex potential energy landscape of the bases, leading to the redistribution of both charge and energy within the molecule via the coupling of vibrational with electronic degrees of freedom. As photoelectron spectroscopy is sensitive to both molecular orbital configurations and vibrational dynamics, Ullrich et al. [12] used Time-Resolved PhotoElectron Spectra (TRPES) to study the electronic relaxation processes in DNA and RNA bases adenine, cytosine, thymine and uracil in a molecular beam. As opposed to other experimental approaches, like for example time-resolved ion yield measurements [107], TRPES

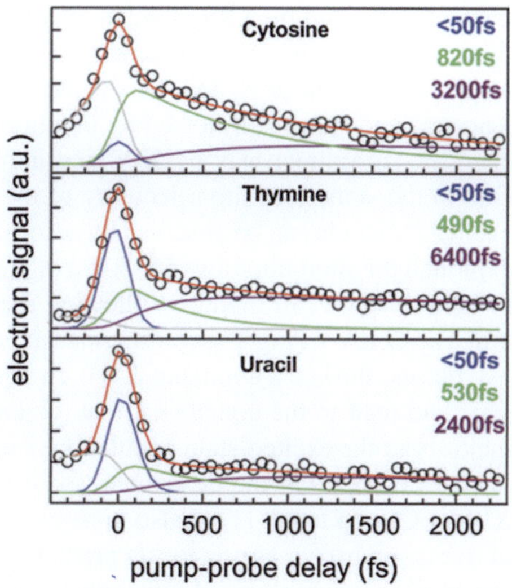

Fig. 9 Comparison of decay traces extracted from two dimensional global fits to TRPES measurements of cytosine, thymine and uracil [12]. *Black circles* represent the energy-integrated photoelectron signal for each channel and the *red line* is the best fit. Four channels are assumed: a Gaussian component (*blue*), a short (*green*) and a long-lived (*purple*) exponential decay, and a probe–pump signal (*grey*). See text for details

allows for direct identification of the states involved in the electronic relaxation process through projection of the excited state dynamics onto cationic states. Hence photoelectron spectroscopy can provide information on the character of the excited state and, since ionisation is always allowed, states appearing as 'dark' in absorption can be probed. In the experiment pump wavelengths between 250 and 277 nm lead to initial excitation of the bright $S_2(\pi\,\pi^*)$ state and ionisation was then produced by a pulse at 200 nm. The instrument response function of 160 fs Full Width at Half Max (FWHM) allowed the determination of time constants down to about 20 % of the cross-correlation FWHM, i.e. 40 fs. The main attention has been placed on adenine and the results show that the initially prepared bright S_2 $(\pi\,\pi^*)$ state decays rapidly (<50 fs) to the S_1 $(n\pi^*)$ state, which has a lifetime of 750 fs. There are also indications for an additional decay pathway consistent with the theoretically predicted $S_3(\pi\,\sigma^*)$. The results for the pyrimidine bases, Fig. 9, displayed a dynamics with a multi-exponential decay: an ultrashort decay <50 fs followed by a slower one in the range of several hundreds of fs (820 fs for cytosine, 490 fs for thymine, 530 fs for uracil) and a ps channel (3.2 ps for cytosine, 6.4 ps for thymine, 2.4 ps for uracil).

These time constants are consistent with calculations of the potential energy surface [108–112], but the assignment of the transients to electronic or nuclear relaxation varies among theoretical models. Approaches based on linear interpolation [110] or minimal energy paths [111] predict a barrierless fast decay to the electronic ground, thus an electronic relaxation. Dynamic simulations [108] predict an indirect path [112] where the initial ultrafast nuclear relaxation into the $\pi\,\pi^*$ state involves the C–O stretch, with the population trapped for picoseconds behind a reaction barrier. This suggests an interpretation of the 100 fs constant as a nuclear relaxation. Ultrafast pump-probe optical spectroscopies due to the reduced Franck–Condon overlap upon vibrational relaxation are unable to disentangle among these models. An alternative is provided by XPS or Auger electron spectroscopies, which can core ionise the molecule at any nuclear geometry. Time-resolved studies in the core shell region needed the advent of X-ray FELs where the evolution of the wavepacket in the valence excited states is followed by monitoring its effects on the inner shell binding energies, in a direct way by XPS and indirectly by Auger electron spectroscopy. Due to the state- and site-selectivity of the core spectroscopy the dynamics of the excited molecule can be observed by a 'local' point of view. A first experiment [113] exploited the short duration (50–80 fs) of the LCLS at the SLAC National laboratory to ionise the O1s in thymine following the excitation of a 70 fs UV pulse of 266 nm. The O KVV Auger spectrum, measured as a function of the delay between the two pulses, showed a dominant 200 fs electronic relaxation of the photoexcited $\pi\ \pi^*$ state and lead to the conclusion that, under the conditions of that experiment, the majority of the excited-state population is not efficiently trapped by an excited-state reaction barrier. On the other hand some preliminary results on the time-resolved XPS of C1s on uracil [114] also measured at LCLS appear to support the existence of a reaction barrier as previously predicted [108]. Thus, still there are many remaining puzzles to completely disentangle the nature of the photostability of nucleobases to UV radiation damage.

Transfer of electronic charge within a single molecule plays a key role in catalysis, DNA damage by ionizing radiation, photosynthesis, respiration. The ability of molecules such as peptides and DNA to act as charge conduits is an intrinsic part of many biological processes. Photodissociation studies with ns UV laser radiation has provided information on the charge transfer, for example in protonated peptides isolated in vacuo, and how, depending on the initial site of photoexcitation, this can lead to a slow and non-hazardous statistical dissociation or a prompt cleavage of peptide bonds [115]. However, only pump-probe experiments can fully elucidate the charge location during the process. In their pioneer works with nanosecond laser in the 1990s Weinkauf and coworkers showed that if an electron is selectively ionised from a chromophore at a terminal end of a peptide, then the location of the charge could be probed using the shift in absorption of the chromophore, so that the passage of the charge through up to 12 sigma bonds in a quadrapeptide was observed [116, 117]. Later experiments were able to extract a 80 fs lifetime for charge transfer from an ionised chromophore to the amine group in 2phenylethyl-N,Ndimethylamine molecule and more recent calculations [118] predicted that, depending on the cationic states contributing to the wavepacket and the conformation of the neutral molecule, the charge migration across the full extent of the molecule could take 5 fs or less. This represents therefore one of the most suited processes to exploit the performances of attosecond pulse trains in the extreme ultraviolet [119]. Calegari et al. [120] investigated ultrafast charge dynamics in the amino acid phenylalanine after prompt ionisation induced by isolated attosecond pulses. A probe pulse then produced a doubly charged molecular fragment by ejection of a second electron, and charge migration manifested itself as a sub-4.5 fs oscillation in the yield of this fragment as a function of pump-probe delay, Fig. 10. This temporal scale is definitely shorter than the vibrational response of the molecule. Numerical simulations of the temporal

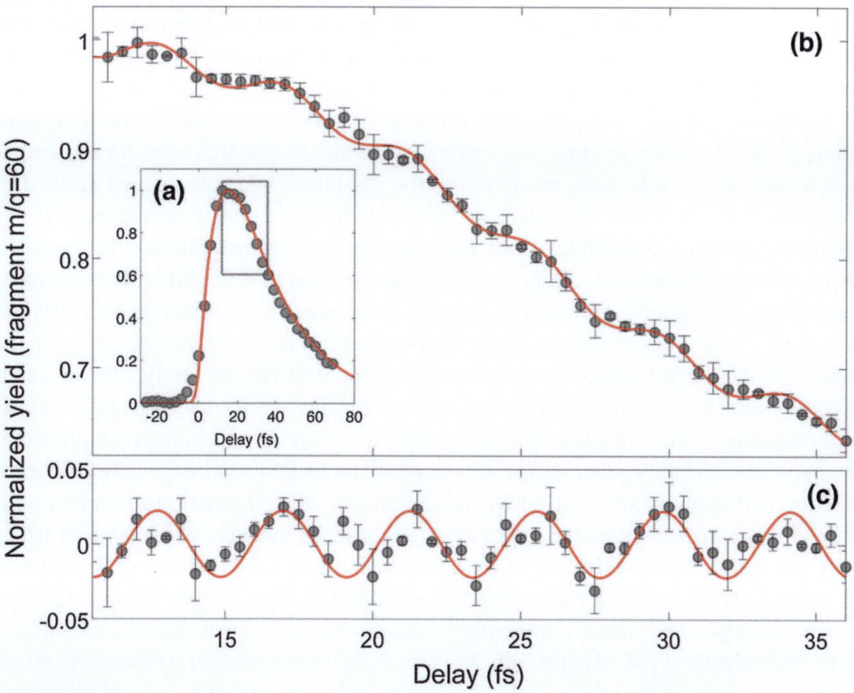

Fig. 10 Pump-probe measurements in phenylalanine amino acid [120]. (*a*) Yield of doubly charged immonium ion (m/z 60) as a function of pump-probe delay, measured with 3 fs temporal steps. The *red line* is a fitting curve with an exponential rise time of 10 fs and an exponential relaxation time of 25 fs. (*b*) Yield of doubly charged immonium ion versus pump-probe delay measured with 0.5-fs temporal steps, within the temporal window shown as dotted box in (*a*). The *red line* is the fitting curve given by the sum of the fitting curve shown in (*a*) and a sinusoidal function of frequency 0.234 PHz (4.3 fs period). (*c*) Difference between the experimental data and the exponential fitting curve displayed in (*a*)

evolution of the electronic wave packet created by the attosecond pulse supported the interpretation of the experimental data in terms of charge migration, resulting from a periodic variation of the charge density around the amine group.

4 Conclusions

In this chapter it has been shown how the electromagnetic radiation sources, mainly synchrotron radiation, combined with electron spectroscopies and mass spectrometry can be used for the characterisation of the electronic structure, the conformations and the dynamic processes that lead to the redistribution of the energy absorbed by molecules of biological interest, such as amino acids, peptides and the building blocks of nucleic acids. The advent of more intense and short pulse sources will enlarge the number of molecules to be investigated, overcoming the limitation of the low density of the beams produced by novel methods.

While most of the work performed up to now has been restricted to thermal evaporation, and therefore to small molecules, a promising research area for the future is towards larger molecules, brought into the gas phase by sophisticated and softer methods. Among others, the state of the art technique to bring unfragmented nucleotides, proteins or peptides from solution into the gaseous phase is represented by the ESI technique [121, 122]. The ESI source overcomes the difficulties of introducing large molecules prepared in solution into a mass spectrometer, which works in a high vacuum environment. Since the commercial development of ESI sources, mass spectrometry has become the most popular tool for the study of very large organic molecules as analytical tool as well as for innovative research applications. The coupling of this versatile ion source with spectroscopic techniques implies that the mass selected ions generated in the gas phase by an ESI source are collected in an ion trap and then excited/ionised by the radiation. The combination of an ESI with powerful fs laser has provided a much better sequence analysis of the protein structure and despite the low density of the target beam, which hampers the achievement of a good signal-to-noise ratio, with the available photon flux in third generation synchrotron sources, some pioneering works using mass spectrometric techniques have been reported [64, 82, 88, 90, 98, 123–125]. Using similar methods, Giuliani et al. [90] produced free protonated and charge state-selected cytochrome ions (consisting of 104 amino acids) and performed VUV photoionisation mass spectrometry. An improvement of the throughput of the sources is needed in order to make them suitable for electron spectroscopies. The high intensity of the new FEL sources will certainly favour the use of ESI sources.

Another subject in fast development is related to the investigation of the effects of the environment in the ionisation/fragmentation of biomolecules. The natural environment is an aqueous solution. Thus sources of dry or hydrated clusters of biomolecules are being developed [126–128]. In this way a solvated environment can be mimicked in the gas phase and its effects on the structure and properties of the biomolecular system investigated.

Furthermore, the new radiation sources will made possible, as described in Sect. 3.4, to follow the evolution of the energy flow in the molecule as well as processes like H migration, molecular dynamics (rotation or bending of bonds, etc.) which can strongly affect the chemical behaviour of these molecules or their fragments.

Acknowledgements Work partially supported by the COST Action Nano-IBCT (CM 1002) and XLIC (CM1204), MAE-CI Italia-Serbia 'A nanoview of radiation-biomatter interaction'.

References

1. Bodaïffa B, Cloutier P, Hunting D, Huels MA, Sanche L (2000) Science 287:1658
2. Fenn JB, Mann M, Meng CK, Wong SF, Whitehouse CM (2003) Science 246 (1989) 64. Fenn Angew JB Chem Int Ed 42:3871
3. Karas M, Bachmann D, Hillenkamp F (1985) Anal Chem 57:2935
4. Golovlev VV, Allman SL, Garrett WR, Taranenko NI, Chen CH (1997) Int J Mass Spectrom 169–170:69
5. http://www.thermocoax.com
6. Ross KJ, Sonntag B (1995) Rev Sci Instrum 66:4409
7. Denifl S, Matejcik S, Gstir B, Hanel G, Probst M, Scheier P, Märk TD (2003) J Phys Chem 118:4107
8. Kopyra J, König-Lehmann C, Szamreja I, Illenberger E (2009) Int J Mass Spectrom 285:131
9. Bergen T, Biquard X, Brenac A, Chandezon F, Huber BA, Jalabert D, Lebius H, Maurel M, Monnand E, Opitz J, Pesnelle A, Pras B, Ristori C, Rocco JC (1999) Rev Sci Instrum 70:3244
10. Lago AF, Countinho LH, Marinho RRT, Naves de Brito A, de Souza GGB (2004) Chem Phys 307:9
11. Tabet J, Eden S, Feil S, Abdoul-Carime H, Farizon B, Farizon M, Ouaskit S, Märk TD (2010) Nucl Instrum Methods Phys Res Sect B 268:2458
12. Levola H, Kooser K, Rachlew E, Nömmiste E, Kukk E (2013) Int J Mass Spectrom 353:7
13. Ullrich S, Schultz T, Zgierski MZ, Stolow A (2004) Phys Chem Chem Phys 6:2796
14. Siegbahn K (ed) (1969) ESCA applied to free molecules. North-Holland, ISBN 0720401607
15. Stöhr J (1992) NEXAFS spectroscopy. Springer Science and Business Media. ISBN 978-3-662-02853-7
16. Wiley WC, McLaren IH (1955) Rev Sci Instrum 25:1150
17. Maclot S (2014) Ph.D. thesis
18. Plekan O, Coreno M, Feyer V, Moise A, Richter R, de Simone M, Sankari R, Prince KC (2008) Phys Scr 78:58105
19. Kukk E, Sankari R, Huttula M, Sankari A, Pilling S, Aksela S (2007) J Electron Spectrosc Relat Phenom 155:141
20. Kimura K, Katsumata S, Achiba Y, Yamazaki T, Iwata S (1981) Handbook of He I photoelectron spectra of fundamental organic molecules. Halsted Press, New York
21. Debies TP, Rabalais JW (1974) J Electron Spectrosc Relat Phenom 3:315
22. O'Keeffe P, Bolognesi P, Casavola AR, Catone D, Zema N, Turchini S, Avaldi L (2009) Mol Phys 107:2025
23. Gerson SH, Worley SD, Bodor N, Kaminski JJ, Flechtner TW (1978) J Electron Spectrosc Relat Phenom 13:421
24. Schwell M, Hochlaf M (2014) Topics of current chemistry. Springer, Berlin, Heidelberg
25. Holland DMP, Potts AW, Karlsson L, Stener M, Decleva P (2011) Chem Phys 390:25
26. Trofimov AB, Schirmer J, Kobychev VB, Potts AW, Holland DP, Karlsson L (2006) J Phys B: At Mol Opt Phys 39:305
27. Zaytseva IL, Trofimov AB, Schirmer J, Plekan O, Feyer V, Richter R, Coreno M, Prince KC (2009) J Phys Chem A 113:15142
28. Plekan O, Feyer V, Richter R, Coreno M, Prince KC (2008) Mol Phys 106:1143
29. Plekan O, Feyer V, Richter R, Coreno M, de Simone M, Prince KC, Carravetta V (2007) J Phys Chem A 111:10998
30. Feyer V, Plekan O, Richter R, Coreno M, Prince KC, Carravetta V (2009) J Phys Chem A 113:10726
31. Wickrama Arachchilage AP, Wang F, Feyer V, Plekan O, Prince KC (2010) J Chem Phys 133:174319
32. Wickrama Arachchilage AP, Wang F, Feyer V, Plekan O, Prince KC (2012) J Chem Phys 136:124301
33. Dampc M, Mielewska B, Siggel-King MRF, King GC, Zubek M (2009) Chem Phys 359:77

34. Castrovilli MC, Bolognesi P, Cartoni A, Catone D, O'Keeffe P, Casavola AR, Turchini S, Zema N, Avaldi L (2014) J Am Soc Mass Spectrom 25:351
35. Feyer V, Plekan O, Richter R, Coreno M, Prince KC (2009) Chem Phys 358:33
36. Maris A, Melandri S, Evangelisti L, Caminati W, Giuliano BM, Plekan O, Feyer V, Richter R, Coreno M, Prince KC (2012) J Electron Spectros Relat Phenom 185:244
37. Ahmed M, Ganesan A, Wang F, Feyer V, Plekan O, Prince KC (2012) J Phys Chem A 116:8653
38. Ahmed M, Wang F, Acres RG, Prince KC (2014) J Phys Chem A 118:3645
39. Feketeova L, Plekan O, Goonewardane M, Ahmed M, Albright AL, White J, O'Hair RAJ, Horsman MR, Wang F, Prince KC (2015) J Phys Chem A 119:9986
40. Giuliani A, Limão-Vieira P, Duflot D, Milosavljevic AR, Marinkovic BP, Hoffmann SV, Mason N, Delwiche J, Hubin-Franskin M-J (2009) Eur Phys J D 51:97
41. Stener M, Decleva P, Holland DMP, Shaw DA (2011) J Phys B: At Mol Opt Phys 44:075203
42. Ferreira da Silva F, Almeida D, Martins G, Milosavljevic AR, Marinkovic BP, Hoffmann SV, Mason NJ, Nunes Y, Garcia G, Limao-Vieira P (2010) Phys Chem Chem Phys 12:6717
43. Potts AW, Holland DMP, Trofimov AB, Schirmer J, Karlsson L, Siegbahn K (2003) J Phys B: At Mol Opt Phys 36:3129
44. Nahon L, Garcia GA, Powis I (2015) J Electron Spectros Relat Phenom 204:322 (and references there in)
45. Tia M, Cunha De Miranda B, Daly S, Gaie-Levrel F, Garcia GA, Powis I, Nahon L (2013) J Phys Chem Lett 4:2698
46. Touboul D, Gaie-Levrel F, Garcia GA, Nahon L, Poisson L, Schwell M, Hochlaf M (2013) J Chem Phys 138:094203
47. Powis I, Rennie EE, Hergenhahn U, Kugeler O, Bussy-Socrate R (2003) J Phys Chem A 107:25
48. Hochlaf M, Pan Y, Lau KC, Majdi Y, Poisson L, Garcia GA, Nahon L, Al-Mogren MM, Schwell M (2015) J Phys Chem A 119:1146
49. Majdi Y, Hochlaf M, Pan Y, Lau K-C, Poisson L, Garcia GA, Nahon L, Al-Mogren MM, Schwell M (2015) J Phys Chem A 119:5951
50. Jochims H-W, Schwell M, Chotin JL, Clemino M, Dulieu F, Baumgärtel H, Leach S (2004) Chem Phys 298:279
51. Richer G, Sandorfy C, Chaer Nascimento MA (1984). J Electron Spectrosc Relat Phenom 34:327
52. Slaughter AR, Banna MS (1988) J Phys Chem 92:2165
53. Feyer V, Plekan O, Richter R, Coreno M, Prince KC, Carravetta V (2008) J Phys Chem A 112:7806
54. Feyer V, Plekan O, Kivimaki A, Prince KC, Moskovskaya TE, Zaytseva IL, Soshnikov DY, Trofimov AB (2011) J Phys Chem A 115:7722
55. Plekan O, Feyer V, Richter R, Coreno M, Vall-llosera G, Prince KC, Trofimov AB, Zaytseva IL, Moskovskaya TE, Gromov EV, Schirmer J (2009) J Phys Chem A 113:9376
56. Feyer V, Plekan O, Richter R, Coreno M, Vall-llosera G, Prince KC, Trofimov AB, Zaytseva IL, Moskovskaya TE, Gromov EV, Schirmer J (2009) J Phys Chem A 113:5736
57. Plekan O, Feyer V, Richter R, Coreno M, de Simone M, Prince KC, Trofimov AB, Gromov EV, Zaytseva IL, Schirmer J (2008) Chem Phys 347:360
58. Itälä E, Ha DT, Kooser K, Huels MA, Rachlew E, Nõmmiste E, Joost U, Kukk E (2011) J Electron Spectrosc Relat Phenom 184:119
59. Lin Yi-Shiue, Lin Huei-Ru, Liu Wei-Lun, Lee Yuan T, Tseng Chien-Ming, Ni Chi-Kung, Liu Chen-Lin, Tsai Cheng-Cheng, Chen Jien-Lian, Wei-Ping Hu (2015) Chem Phys Lett 636:146
60. Itälä E, Kooser K, Rachlew E, Levola H, Ha DT, Kukk E (2015) J Chem Phys 142:194303
61. Milosavljevic AR, Cerovski VZ, Canon F, Rankovic ML, N. Škoro, Nahon L, Giuliani A (2014) J Phys Chem Lett 5:1994
62. Zhang W, Carravetta V, Plekan O, Feyer V, Richter R, Coreno M, Prince KC (2009) J Chem Phys 131:035103
63. Plekan O, Feyer V, Richter R, Coreno M, de Simone M, Prince KC, Carravetta V (2007) Chem Phys Lett 442:429

64. Marinho RRT, Lago AF, Homem MGP, Coutinho LH, de Souza GGB, Naves de Brito A (2006) Chem Phys 324:420
65. Gonzalez-Magaña O, Reitsma G, Tiemens M, Boschman L, Hoekstra R, Schlathölter T (2012) J Phys Chem A 116:10745
66. Milosavljević AR, Nicolas C, Ranković ML, Canon F, Miron C, Giuliani A (2015) J Phys Chem Lett 6:3132
67. Bolognesi P, Mattioli G, O'Keeffe P, Feyer V, Plekan O, Ovcharenko Y, Prince KC, Coreno M, Amore Bonapasta A, Avaldi L (2009) J Phys Chem A 113:48
68. Bolognesi P, O'Keeffe P, Ovcharenko Y, Coreno M, Avaldi L, Feyer V, Plekan O, Prince KC, Zhang W, Carravetta V (2010) J Chem Phys 133:034302
69. Plekan O, Feyer V, Richter R, Moise A, Coreno M, Prince KC, Zaytseva IL, Moskovskaya TE, Soshnikov DY, Trofimov AB (2012) J Phys Chem A 116:5653
70. Ha DT, Huels MA, Huttula M, Urpelainen S, Kukk E (2011) Phys Rev A 84:033419
71. Watson JD, Crick F (1953) Nature 171:964
72. Nivethaa EAK, Dhanavel S, Narayanan V, Vasuc CA, Stephen A (2015) RSC Adv 5:1024 (and reference therein)
73. Wang B, Feng W, Chai Z, Zhao Y (2015) Sci China 58:768
74. Pilling S, Lago AF, Coutinho LH, De Castilho RB, De Souza GGB, De Brito AN (2007) Rapid Commun Mass Sp 21:3646
75. Pilling S, Andrade DPP, de Castilho RB, Cavasso-Filho RL, Lago AF, Coutinho LH, de Souza GGB, Boechat-Roberty HM, Naves de Brito A (2008) Org Matter Space Proc IAU Symp No 251:371
76. de Souza GGB, Coutinho LH, Nunez C, Bernini R, Castilho RB, Lago AF (2007) J Phys: Conf Ser 88:012005
77. Levola H, Kooser K, Itälä E, Kukk E (2014) Int J Mass Spectrom 370:96
78. Barc B, Ryszka M, Poully J-C (2014) Al Maalouf EJ, el Otell Z, Tabet J, Parajuli R, van der Burgt PJM, Limão-Vieira P, Cahillane P, Dampc M, Mason NJ, Eden S. Int J Mass Spectr 194:365–366
79. Barc B, Ryszka M, Spurrell J, Dampc M, Limão-Vieira P, Parajuli R, Mason NJ, Eden S (2013) J Chem Phys 139:244311
80. Barbier B, Chabin A, Chaput D, Brack A (1998) Planet Space Sci 46:391
81. Boillot F, Chabin A, Bure C, Venet M, Belsky A, Bertrand-Urbaniak M, Delmas A, Brack A, Barbier B (2002) Orig Life Evol Biosph 32:359
82. González-Magaña O, Reitsma G, Bari S, Hoekstra R, Schlathölter T (2012) Phys Chem Chem Phys 14:4351
83. Milosavljevic AR, Nicolas C, Lemaire J, Dehon C, Thissen R, Bizau J-M, Refregiers M, Nahon L, Giuliani A (2011) Phys Chem Chem Phys 13:15432
84. Brunet C, Antoine R, Allouche A-R, Dugourd P (2011) J Phys Chem A 115:8933
85. Bari S, Gonzalez-Magaña O, Reitsma G, Werner J, Schippers S, Hoekstraand R, Schlathölter T (2011) J Chem Phys 134:024314
86. Milosavljević AR, Nicolas C, Gil JF, Canon F, Réfrégiers M, Nahon L, Giuliani A (2012) J Synch Rad 19:174
87. Giuliani A, Milosavljević AR, Canon F, Nahon L (2014) Mass Spectrom Rev 33:424
88. Canon F, Milosavljević AR, van der Rest G, Réfrégiers M, Nahon L, Sarni-Manchado P, Cheynier V, Giuliani A (2013) Angew Chem 125:8535
89. Canon F, Milosavljević AR, Van Der Rest G, Réfrégiers M, Nahon L, Sarni-Manchado P, Cheynier V, Giuliani A (2013) Angew Chem Int Ed 52:8377
90. Giuliani A, Milosavljević AR, Hinsen K, Canon F, Nicolas C, Réfrégiers M, Nahon L (2012) Angew Chem 124:9690
91. Giuliani A, Milosavljević AR, Hinsen K, Canon F, Nicolas C, Réfrégiers M, Nahon L (2012) Angew Chem Int Ed 51:9552
92. Canon F, Milosavljević AR, Nahon L, Giuliani A (2015) Phys Chem Chem Phys 17:25725
93. Schwell M, Jochims H-W, Baumgärtel H, Leach S (2008) Chem Phys 353:145
94. Jochims H-W, Schwell M, Baumgärtel H, Leach S (2005) Chem Phys 314:263

95. Gaie-Levrel F, Garcia GA, Schwell M, Nahon L (2011) Phys Chem Chem Phys 13:7024
96. Schwell M, Jochims H-W, Baumgärtel H, Dulieuc F, Leach S (2006) Planet Space Sci 54:1073
97. Bolognesi P, O'Keeffe P, Avaldi L (2012) Soft X-ray interaction with organic molecules of biological interest. In: Garcia Gomez-Tejedor G, Fuss MC (eds) Radiation damage in biomolecular systems. Springer Science + Business Media B.V, pp 165
98. Bolognesi P, Kettunen JA, Cartoni A, Richter R, Tosic S, Maclot S, Rousseau P, Delaunay R, Avaldi L (2015) Phys Chem Chem Phys 17:24063
99. Milosavljević AR, Canon F, Nicolas C, Miron C, Nahon L, Giuliani A (2012) J Phys Chem Lett 3:1191
100. Itälä E, Ha DT, Kooser K, Nõmmiste E, Joost U, Kukk E (2011) Int J Mass Sp 306:82
101. Itälä E, Kooser K, Rachlew E, Huels MA, Kukk E (2014) J Chem Phys 140:234305
102. Itälä E, Huels MA, Rachlew E, Kooser K, Hägerth T, Kukk E (2013) J Phys B: At Mol Opt Phys 46:215102
103. Ha DT, Wang Y, Alcamí M, Itälä E, Kooser K, Urpelainen S, Huels MA, Kukk E, Martín F (2014) J Phys Chem A 118:1374
104. Miller DL, Weinstock MA (1994) J Am Acad Dermatol 30:774
105. Kraemer KH (1997) 94:11
106. Young AR (1997) Br J Clin Pract 89:10
107. Crespo-Hernandez CE, Cohen B, Hare PM, Kohler B (2004) Chem Rev 104:1977
108. Canuel C, Mons M, Piuzzi F, Tardivel B, Dimicoli I, Elhanine M (2005) J Chem Phys 122:074316
109. Hudock HR, Levine BG, Thompson AL, Satzger H, Townsend D, Gador N, Ullrich S, Stolow A, Martínez TJ (2007) J Phys Chem A 111:8500
110. Szymczak JJ, Barbatti M, Soo Hoo JT, Adkins JA, Windus TL, Nachtigallová D, Lischka H (2009) J Phys Chem A 113:12686
111. Perun S, Sobolewski AL, Domcke WJ (2006) J Phys Chem A 110:13238
112. Merchan M, González-Luque R, Climen T, Serrano-Andrés L, Rodríguez E, Reguero M, Peláez D (2006) J Chem Phys B 110:26471
113. Asturiol D, Lasorne B, Robb MA, Blancafort LJ (2009) Phys Chem A 113:10211
114. McFarland BK, Farrell JP, Miyabe S, Tarantelli F, Aguilar A, Berrah N, Bostedt C, Bozek JD, Bucksbaum PH, Castagna JC, Coffee RN, Cryan JP, Fang L, Feifel R, Gaffney KJ, Glownia JM, Martinez TJ, Mucke M, Murphy B, Natan A, Osipov T, Petrović VS, Schorb S, Schultz Th, Spector LS, Swiggers M, Tenney I, Wang S, White JL, White W, Gühr M (2014) Nature. Communication 5:4235
115. Bolognesi P et al (2015) J Phys Conf Ser 635:112062
116. Skinnerup Byskov C, Jensen F, Jørgensenc TJD, Brøndsted Nielsen S (2014) Phys Chem Chem Phys 16:15831
117. Weinkauf R, Aicher P, Wesley G, Grotemeyer J, Schlag EW (1994) J Phys Chem 98:8381
118. Weinkauf R, Schanen P, Metsala A, Schlag EW, Bürgle M, Kessler H (1996) J Phys Chem 100:18567
119. Lünnemann S, Kuleff AI, Cederbaum LS (2013) Chem Phys 414:100
120. Krausz F, Ivanov M (2009) Rev Mod Phys 81:163
121. Calegari F, Ayuso D, Trabattoni A, Belshaw L, De Camillis S, Anumula S, Frassetto F, Poletto L, Palacios A, Decleva P, Greenwood JB, Martín F, Nisoli M (2014) Science 346:336
122. Yamashita M, Fenn JB (1984) J Phys Chem 88:4451
123. Fenn JB, Mann M, Meng CK, Wong SF, Whitehouse CM (1989) Science 246:64
124. Reitsma G, Boschman L, Deuzeman MJ, González-Magaña O, Hoekstra S, Cazaux S, Hoekstra R, Schlathölter T (2014) Phys Rev Lett 113:053002
125. Milosavljević AR, Giuliani A, Nicolas C, Gil J-F, Lemaire J, Réfrégiers M, Nahon L (2010) J Phys: Conf Ser 257:012006
126. Milosavljević AR, Cerovski VZ, Canon F, Nahon L, Giuliani A (2013) Angew Chem Int Ed 52:7286
127. Maclot S, Capron M, Maisonny R, Ławicki A, Méry A, Rangama J, Chesnel J-Y, Bari S, Hoekstra R, Schlatholter T, Manil B, Adoui L, Rousseau P, Huber BA (2011) Chem Phys Chem 12:930

128. Zimmermann U, Malinowski N, Näher U, Franck S, Martin TP (1994) Z Physik D 31:85
129. Khistyaev K, Golan A, Bravaya KB, Orms N, Krylov AI, Ahmed M (2013) J Phys Chem A 117:6789

Irradiation-Induced Processes with Atomic Clusters and Nanoparticles

Alexey Verkhovtsev, Andrei V. Korol and Andrey V. Solov'yov

Abstract This chapter gives an overview of theoretical and computational studies of physical phenomena manifesting themselves in photon, electron and ion collisions with atomic clusters and nanoparticles (NPs). The emphasis is made on ion and electron scattering as well as photoabsorption of metal NPs which are of current interest in application in cancer treatments with ionizing radiation. Although the number of reports on dose enhancement and radiosensitization due to metal NPs has been rapidly increasing during the past years, physical mechanisms of enhanced production of secondary electrons and reactive species due to sensitizing NPs are still a debated issue and require thorough investigation. In this chapter, we elucidate the essential role of collective electron excitations in the formation of electron emission spectra of metal clusters and NPs. These effects appear also in other types of nanoscale systems, such as carbon-based NPs. We also briefly overview a number of recent Monte Carlo-based studies devoted to the investigation of radiosensitization and dose enhancement effects for proton irradiation combined with metal NPs.

A. Verkhovtsev (✉)
Instituto de Fisica Fundamental, Consejo Superior de Investigaciones Cientificas (CSIC),
Serrano 113-bis, 28006 Madrid, Spain
e-mail: verkhovtsev@iff.csic.es

A. Verkhovtsev · A.V. Korol
MBN Research Center,
60438 Frankfurt am Main, Germany
e-mail: korol@th.physik.uni-frankfurt.de

A.V. Korol
Department of Physics, St. Petersburg State Maritime Technical University,
Leninskii Prospekt 101, 198262 St. Petersburg, Russia

A.V. Solov'yov
MBN Research Center at Frankfurter Innovationszentrum Biotechnologie,
Altenhöferallee 3, 60438 Frankfurt am Main, Germany
e-mail: solovyov@mbnresearch.com

A.V. Solov'yov
A.F. Ioffe Physical-Technical Institute, Polytekhnicheskaya Ul. 26,
194021 Saint Petersburg, Russia

A.V. Solov'yov (ed.), *Nanoscale Insights into Ion-Beam Cancer Therapy*,
DOI 10.1007/978-3-319-43030-0_7

1 Introduction

At present, there is a vivid scientific interest in studying the interaction of nanoscale systems, such as atomic clusters and nanoparticles (NPs), with biological media because of the large number of possible applications in nanomedicine [1, 2]. One of the promising ideas is the use of metal-based NPs or small atomic clusters in cancer treatments with ionizing radiation [3–8]. It has been suggested that such NPs, being delivered to the tumor region, can act as radiosensitizers. They may locally enhance the radiation damage of the tumor cells relative to normal tissues thereby increasing the efficiency of treatments with ionizing radiation. Understanding and exploiting the nanoscale processes that drive physical, chemical, and biological effects induced by ionizing radiation in combination with radiosensitizing NPs is within the scope of an ongoing international project, entitled "Advanced Radiotherapy, Generated by Exploiting Nanoprocesses and Technologies (ARGENT)", supported by the European Commission [9]. Activities undertaken within this project are described in more detail in Chap. 12 of this book.

Currently investigated NPs are generally composed of metals or metal-based materials that can strongly absorb and re-emit radiation energy, resulting in a local radiation dose increase when they are accumulated in tumors [10–12]. Such nanoagents delivered to the tumor region can boost the production of secondary electrons, free radicals, and other reactive species [13–15], which are formed as a result of energy deposition by ionizing radiation in a biological medium [16, 17]. Of particular research interest is the production of low-energy electrons which have the kinetic energy from a few electronvolts to several tens of electronvolts. It is currently established that the secondary electrons of such low energies play an important role in the nanoscale mechanisms of biological damage resulting from ionizing radiation [18, 19].

After the first experimental evidence of radiosensitization by gold NPs [4], a number of follow-up experiments with platinum, silver and gadolinium-based NPs [20–24] demonstrated an increase of cancer cell killing during the X-ray irradiation. Several experiments were performed also with the combination of NPs and ion beams [5, 25–27]. Experiments performed at the molecular scale (using plasmid DNA as a probe) have shown that the addition of platinum NPs during irradiation with C^{6+} and He^{2+} ions amplifies the induction of important DNA lesions, such as double strand breaks [5]. Very recently, similar effects were also observed with small platinum and gadolinium-based NPs irradiated with protons [28].

Despite an increased number of studies on radiosensitization effects of metal NPs, the physical mechanisms of an enhanced yield of secondary electrons emitted from these nanoagents and the subsequent enhancement of the reactive species production are still not entirely understood. For photon beams, the effects of gold NPs have been explained recently by microscopic mechanisms [6], however it is still not established whether similar mechanisms play the main role in the irradiation with ions. This chapter gives an overview of theoretical studies of physical phenomena manifesting themselves in photon, electron and ion collisions with atomic clusters and NPs. We emphasize the essential role of collective electron excitations in strong electron emission from these systems. In particular, we describe recent theoretical and computational studies of the physical mechanisms of the enhanced production of secondary electrons from sensitizing NPs under ionizing irradiation. We describe recent achievements in the analysis of electron production by NPs under ion impact, while irradiation with photons is also briefly discussed. The main focus is made on gold, platinum, silver and gadolinium NPs, which are of current interest [5–7, 22, 25] for radiotherapy applications.

2 Monte Carlo Simulations of Dose Enhancement and Radiosensitization Due to Nanoparticles

Several hypotheses have been proposed recently to understand microscopic mechanisms of radiosensitization by gold NPs irradiated with X-rays. Monte Carlo simulations revealed that the nanoscale dose distribution around gold NPs is highly non-uniform, so that very high doses are deposited in the nearest region around the NP. The dose enhancement (i.e., the increase in energy deposited in the target volume) on the nanoscale was explained by the large number of Auger electrons produced following ionization in the elements with large atomic number [6, 11, 29]. A similar explanation was also given for nanoscale effects arising in the presence of gadolinium-based NPs irradiated with X-rays [30]. It was stated that Auger electrons deposit locally their energy in a region of up to a few tens of nanometers around the NP, leading to highly localized nanoscale doses.

Despite extensive research on NP sensitization under photon irradiation, much smaller number of theoretical and computational studies have been carried out for ion beams. Ion-beam cancer therapy (IBCT) is currently considered as one of the most promising modern treatment techniques [16, 31, 32] because it allows one to deliver a higher dose to the target region, as compared to conventional radiotherapy with X-rays, and also to minimize the exposure of healthy tissue to radiation [32]. Very recently, several Monte Carlo-based studies have evaluated the local dose enhancement for proton irradiation combined with metal NPs [33–36]. The first study of this kind was presented in Ref. [33] where electron emission from 4 and 44 nm NPs composed of gold, platinum, silver, gadolinium, and iron, was investigated by means of the track structure Monte Carlo code TRAX [37]. In the simulations, the NPs were irradiated with 2, 80 and 300 MeV monoenergetic proton beams. A data set of elastic and inelastic (direct ionization, production of Auger electrons, and excitations) cross sections was used in order to track low-energy electrons in the target materials down to energies of a few eV. The simulations demonstrated that the maximal dose enhancement is achieved in the case of platinum and gold NPs, being in the range from 1.9 to 2.5 depending on the NP size. The estimated range of the excess dose was about 5 nm from the surface of small NPs and about 20 nm from the large ones. It was stated that Auger electrons and Auger cascades make a significant contribution to the observed dose enhancement on a nanoscopic level.

In Ref. [34], Monte Carlo simulations were performed to compare the sensitization effect of a 50 nm gold NP using photon and proton beams. The simulations were carried out using the TOPAS (TOol for PArticle Simulation) tool [38]. It was demonstrated that the mechanism by which gold NPs can lead to dose enhancement is different in the case of photon and proton irradiation. For instance, it was shown that for the same amount of energy absorbed inside the NP the difference in the doses deposited by secondary electrons within several nanometers from the NP surface does not exceed 20 % for different radiation sources. However, secondary electrons produced by kilovoltage photons cause a dose enhancement an order of magnitude higher than that caused by protons 10 μm away from the NP surface. Thus, it was concluded that the sensitization effect is radiation type dependent, so that proton therapy can only be enhanced if the NPs and located in close proximity to the biological target being effectively internalized in the cells.

In the subsequent work [35] performed by the same group, the authors adopted the similar methodology to study biological effects of gold NPs radiosensitization, depending on their size and localization inside a cell. The simulations were performed for the NPs ranging from 2 to 50 nm in diameter. For the same mass of gold NPs in the cells, the maximal sensitization effect was observed for smaller systems. This was explained by a higher fraction of low-energy electrons escaping from the 2 nm NPs and thus contributing to the local dose enhancement. As a result of the simulations, it was also concluded that proton radiotherapy can be significantly enhanced with gold NPs only if these nanoagents are internalized into cells. Because of the shorter range of secondary electrons generated by protons, as compared to kilovoltage photons [34], gold NPs cannot contribute to direct DNA damage being located in the extracellular medium. Thus, the cellular uptake is crucial to bring NPs in close proximity to the cell nucleus.

In Ref. [36], Monte Carlo simulations were performed to assess the NP-induced dose enhancement under proton irradiation for several conditions, including several NP types and sizes (gold and gadolinium NPs with 4 and 50 nm in diameter), various distances between the radiation source and the NP, and various source sizes. The simulation results demonstrated that the number of electrons produced by a gold NP is 2–3 times higher than that emitted from a Gd NP of the same size and exceeds by an order of magnitude that emitted from equivalent volume of water. The contribution of Auger electrons to the obtained dose distributions was found to be insignificant (around 1 % at a few nm distance from the NP surface), that agreed with the results of Ref. [34] but contradicted with the results of Ref. [33]. This discrepancy was explained by diverse physical models used in different simulations. The calculated dose enhancement ratio was equal to 1.4–1.7 for gold and 1.1–1.2 for gadolinium, depending on the NP size and geometry of the simulation. As a result of this study, it was concluded that more realistic configurations (i.e., when the proton source is located not at the NP surface but at some distance from it) lead to a very small increase in dose deposition that is mostly located within 1–3 nm from the NP surface.

Most of the above described simulations have been carried out using a 'Geant4-Penelope' physics model for metals, which tracks electrons down to the energy of 100 eV [39]. A similar cutoff exists in other models for tracking low-energy electromagnetic interactions in Geant4 [39]. Thus, the simulation of secondary electron production from metal NPs by means of this approach is limited to the electrons with kinetic energy greater than 100 eV. Microdosimetry models included into the 'Geant4-DNA' physics package allow one to analyze the ionization and excitation spectra in water down to 7.4 eV but are not adapted for studying the production of low-energy secondary electrons in metal materials. Because of a lack of an accurate description of low-energy electrons in most of Monte Carlo simulations, many important physical processes involving these particles may be missing, and it may affect directly the outcome of the simulations. Important phenomena which are not accounted for in most of such simulations is the formation and decay of collective electron excitations which play a significant role in the ionization and excitation of metallic systems. These phenomena are addressed in detail further in this chapter.

3 Collective Electron Excitations in Metal Clusters and Nanoparticles

Metallic clusters and NPs are characterized by a full delocalization of their valence electrons. To some extent this feature is also valid for carbon fullerenes and polycyclic aromatic hydrocarbons (PAHs), where the delocalization of electrons takes place within aromatic rings. Due to the presence of highly mobile delocalized valence electrons, all these systems are highly polarizable. Collective oscillations of delocalized electrons can be excited by a photon or a charged projectile (e.g., an electron or an ion). Such collective excitations are known as plasmons [40, 41].

When considering electron, photon and ion collisions with metal clusters and NPs as well as fullerenes, the delocalized valence electrons often play the most important role in the formation of the cross sections of various collision processes. Plasmon excitations appear as prominent resonances in the cross sections, and the position of the resonance peak depends strongly on the type of a system [42–45]. Investigation of electron collisions with metal clusters and fullerenes in a gas phase has led to manifestation of the interference and diffraction phenomena [46–51], the role of surface and volume plasmon excitations in the formation of electron energy loss spectra [48–54], as well as the total inelastic scattering [48, 49], photoabsorption [55] and photoionization cross sections [56, 57]. All these phenomena are described below in this section.

3.1 Diffraction of Fast Electrons on Atomic Clusters

The phenomenon of elastic scattering of fast electrons on metal clusters and fullerenes appears because the ionic density distribution in a cluster is typically characterized by a rigid border. The presence of a surface in a cluster results in the specific oscillatory behavior of the electron elastic scattering cross sections, which can be interpreted in terms of electron diffraction by the cluster surface [48, 51]. The detailed theoretical treatment of the diffraction phenomena arising in electron scattering on metal clusters and fullerenes was given in Refs. [48–50]. Experimentally, diffraction in electron elastic scattering cross sections on gas-phase fullerenes was observed for the first time in Ref. [51].

Let us explain the physical nature of the diffraction phenomena arising in elastic electron–cluster scattering. For the sake of simplicity, we consider atomic clusters as spherically symmetric systems with a uniform electron density distribution; this model is well applicable, e.g., to highly-symmetric 'magic'-number metal clusters or carbon fullerenes. The cross section of elastic scattering of a fast electron on a cluster in the first Born approximation (see, e.g., Ref. [58]) reads as (in this chapter we use the atomic system of units, $m_{\rm e} = |e| = \hbar = 1$):

$$\frac{{\rm d}\sigma}{{\rm d}\Omega_{{\bf p}_2}} = \frac{4}{q^4}\,F(q)^2 \,. \tag{1}$$

Here, $F(q)$ is the form-factor of the cluster, $q = |{\bf p}_2 - {\bf p}_1|$ is the momentum transfer, with ${\bf p}_1$, ${\bf p}_2$ being the momenta of the electron in the initial and the final state, respectively, and ${\rm d}\Omega_{{\bf p}_2}$ denotes the differentiation over the solid angle of the scattered electron. The magnitude of q^2 is related to the scattering angle $\theta = \widehat{{\bf p}_1\,{\bf p}_2} \ll 1$ rad via:

$$q^2 = p_1^2 + p_2^2 - 2p_1p_2\cos\theta = 2p_1^2(1-\cos\theta) \approx p_1^2\theta^2 \,. \tag{2}$$

The form factor of the target, $F(q)$, can be expressed as product of the form factor of the atomic concentration, $n(q)$, and the form factor of a single atom, $F_{\rm A}(q)$:

$$F(q) = F_{\rm A}(q)\sum_j e^{{\rm i}{\bf q}{\bf r}_j} = F_{\rm A}(q)\,n(q) \tag{3}$$

where the summation is performed over all coordinates, ${\bf r}_j$, of all atoms in the cluster. The applicability of this approximation has been examined in Ref. [51] for metal clusters and fullerenes.

The form factor of the atomic concentration, $n(q)$, depends on the geometry of the cluster. In the case of the *metal cluster*, assuming a homogeneous distribution of atoms in the volume of the cluster of the radius R, one derives

$$n(q) = 3N\left[\frac{\sin(qR)}{(qR)^3} - \frac{\cos(qR)}{(qR)^2}\right] = 3N\,\frac{j_1(qR)}{qR} \,, \tag{4}$$

where $j_1(qR)$ is the spherical Bessel function of the first order and N is the number of atoms in the cluster [48]. The simplest approximation for the description of *a fullerene* is to assume that carbon atoms are uniformly distributed on the surface of a sphere of the radius R. In this case, one derives

$$n(q) = N\,\frac{\sin(qR)}{qR}\ . \tag{5}$$

This form-factor oscillates with the period $q = 2\pi/R \ll 1$. These oscillations form the diffraction pattern of the differential cross section (1) which possesses a series of diffraction maxima and minima whose positions are mainly determined by the radius of the target.

Figure 1 presents the dependence of the cross section $\mathrm{d}\sigma/\mathrm{d}\Omega_{\mathbf{p}_2}$ on the scattering angle θ for elastic collision of a 809-eV electron with the C_{60} fullerene [51]. The figure shows that the cross section possesses a series of diffraction maxima and minima. Experimental data points obtained in Ref. [51] in the two sets of measurements are illustrated by open and closed circles. The cross section dependence obtained theoretically is shown by a solid curve. Experimental data have been normalized to the theoretical cross section at the second diffraction maximum ($\theta = 5°$). Figure 1 shows quite a good agreement of the experimental and the theoretical results in position of the first and the second maxima. The entire pattern of the differential cross section obtained theoretically is very similar to that measured in experiment.

In the vicinity of the diffraction maxima at $\theta < 10°$ the cross section greatly exceeds the elastic scattering cross section on the equivalent number of isolated atoms because of the coherent interaction of the projectile electron with the fullerene sphere.

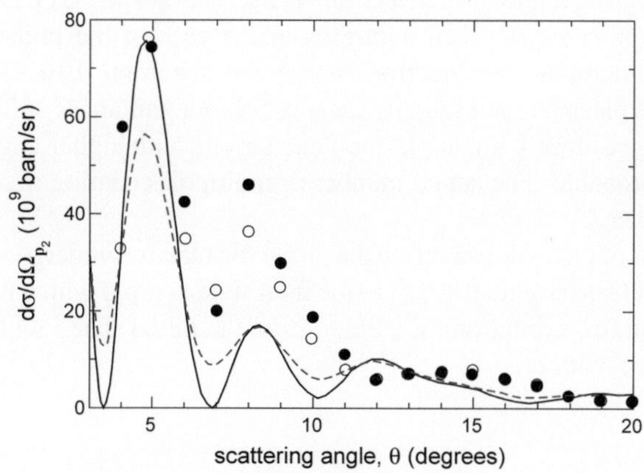

Fig. 1 Experimental (*symbols*) and theoretical (*solid curve*) angular dependencies of the differential elastic scattering cross section in collision of a 809 eV electron with the C_{60} molecule. Symbols correspond to the two independent sets of measurements. *Dashed curve* is the cross section for the mixture containing 60 % of C_{60} and 40 % of equivalent isolated carbon atoms. The figure is adapted from Ref. [51] with editorial modifications

In the region $\theta > 10°$, where $q > 1$, the projectile electron scatters on individual carbon atoms of the fullerene rather than on the entire molecule. Therefore diffraction features of the cross section in the region $\theta < 10°$ are much more pronounced than in the region $\theta > 10°$.

In the region $\theta < 10°$, where $q < 1$, the theoretical cross section has zeros while the experimental one does not. The presence of zeros at $q \approx \pi k/R < 1$ where k is integer, in the theoretical curve is the consequence of the coherent scattering of electron on the fullerene sphere.[1] However, in experiment, zeros in the cross section can disappear because of the presence of carbon atoms or some other impurities in the gas cell. Figure 1 also shows the differential cross section for the mixture containing 60 % of C_{60} and 40 % of isolated carbon atoms (the dashed curve). The differential electron elastic scattering cross section on single carbon atoms does not have diffraction oscillations and thus it forms the smooth background removing zeroes in the angular dependence of the cross section.

3.2 Inelastic Scattering of Electrons on Atomic Clusters

Plasmon excitations in metal clusters and fullerenes have been intensively studied during the past several decades, see, e.g., Refs. [40, 59–66]. They were observed in photoabsorption experiments with metal clusters [42, 43] and in photoionization studies with fullerenes [44, 67–70].

In the photoionization experiments, only the dipole plasmon excitation mode (the angular momentum $l = 1$) can be effectively probed [71, 72]. Electron collective modes with higher angular momenta can be studied in the fast electron–cluster collisions if the scattering angle of the electron is large enough [48–51]. Dipole plasmon resonances of the same physical nature as in the case of the photoabsorption or photoionization dominate the electron energy loss spectrum (EELS) if the scattering angle of the electron, and thus its transferred momentum, is sufficiently small. With increasing scattering angle plasmon excitations with higher angular momenta become more probable. The actual number of multipoles coming into play depends on the cluster size.

In the process of inelastic scattering the projectile electron undergoes the transition from the initial electron state $(\varepsilon_1, \mathbf{p}_1)$ to the final state $(\varepsilon_2, \mathbf{p}_2)$ which is accompanied by the ionization (or, excitation) of a target from the initial state i with the energy ε_i to the final state f with ε_f.

---

[1] In the region $\theta > 10°$, where $q > 1$, the process of elastic scattering on the fullerene shell with the subsequent excitation of surface multipole plasmons becomes dominating. This process is described by the formulas of the second Born approximation which was used to correct the calculated cross section at large values of transferred momentum.

The matrix element, M, which defines the amplitude of the inelastic scattering is given by

$$M = \left\langle f, 2 \left| \sum_a \frac{1}{|\mathbf{r} - \mathbf{r}_a|} \right| 1, i \right\rangle$$
$$= \sum_a \int \psi_2^{(-)*}(\mathbf{r}) \psi_f^*(\{\mathbf{r}_a\}) \frac{1}{|\mathbf{r} - \mathbf{r}_a|} \psi_i(\{\mathbf{r}_a\}) \psi_1^{(+)}(\mathbf{r}) \{d\mathbf{r}_a\} d\mathbf{r} \,, \tag{6}$$

where $\{\mathbf{r}_a\} = \mathbf{r}_1 \ldots \mathbf{r}_N$ are the position vectors of the delocalized electrons in the target, $\mathbf{r}$ is the position vector of the projectile, $\psi_1^{(+)}(\mathbf{r})$ and $\psi_2^{(-)}(\mathbf{r})$ stand for the initial- and the final state wave functions of the projectile, respectively. Superscripts $(+)$ and $(-)$ indicate that asymptotic behavior of the wave functions is 'plane wave + outgoing spherical wave' and 'plane wave + incoming wave', respectively.

The matrix element can be written as follows:

$$M = \int \frac{4\pi}{q^2} \frac{d\mathbf{q}}{(2\pi)^3} \langle 2 \left| e^{-i\mathbf{q}\cdot\mathbf{r}} \right| 1 \rangle \left\langle f \left| \sum_a e^{i\mathbf{q}\cdot\mathbf{r}_a} \right| i \right\rangle , \tag{7}$$

where $\mathbf{q} = \mathbf{p}_1 - \mathbf{p}_2$ is the transferred momentum.

If the velocity of a projectile is high and significantly exceeds the characteristic velocities of delocalized electrons in the target, the first Born approximation is applicable [48]. Within this approximation the initial and the final states of the incident electron can be described by plane waves:

$$\psi_1^{(+)}(\mathbf{r}) = e^{i\mathbf{p}_1\cdot\mathbf{r}} \,, \qquad \psi_2^{(-)}(\mathbf{r}) = e^{i\mathbf{p}_2\cdot\mathbf{r}} \,. \tag{8}$$

Then, the amplitude of the process reduces to

$$M = \frac{4\pi}{q^2} \left\langle f \left| \sum_a e^{i\mathbf{q}\cdot\mathbf{r}_a} \right| i \right\rangle_{\mathbf{q}=\mathbf{p}_1-\mathbf{p}_2} . \tag{9}$$

The magnitude q^2 is related to the scattering angle as $q^2 \approx p_1^2\theta^2$ under the assumption that the energy loss $\omega = \varepsilon_1 - \varepsilon_2$ is small, $\omega \ll \varepsilon_1$ (which implies $p_1 \approx p_2$) and the scattering angle is small, $\theta \ll 1$ rad.

Performing the multipole expansion of the exponential factors in (9) (see, e.g., Ref. [73]), one obtains:

$$M = 4\pi \sum_{lm} i^l \, Y_{lm}^*(\mathbf{q}) \left\langle f \left| \sum_a \phi_l(r_a) Y_{lm}(\mathbf{r}_a) \right| i \right\rangle , \tag{10}$$

where the notation

$$\phi_l(r) = 4\pi \frac{j_l(qr)}{q^2} \tag{11}$$

is introduced and j_l is a spherical Bessel function of the order l.

Let us consider a general expression for the cross section of the scattering process:

$$\mathrm{d}\sigma = \frac{2\pi}{p_1}\,\delta(\omega_{fi} - \omega) \sum_{\mathrm{pol}_f} \overline{\sum_{\mathrm{pol}_i}} |M|^2 \frac{\mathrm{d}\mathbf{p}_2}{(2\pi)^3}\,\mathrm{d}\rho_f\,. \tag{12}$$

Here $\omega_{fi} = \varepsilon_f - \varepsilon_i$, the sign $\sum_{\mathrm{pol}_f}$ denotes the summation over the projection of the final state f orbital momentum, whereas $\overline{\sum}_{\mathrm{pol}_i}$ denotes the averaging over the projections of the initial state orbital momentum, and $\mathrm{d}\rho_f$ is the density of final states of the target.

Substituting the scattering amplitude (10) into Eq. (12), one derives the doubly differential cross section:

$$\frac{\mathrm{d}^2\sigma}{\mathrm{d}\varepsilon_2 \mathrm{d}\Omega_{\mathbf{p}_2}} = \frac{1}{\pi}\frac{p_2}{p_1} \sum_{lm} \int \left| \left\langle f \left| \sum_a V_{lm}(\mathbf{r}_a) \right| i \right\rangle \right|^2 \delta(\omega_{fi} - \omega)\,\mathrm{d}\rho_f\,, \tag{13}$$

where

$$V_{lm}(\mathbf{r}) = \phi_l(r) Y_{lm}(\mathbf{r}) \tag{14}$$

is the multipolar potential of the fast projectile, $\mathrm{d}\Omega_{\mathbf{p}_2}$ denotes the differentiation over the solid angle of the scattered electron and sign $\int \mathrm{d}\rho_f$ means the summation over the final states (which includes the summation over the discrete spectrum and the integration over the continuous spectrum).

3.3 Plasmon Resonance Approximation

An effective tool for evaluation of the contribution of plasmon excitations to the excitation and ionization spectra of different atomic clusters and NPs is based on the plasmon resonance approximation (PRA) [48, 50, 55, 72]. This approach postulates that the dominating contribution to the cross section in the vicinity of the plasmon resonance frequency comes from collective electron excitations, while single-particle effects give a much smaller contribution. Hence, one can neglect single-particle excitations when calculating the matrix element in Eqs. (12) and (13). During the past decades, this approach has provided a clear physical explanation of the resonant-like structures in photoionization spectra [55, 57, 74] and differential inelastic scattering cross sections [48, 50, 53, 54, 74–76] of small metallic clusters and nanoparticles, as well as carbon fullerenes by the photon, electron and ion impact. It was also applied [55, 77–79] to describe the dynamic response of alkali and noble metal clusters in the processes of radiative electron capture, polarization bremsstrahlung, and multiphoton excitation.

According to Kubo linear response theory [48, 80], the integral on the right-hand side of Eq. (13) can be related to the variation of electron density caused by an external electric field, and the following substitution can be performed:

$$\int \left| \left\langle f \left| \sum_a V_{lm}(\mathbf{r}_a) \right| i \right\rangle \right|^2 \delta(\omega_{fi} - \omega) \mathrm{d}\rho_f \;\rightarrow\; \frac{1}{\pi} \mathrm{Im} \int V_{lm}^*(\mathbf{r}) \delta\rho_l(\omega, q; \mathbf{r}) \mathrm{d}\mathbf{r} \,. \quad (15)$$

Here, $\delta\rho_l(\omega, q; \mathbf{r})$ is the density variation due to the exposure of the system to the multipolar potential $V_{lm}(\mathbf{r})$. In a general case, this variation depends on the transferred energy ω, transferred momentum q and the position vector $\mathbf{r}$.

Using (15) in (13), the doubly differential cross section acquires the form

$$\frac{\mathrm{d}^2\sigma}{\mathrm{d}\varepsilon_2 \mathrm{d}\Omega_{\mathbf{p}_2}} = \frac{1}{\pi^2} \frac{p_2}{p_1} \sum_l \mathrm{Im}\Big[I_l(\omega, q) \Big] \,, \quad (16)$$

where

$$I_l(\omega, q) = \sum_m \int V_{lm}^*(\mathbf{r}) \delta\rho_l(\omega, q; \mathbf{r}) \mathrm{d}\mathbf{r} \,. \quad (17)$$

In the general case, an atomic cluster can be considered as a spherically symmetric system where the charge is distributed homogeneously between two concentric spheres. The width of a spherical shell is defined as $\Delta R = R_2 - R_1$ where R_1, R_2 are the inner and the outer radii of the system, respectively. The equilibrium electron density distribution $\rho_0(r)$ is expressed via the number N_e of delocalized electrons and the system's volume V:

$$\rho_0 = \begin{cases} N_e/V & \text{for } R_1 \le r \le R_2 \\ 0 & \text{if otherwise} \,. \end{cases} \quad (18)$$

The volume of the spherical shell reads as

$$V = \frac{4\pi}{3}\left(R_2^3 - R_1^3\right) = \frac{4\pi}{3} R_2^3 \left(1 - \xi^3\right) , \quad (19)$$

where $\xi = R_1/R_2 \le 1$ is the ratio of the inner to the outer radii. Such a "jellium"-shell representation has been successfully utilized for the description of plasmon formation in carbon fullerenes [53, 54, 57] and fullerene-like hollow gold clusters (see Ref. [75] and Sect. 4). For such a geometry, the density variation can be written in the following form

$$\delta\rho_l(r) = \delta\rho_l^{(\mathrm{v})}(r)\Theta(r - R_1)\Theta(R_2 - r) + \sigma_l^{(1)}\delta(r - R_1) + \sigma_l^{(2)}\delta(r - R_2) \,, \quad (20)$$

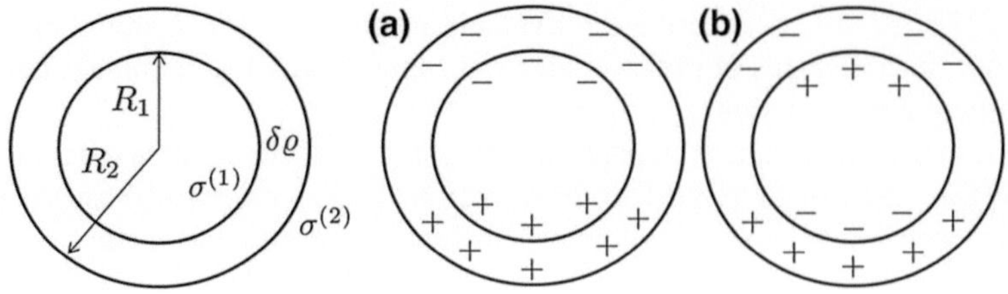

Fig. 2 *Left panel*: Representation of a target system (e.g., a fullerene) as a spherical shell of a width $\Delta R = R_2 - R_1$. Variation of the surface charge densities, $\sigma^{(1,2)}$, and the volume charge density, $\delta\rho^{(v)}$, is also shown. *Right panel*: Representation of the symmetric (**a**) and the antisymmetric (**b**) modes of the surface plasmon. The figure is adapted from Ref. [72] with minor editorial modifications

where $\delta\rho_l^{(v)}(r)$ describes the volume density variation arising inside the shell, and $\sigma_l^{(1,2)}$ are variations of the surface charge densities at the inner and the outer surfaces of the shell, respectively (see the left panel of Fig. 2). In this expression, $\Theta(x)$ and $\delta(x)$ are the Heaviside step function and the delta-function, respectively. The volume density variation causes the formation of the volume plasmon, while the variations of the surface densities correspond to two surface plasmon modes, the so-called symmetric and antisymmetric ones. In the symmetric mode the charge densities of the two surfaces oscillate in phase, while in the antisymmetric mode they are out of phase (see the right panel of Fig. 2). The volume plasmon appears due to compression of the electron density inside the volume of the shell, therefore it does not interfere with either of the surface plasmon modes [72]. The formation of the volume plasmon in the electron impact ionization of metal clusters and carbon fullerenes was revealed in Refs. [52–54]. The model accounting for the contribution of different plasmon modes was successfully utilized to describe the experimentally observed variation of the electron energy loss spectra of C_{60} in collision with fast electrons [53, 54].

The spherical-shell model defined by Eqs. (18) and (19) is applicable for any spherically symmetric system with an arbitrary value of the ratio ξ. Supposing $\xi = 0$ (i.e., $R_1 \to 0$), one obtains a model of a metallic cluster/NP. As a result, the system is treated not as a "jellium" hull but as a full sphere, where the electron density is uniformly distributed inside the sphere of a radius R [49, 52]. In this case, the electron density variation on the surface and in the volume of the cluster or NP leads to the formation of the surface (symmetric mode) and the volume plasmon, respectively, while the antisymmetric surface plasmon mode does not contribute to the cross section. A detailed explanation of this phenomenon can be found in Ref. [72].

Presenting the multipole variation of the electron density, $\delta\rho_l(\omega, q; \mathbf{r})$, as a sum of three contributions (see Eq. (20)), using the explicit expression for the multipolar potential $V_{lm}(r)$ (see Eqs. (11) and (14)), and performing some algebraic transformations, one comes to the formula for the differential inelastic scattering cross section with no damping of plasmon oscillations [72]:

$$\frac{d^2\sigma}{d\varepsilon_2 d\Omega_{\mathbf{p}_2}} = \frac{2}{\pi}\frac{R_2}{q^4}\frac{p_2}{p_1}\mathrm{Im}\sum_l \left[\frac{\omega_p^2\, V_l(q)}{\omega^2-\omega_p^2} + \frac{\omega_{1l}^2\, S_{1l}(q)}{\omega^2-\omega_{1l}^2} + \frac{\omega_{2l}^2\, S_{2l}(q)}{\omega^2-\omega_{2l}^2}\right]. \tag{21}$$

In this expression, ω_p is the volume plasmon frequency associated with the ground-state electron density ρ_0 of N_e electrons,

$$\omega_p = \sqrt{4\pi\rho_0} = \sqrt{\frac{3N_e}{R_2^3 - R_1^3}}, \tag{22}$$

and ω_{1l}, ω_{2l} are the frequencies of the two surface plasmon modes of multipolarity l:

$$\omega_{jl} = \left(1 \mp \frac{1}{2l+1}\sqrt{1+4l(l+1)\xi^{2l+1}}\right)^{1/2}\frac{\omega_p}{\sqrt{2}}, \quad (j=1,2) \tag{23}$$

where '$-$' and '$+$' stand for the symmetric ($j = 1$) and antisymmetric ($j = 2$) mode, respectively. The volume plasmon frequency, ω_p, is independent of l as it follows from Eq. (22). Functions $V_l(q)$, $S_{1l}(q)$ and $S_{2l}(q)$ are the diffraction factors depending on the transferred momentum q. They determine the relative significance of the multipole plasmon modes in various ranges of the projectile's scattering angles. The dominant contribution of different multipole modes results in a significant angular dependence for the differential electron energy loss spectrum [54]. Explicit expressions for these functions are given in Ref. [72].

Plasmons decay from the collective excitation mode to the incoherent sum of single-electron excitations; therefore, it is essential to account for the damping of plasmon oscillations. This can be done by introducing the finite widths, $\Gamma_l^{(v)}$ and $\Gamma_{jl}^{(s)}$ ($j = 1, 2$) of the volume and surface plasmon resonances, respectively, and making the following substitutions in the right-hand side of Eq. (21):

$$\frac{1}{\omega^2-\omega_{jl}^2} \to \frac{1}{\omega^2-\omega_{jl}^2 + i\omega\Gamma_{jl}^{(s)}}, \qquad \frac{1}{\omega^2-\omega_p^2} \to \frac{1}{\omega^2-\omega_p^2 + i\omega\Gamma_l^{(v)}}. \tag{24}$$

The final formula for the differential inelastic scattering cross section which accounts for three plasmons and with damping included is:

$$\frac{d^2\sigma}{d\varepsilon_2 d\Omega_{\mathbf{p}_2}} = \frac{d^2\sigma^{(v)}}{d\varepsilon_2 d\Omega_{\mathbf{p}_2}} + \frac{d^2\sigma^{(s_1)}}{d\varepsilon_2 d\Omega_{\mathbf{p}_2}} + \frac{d^2\sigma^{(s_2)}}{d\varepsilon_2 d\Omega_{\mathbf{p}_2}}, \tag{25}$$

where

$$
\begin{aligned}
\frac{d^2\sigma^{(v)}}{d\varepsilon_2 d\Omega_{\mathbf{p}_2}} &= \frac{2R_2p_2}{\pi q^4 p_1}\,\omega \sum_l \frac{\omega_p^2\,\Gamma_l^{(v)}\,V_l(q)}{\left(\omega^2-\omega_p^2\right)^2+\omega^2\left(\Gamma_l^{(v)}\right)^2}\\
\frac{d^2\sigma^{(s_1)}}{d\varepsilon_2 d\Omega_{\mathbf{p}_2}} &= \frac{2R_2p_2}{\pi q^4 p_1}\,\omega \sum_l \frac{\omega_{1l}^2\,\Gamma_{1l}^{(s)}\,S_{1l}(q)}{\left(\omega^2-\omega_{1l}^2\right)^2+\omega^2\left(\Gamma_{1l}^{(s)}\right)^2}\\
\frac{d^2\sigma^{(s_2)}}{d\varepsilon_2 d\Omega_{\mathbf{p}_2}} &= \frac{2R_2p_2}{\pi q^4 p_1}\,\omega \sum_l \frac{\omega_{2l}^2\,\Gamma_{2l}^{(s)}\,S_{2l}(q)}{\left(\omega^2-\omega_{2l}^2\right)^2+\omega^2\left(\Gamma_{2l}^{(s)}\right)^2}\,.
\end{aligned}
\tag{26}
$$

The cross section $d^2\sigma/d\varepsilon_2 d\Omega_{\mathbf{p}_2}$ can also be written in terms of the energy loss $\omega = \varepsilon_1 - \varepsilon_2 \equiv \Delta\varepsilon$ of the incident projectile of energy ε_1. Integration of $d^2\sigma/d\Delta\varepsilon\, d\Omega_{\mathbf{p}_2}$ over the solid angle leads to the single differential cross section:

$$
\frac{d\sigma}{d\Delta\varepsilon} = \int d\Omega_{\mathbf{p}_2} \frac{d^2\sigma}{d\Delta\varepsilon\, d\Omega_{\mathbf{p}_2}} = \frac{2\pi}{p_1p_2}\int_{q_{\min}}^{q_{\max}} q\, dq \frac{d^2\sigma}{d\Delta\varepsilon\, d\Omega_{\mathbf{p}_2}}\,. \tag{27}
$$

As follows from the expressions presented above, the PRA relies on a few parameters, which include the oscillator strength of the plasmon excitation, position of the plasmon resonance peak and its width. The choice of these parameters can be justified by comparing the model-based spectra either with experimental data or with the results of more advanced calculations. Note that values of the plasmon width cannot be obtained directly on the basis of the utilized model. A precise calculation of the widths can be performed by analyzing the decay of the collective excitation mode into the incoherent sum of single-electron excitations. This process should be considered within the quantum-mechanical framework [52] and cannot be treated within the classical physics framework, as the PRA does. In Ref. [52], such an analysis was made to obtain the values of the surface and the volume plasmon width for a Na_{40} cluster. This analysis is discussed in the following section.

3.4 Contribution of the Surface and Volume Plasmon Excitations to the formation of Electron Energy Loss Spectra of Metal Clusters

Damping of the plasmon oscillations is related to the decay of the collective electron excitations to the single-particle ones similar to the mechanism of Landau damping in infinite electron gas. Frequencies of the surface plasmon excitations in neutral metal clusters lie in the vicinity of the ionization threshold. For instance, in small sodium clusters, they are below the ionization potential, and single-particle excitations in the vicinity of the surface plasmon resonance have therefore the discrete spectrum. In this case, the width of a surface plasmon excitation caused by the Landau damping should be treated as the width of the distribution of the oscillator strengths in the vicinity of the resonance. The problem of the formation of the surface plasmon resonance widths in clusters was studied in a number of works [52, 81–83].

The resonance frequencies of volume plasmon excitations in metal clusters are typically located above the ionization threshold. This means that the volume plasmon excitations are quasi-stable and have the real channel of the Landau damping leading to the ionization of the cluster [52]. Thus, the process of inelastic scattering in the region of transferred energies above the ionization threshold can be described as follows. The projectile particle induces the oscillations of the electron density in the cluster; in turn, they cause oscillations of the electric field which result in the ionization of the cluster. The similar scenario takes place with damping of the surface plasmon resonances in fullerenes [84], which also decay via the autoionization channel.

The differential cross section of the electron inelastic scattering on metal clusters obtained in the PRA with accounting for both surface and volume excitations [52] reads as:

$$\begin{aligned}\frac{d^2\sigma}{d\varepsilon_2 d\Omega_{\mathbf{p}_2}} &= \frac{4Rp_2}{\pi q^4 p_1}\sum_l (2l+1)^2 j_l^2(qR)\frac{\omega_l^2\,\omega\,\Gamma_l^{(s)}}{(\omega^2-\omega_l^2)^2+\omega^2\left(\Gamma_l^{(s)}\right)^2} \\ &+ \frac{2R^3p_2}{\pi q^2 p_1}\sum_l (2l+1)\frac{\omega_p^2\,\omega\,\Gamma_l^{(v)}}{(\omega^2-\,\omega_p{}^2)^2+\omega^2\left(\Gamma_l^{(v)}\right)^2} \\ &\times\left[j_l^{\prime 2}(qR) - j_{l+1}(qR)j_{l-1}(qR) - \frac{2}{qR}j_{l+1}(qR)j_l(qR)\right],\end{aligned} \tag{28}$$

where $\omega_p = \sqrt{3N_e/R^3}$ is the volume plasmon resonance frequency and $\omega_l = \sqrt{l/(2l+1)}\omega_p$ is the frequency of a surface plasmon excitation with the angular momentum l, $\Gamma_l^{(v)}$ and $\Gamma_l^{(s)}$ are the corresponding widths. The cross section (28) is similar to the expression obtained in Ref. [85] for electron scattering on small metal particles by means of classical electrodynamics.

According to Ref. [52], the width of the surface plasmon resonance in the PRA is equal to:

$$\Gamma_l^{(s)} = \frac{4\pi\,\omega_l}{(2l+1)R}\sum_{\nu,\mu}\left|\left\langle\psi_\mu\left|\varphi_l^{(s)}(\mathbf{r})\right|\psi_\nu\right\rangle\right|^2\delta(\omega_l-\varepsilon_\mu+\varepsilon_\nu), \tag{29}$$

where $\varphi_l^{(s)}(\mathbf{r}) = \sum_m\left[(r/R)^l\theta(R-r)+(R/r)^{(l+1)}\theta(r-R)\right]Y_{lm}(\mathbf{n})$. Note that the same expression was obtained in earlier studies [81, 82] using other methods. Evaluation of the expression (29) for sufficiently large clusters leads to the well-known result for the Landau damping of the surface plasmon oscillations, $\Gamma_l^{(s)} = 3l\,v_F/R$, where v_F is the velocity of the cluster electrons on the Fermi surface [82].

In the PRA, one can also determine the autoionization width of the volume plasmon resonance [52], which is equal to:

$$\Gamma_l^{(\mathrm{v})} = \frac{8\pi^2\omega_\mathrm{p}}{q^2R^3} \frac{\sum_\nu \int \left| \left\langle \psi_\mu \left| \varphi_l^{(\mathrm{v})}(\mathbf{r}) \right| \psi_\nu \right\rangle \right|^2 \delta(\omega_\mathrm{p} - \varepsilon_\mu + \varepsilon_\nu) \mathrm{d}\mu}{j_l^2(qR) - j_{l+1}(qR)j_{l-1}(qR) - \frac{2}{qR} j_{l+1}(qR) j_l(qR)} , \quad (30)$$

where $\varphi_l^{(\mathrm{v})}(\mathbf{r}) = \sum_m \left[j_l(qr) - j_l(qR)(r/R)^l \right] \theta(R-r) Y_{lm}(\mathbf{n})$. The summation is performed over the occupied single-electron states ν and the integration is performed over the electronic states μ of the continuous spectrum.

The projectile particle excites simultaneously numerous modes of the volume plasmon. The sum of the potentials of all the modes gives the resulting potential $\varphi_l^{(\mathrm{v})}(\mathbf{r})$. It is essential that all normal modes of the volume plasmon have the same resonance frequency ω_p, but the excitation probability for these modes depends on the kinematics of collision. This leads to the dependence of the volume plasmon potential $\varphi_l^{(\mathrm{v})}(\mathbf{r})$ upon the transferred momentum. The oscillations of the volume plasmon potential result in the ionization of the cluster, which probability and the volume plasmon resonance width depend on transferred momentum q. However, the numerical analysis [52] showed that the dependence of $\Gamma_l^{(\mathrm{v})}$ on q is rather weak in the region of $q \ll 1$, where collective electron oscillations mainly take place. Therefore, the volume plasmon resonance width with the given l can be approximated by the limiting value following from Eq. (30) at $q = 0$:

$$\Gamma_l^{(\mathrm{v})} = (2l+5) \frac{\pi^2 \omega_\mathrm{p}}{R} \sum_\nu \int \left| \left\langle \psi_\mu \left| \varphi_l^{(\mathrm{v})}|_{q=0}(\mathbf{r}) \right| \psi_\nu \right\rangle \right|^2 \delta(\omega_\mathrm{p} - \varepsilon_\mu + \varepsilon_\nu) \mathrm{d}\mu \, . \quad (31)$$

Figure 3 shows the dependence of the autoionization width $\Gamma_l^{(\mathrm{v})}$ on the transferred momentum q for the volume plasmon modes, which provide significant contribution to the EELS. The width of the dipole, quadrupole and octupole volume plasmon resonances has been calculated according to Eq. (30). The transferred momentum q plays the role of the wave vector for the volume plasmon excitations. All three plasmon modes have the similar dependence of $\Gamma_l^{(\mathrm{v})}$ upon q. The width grows slowly in the region of small q and it decreases rapidly at larger q. In the latter region, the probability of volume plasmon excitation by the incoming electron is correspondingly reduced. Note that the wave length of a collective electron oscillation should be larger than the inter-electronic distance in the cluster, i.e. plasmon wave vector should be smaller than the Fermi momentum of cluster electrons. In the region $q < 0.5$, where the latter condition is fulfilled, the dependence of $\Gamma_l^{(\mathrm{v})}$ upon q is rather weak, and the resonance width can be approximated by the following values: $\Gamma_1^{(\mathrm{v})} \simeq 0.5\omega_\mathrm{p}$, $\Gamma_2^{(\mathrm{v})} \simeq 0.3\omega_\mathrm{p}$, and $\Gamma_3^{(\mathrm{v})} \simeq 0.23\omega_\mathrm{p}$. Contrary to surface plasmons, the autoionization width of a volume plasmon decreases with the growth of the angular momentum.

Figure 4 shows the EELS of a sodium Na_{40} cluster in collision with a 50-eV electron at the scattering angle $\theta = 9°$. The figure illustrates the region of transferred

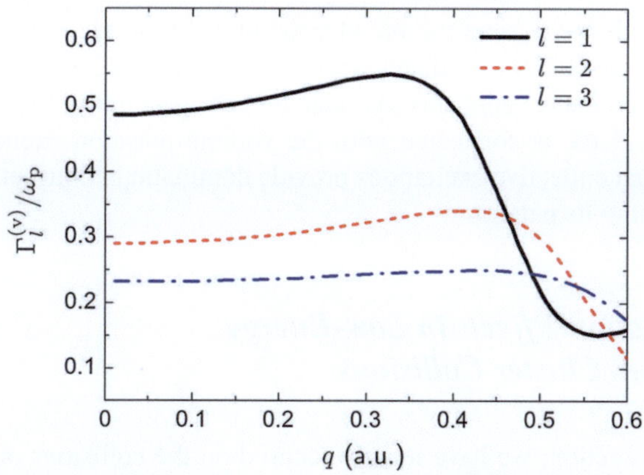

Fig. 3 Autoionization width $\Gamma_l^{(v)}$ of the dipole ($l = 1$), quadrupole ($l = 2$) and the octupole ($l = 3$) volume plasmon excitations as a function of transferred momentum q. The figure is adapted from Ref. [52] with editorial modifications

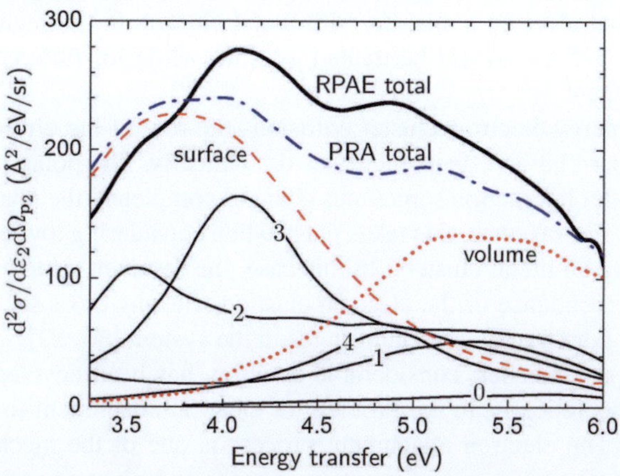

Fig. 4 Differential cross section $d^2\sigma/d\varepsilon_2 d\Omega_{\mathbf{p}_2}$ as a function of the transferred energy ω calculated for the collision of a 50 eV electron with a Na_{40} cluster for the scattering angle $\theta = 9°$ [52]. *Solid lines* represent the RPAE results (see the text for further details). Contributions of the surface and the volume plasmons calculated in the PRA (28) are shown by *dashed* and *dotted lines*, respectively. *Dashed-dotted line* represents the sum of these two contributions

energy above the ionization potential, $\omega > 3.3$ eV, where volume plasmon modes become significant. Solid curves illustrate the spectrum calculated using the random phase approximation with exchange (RPAE) method [49, 71]—an *ab initio* method which takes into account many-electron correlations in a many-particle system. In Fig. 4, the thick curve corresponds to the total EELS calculated with RPAE, while thin curves show various partial contributions corresponding to different angular momenta ($l = 0$ to 4). These curves are marked by numbers. The partial contribution to the EELS with $l < 3$ have the broad maximum in the vicinity of $\omega \simeq 5.1$ eV. Comparison of the EELS calculated within the *ab initio* RPAE approach and the PRA (dash-dotted curve) confirms that idea that the peculiarity in the EELS in the vicinity of $\omega \sim 5$ eV is connected with the volume plasmon excitation. Figure 4 demonstrates that collective excitations provide dominating contribution to the total EELS determining its pattern.

3.5 *Polarization Effects in Low-Energy Electron–Cluster Collisions*

In the previous section, we have mainly focused on the collisions of fast electrons with metal clusters and fullerenes. When considering low-energy electron–cluster collisions, i.e., when the velocity of the projectile is lower or comparable with characteristic velocities of the delocalized cluster electrons, polarization effects come into play [71]. In Ref. [48], on the basis of the Born theory of electron–cluster collisions, it was shown that electron collisions with metal clusters in the region of collision energies below 3–5 eV should be treated as slow, while for fullerenes, the region extends up to 30 eV.

In the low-energy electron–cluster collisions the role of the cluster polarization and exchange-correlation effects increases dramatically. The polarization potential of electron–cluster interaction sometimes changes completely the qualitative picture of the collision. For instance, this takes place when considering low-energy electron elastic scattering on metal clusters. In this case, the resonant structures can appear in the energy dependence of the electron elastic scattering cross section due to the presence of the bound or quasi-bound states in the system [86, 87].

During the past decades, considerable attention has been devoted, both experimentally and theoretically, to the problem of electron attachment to metal clusters and fullerenes. The electron attachment process is one of the mechanisms which leads to the negative cluster ion formation in gases and plasmas and thus it attracts the interest of numerous researchers. Low-energy electron–fullerene scattering was studied in Refs. [88–91]. For metal clusters, the electron attachment problem has been the subject of the intensive experimental [92–96] and theoretical [77, 97–100] investigations. Below, this problem is discussed in more details.

The very simple picture of attachment is described in many textbooks (see, e.g., Ref. [101]). Let us assume that there exists a Langevin attractive potential of the form

$$V(r) = -\frac{\alpha}{2r^4} \tag{32}$$

outside the cluster radius. The constant α is the static polarizability of the cluster. One can then show that there is an orbiting cross section,

$$\sigma = \left(\frac{2\pi^2\alpha}{\varepsilon}\right)^{1/2}, \tag{33}$$

which sets an upper limit bound to the attachment cross section (the so-called Langevin limit). Here, ε is the kinetic energy of the projectile electron. This simple treatment, if valid, would explain the behavior of the cross section in the vicinity of the threshold.

It is known that metal clusters possess a high polarizability (see, e.g., Ref. [59]); hence, large capture cross sections are anticipated. However, simple attempts to account for attachment by using the static polarizability α are not in accordance with observations [93]. The great weakness of the Langevin model is the treatment of α as an approximate constant. In fact, it possesses a complicated energy dependence due to the dynamical polarizability of the metallic cluster.

The possibility of resonances in the capture cross section was considered theoretically in Refs. [77, 97]. It was demonstrated that low-energy electrons can excite a collective plasmon resonance within the metal cluster in the electron attachment process as a result of a strong dipole deformation of the charge density of the cluster. Later this idea was commented in the context of the measurements performed in Ref. [92], although no clear evidence of the resonant behavior was found. The total inelastic scattering cross sections measured in Ref. [92] included attachment as only one of several possible contributing channels.

The resonant electron attachment mechanism was called in Refs. [77, 97] a "polarizational capture" in analogy with the similar mechanism known in the theory of bremsstrahlung (see, e.g., Ref. [79] and references therein). An important consequence of the polarization mechanism is that the low-energy electron falls into the target and the probability of this process is enhanced. Since the process as a whole is resonant, the enhancement is greatest for energies rather close to the plasmon resonance in the dynamic polarizability of the cluster.

In the attachment process, the electron loses its excess energy. Emission of the photon via the polarizational mechanism, is one of the possible channels of the energy loss [63, 77, 78, 97, 98, 102, 103]. The energy of the electron can also be transferred to the excitations of the ionic background of the cluster [104], which may lead to increase of its vibrations and final fragmentation. In spite of the significant physical difference between various channels of the electron energy loss, they have one important common feature: the energy is transferred to the system via the plasmon excitation. Therefore, calculating the total electron attachment cross section including all possible channels of the electron energy loss in the system, one obtains [98, 99] qualitatively similar dependence of the cross section as it was obtained initially for the radiative channel of electron energy loss [77].

In Ref. [77] the attachment cross section has been calculated within the jellium model in a scheme which holds best if the kinetic energy of the electrons is somewhat higher than the energy of the resonance. Also, it was assumed that the attached ion is created in the ground state. As a useful step in simplifying the calculation, a Kramers-Kronig transformation procedure was introduced to compute the polarizability from the absorption coefficient, thereby circumventing the need for full ab initio calculations. Within this approximate scheme, it was found [77, 97] that the resonant attachment cross section dominates over the non-resonant one by a factor of about 10^3–10^4 near resonance, and is therefore a very significant pathway for electrons of low enough energy.

In Refs. [98, 99], the earlier theoretical work on attachment was extended by including the following improvements: (a) all possible channels of the electron attachment were included and the total cross section of the process was calculated rather than analyzing a particular single channel; no assumption that the system can only return to its ground state had been made; (b) theoretical approximation was used to treat electron energies not only in the resonance region, but also throughout the range of interest; (c) an RPAE calculation of the dynamical polarizability was performed along with the corresponding electron attachment cross sections on the basis of the consistent many-body theory with the use of the Hartree-Fock jellium model wave function; (d) calculations were performed for both neutral and charged cluster targets; (e) the polarization effect on the incoming particle as well as collective excitations of different multipolarity in the target electron system were taken into account; (f) Dyson's equation was used to reduce the problem of the interaction of an extra electron with a many-electron target system to a quasi-single-particle problem in a similar way as it was done for negative atomic ions calculations [105].

An example of such a calculation is shown in Fig. 5a. This plot represents the total and partial electron capture cross sections calculated for neutral potassium K_8 cluster. The inset demonstrates the photoabsorption spectrum of K_8. In Ref. [99] this calculation was performed in various approximations outlined above. It was found that the resonance pattern in the electron capture cross section for the K_8 cluster turns out to be similar in various approaches, although for some other sodium and potassium clusters it is more sensitive to the approximations made [99]. The plasmon resonance in the electron capture cross section is shifted on the value of energy of the attached electron as compared to the photoabsorption case shown in the inset.

Experimental evidence for the resonance enhancement of the cross sections of electron attachment process has been obtained in Ref. [94]. The experimental points from the cited paper are shown in Fig. 5b. Comparison of the two panels of Fig. 5 indicates the reasonable agreement between the predictions of theory and the experimental results. However, more precise measurements would be desirable to resolve the more detailed structures in the electron attachment cross sections.

The plasmon resonance enhanced mechanism of electron attachment considered above is typical for metal clusters rather than for carbon fullerenes. Although fullerenes have many similarities in the properties with metal clusters and also possess the plasmon resonances, the energies of these resonances are much higher ($\sim$7 eV and $\sim$20 eV) and thus cannot be reached at low kinetic energies of the projectile electron.

3.6 Relaxation of Electronic Excitations in Metal Clusters

As described above, the plasmon resonances in metal clusters may lie below the ionization thresholds, i.e., in the region of the discrete spectrum of electron excita-

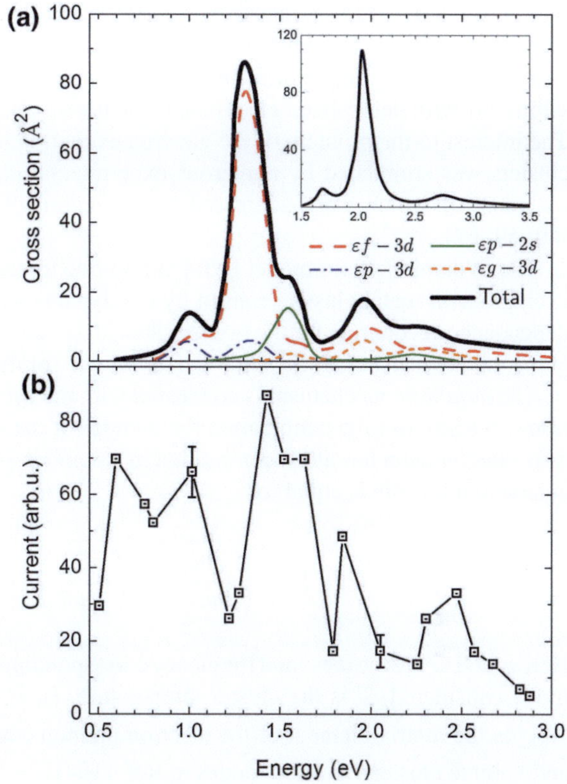

Fig. 5 *Panel a* shows total and partial electron capture cross sections in the vicinity of the plasmon resonance, calculated for a neutral potassium K_8 cluster [99]. The *inset* shows the photoabsorption spectrum of K_8. *Panel b* shows experimental evidence for the resonance enhancement of the electron attachment cross section. The experimental points are from Ref. [94]

tions [61]. This fact rises an interesting physical problem about the eigenwidths of these electronic excitations which possess large oscillator strengths and form the plasmon resonances. Knowledge of these widths is necessary for the complete description of the electron energy loss spectra, electron attachment, polarization bremsstrahlung and photoabsorption cross sections in the vicinity of the plasmon resonances and the description of their dependence on the cluster temperature. The dependence of the plasmon resonance photoabsorption patterns of metal clusters on temperature has been studied experimentally in Ref. [106].

In metal clusters, the origination of the electron excitation widths is mainly connected with the dynamics of the ionic cluster core [104, 107–112]. Let us focus on the influence of the dynamics of ions on the motion of delocalized electrons in metal clusters and discuss it on the basis of the dynamic jellium model suggested in Ref. [107] and developed further in Refs. [104, 113]. This model generalizes the static jellium model [114–116] which treats the ionic background of an atomic cluster as frozen by taking into account vibrations of the ionic background near the equilibrium point. The dynamic jellium model treats simultaneously the vibration modes of the ionic jellium background, the quantized electron motion and the interaction between the electronic and the ionic subsystems. In Ref. [107], the dynamic jellium model was applied for a consistent description of the physical phenomena arising from the oscillatory dynamics of ions.

An important example of the effect, originating from the interaction of the ionic vibrations with delocalized electrons, is the broadening of electron excitation lines. The interest to the problem of the electron excitation linewidths formation in metal clusters was stimulated by numerous experimental data on photoabsorption spectra, most of which were addressed to the region of dipole plasmon resonances [40, 59, 62].

The dynamic jellium model [107] allows one to calculate widths of the electron excitations in metal clusters caused by the dynamics of ions and their temperature dependence accounting for the two mechanisms of the electron excitation line broadening, namely, an *adiabatic* and a *non-adiabatic* (or dynamic) ones.

The *adiabatic* mechanism is connected with the averaging of the electron excitation spectrum over the temperature fluctuations of the ionic background in a cluster. This phenomenon has also been studied in a number of papers [104, 107–112]. The adiabatic linewidth is equal to

$$\Gamma = \sqrt{\frac{4\ln 2}{m\Omega}\,\mathrm{cth}\left(\frac{\Omega}{2k_{\mathrm{B}}T}\right)}\left|\hat{V}_{\mathrm{nn}}\right| . \tag{34}$$

Here m and Ω are the mass and frequency corresponding to the generalized oscillatory mode considered, T is the cluster temperature, k_{B} is the Bolzmann constant, and $\left|\hat{V}_{\mathrm{nn}}\right|$ is the matrix element of the electron phonon coupling, calculated for surface and volume cluster vibration modes in Ref. [104].

The mechanism of *non-adiabatic* electron excitation line broadening has been considered for the first time in Refs. [104, 107]. This mechanism originates from the real multiphonon transitions between the excited electron energy levels. Therefore, the dynamic linewidths characterize the real lifetimes of the electronic excitations in a cluster.

According to Ref. [107], the probability of a multiphonon transition from an excited cluster state with electronic and phononic quantum numbers n and N to all possible states (n', N') is equal to

$$\Gamma = \frac{2\pi}{\Omega}\,|A|^2 = \Omega\left|\frac{H^2_{\mathrm{n'n}}}{v(q_0)(V_{\mathrm{n'n'}} - V_{\mathrm{nn}})}\right| e^{2(\varphi_{\mathrm{n'}} - \varphi_{\mathrm{n}})} . \tag{35}$$

Here $H_{\mathrm{n'n}}$ is the half-distance between the electron energy levels $\varepsilon_{\mathrm{n}}(q)$ and $\varepsilon_{\mathrm{n'}}(q)$ in the tangent point,

$$v(q_0) = \sqrt{\Omega\left[l^2 - 2S(2N - l - 1) + S^2\right]/2S} \tag{36}$$

is the ion velocity in the tangent point, which is expressed via the number of emitted phonons $l = N' - N$, where N' and N are the phonon numbers, and the parameter

$S = \left|\hat{V}_{\rm nn} - \hat{V}_{\rm n'n'}\right|^2 / 2m\Omega^3$; $\varphi_{\rm n}$, $\varphi_{\rm n'}$ are the phases of ionic motion, arising from the distance between the turning points and the tangent point, being equal to

$$\varphi_{\rm n} = \frac{Z_{\rm n}\sqrt{Z_{\rm n}^2 - 2N - 1}}{2} + \frac{2N+1}{4}\ln\left(\frac{Z_{\rm n} + \sqrt{Z_{\rm n}^2 - 2N - 1}}{2N+1}\right), \quad (37)$$

where $Z_{\rm n} = (l - S)/\sqrt{2S}$. The expression for $\varphi_{\rm n'}$ is the same, but the parameter $Z_{\rm n'}$ is equal to $Z_{\rm n'} = (l + S)/\sqrt{2S}$.

The adiabatic broadening mechanism explains the temperature dependence of the photoabsorption spectra in the vicinity of the plasmon resonance via the coupling of the dipole excitations in a cluster with the quadrupole deformation of the cluster surface. The photoabsorption spectra were calculated within the framework of deformed jellium model using either the plasmon pole approximation [108, 109] or the local density approximation [110–112, 117, 118]. In Ref. [112], the octupole deformation of the cluster surface was taken into account. It was demonstrated that the octupole deformation increases the Landau damping as a result of breaking the selection rule, leading to a mixture of the dipole and the quadrupole electronic excitations. Via this mechanism, the octupole deformations of the cluster surface provide the dominating contribution to the thermal broadening of electron excitation lines in small metal clusters.

In Refs. [104, 107] both the adiabatic and non-adiabatic linewidths of electron excitations in the vicinity of the plasmon resonance caused by coupling of electrons with various ionic vibration modes have been calculated. The behavior of the adiabatic and non-adiabatic linewidths as a function of temperature is shown in Fig. 6. The non-adiabatic linewidths characterize the real lifetimes of cluster electron excitations. Naturally, the non-adiabatic widths turn out to be much smaller than the adiabatic ones due to the slow motion of ions in the cluster. However, the adiabatic linewidths do not completely mask the non-adiabatic ones because the two types of widths manifest themselves differently. The adiabatic broadening determines the pattern of the photoabsorption spectrum in the linear regime. The non-adiabatic linewidths are important for the processes, in which the real lifetime of electron excitations and the electron-ion energy transfer are essential. The information about the non-adiabatic electron-phonon interactions in clusters is necessary for the description of electron inelastic scattering on clusters [48–50, 52], including the processes of electron attachment [77, 98, 99], the non-linear photo-absorption and bremsstrahlung [63, 78, 97, 102], the problem of cluster stability and fission. The non-adiabatic linewidths determined by the probability of multiphonon transitions are also essential for the treatment of the relaxation of electronic excitations in clusters and the energy transfer from the excited electrons to ions, which occurs after the impact- or photoexcitation of the cluster.

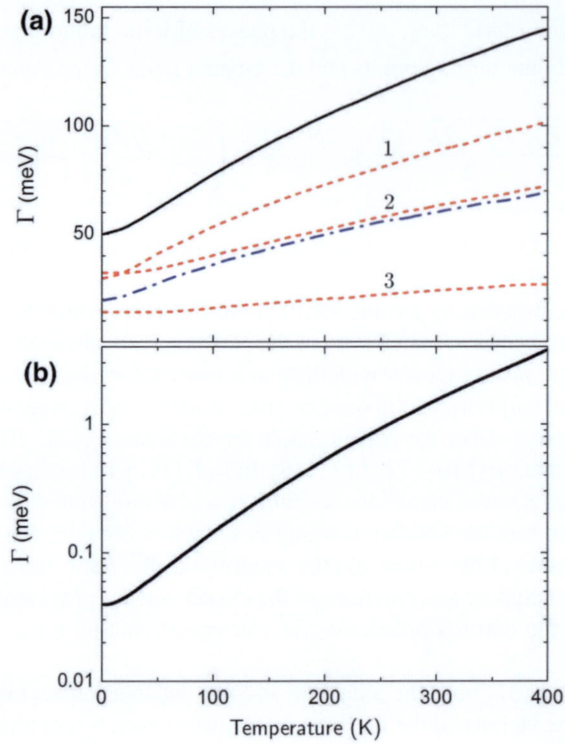

Fig. 6 *Panel a* shows temperature dependence of the adiabatic linewidth Γ, calculated according to (34) for the dipole electron excitation with the energy $\omega_{\rm n} = 3.013$ eV in the Na_{40} cluster. *Dashed curves* labeled as 1, 2 and 3 show the adiabatic width corresponding to the electron coupling with the three first volume vibration modes, respectively. *Dashed-dotted curve* shows the adiabatic width arising from the electron coupling with surface vibrations of the cluster. *Solid curve* shows the total adiabatic linewidth. *Panel b* shows temperature dependence of the non-adiabatic width calculated according to Eq. (35) for the dipole excitation with the energy $\omega_{\rm n} = 3.013$ eV in the Na_{40} cluster. The figures are adapted from Refs. [104, 107] with minor editorial modifications

In Refs. [104, 107] the role of the volume and the surface vibrations of the ionic cluster core in the formation of the electron excitation linewidths was investigated. It was demonstrated that the volume and surface vibrations provide comparable contributions to the adiabatic linewidths, but the surface vibrations are much more essential for the non-adiabatic multiphonon transitions than the volume ones.

4 Collective Electron Excitations as a Source of Strong Low-Energy Electron Emission from Metal Nanoparticles

As described in the previous section, an important mechanism of ionization or excitation of metallic clusters and NPs, as well as some other nanoscale systems, relies on the formation of plasmons—collective excitations of delocalized valence electrons that are induced by an external electric field [40, 41]. These excitations appear as prominent resonances in the ionization/excitation spectra of various atomic and molecular clusters, and the position of the resonance peak depends strongly on the type of a system. In the case of metallic nanosystems, a typical energy of the plasmon excitations is about several electronvolts, so the resonance peak is located in the vicinity of the ionization threshold [42, 43].

In the recent Monte Carlo simulation [33], the authors included the contribution of plasmon excitations when calculating the spectra of electron emission from metal NPs under proton impact. However, having accounted for the volume plasmon only, they concluded that the plasmon excitation does not play an important role in the process of electron emission, contributing much less to the overall cross sections than individual excitations. On the contrary, the recent studies [74, 75] revealed that a significant increase in the number of emitted electrons due to irradiation of noble metal NPs by fast protons comes from the two distinct types of collective electron effects, namely excitation of delocalized valence electrons in a NP (plasmons) and that of *d* electrons in individual atoms (atomic giant resonances).

The contribution of the plasmon excitations was evaluated by means of the PRA (see Sect. 3.3). In Ref. [75], parameters of the utilized model approach were justified by calculating photoabsorption spectra of several three-dimensional gold clusters. The PRA-based spectra were compared with those obtained by means of a more advanced method, namely by time-dependent density-functional theory (TDDFT) [119]. To evaluate the contribution of individual atomic excitations, an analytical expression was introduced, which relates the cross section of photoionization with that of inelastic scattering in the dipole approximation. In Ref. [74], this methodology was applied for different metal NPs, which have been proposed as sensitizers in radiation therapy applications.

4.1 *Collective Electron Excitations in Gold Clusters Under Photon Impact*

Figure 7a presents the photoabsorption spectra of the Au_{18}, Au_{20}, Au_{32} and Au_{42} clusters calculated by means of TDDFT for the photon energy up to 60 eV [74, 75]. The spectra, having a similar profile, are characterized by a low-energy peak located below 10 eV and by a broad feature with a maximum at about 25 eV. The analysis performed in Refs. [74, 75] has revealed that the high-energy feature is the atomic

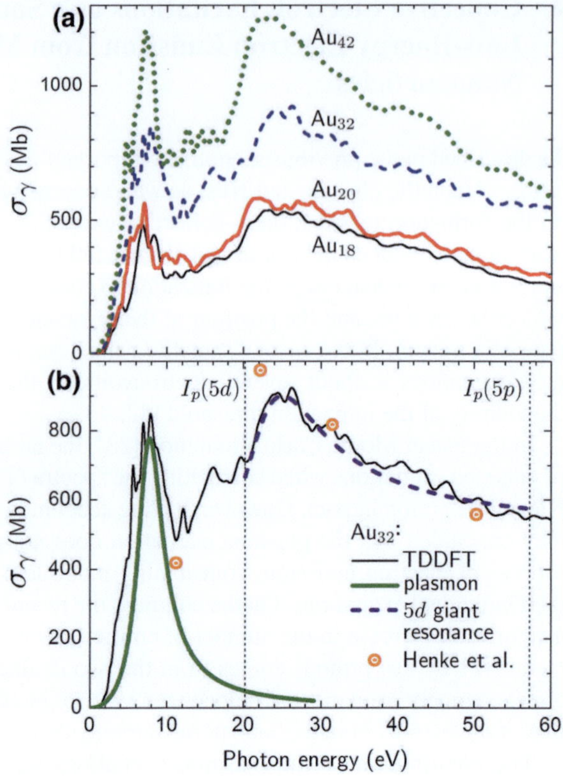

Fig. 7 *Panel a* shows the photoabsorption cross section of the Au_{18}, Au_{20}, Au_{32} and Au_{42} clusters calculated within the TDDFT framework. *Panel b* shows contribution of the plasmon-type excitation and the 5*d* giant atomic resonance in the photoabsorption cross section of Au_{32}. *Symbols* represent the data for atomic gold [120], multiplied by the number of atoms in the cluster. *Vertical lines* mark the 5*d* and 5*p* ionization thresholds in the atom of gold. The figures are adapted from Refs. [74, 75] with minor editorial changes

giant resonance formed due to the excitation of electrons in the 5*d* atomic shell. The integration of the oscillator strength from 20.2 eV (ionization threshold of the 5*d* shell in a single atom of gold) up to 57.2 eV (the 5*p* shell ionization threshold [120]), indicated that about eight localized *d*-electrons contribute to the excitation of the 5*d* shell forming the broad peak in the spectrum.

The low-energy peak is due to the plasmon-type excitation, which involves some fraction of *s* and *d* electrons delocalized over the whole cluster. The delocalization comes from a partial hybridization of the 6*s* and 5*d* atomic shells. Figure 7b shows the TDDFT-based photoabsorption spectrum of a highly-symmetric icosahedral Au_{32} cluster with the diameter of about 0.9 nm (thin solid curve). The results of the calculation are compared to the X-ray absorption data for atomic gold [120], multiplied by the number of atoms in the cluster. The integration of the spectrum of Au_{32} up to 11.2 eV, that is the energy at which the first dip after the resonance peak is observed in the TDDFT spectrum, revealed that about 1.5 electrons from each atom contribute to the collective plasmon-type excitation. On the basis of this analysis, it was stated [74, 75] that the total photoabsorption spectra of gold clusters in the energy region up to 60 eV can be approximated by the sum of the plasmon contribution and that of the 5*d* electron excitations in individual atoms, $\sigma_\gamma \approx \sigma_{pl} + \sigma_{5d}$.

4.2 Collective Electron Excitations in Small Gold Nanoparticles Under Charge Particle Impact

Similar to the photoionization, the two distinct types of collective electron excitations appear in the process of charge particle impact ionization. The single differential inelastic scattering cross section of a fast projectile in collision with a cluster/NP is then given by

$$\frac{\mathrm{d}\sigma}{\mathrm{d}\Delta\varepsilon} \approx \frac{\mathrm{d}\sigma_{\mathrm{pl}}}{\mathrm{d}\Delta\varepsilon} + \frac{\mathrm{d}\sigma_{\mathrm{at}}}{\mathrm{d}\Delta\varepsilon}\,, \tag{38}$$

where $\Delta\varepsilon = \varepsilon_1 - \varepsilon_2$ is the energy loss of the incident projectile of energy ε_1, $\mathbf{p}_1$ and $\mathbf{p}_2$ are the initial and the final momenta of the projectile, and $\Omega_{\mathbf{p}_2}$ is its solid angle. The cross sections $\mathrm{d}\sigma_{\mathrm{pl}}$ and $\mathrm{d}\sigma_{\mathrm{at}}$ denote the contributions of the plasmon and individual atomic excitations, respectively.

The contribution of the plasmon excitations to the ionization cross section has been described by means of PRA. In this approach, the double differential cross section $\mathrm{d}^2\sigma/\mathrm{d}\Delta\varepsilon\,\mathrm{d}\Omega_{\mathbf{p}_2}$ for a spherical NP is defined as a sum of the surface (s) and the volume (v) plasmon terms, which are constructed as a sum over different multipole contributions corresponding to different values of the angular momentum l (see Sect. 3.3):

$$\begin{aligned} \frac{\mathrm{d}^2\sigma^{(\mathrm{s})}}{\mathrm{d}\Delta\varepsilon\,\mathrm{d}\Omega_{\mathbf{p}_2}} &\propto \sum_l \frac{\omega_l^{(\mathrm{s})2}\,\Gamma_l^{(\mathrm{s})}}{\left(\omega^2 - \omega_l^{(\mathrm{s})2}\right)^2 + \omega^2\left(\Gamma_l^{(\mathrm{s})}\right)^2} \\ \frac{\mathrm{d}^2\sigma^{(\mathrm{v})}}{\mathrm{d}\Delta\varepsilon\,\mathrm{d}\Omega_{\mathbf{p}_2}} &\propto \sum_l \frac{\omega_{\mathrm{p}}^2\,\Gamma_l^{(\mathrm{v})}}{\left(\omega^2 - \omega_{\mathrm{p}}^2\right)^2 + \omega^2\left(\Gamma_l^{(\mathrm{v})}\right)^2}\,. \end{aligned} \tag{39}$$

Here $\omega_l^{(\mathrm{s})} = \sqrt{l/(2l+1)}\,\omega_{\mathrm{p}}$ is the frequency of the surface plasmon of the multipolarity l, $\omega_{\mathrm{p}} = \sqrt{3N_{\mathrm{e}}/R^3}$ is the volume plasmon frequency, and the quantities $\Gamma_l^{(i)}$ ($i = \mathrm{s, v}$) are the corresponding widths. Only the dipole ($l = 1$), quadrupole ($l = 2$) and octupole ($l = 3$) terms were taken into account in this analysis. Excitations with larger l have a single-particle rather than a collective nature [71], thus not contributing to the plasmon formation. The cross sections (39) were obtained within the first Born approximation which is applicable for the collision of a NP with a fast heavy projectile.

As mentioned above, the PRA relies on a few parameters, which include the oscillator strength of the plasmon excitation, position of the peak and its width. In the dipole case, these were validated by fitting the TDDFT-based spectra of several three-dimensional gold clusters to those calculated within the model approach (see Sect. 4.1).

Figure 8a shows the cross section $\mathrm{d}\sigma_{\mathrm{pl}}/\mathrm{d}\Delta\varepsilon$ calculated for a 1 nm gold NP irradiated by fast protons of different incident energies as indicated. The figure demonstrates that the amplitude and the shape of the plasmon resonance depend strongly on the kinetic energy of the projectile. At high velocities, the dipole contribution dom-

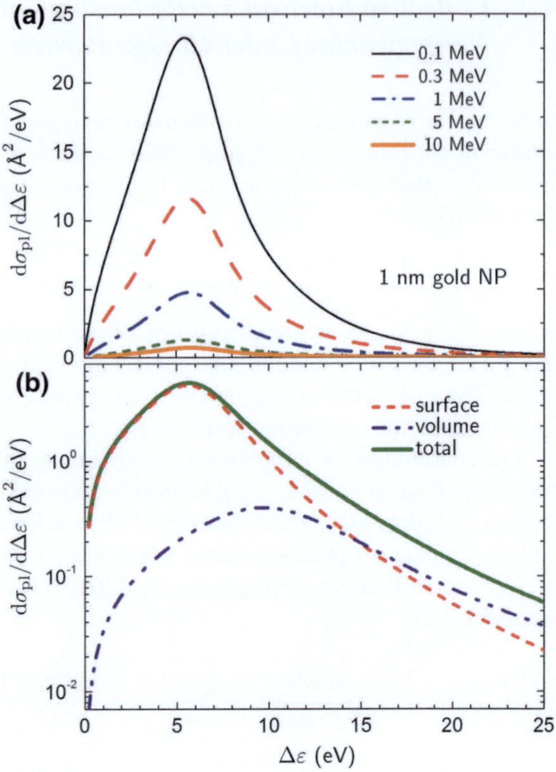

Fig. 8 *Panel a* shows contribution of the plasmon excitations to the single differential cross section, $d\sigma_{pl}/d\Delta\varepsilon$, of 1 nm gold NP irradiated by fast protons of different incident energies as a function of the energy loss. *Panel b* shows contribution of different plasmon excitations to the resulting cross section $d\sigma_{pl}/d\Delta\varepsilon$ of a 1 nm gold NP irradiated by a 1 MeV proton

inates over the higher multipole contributions, since the dipole potential decreases slower at large distances than the higher multipole potentials. Figure 8b illustrates the contribution of different plasmon modes to the spectrum of a 1 nm gold NP irradiated by a 1 MeV proton. The main contribution to the cross section in the region of low-energy transfer comes from the surface plasmon, which exceeds that of the volume plasmon by more than an order of magnitude. Thus, the leading mechanism of low-energy electron production by gold NPs is related to the surface plasmon which has not been accounted for in the Monte Carlo simulations [33].

The *d* electrons in the atoms of noble metals play a dominant role at the excitation energies from approximately 20 to 60 eV (see Fig. 7 for the case of gold). For distant collisions, i.e., when the impact parameter exceeds the radius R_{at} of the atomic shell, the ionization spectra of the gold atoms are dominated by the dipole term [58]. Comparison of the cross sections of photoionization, σ_γ, and the dipole term of inelastic scattering, $d\sigma_{at}/d\Delta\varepsilon$, calculated in the Born approximation, leads to the following expression [74, 75]:

$$\frac{d\sigma_{at}}{d\Delta\varepsilon} = \frac{2c}{\pi\omega v_1^2}\sigma_\gamma \ln\left(\frac{v_1}{\omega R}\right), \tag{40}$$

where $\omega = \varepsilon_1 - \varepsilon_2$ is the energy transfer and v_1 is the projectile velocity. Equation (40), obtained within the so-called "logarithmic approximation", assumes that the main contribution to the cross section $\mathrm{d}\sigma_{\mathrm{at}}/\mathrm{d}\Delta\varepsilon$ comes from the region of large distances, $R_{\mathrm{at}} < r < v_1/\omega$. This relation has the logarithmic accuracy which implies that the logarithmic term dominates the cross section while all non-logarithmic terms are neglected [79]. Making an estimate for the gold atoms, it was assumed that $\omega \approx 1$ a.u. which corresponds to the maximum of the 5*d* giant resonance in gold [120], $v_1 \approx 6.3$ a.u. for a 1 MeV proton, and the electron shell radius $R_{5d}(\mathrm{Au}) \approx 2$ a.u. The interaction of the incident projectile with the NP leads to the formation of a giant resonance not in all atoms of the system but only in those located within the impact parameter interval from $r_{\min} \simeq R_{5d}$ to $r_{\max} \simeq v_1/\omega$. This estimate indicates that the 5*d* giant resonance is formed in about one third of atoms of the NP.

To quantify the production of secondary electrons in collision with the NPs, the cross section $\mathrm{d}\sigma/\mathrm{d}\Delta\varepsilon$ (38) is redefined as a function of the kinetic energy E of the electrons: $E = \Delta\varepsilon - I_p$, where I_p is the ionization threshold of the system. The cross section $\mathrm{d}\sigma/\mathrm{d}E$ can be related to the probability to produce N electrons with kinetic energy within the interval $\mathrm{d}E$, emitted from a segment $\mathrm{d}x$ of the trajectory, via [16]:

$$\frac{\mathrm{d}^2 N(E)}{\mathrm{d}x\,\mathrm{d}E} = n\frac{\mathrm{d}\sigma}{\mathrm{d}E}\,, \tag{41}$$

where n is the atomic density of the target.

Figure 9 shows the relative enhancement of the electron yield from a 1 nm gold NP compared to an equivalent volume of pure water. The data for the gold NP have been normalized to the spectrum for liquid water [121]. The dashed line shows the contribution of the plasmon excitations to the electron yield, while the dash-dotted line presents the contribution from the atomic 5*d* giant resonance, estimated using

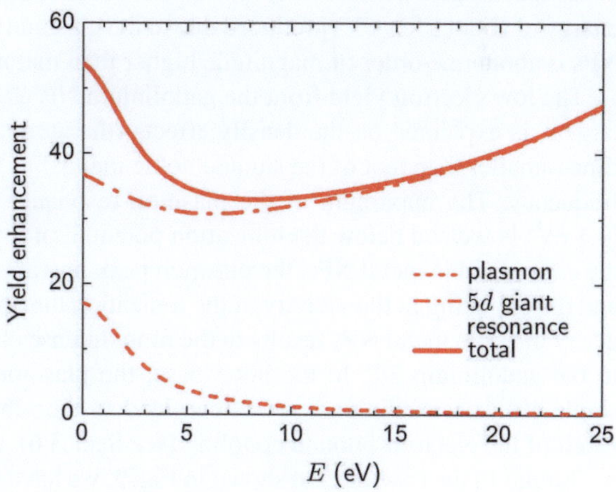

Fig. 9 Electron yield enhancement from the 1 nm gold NP as compared to an equivalent volume of pure water [121]. The *dashed* and *dash-dotted lines* show the contribution of the plasmons and the atomic 5*d* excitations, respectively. *Solid line* illustrates the resulting enhancement

Eq. (40). Making this estimate, we have assumed that the ionization cross sections of individual atoms are dominated by the dipole excitation. Contribution of quadrupole and higher multipole terms will lead to an increase in the number of emitted electrons but their relative contribution will be not as large as that from the dipole excitation. Accounting for the plasmon contribution leads to a significant additional increase in the number of 1–5 eV emitted electrons as compared to the pure water. Due to the collective electron excitations arising in the 1 nm gold NP, it can thus produce up to 50 times larger number of the low-energy electrons comparing to the equivalent volume of pure water medium. The enhancement of the secondary electron yield may increase the number of hydroxyl radicals [15] which recombinate with each other to form more stable hydrogen peroxide H_2O_2 [122]. The latter can propagate for large distances inside the cell and deliver damage to the DNA in the cell nucleus even if the NPs are localized in other cell compartments [25]. Thereby, the results of the analysis performed in Refs. [74, 75] have indicated that the decay of the collective electron excitations in gold NPs is an important mechanism of enhancement of the yield of secondary species. However, more investigation is needed to acquire the complete understanding of all physical and chemical processes involved into the process of radiosensitization due to the NPs.

4.3 Electron Production by Different Metal Nanoparticles

A similar analysis was performed for other metal systems, which are of current interest for cancer treatment with sensitizing NPs. The electron production due to collective electron excitations in small NPs composed of gold, platinum, silver, and gadolinium was analyzed in Ref. [74]. Figure 10 presents the number of electrons per unit length per unit energy produced via the plasmon excitation mechanism by the 1 nm spherical metal NPs due to 1 MeV proton irradiation. Comparative analysis of the spectra demonstrates that the number of low-energy electrons (with the kinetic energy of about a few eV) produced due to the plasmon excitations in the noble metal NPs is about one order of magnitude higher than that by liquid water.

The low electron yield from the gadolinium NP, as compared to the noble metal targets, is explained by the density effects (the atomic density of Gd is about two times smaller than that of the studied noble metals) as well as by the lower plasmon frequency. The maximum of the plasmon resonance peak in the gadolinium NP (4.1 eV) is located below the ionization potential of the system ($\sim$5.0 eV) [123]. In the case of noble metal NPs, the plasmon peak maxima are in the range between 5.5 and 6.0 eV, being in the vicinity of the ionization thresholds. Therefore, the plasmon decay in noble metal NPs results in the more intense electron emission as compared to the gadolinium NP. In the latter case, the plasmon will mostly decay into the single-electron excitations, which can lead to the vibration of the ionic core as a result of the electron-phonon coupling (see Sect. 3.6).

Similar to the case of gold shown in Fig. 9, we have estimated the total number of electrons produced due to the collective excitations in the metal NPs by accounting for

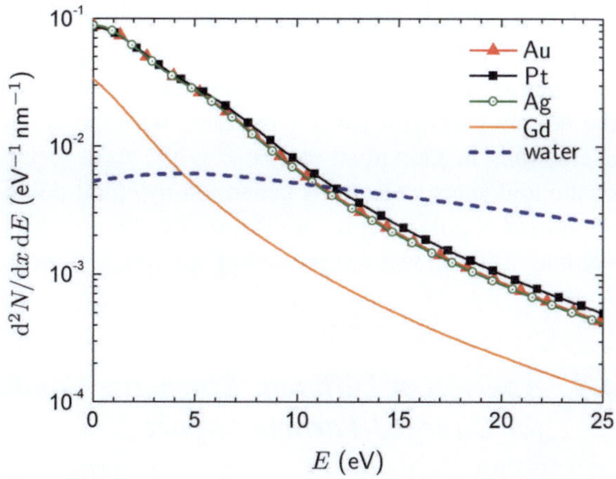

Fig. 10 Number of electrons per unit length per unit energy produced via the plasmon excitations in the 1 nm Au, Pt, Ag and Gd NPs irradiated by a 1 MeV proton. *Dashed curve* represents the number of electron generated from the equivalent volume of water [121]

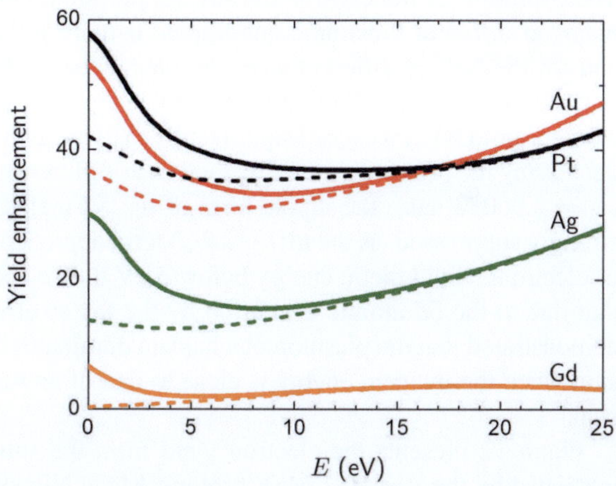

Fig. 11 Yield enhancement from the 1 nm metallic NPs. *Dashed lines* show the contribution of individual atomic excitations. *Solid lines* show the resulting contribution with an account of the plasmons. The figure is adapted from Ref. [74] with minor editorial changes

the contribution of excitations in individual atoms. Figure 11 demonstrates the relative enhancement of the electron yield from the considered NPs as compared to pure water. This quantity was obtained by summing up the contribution of the plasmons and individual atomic excitations. Contrary to the noble metals, the Gd atom has a single electron in the 5*d* shell. Thus, there is no atomic giant resonance in the ionization spectrum of Gd in the 20–60 eV range, and the spectrum is characterized by a narrow peak at $\omega \approx 1.2$ a.u., formed due to ionization of the 5*p* shell. The dashed lines present the contribution of the atomic giant resonances (5*d* in Au and Pt, and 4*d* in Ag) as well as the total $5p + 5d$ contribution in Gd, estimated using Eq. (40). The solid line is the sum of the excitations in individual atoms and the plasmons. The significant yield enhancement arises in those nanoparticles whose constituent atoms possess the giant resonance, contrary to case of gadolinium which has a single 5*d* electron.

The presented results demonstrate that accounting for the plasmon contribution leads to a significant increase of the 1–10 eV electron yield. For higher electron energies (of a few tens of eV), the main contribution to the electron yield arises from the atomic giant resonance associated with the collective excitation of outer-shell *d* electrons in individual atoms of a NP. As a result of these effects, the number of the low-energy electrons generated by small noble metal NPs (especially, those composed of gold and platinum) significantly exceeds that produced by an equivalent volume of liquid water representing a biological medium.

4.4 Analysis of Different Kinematic Conditions for Charged-Particle Impact

The evolution of the contribution of the plasmon and the giant resonance mechanisms at different kinematic conditions, namely for different projectile velocities and for the NPs of different sizes, was analyzed in Ref. [75]. It was demonstrated that the plasmon contribution to the low-energy (of about 1–10 eV) electron yield from the gold NP can exceed significantly that due to the atomic giant resonance with decreasing the projectile's energy. Indeed, as follows from Eq. (40), at $\varepsilon_1 = 0.1$ MeV ($v_1 = 2.00073$ a.u.), the dipole term of the $5d$ inelastic scattering cross section is strongly suppressed, as the $\ln(v_1/\omega R_{5d})$ term approaches zero. In this case, the yield of electrons with kinetic energy below 5 eV due to the plasmon excitation exceeds that due to the $5d$ atomic excitation by the factor of about 10^3 [75]. This analysis demonstrated that the plasmon mechanism dominates the low-energy electron emission when the incident energy is close to that of an ion in the vicinity of the Bragg peak.

Figure 12 presents the electron yield from the solid gold NPs of different size irradiated by the 1 and 0.1 MeV protons. Metal NPs of this size range were studied recently in relation to the radiotherapies with charged ions [5, 25]. At certain conditions, the contribution of the plasmon excitations saturates, so that larger NPs emit a smaller number of electrons via the plasmon damping mechanism. It was shown previously [48] that the dipole mode of the plasmon excitations arising in a NP gives the dominating contribution to the ionization cross section when the characteristic collision distance exceeds significantly the NP size, $v_1/\omega \gg D/2$, where D is the NP diameter. At large collision distances, the dipole contribution dominates over the higher multipole contributions. Terms with higher l become significant only in the case when the collision distances become comparable with the cluster size. This means that for a given incident energy the plasmon mechanism of electron production will be efficient for relatively small NPs, while the dipole plasmon mode will be suppressed for larger D. A small increase in the number of 5 eV electrons produced by larger NPs as compared to the smaller ones is the result of an increased role of the volume plasmon due to the increased volume/surface ratio. A similar scenario holds for other incident velocities [75].

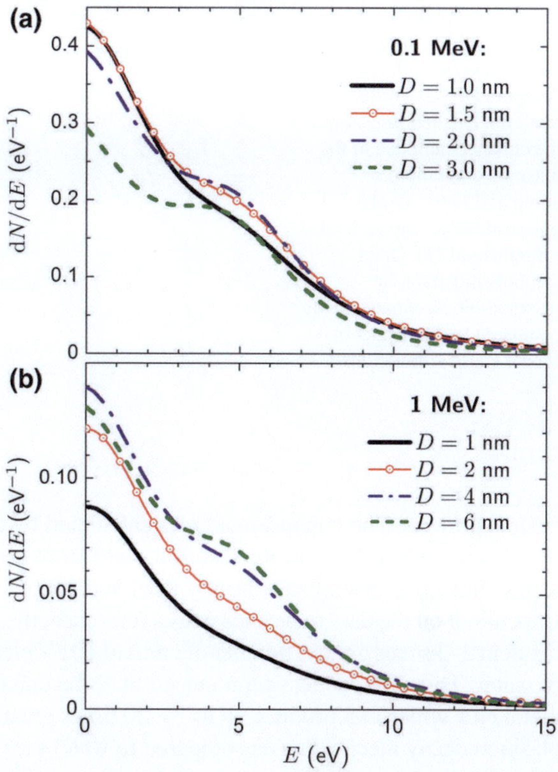

Fig. 12 Number of electrons per unit energy produced via the plasmon excitation mechanism in the solid gold NPs of different size irradiated by the protons of different kinetic energy: 0.1 MeV (**a**) and 1 MeV (**b**). The figures are adapted from [75] with minor editorial changes

5 Enhanced Low-Energy Electron Production by Carbon Nanoparticles

As it was outlined throughout this chapter, excitation of plasmons by an external electric field is a characteristic feature of not only metallic but also, to some extent, of carbon nanoscale systems. For instance, it is well established that plasmon excitations dominate the spectra of photon- and charge-particle impact ionization of fullerenes and polycyclic aromatic hydrocarbons [44, 45, 51, 53, 57, 124, 125].

A recent work [76] presented the spectra of secondary electrons ejected from a carbon NP composed of fullerite, a crystalline form of C_{60} fullerene, irradiated by fast protons. It was demonstrated that the decay of plasmons excited in carbon NPs also enhances the production of low-energy secondary electrons in a biological medium. The contribution of plasmon excitations to the electron production was evaluated by means of the PRA, and the results of these calculations were compared to the model calculations based on the dielectric formalism [121] and Monte Carlo simulations, carried out for pure water medium and for the medium with an embedded NP.

Figure 13 presents the relative enhancement of the electron yield from a 50 nm carbon NP as compared to the equivalent volume of water due to irradiation with

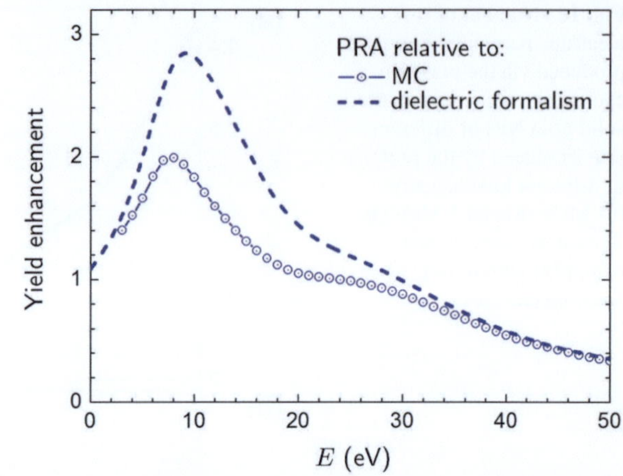

Fig. 13 Electron yield enhancement from the 50 nm carbon nanoparticle as compared to pure water medium. *Dashed line* shows the enhancement due to the plasmon excitations as compared to the results obtained within the dielectric formalism [121]. Open symbols illustrate the plasmon-based enhancement compared to the results of Monte Carlo simulations

1 MeV protons. The enhancement was calculated by comparing the contribution of the plasmon excitations to the electron yield from pure water calculated by means of the dielectric formalism (dashed line) and Monte Carlo simulations (symbols). Depending on the data to be chosen as a reference, the collective electron excitations result in 2–3 times greater number of emitted 10 eV electrons as compared to the case of water. This effect is less pronounced than the enhancement done by small noble metal NPs which can produce up to 15–20 times greater number of electrons via the plasmon decay mechanism as compared to water (see Sect. 4.3). On the other hand, this enhancement results in an excessive emission of the very low-energy electrons of about a few eV, while the carbon-based NP can enhance the yield of more energetic electrons. On this basis, an idea was proposed [76] about considering novel metal-organic sensitizing NPs, where collective excitations will arise in both parts of the system. A proper choice of the constituents may allow for tuning the position of the resonance peaks in the ionization spectra of such systems and, subsequently, for covering a broader kinetic energy spectrum of electrons emitted from such NPs.

6 Conclusions

We presented an overview of theoretical and computational studies of physical phenomena related to the formation and the decay of collective electron excitations in atomic clusters and nanoparticles. These phenomena have a common physical nature and manifest themselves in collision with photons and charged projectiles like electrons, protons or heavier ions. We also made a brief overview of recent Monte Carlo-based studies which have been devoted to the investigation of radiosensitization and dose enhancement effects for proton irradiation combined with metal nanoparticles which are of current interest for radiation therapy applications. Because of a lack of the description of collective electron effects in most of Monte Carlo simulations, many important physical phenomena may be missing. In the presented overview, we emphasized the role of collective electron excitations in various collision processes with atomic clusters and nanoparticles. An accurate description of these phenomena may lead to a better and more complete understanding of the physical picture related to the nanoscale mechanisms of radiation damage in the presence of nanoagents.

Acknowledgements We acknowledge the financial support received from the European Union Seventh Framework Programme (PEOPLE-2013-ITN-ARGENT project) under grant agreement no. 608163.

References

1. Salata OV (2004) Applications of nanoparticles in biology and medicine. J Nanobiotechnol 2:3
2. Murthy SK (2007) Nanoparticles in modern medicine: state of the art and future challenges. Int J Nanomed 2:129–141
3. Herold DM, Das IJ, Stobbe CC, Iyer RV, Chapman JD (2000) Gold microspheres: a selective technique for producing biologically effective dose enhancement. Int J Radiat Biol 76:1357–1364
4. Hainfeld JJ, Slatkin DN, Smilowitz HM (2004) The use of gold nanoparticles to enhance radiotherapy in mice. Phys Med Biol 49:N309–N315
5. Porcel E, Liehn S, Remita H, Usami N, Kobayashi K, Furusawa Y, Le Sech C, Lacombe S (2010) Platimun nanoparticles: a promising material for future cancer therapy? Nanotechnology 21:085103
6. McMahon SJ et al (2011) Biological consequences of nanoscale energy deposition near irradiated heavy atom nanoparticles. Sci Rep 1:8; Corrigendum: ibid. **3**, 1725 (2013)
7. Zhang X-D et al (2015) Ultrasmall glutathione-protected gold nanoclusters as next generation radiotherapy sensitizers with high tumor uptake and high renal clearance. Sci Rep 5:8669
8. McQuaid HN et al (2016) Imaging and radiation effects of gold nanoparticles in tumour cells. Sci Rep 6:19442
9. FP7 Initial Training Network Project "Advanced Radiotherapy, Generated by Exploiting Nanoprocesses and Technologies" (ARGENT), http://www.itn-argent.eu
10. Ali H, van Lier JE (1999) Metal complexes as photo- and radiosensitizers. Chem Rev 99:2379–2450
11. Butterworth KT, McMahon SJ, Currell FJ, Prise KM (2012) Physical basis and biological mechanisms of gold nanoparticle radiosensitization. Nanoscale 4:4830–4838
12. Kobayashi K, Usami N, Porcel E, Lacombe S, Le Sech C (2010) Enhancement of radiation effect by heavy elements. Mutat Res 704:123–131
13. Xiao F, Zheng Y, Cloutier P, He Y, Hunting D, Sanche L (2011) On the role of low-energy electrons in the radiosensitization of DNA by gold nanoparticles. Nanotechnology 22:465101
14. Zheng Y, Hunting DJ, Ayotte P, Sanche L (2008) Radiosensitization of DNA by gold nanoparticles irradiated with high-energy electrons. Radiat Res 169:19–27; Erratum: ibid. **169**, 481–482 (2008)
15. Sicard-Roselli C et al (2014) A new mechanism for hydroxyl radical production in irradiated nanoparticle solutions. Small 10:3338–3346
16. Surdutovich E, Solov'yov AV (2014) Multiscale approach to the physics of radiation damage with ions. Eur Phys J D 68:353
17. Garcia Gomez-Tejedor G, Fuss MC (ed) (2012) Radiation damage in biomolecular systems. Springer Science+Business Media B.V

18. Boudaiffa B, Cloutier P, Hunting D, Huels MA, Sanche L (2000) Resonant formation of DNA strand breaks by low-energy (3 to 20 eV) electrons. Science 287:1658–1660
19. Huels MA, Boudaiffa B, Cloutier P, Hunting D, Sanche L (2003) Single, double, and multiple double strand breaks induced in DNA by 3–100 eV electrons. J Am Chem Soc 125:4467–4477
20. Jain S et al (2011) Cell-specific radiosensitization by gold nanoparticles at megavoltage radiation energies. Int J Radiat Oncol Biol Phys 79:531–539
21. Jain S, Hirst DG, O'Sullivan JM (2012) Gold nanoparticles as novel agents for cancer therapy. Br J Radiol 85:101–113
22. Liu P, Huang Z, Chen Z, Xu R, Wu H, Zang F, Wang C, Gu N (2013) Silver nanoparticles: a novel radiation sensitizer for glioma? Nanoscale 5:11829–11836
23. Luchette M, Korideck H, Makrigiorgos M, Tillement O, Berbeco R (2014) Radiation dose enhancement of gadolinium-based AGuIX nanoparticles on HeLa cells. Nanomed Nanotechnol 10:1751–1755
24. Miladi I et al (2015) Combining ultrasmall gadolinium-based nanoparticles with photon irradiation overcomes radioresistance of head and neck squamous cell carcinoma. Nanomed Nanotechnol 11:247–257
25. Porcel E et al (2014) Gadolinium-based nanoparticles to improve the hadrontherapy performances. Nanomed Nanotechnol 10:1601–1608
26. Kim J-K et al (2010) Therapeutic application of metallic nanoparticles combined with particle-induced X-ray emission effect. Nanotechnology 21:425102
27. Polf JC, Bronk LF, Driessen WHP, Arap W, Pasqualini R, Gillin M (2011) Enhanced relative biological effectiveness of proton radiotherapy in tumor cells with internalized gold nanoparticles. Appl Phys Lett 98:193702
28. Schlathölter T et al (2016) Improving proton therapy by metal-containing nanoparticles: nanoscale insights. Int J Nanomed 11:1549–1556
29. McMahon SJ, Paganetti H, Prise KM (2016) Optimising element choice for nanoparticle radiosensitisers. Nanoscale 8:581–589
30. Sancey L et al (2014) The use of theranostic gadolinium-based nanoprobes to improve radiotherapy efficacy. Br J Radiol 87:20140134
31. Baccarelli I, Gianturco FA, Scifoni E, Solov'yov AV, Surdutovich E (2010) Molecular level assessments of radiation biodamage. Eur Phys J D 60:1–10
32. Schardt D, Elsässer T, Schulz-Ertner D (2010) Heavy-ion tumor therapy: physical and radiobiological benefits. Rev Mod Phys 82:383–425
33. Wälzlein C, Scifoni E, Krämer M, Durante M (2014) Simulations of dose enhancement for heavy atom nanoparticles irradiated by protons. Phys Med Biol 59:1441–1458
34. Lin Y, McMahon SJ, Scarpelli M, Paganetti H, Schuemann J (2014) Comparing gold nanoparticle enhanced radiotherapy with protons, megavoltage photons and kilovoltage photons: a Monte Carlo simulation. Phys Med Biol 59:7675–7689
35. Lin Y, McMahon SJ, Paganetti H, Schuemann J (2015) Biological modeling of gold nanoparticle enhanced radiotherapy for proton therapy. Phys Med Biol 60:4149–4168
36. Martinez-Rovira I, Prezado Y (2015) Evaluation of the local dose enhancement in the combination of proton therapy and nanoparticles. Med Phys 42:6703–6710
37. Krämer M, Kraft G (1994) Calculations of heavy-ion track structure. Radiat Environ Biophys 33:91–109
38. Perl J, Shin J, Schümann J, Faddegon B, Paganetti H (2012) TOPAS: an innovative proton Monte Carlo platform for research and clinical applications. Med Phys 39:6818–6837
39. GEANT4, LEEPWG–Low Energy Electromagnetic Physics Working Group (2013). http://geant4.in2p3.fr/2013/resources/L11-EMLowE.pdf
40. Kreibig U, Vollmer M (1995) Optical properties of metal clusters. Springer, Berlin-Heidelberg
41. Dinh PM, Reinhard P-G, Suraud E (2013) An introduction to cluster science. Wiley, 2013
42. Bréchignac C, Cahuzac Ph, Carlier F, Leygnier J (1989) Collective excitation in closed-shell potassium cluster ions. Chem Phys Lett 164:433–437
43. Selby K, Vollmer M, Masui J, Kresin V, de Heer WA, Knight WD (1989) Surface plasma resonances in free metal clusters. Phys Rev B 40:5417–5427

44. Hertel IV, Steger H, de Vries J, Weisser B, Menzel C, Kamke B, Kamke W (1992) Giant plasmon excitation in free C_{60} and C_{70} molecules studied by photoionization. Phys Rev Lett 68:784–787
45. Ling Y, Lifshitz C (1996) Plasmon excitation in polycyclic aromatic hydrocarbons studied by photoionization. Chem Phys Lett 257:587–591
46. Liebsch T et al (1995) Angle-resolved photoelectron spectroscopy of C_{60}. Phys Rev A 52:457–464
47. Liebsch T et al (1996) Photoelectron spectroscopy of free fullerenes. J Electron Spectrosc Relat Phenom 79:419–422
48. Gerchikov LG, Connerade JP, Solov'yov AV, Greiner W (1997) Scattering of electrons on metal clusters and fullerenes. J Phys B: At Mol Opt Phys 30:4133–4161
49. Gerchikov LG, Ipatov AN, Solov'yov AV (1997) Many-body treatment of electron inelastic scattering on metal clusters. J Phys B: At Mol Opt Phys 30:5939–5959
50. Gerchikov LG, Ipatov AN, Solov'yov AV, Greiner W (1998) Excitation of multipole plasmon resonances in clusters by fast electron impact. J Phys B: At Mol Opt Phys 31:3065–3077
51. Gerchikov LG, Efimov PV, Mikoushkin VM, Solov'yov AV (1998) Diffraction of fast electrons on the fullerene C_{60} molecule. Phys Rev Lett 81:2707–2710
52. Gerchikov LG, Ipatov AN, Polozkov RG, Solov'yov AV (2000) Surface and volume plasmon excitation in electron inelastic scattering on metal clusters. Phys Rev A 62:043201
53. Verkhovtsev AV, Korol AV, Solov'yov AV, Bolognesi P, Ruocco A, Avaldi L (2012) Interplay of the volume and surface plasmons in the electron energy loss spectra of C_{60}. J Phys B: At Mol Opt Phys 45:141002
54. Bolognesi P, Avaldi L, Ruocco A, Verkhovtsev A, Korol AV, Solov'yov AV (2012) Collective excitations in the electron energy loss spectra of C_{60}. Eur Phys J D 66:254
55. Connerade J-P, Solov'yov AV (2002) Formalism for multiphoton plasmon excitation in jellium clusters. Phys Rev A 66:013207
56. Ivanov VK, Kashenock GYu, Polozkov RG, Solov'yov AV (2001) Photoionization cross sections of the fullerenes C_{20} and C_{60} calculated in a simple spherical model. J Phys B: At Mol Opt Phys 34:L669–L677
57. Verkhovtsev AV, Korol AV, Solov'yov AV (2013) Quantum and classical features of the photoionization spectrum of C_{60}. Phys Rev A 88:043201
58. Landau LD, Lifshitz EM (1976) Quantum Mechanics: non-relativistic theory. 3rd edn. Course of Theoretical Physics, vol. 3. Butterworth-Heinemann
59. de Heer WA (1993) The physics of simple metal clusters: experimental aspects and simple models. Rev Mod Phys 65:611–676
60. Brack M (1993) The physics of simple metal clusters: self-consistent jellium model and semiclassical approaches. Rev Mod Phys 65:677–732
61. Bréchignac C, Connerade JP (1994) Giant resonances in free atoms and in clusters. J Phys B: At Mol Opt Phys 27:3795–3828
62. Haberland H (ed) (1994) Clusters of atoms and molecules, theory, experiment and clusters of atoms. Springer Series in Chemical Physics, vol. 52. Springer, Berlin, Heidelberg, New York
63. Korol AV, Solov'yov AV (1997) Polarizational bremsstrahlung of electrons in collisions with atoms and clusters. J Phys B: At Mol Opt Phys 30:1105–1150
64. Alasia F, Broglia RA, Roman HE, Serra L, Colo G, Pacheco JM (1994) Single-particle and collective degrees of freedom in C_{60}. J Phys B: At Mol Opt Phys 27:L643–L650
65. Madjet M, Guet C, Johnson WR (1995) Comparative study of exchange-correlation effects on the electronic and optical properties of alkali-metal clusters. Phys Rev A 51:1327–1339
66. Campbell EE, Rohmund F (2000) Fullerene reactions. Rep Prog Phys 63:1061–1109
67. Berkowitz J (1999) Sum rules and the photoabsorption cross sections of C_{60}. J Chem Phys 111:1446–1453
68. Reinköster A, Korica S, Viefhaus J, Godenhusen K, Schwarzkopf O, Mast M, Becker U (2004) The photoionization and fragmentation of C_{60} in the energy range 26–130 eV. J Phys B: At Mol Opt Phys 37:2135–2144

69. Scully SWJ et al (2005) Photoexcitation of a volume plasmon in C_{60} ions. Phys Rev Lett 94:065503
70. Baral KK et al (2016) Photoionization and photofragmentation of the C_{60}^{+} molecular ion. Phys Rev A 93:033401
71. Solov'yov AV (2005) Plasmon excitations in metal clusters and fullerenes. Int J Mod Phys B 19:4143–4184
72. Verkhovtsev AV, Korol AV, Solov'yov AV (2012) Formalism of collective excitations in fullerenes. Eur Phys J D 66:253
73. Varshalovich DA, Moskalev AN, Khersonskii VK (1988) Quantum theory of angular momentum. World Scientific Publishing, Singapore
74. Verkhovtsev AV, Korol AV, Solov'yov AV (2015) Revealing the mechanism of the low-energy electron yield enhancement from sensitizing nanoparticles. Phys Rev Lett 114:063401
75. Verkhovtsev AV, Korol AV, Solov'yov AV (2015) Electron production by sensitizing gold nanoparticles irradiated by fast ions. J Phys Chem C 119:11000–11013
76. Verkhovtsev A, McKinnon S, de Vera P, Surdutovich E, Guatelli S, Korol AV, Rosenfeld A, Solov'yov AV (2015) Comparative analysis of the secondary electron yield from carbon nanoparticles and pure water medium. Eur Phys J D 69:116
77. Connerade J-P, Solov'yov AV (1996) Radiative electron capture by metallic clusters. J Phys B: At Mol Opt Phys 29:365–375
78. Gerchikov LG, Ipatov AN, Solov'yov AV (1998) Many-body treatment of the photon emission process in electron-clusters collisions. J Phys B: At Mol Opt Phys 31:2331–2341
79. Korol AV, Solov'yov AV (2014) Polarization Bremsstrahlung, Springer Series on Atomic, Optical, and Plasma Physics, vol 80. Springer, Berlin Heidelberg
80. Kubo R (1962) Electronic properties of metallic fine particles. I. J Phys Soc Jpn 17:975–986
81. Lushnikov AA, Simonov AJ (1974) Surface plasmons in small metal particles. Z Phys 270:17–24
82. Yannouleas C, Broglia RA (1992) Landau damping and wall dissipationin large metal clusters. Ann Phys 217:105–141
83. Yannouleas C (1998) Microscopic description of the surface dipole plasmon in large Na_N clusters ($950 \le N \le 12050$). Phys Rev B 58:6748–6751
84. Bertsch GF, Bulgac A, Tomanek D, Wang Y (1992) Collective plasmon excitations in C_{60} clusters. Phys Rev Lett 67:2690–2693
85. Lushnikov AA, Simonov AJ (1975) Excitation of surface plasmons in metal particles by fast electrons and x rays. Z Phys B 21:357–362
86. Ipatov AN, Ivanov VK, Agap'ev BD, Eckardt W (1998) Exchange and polarization effects in elastic electron scattering by metallic clusters. J Phys B: At Mol Opt Phys 31:925–934
87. Descourt P, Farine M, Guet C (2000) Many-body approach of electron elastic scattering on sodium clusters. J Phys B: At Mol Opt Phys 33:4565–4574
88. Lezius M, Scheier P, Märk TD (1993) Free electron attachment to C_{60} and C_{70}. Chem Phys Lett 203:232–236
89. Huang J, Carman HS Jr, Compton RN (1995) Low-energy electron attachment to C_{60}. J Phys Chem 99:1719–1726
90. Elhamidi O, Pommier J, Abouaf R (1997) Low-energy electron attachment to fullerenes C_{60} and C_{70} in the gas phase. J Phys B: At Mol Opt Phys 30:4633–4642
91. Ptasinska S et al (2006) Electron attachment to higher fullerenes and to $Sc_3N@C_{80}$. J Phys Chem A 110:8451–8456
92. Kasperovich V, Tikhonov G, Wong K, Brockhaus P, Kresin V (1999) Polarization forces in collisions between low-energy electrons and sodium clusters. Phys Rev A 60:3071–3075
93. Kresin V, Guet C (1999) Long-range polarization interactions of metal clusters. Philos Mag B 79:1401–1411
94. Sentürk S, Connerade JP, Burgess DD, Mason NJ (2000) Enhanced electron capture by metallic clusters. J Phys B: At Mol Opt Phys 33:2763–2774
95. Rabinovitch R, Xia C, Kresin VV (2008) Evaporative attachment of slow electrons to alkali-metal nanoclusters. Phys Rev A 77:063202

96. Rabinovitch R, Hansen K, Kresin VV (2011) Slow electron attachment as a probe of cluster evaporation processes. J Phys Chem A 115:6961–6972
97. Connerade JP, Solov'yov AV (1996) Giant resonances in photon emission spectra of metal clusters. J Phys B: At Mol Opt Phys 29:3529–3547
98. Ipatov A, Connerade J-P, Gerchikov LG, Solov'yov AV (1998) Electron attachement to metallic clusters. J Phys B: At Mol Opt Phys 31:L27–L34
99. Connerade J-P, Gerchikov LG, Ipatov AN, Solov'yov AV (1999) Polarization effects in electron attachement to metallic clusters. J Phys B: At Mol Opt Phys 32:877–894
100. Hervieux P-A, Madjet ME, Benali H (2002) Capture of low-energy electrons by simple closed-shell metal clusters. Phys Rev A 65:023202
101. Massey HSW (1979) Atomic and molecular collisions. Taylor and Francis, London
102. Gerchikov LG, Solov'yov AV (1997) Photon emission in electron-cluster collision in the vicinity of plasmon resonance. Z Phys D 42:279–287
103. Amusia MYa, Korol AV (1994) On the continuous spectrum electromagnetic radiation in electron-fullerene collisions. Phys Lett A 186:230–234
104. Gerchikov LG, Ipatov AN, Solov'yov AV, Greiner W (2000) Non-adiabatic electron-ion coupling in dynamical jellium model for metal clusters. J Phys B: At Mol Opt Phys 33:4905–4926
105. Chernysheva LV, Gribakin GF, Ivanov VK, Kuchiev MYu (1988) Many-body calculation of negative ions using the Dyson equation. J Phys B: At Mol Opt Phys 21:L419–L425
106. Ellert Ch, Schmidt M, Schmitt M, Reiners Th, Haberland H (1995) Temperature dependence of the optical response of small, open shell sodium clusters. Phys Rev Lett 75:1731–1734
107. Gerchikov LG, Solov'yov AV, Greiner W (1999) Dynamical jellium model for metallic clusters. Int J Mod Phys E 8:289–298
108. Pacheco JM, Broglia RA (1989) Effect of surface fluctuations in the line shape of plasma resonances in small metal clusters. Phys Rev Lett 62:1400–1402
109. Bertsch GF, Tomanek D (1989) Thermal line broadening in small metal clusters. Phys Rev B 40:2749–2751
110. Penzar Z, Ekardt W, Rubio A (1990) Temperature effects on the optical absorption of jellium clusters. Phys Rev B 42:5040–5045
111. Montag B, Reinhard P-G, Meyer J (1994) The structure-averaged jellium model for metal clusters. Z Phys D 32:125–136
112. Montag B, Reinhard PG (1995) Width of the plasmon resonance in metal clusters. Phys Rev B 51:14686–14692
113. Lyalin AG, Semenov SK, Cherepkov NA, Solov'yov AV, Greiner W (2000) Hartree-Fock deformed jellium model for metal clusters. J Phys B: At Mol Opt Phys 33:3653–3664
114. Ekardt W (1985) Collective multipole excitations in small metal particles: critical angular momentum l^{cr} for the existence of collective surface modes. Phys Rev B 32:1961–1970
115. Guet C, Johnson WR (1992) Dipole excitations of closed-shell alkali-metal clusters. Phys Rev B 45:11283–11287
116. Kharchenko VA, Ivanov VK, Ipatov AN, Zhizhin ML (1994) Size dependence of electronic structure and adiabatic type of collective vibration in small metal clusters. Phys Rev A 50:1459–1464
117. Wang Y, Lewenkopf C, Tomanek D, Bertsch G, Saito S (1993) Collective electronic excitations and their damping in small alkali clusters. Chem Phys Lett 205:521–528
118. Pacheco JM, Schöne W-D (1997) Shape phase transitions in the absorption spectra of atomic clusters. Phys Rev Lett 79:4986–4989
119. Runge E, Gross EKU (1984) Density-functional theory for time-dependent systems. Phys Rev Lett 52:997–1000
120. Henke BL, Gullikson EM, Davis JC (1993) X-ray interactions: photoabsorption, scattering, transmission, and reflection at $E = 50 - 30,000$ eV, $Z = 1 - 92$. At Data Nucl Data Tables 54:181–342
121. de Vera P, Garcia-Molina R, Abril I, Solov'yov AV (2013) Semiempirical model for the ion impact ionization of complex biological media. Phys Rev Lett 110:148104

122. LaVerne J (1989) Radical and molecular yields in the radiolysis of water with carbon ions. Radiat Phys Chem 34:135–143
123. Yuan HK, Chen H, Tian CL, Kuang AL, Wang JZ (2014) Density functional calculations for structural, electronic, and magnetic properties of gadolinium-oxide clusters. J Chem Phys 140:154308
124. Schüler M, Berakdar J, Pavlyukh Y (2015) Disentangling multipole contributions to collective excitations in fullerenes. Phys Rev A 92:021403(R)
125. Biswas S, Tribedi LC (2015) Plasmon-mediated electron emission from the coronene molecule under fast ion impact. Phys Rev A 92:060701(R)

On the Quantum Description of Irradiation Dynamics in Systems of Biological Relevance

P.M. Dinh, L. Bouëssel du Bourg, C.-Z. Gao, Bin Gu, L. Lacombe, M. McAllister, M. Smyth, G. Tribello, M. Vincendon, J. Kohanoff, P.-G. Reinhard, L. Sanche and E. Suraud

Abstract The two main products of ionizing radiation in biological tissues are electrons and radicals. The numerous secondary electrons are generated by ionisation in the molecules in the vicinity of DNA and are produced with a mean energy about 10 eV. These low-energy electrons can lead to DNA strand breaks via dissociative electron attachment and other mechanisms. The modelling of these phenomena requires, on the one hand, an explicit quantum description of the electrons of the target molecule (typically, a subunit of a DNA strand), and on the other hand, a realistic account of the DNA environment. This chapter reviews theoretical and computational approaches that have allowed us to study electron dynamics (excitation, ionization, transport and localization) in systems of biological interest.

P.M. Dinh (✉) · C.-Z. Gao · L. Lacombe · M. Vincendon · E. Suraud
Laboratoire de Physique Théorique, Université Paul Sabatier and CNRS,
118 Route de Narbonne, 31062 Toulouse, France
e-mail: dinh@irsamc.ups-tlse.fr

C.-Z. Gao
e-mail: Congzhang.Gao@irsamc.ups-tlse.fr

L. Lacombe
e-mail: lacombe@irsamc.ups-tlse.fr

M. Vincendon
e-mail: Marc.Vincendon@irsamc.ups-tlse.fr

E. Suraud
e-mail: suraud@irsamc.ups-tlse.fr

L.B. du Bourg · B. Gu · M. McAllister · M. Smyth · G. Tribello · J. Kohanoff
Atomistic Simulation Centre, Queen's University Belfast, Belfast Bt7 1nn, UK
e-mail: lila.bouesseldubourg@univ-rennes1.fr

B. Gu
e-mail: gubin@nuist.edu.cn

M. McAllister
e-mail: mmcallister19@qub.ac.uk

M. Smyth
e-mail: Maeve.Smyth2@wales.nhs.uk

A.V. Solov'yov (ed.), *Nanoscale Insights into Ion-Beam Cancer Therapy*,
DOI 10.1007/978-3-319-43030-0_8

1 Introduction

Radiation damage of biological tissue is a field of strong current interest due to its importance both within the medical context and in relation to Earth and space radiation. The response of the system to radiation is primarily of electronic nature and electrons thus play a key role as the doorway to all subsequent dynamical scenarios. A basic feature is here the optical response corresponding to electronic oscillations, which defines the coupling between radiation and matter in a large variety of dynamical situations from gentle to strong perturbations, involving lasers [1] or charged projectiles [2]. Equally important in energetic irradiation processes is electron transport, particularly electron emission. As typical examples for the relevance of electronic emission, one can cite the many studies on irradiation of clusters by short and intense laser pulses [1], providing invaluable information especially through energy (Photo Electron Spectra, PES [3]) and, more recently, angle-resolved [4] distributions of emitted electrons (Photo Angular Distributions, PAD). Such quantities, namely energy and/or angular resolved cross sections of emitted electrons also begin to be investigated in the case of irradiation via swift charged projectiles, a situation closer to potential applications in a biological context. Indeed the response of DNA basis has already been analysed that way [5, 6]. In turn, DNA damage due to secondary electrons generated by primary irradiation of the surrounding medium or DNA itself also provides a remarkable example, where a microscopic understanding of radiation damage in biological systems will only be achieved when including such complex non-linear electronic effects.

G. Tribello
e-mail: g.tribello@qub.ac.uk

J. Kohanoff
e-mail: j.kohanoff@qub.ac.uk

L.B. du Bourg
Chimie Thorique Inorganique, UMR CNRS 6226 - Institut des Sciences Chimiques de Rennes Université de Rennes 1, Campus de Beaulieu - Bt. 10B - Case 1009, 35042 Rennes Cedex, France

B. Gu
Department of Physics, Nanjing University of Information Science and Technology, Nanjing 21004, China

M. Smyth
Velindre Cancer Centre, Cardiff, Velindre Road CF14 2TL, UK

P.-G. Reinhard
Institut Für Theoretische Physik, Universität Erlagen, Staudstrasse 7, D91058 Erlangen, Germany
e-mail: Paul-Gerhard.Reinhard@fau.de

L. Sanche
Département de médecine nucléaire et de Radiobiologie, Université de Sherbrooke, 3001, 12e Avenue Nord, Sherbrooke, QC J1H 5N4, Canada
e-mail: leon.sanche@USherbrooke.ca

In a biological environment, the two main products of ionizing radiation are electrons and radicals. Both have the potential to cause damage to biological molecules, in particular DNA, thus leading, in the long run, to the inability of cells to replicate, and hence to cell death. For a long time, it was assumed that the main actors were the radicals. However, in 2000, a seminal paper by the group of L. Sanche showed experimentally that Low-Energy Electrons (LEE) produced by ionization can also lead to bond breaking in biomolecules, in particular to strand breaks in plasmid DNA [7]. These results prompted, during the past 15 years, a significant amount of both experimental [8–12] and theoretical [13–15] works aiming at elucidating the role played by LEE in DNA damage. In the past few years, this research has moved on from idealized systems, e.g. isolated fragments like nucleic acids, nucleotides and even plasmid DNA, towards a more realistic environment reminiscent of physiological conditions. This requires to take into account the aqueous solution containing ions and the presence of amino acids from the histones in chromatin. Moreover, it also requires to study these phenomena in the presence of thermal fluctuations, as all processes happen at room temperature. Figure 1 shows a schematic view of the irradiation process in a biological medium. Electromagnetic radiation in the form of X- or γ-rays as well as high-energy ions produce mainly ionization of the medium generating LEE and radicals.

While radicals are extremely important and interesting, in the present paper we have chosen to focus on the dynamics of LEE only and its impact on DNA damage. The long-term societal importance of these studies is obvious, especially in relation to oncology. There is by now a significant body of research devoted to the relation between cancer treatment and LEE [16–22]. The microscopic understanding of underlying mechanisms, however, is still in its infancy and requires dedicated efforts, particularly from a modelling point of view, in order to better understand ongoing experiments. The aim of this contribution is to describe some developments in the microscopic quantum description of irradiation processes, addressing in particular the role of electrons. We shall thus review a body of theoretical works based

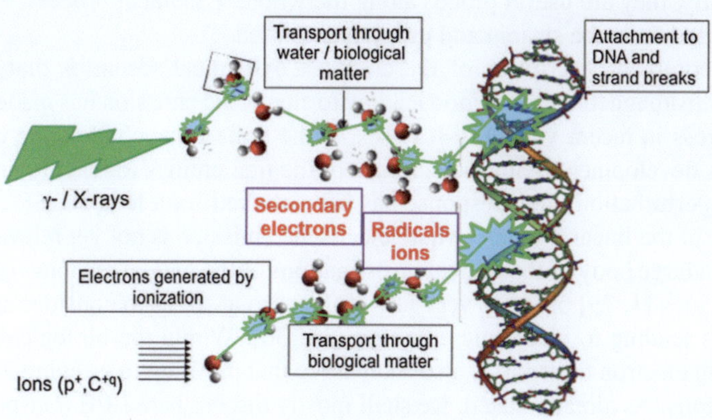

Fig. 1 Schematic view of the processes involved in the irradiation of biological media

on quantum-mechanical simulations that focus on electron dynamics (ionization and transport) and on physical and chemical processes like electron localization and bond breaking in a realistic, explicit environment.

Paradoxically, an atomic-scale characterisation may be easier to achieve via computer simulations. In fact, contrary to gas-phase systems, where mass spectrometry is an invaluable tool to study the distribution of fragments upon electron [23] or ion irradiation [24], the presence of the aqueous environment complicates the extraction and identification of fragmentation products. A large fraction of the experimental information on single and double strand breaks is obtained by gel electrophoresis, which is sensitive to the mass and geometric structure of the products. Moreover, to obtain a sufficiently strong signal, experiments have generally been conducted at low temperatures and in lyophilized samples, with a small content of structural waters. Only quite recently experiments have been conducted in a solvated environment [25]. In between, some insight on the role of the environment was obtained through studies of fragmentation in microsolvated clusters, where gas-phase techniques are still applicable [26, 27].

Electrons are usually the first constituents to respond to an electromagnetic pulse. Strong excitations lead to an immediate ionization of the system, often with dramatic long-time effects like dissociation, Coulomb explosion, or thermal spikes. This leads to electronic transport and possible indirect effects on neighbouring species. A typical example of indirect effects is provided by Dissociative Electron Attachment (DEA) in biological systems where primarily (low-energy) emitted electrons attach resonantly to target biological molecules transferring electronic energy into vibrations, eventually leading to the break up of the molecule. Emitted electrons can also provide a valuable insight into reaction pathways, once they have been properly detected. Typical examples here are PES and PAD. Moreover, Time-Resolved (TR) PES and PAD have been recorded in molecules and more recently in clusters. Electrons are thus leading players at all stages of an excitation of a system subject to an electromagnetic perturbation, e.g. irradiation. They provide the first response at short time scales, which is more or less quickly coupled to other degrees of freedom. Additionally, they are useful probes along the whole dynamical process, especially when emitted from the system and properly recorded.

The theoretical description of the complex dynamical scenarios that follow a strong electromagnetic perturbation leading to electronic emission has made remarkable progress in recent years [1]. It raises, at the same time, challenging questions for further developments. Since electrons are the first entities to react to an electromagnetic perturbation, their response has been studied for a long time [1, 28, 29], especially in the linear domain (where electronic emission is not yet relevant). This has led to a large body of theoretical investigations of the optical response of clusters and molecules [1, 29] but only very few of these go as far as to consider non-linear excitations leading to electronic emission [28, 30]. Within the biological context, the ensuing electron transport is also a key issue that motivates a growing number of investigations. As already stated, we shall mostly discuss here LEE transport.

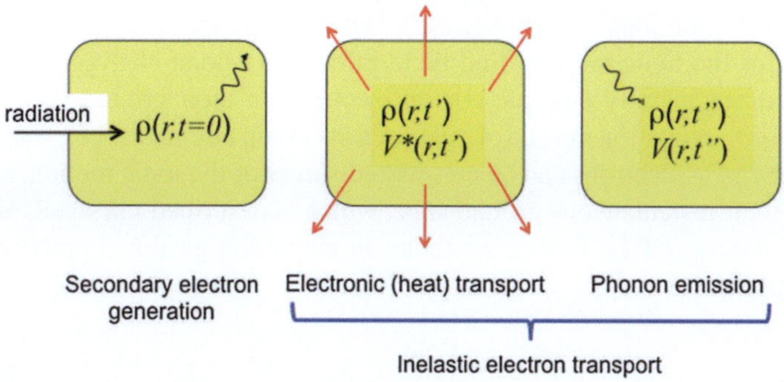

Fig. 2 Physico-chemical stages in the biological irradiation process. In the first stage, radiation ionizes the medium producing secondary electrons and radicals. Within time-dependent DFT (see Sect. 2.1) this is translated into a time-dependent electronic density $\rho(\mathbf{r}, t)$. When electrons are not in the ground state, the interaction potential is modified to $V^*(\mathbf{r}, t)$, which also evolves in time together with the density. Next, secondary electrons move through the medium losing energy via electronic, vibrational and rotational excitations. These energy losses can occur via direct scattering or through the formation of transient anions. When the latter are dissociative states bond scission can occur, if the anion's lifetime is of the order of or greater than a vibrational period of the target molecule, this process is called dissociative electron attachment (DEA). All of these phenomena can be seen as components of electronic heat transport. Once stabilized in a molecule, the excess electron usually weakens bonds and lowers energy barriers, thus facilitating bond cleavage via thermal activation

LEE are transported through the aqueous medium, interacting inelastically with, e.g., water molecules, and transferring part of their kinetic energy to the medium via electronic and vibrational excitation [11]. As a consequence of this energy loss, LEE are driven into a state that is often known as *pre-solvated*. At this stage, either they become fully solvated in the medium or they interact with DNA or other species as amino-acids producing chemical modifications, e.g. the cleavage of some bond. Figure 2 shows a scheme of the physical and chemical stages involved in the biological irradiation process, using some basic theoretical concepts such as the time-dependent electronic density, which we shall present and use below.

2 Methodological Considerations

In this section, we describe the theoretical framework underlying the simulations that will be presented in Sect. 3. The electronic excitation and ionization processes require the description of the intrinsic dynamics of the electrons, independently of the motion of the nuclei (actually the ionic cores containing the nucleus and core electrons which are frequently called *ions*). In the approach presented in this paper, this is achieved via time-dependent density-functional theory (TDDFT). If, within this time scale, the ions also move, then TDDFT is supplemented with molecular dynamics for the ionic motion. End-point effects like bond cleavage are chemical processes that generally take place with electrons in their ground state, although electronic (de)-excitation may be of relevance in certain circumstances, such as DEA. In all other cases, electrons can be considered *slaves* of the ionic motion, remaining always in their instantaneous ground state, which is described via standard density-functional theory (DFT). Below, we focus on methodological and practical aspects of these approaches.

2.1 Electron Dynamics in Model Systems Within (TD)DFT

Whatever the framework of macroscopic modeling of the environment may be, at the deepest level of description, we focus on the detailed dynamics of a small subsystem, an acceptably small molecule or a well-confined active region of a macromolecule. In most cases, this also embraces electronic dynamics which, in turn, requires a quantum mechanical description. In this section, we focus on the description of a relatively small quantum mechanical subsystem within a framework of hierarchical approaches, describing a typical theoretical framework and presenting some results on dynamical calculations of covalent molecules.

TDDFT is a widely used and very efficient tool for that purpose, usually treated at the level of the time-dependent local-density approximation (TDLDA). It describes statics and dynamics of the electron cloud in terms of single-particle (s.p.) wave functions where the expensive exchange and correlation effects are incorporated into a functional of local density alone, see [31, 32] and Sect. 2.1.1. The scheme is robust and efficient. Actually, there exist many refinements and extensions in the literature to mention, e.g., additional terms depending on the gradient of density (generalised gradient approximations or GGA) [33] or the self-interaction correction (SIC) [34]. The latter will be addressed briefly in Sect. 2.1.2.

The coupling to ionic motion is implemented in different ways. Most widely used is the Born-Oppenheimer (sometimes also called adiabatic) approximation, where the electronic ground state is computed for fixed ionic configuration from which one deduces the force fields for ionic dynamics. Strong excitations require to go beyond. The next step is to propagate electrons and ions simultaneously, the electrons quantum mechanically by TDLDA and the ions by classical molecular dynamics, together called TDLDA-MD [29] or Ehrenfest dynamics. A limitation of this approach, even if fully dynamical, is that it describes only an average ionic motion and cannot account for the splitting of ionic trajectories induced by electronic quantum jumps. Similarly, since ions see only the mean field of the electrons via the electronic density, there is no information about individual collisions and thus phenomena like spontaneous phonon emission are precluded. To restore such effects one has to resort to other methods *beyond* Ehrenfest. One such method is surface hopping [35, 36], which consists of augmenting Born-Oppenheimer dynamics with stochastic jumps between (electronic) Born-Oppenheimer surfaces. Another, computationally more expensive, approach is Correlated Electron-Ion Dynamics (CEID) [37, 38] where the ionic motion is treated approximately by a perturbative expansion in powers of quantum fluctuations about the mean trajectory. In the following, we will address only TDLDA-MD and pure TDLDA calculations, for isolated systems or in the presence of an environment.

2.1.1 TDLDA and TDLDA-MD

In the Kohn-Sham formulation [39–41], (TD)DFT describes a system in terms of a set of occupied s.p. wave functions $\{\varphi_i\}$. It is a variational formulation which is fully defined from its starting point, via the following expression for the total energy

$$E^{(\mathrm{LDA})} = E_{\mathrm{kin}}[\{\varphi_\alpha\}] + E_{\mathrm{ion}} + E_{\mathrm{ext}} + E_{\mathrm{Cxc}}[\rho] \quad , \tag{1a}$$

$$\rho(\mathbf{r},t) = \sum_{j=1}^{N} \left|\varphi_j(\mathbf{r},t)\right|^2 \quad , \tag{1b}$$

where $\rho(\mathbf{r},t)$ is the total electronic density, i.e. the probability of finding an electron at position $\mathbf{r}$ and at time t. The kinetic energy E_{kin} carries detailed information on the s.p. wave functions which serves to maintain quantum shell effects, e.g. electronic shells in atoms. The energy E_{ion} stands for the field from the ionic background (usually ionic pseudo-potentials representing nuclei and core electrons). The energy E_{ext} describes the impact of a time-dependent external perturbation as, e.g., from a laser or a bypassing ion. The term E_{Cxc} comprises all electronic interactions, the Coulomb-Hartree term (C), the Coulomb exchange term (x), and a correlation term (c). The Coulomb-Hartree term depends bi-linearly on $\rho(\mathbf{r})$. The other two terms are mapped onto a pure $\rho(\mathbf{r})$-dependence by virtue of the LDA, while the GGA also involves a dependence on $\nabla\rho(\mathbf{r})$. Variation with respect to φ_i^* yields in standard manner the Kohn-Sham equations:

$$\left.\begin{matrix} \mathrm{i}\hbar\partial_t\varphi_i \\ \varepsilon_i\varphi_i \end{matrix}\right\} = \hat{h}^{\mathrm{LDA}}\varphi_i \quad , \quad \hat{h}^{\mathrm{LDA}} = \frac{\hat{p}^2}{2m} + u_{\mathrm{ext}} + u^{\mathrm{LDA}}[\rho] \quad , \tag{2a}$$

where the LDA potential $u^{\mathrm{LDA}}[\rho]$ is obtained as the functional derivative with respect to the local density, i.e.

$$u^{\mathrm{LDA}}[\rho(\mathbf{r},t)] = \left.\frac{\delta E_{\mathrm{Cxc}}}{\delta\rho}\right|_{\rho=\rho(\mathbf{r},t)} \quad . \tag{2b}$$

The ion-electron interaction is mediated by pseudo-potentials

$$\hat{V}_{\text{ion}} = \sum_{I=1}^{N_{\text{ion}}} \hat{V}^{(\text{PsP})}(\mathbf{R}_I) \quad , \tag{3}$$

where $\hat{V}_{\text{ion}}$ is a, generally, non-local operator acting in the Hilbert space of the electronic s.p. orbitals. It depends on the position of the corresponding ion. The ion-electron energy in (1b) is then

$$E_{\text{ion}} = \sum_{i=1}^{N} \sum_{I=1}^{N_{\text{ion}}} \langle \varphi_i | \hat{V}^{(\text{PsP})}(\mathbf{R}_I) | \varphi_i \rangle \quad . \tag{4}$$

The action on the electrons is accounted for in the Kohn-Sham equations (2). The first line in the left equation in (2) is for TDDFT, and the second for static DFT.

The back-effect on the ions contributes to the classical, Hamilton-like equations of motion:

$$\partial_t \mathbf{P}_I = -\nabla_{\mathbf{R}_I} \left[E_{\text{pot,ion}}(\mathbf{R}_1, ..., \mathrm{R}_{N_{\text{ion}}}) + E_{\text{ion}}(\mathbf{R}_1, ..., \mathrm{R}_{N_{\text{ion}}}) \right] \quad , \tag{5a}$$

$$\partial_t \mathbf{R}_I = \frac{\mathbf{P}_I}{M_I} \quad , \tag{5b}$$

where $E_{\text{pot,ion}}$ is the potential energy due to the, usually Coulomb, interaction between ions. The simultaneous propagation of this equation together with the dynamical Kohn-Sham equations constitutes TDLDA-MD or Ehrenfest dynamics. Notice that the electronic description is not restricted to TDLDA. Extensions such as the GGA or SIC can equally well be employed.

2.1.2 Self-interaction Correction (SIC)

The effective mean-field potential u^{LDA} depends on the density (1b), which contains a sum over all occupied states. This includes the state φ_i on which the mean-field Hamiltonian acts in Eq. (2a). Thus the state φ_i is, unphysically, interacting with itself which leads to what is called the self-interaction error. This is partially cured for many purposes by the GGA [33], although both LDA and GGA produce an incorrect asymptotic behavior of u^{LDA}. A proper asymptotics is achieved by augmenting the energy-density functional with a SIC [34] which reads

$$E_{\text{Cxc}} \longrightarrow E_{\text{Cxc}}[\rho] \underbrace{- \sum_j E_{\text{Cxc}}[\rho_j]}_{E_{\text{SIC}}} \quad , \quad \rho_j(\mathbf{r}) = |\varphi_j(\mathbf{r})|^2 \quad . \tag{6}$$

At variance with Hartree-Fock, which is self-interaction free by construction, this SIC functional is not invariant under unitary transformations because it refers explicitly to a functional of the s.p. densities ρ_j. Since the SIC Hamiltonian is state-dependent, orthonormality between the s.p. orbitals is not automatically guaranteed. In order to retrieve this property, i.e. $\langle\varphi_i|\varphi_j\rangle = \delta_{ij}$, orthonormality has to be enforced via a Lagrangian multiplier λ_{ij} for each overlap, thus augmenting the variational energy expression with the term $-\sum_{ij}\lambda_{ij}\langle\varphi_i|\varphi_j\rangle$.

Variation of the augmented functional then yields the coupled equations:

$$\left.\begin{array}{r}\mathrm{i}\hbar\partial_t\varphi_i\\ \varepsilon_i\varphi_i\end{array}\right\} = \hat{h}^{\mathrm{SIC}}\varphi_i - \sum_{j\neq i}\varphi_j\lambda_{ji} \quad ,$$

$$\hat{h}^{\mathrm{SIC}} = \hat{h}^{\mathrm{LDA}} - \sum_i u_i|\varphi_i\rangle\langle\varphi_i| \quad , \quad u_i = u^{\mathrm{LDA}}[\rho_i] \quad , \tag{7a}$$

$$0 = \langle\varphi_i|u_i - u_j|\varphi_j\rangle \quad . \tag{7b}$$

It is recommended to employ complex wave functions already at the stage of stationary solutions because the SIC energy often takes advantage of the complex degrees-of-freedom when minimizing the energy [42]. Unlike the standard Kohn-Sham scheme, two equations emerge from the variation. The second one, Eq. (7b), which emerges from the orthonormalization constraint, is new. This is often called localization condition because it tends to localize the φ_i. Alternatively, the notion of symmetry condition is used because it is related to symmetry (hermiticity) of the matrix of Lagrangian multipliers λ_{ij}. The first equation (7a) looks deceivingly similar to the Kohn-Sham equation. However, we face now the inconvenience that $\hat{h}_{\mathrm{SIC}}$ is not hermitian. It is the symmetry condition (7b) which guarantees $\langle\varphi_i|\hat{h}^{\dagger}_{\mathrm{SIC}} - \hat{h}_{\mathrm{SIC}}|\varphi_j\rangle = 0$ for all $i, j = 1...N$, such that hermiticity holds at least within the occupied states. Unfortunately, the full solution of the coupled equations (7) in terms of the set $\{\varphi_i, i = 1...N\}$ is inconvenient in the static case and intractable in dynamics. The way out of this dilemma is to use two equivalent sets of s.p. states, $\{\varphi_i, i = 1...N\}$ and $\{\psi_\alpha, i = 1...N\}$, connected by a unitary transformation

$$\varphi_i = \sum_{\alpha=1}^{N}\mathcal{U}_{i\alpha}\psi_\alpha \quad . \tag{7c}$$

The set $\{\varphi_i, i = 1...N\}$ is called the localizing set. It serves to define the SIC potential and to fulfill the symmetry condition. The set $\{\psi_\alpha, i = 1...N\}$ is designed to solve the mean-field equation, either static or dynamic. The 2set-SIC equations thus become:

$$\left.\begin{array}{r}\mathrm{i}\hbar\partial_t\psi_\alpha\\ \varepsilon_\alpha\psi_\alpha\end{array}\right\} = \hat{h}^{\mathrm{SIC}}\psi_\alpha \ , \tag{7d}$$

$$\hat{h}^{\mathrm{SIC}} = \hat{h}^{\mathrm{LDA}} - \sum_i u_i|\varphi_i\rangle\langle\varphi_i| \quad , \quad u_i = u^{\mathrm{LDA}}[\rho_i] \ , \tag{7e}$$

together with the unchanged Eqs. (7a) and (7b). The set $\{\psi_\alpha, i = 1...N\}$ is the diagonalizing (for statics) or propagating (for dynamics) set. It is treated in standard fashion as for any conventional (time-dependent) Kohn-Sham equation. The localizing set $\{\varphi_i, i = 1...N\}$ is found by determining the coefficients $\mathcal{U}_{i\alpha}$ of the unitary transformation (7c) amongst the occupied states such that, for given ψ_α's, the φ_i's meet the symmetry condition (7b). Notice that the energy (6) depends on the localizing set. However, thanks to the symmetry condition, this set is unique. For further details of this 2set-SIC scheme, see [43, 44].

Although the full SIC treatment has become manageable with the 2set-SIC scheme, there are many situations that allow a simpler and more robust approach, such as the average-density SIC (ADSIC) [45]. This scheme augments the LDA functional by:

$$E_{\rm Cxc} \longrightarrow E_{\rm Cxc}[\rho] - E_{\rm ADSIC} \quad , \quad E_{\rm ADSIC} = N_\uparrow E^{\rm LDA}\left[\frac{\rho_\uparrow}{N_\uparrow}\right] + N_\downarrow E^{\rm LDA}\left[\frac{\rho_\downarrow}{N_\downarrow}\right] \quad (8)$$

where $\rho_\uparrow$ and $N_\uparrow$ (resp. $\rho_\downarrow$ and $N_\downarrow$) are the density and the total number of electrons with spin up (resp. spin down). This $E_{\rm ADSIC}$ is an approximate functional of the local density and can be treated the same way as the LDA functional. This simplifies dramatically both the static solution and the time propagation, and yet delivers already the correct asymptotic behavior of the mean-field. Although the averaging over all s.p. densities looks too crude, ADSIC performs surprisingly well for many different bond types [42]. Problems appear however if a molecule mixes elements with very different bonding properties. An example are $Na(H_2O)_n$ complexes discussed in Sect. 3.1.

Another viable SIC framework was proposed by d'Avezac et al. [46] with the aim of counteracting another problem introduced by the incomplete cancellation of the self-interaction in the LDA and GGA functionals, namely their excessive delocalization. This approach starts from Perdew and Zunger's correction [34] as in Eq. (6), but the correction $E_{\rm SIC}$ is modified in the following way:

$$E_{\rm SIC} = -aE_{\rm C}[m] - bE_{\rm xc}[m; 0] \quad , \quad (9)$$

where $m = \rho_\uparrow - \rho_\downarrow$ is the spin density of the system, and a and b are empirical scaling parameters for the Coulomb and exchange-correlation energies [47]. The delocalization problem is more important for unpaired electrons or holes than for doubly occupied orbitals. Therefore, to make this approach more efficient, the SIC is applied only to the singly-occupied molecular orbital (SOMO), while the double-occupied orbitals are left uncorrected. Since this approach can easily lead to spin contamination, a restricted open-shell Kohn-Sham (ROKS) approach is used, where the SIC is applied only to the open shell. Hence, m coincides with the density of the unpaired electron, which in the present work refers to the excess electron. There have been several proposals for the scaling parameters. We have carried out our own determination based on reproducing the MP2 electronic density for a thymine dimer anion, obtaining the optimal values $a = 0.6$ and $b = 0.4$ [48].

A third approach for correcting the self-interaction error is based on the optimized effective potential (OEP) method [49]. Here the effective Kohn-Sham potential is not state-dependent and the density is invariant under unitary transformations, thus overcoming the main limitations of the two methods above. This comes at the expense of having to solve the OEP equations. While this is an interesting route, we will not pursue it any further in what follows, and refer the reader to the relevant literature [50].

2.1.3 Beyond TDLDA: Electron Attachment as an Example

TDLDA is designed to optimize the average properties of dynamical evolution. It is not well suited for reaction channels which deviate significantly from the average path, and even less appropriate for describing rare reactions. Such situations require a description allowing for a mix of different mean-field configurations. This is done in multi-configuration TDHF [51] that, however, is extremely expensive. The propagation can be simplified when reducing once in a while the involved coherently correlated state into an incoherent ensemble of mean field states. This leads to mean-field propagation augmented by stochastic jumps between different mean fields [52], for a recent example, see [53]. The case of rare reaction channels becomes particularly simple in that respect. The jump probability is so low that it suffices to consider one jump from the mean-field trajectory to the final state. We consider this situation for the example of electron attachment [54]. The initial state $|\Phi(0)\rangle$ consists of a molecule in its ground state (of energy E_0) and an incoming electron wave packet, denoted by $|\Phi_{\text{in}}\rangle$ with a certain momentum (corresponding to an energy E_{in}) and impact parameter. This state is propagated by TDLDA. The mean-field path describes predominantly elastic electron scattering. Now we are interested in final states where the impinging electron merges with the molecule leading to a final state

$$|\Psi_n\rangle \approx |\Phi_{pp'h}\rangle \quad , \quad E_n \approx E_0 + \varepsilon_p + \varepsilon_{p'} - \varepsilon_h \quad , \tag{10}$$

out of the excitation spectrum of the new molecule (the ground state is inaccessible directly for reasons of energy conservation). We approximate it by a 2-particle-1-hole($2p1h$) state with respect to the ground state $|\Phi_0\rangle$ of the original target molecule where the excitation energy E_n can be expressed conveniently in terms of s.p. energies ε_i. Following time-dependent perturbation theory, the attachment probability becomes

$$\mathscr{P}_{\text{att}}(t) = \frac{2\pi}{\hbar} \int_0^t \mathrm{d}\tau \sum_{n=1}^{N_{2p1h}} \left| \langle \Psi_n | \hat{V}_{\text{coll}} | \Phi_{\text{in}}(\tau) \rangle \right|^2 \delta(E_n - E_{\text{in}} - E_0) \quad . \tag{11a}$$

Here, V_{coll} is the residual interaction for the collision channel, and N_{2p1h} the number of $2p1h$ transitions fulfilling the energy conservation imposed by the Dirac function.

An exact Dirac δ-function is not applicable in a finite system with a discrete excitation spectrum. We associate in practice a finite-width δ-function

$$\delta(E_n - E_{\rm in} - E_0) \longrightarrow \delta_{\Delta E_{\rm in}}(E_n - E_{\rm in} - E_0) \quad , \quad \Delta E_{\rm in} = \frac{3\hbar^2}{8m\,\Delta r^2} \tag{11b}$$

where the width $\Delta E_{\rm in}$ corresponds to the energy width of the impinging wave packet and thus to the energetic resolution of the setup.

2.1.4 QM/MM

At the level of detail allowing to address electron dynamics fully microscopically, most TDDFT computations were performed for isolated (gas phase) systems. As already pointed out the impact of the environment is a crucial issue and needs a specific treatment. This may be done, to some extent, at a fully quantum level (see examples in Sect. 3.2) but an alternative route consists in describing the environment via a simplified (classical) approach within a scheme usually denominated Quantum Mechanical/Molecular Mechanical (QM/MM). This is a typical example of a hierarchical strategy. The idea of QM/MM is to treat quantally the active zone of interest (typically the region subject to irradiation in our specific context) and couple this "active" quantum subsystem to an environment treated at classical level. This type of methods has been especially developed and used in biochemistry [55, 56], as well as in surface chemistry [57–59]. Standard QM/MM approaches mostly deal with geometry optimization or Born-Oppenheimer MD, but are not tuned to electron emission and transport. We thus developed a specific modelling allowing to treat in a QM/MM picture the irradiation (and ensuing ionization and electron transport) in a minimal framework.

In this model, the QM part describes a Na cluster while a polarizable substrate (rare gas, MgO) is treated in an MM fashion. The novelty with respect to a standard QM/MM relies on the fact the substrate atoms possess two explicit dynamical degrees of freedom, namely the atomic cores and the electronic dipoles, allowing a dynamical (explicitly time-dependent) polarization of the substrate. This provides the minimal requirement of account for polarization effects of the environment, in particular the ones due to travelling electrons following ionization. The latter model has been applied to metal clusters in contact with an insulating environment [60]. Applications to other quantum/classical combinations, in particular in relation to a biological environment, are possible but have not yet considered explicitly with this model.

2.1.5 Some Practical Considerations

The TDLDA calculations presented above have been carried out by means of an implementation of the equations on a grid in coordinate space. The ground state of the system is determined iteratively using a damped gradient method. The time propagation is performed using a T-V splitting. The quantitative analysis of ionization requires the use of absorbing boundary conditions which are presently implemented in terms of a mask function. Details on the numerics can be found in [28, 29].

Such TDLDA and TDLDA-MD calculations give access to several observables both at the electronic and ionic level. We will show here both kinds of observables, presenting: (i) a fragment analysis (Sect. 3.1.1) basically relying on total ionization and ion dynamics, (ii) optical response (Sect. 3.1.2, obtained as Fourier transform of electronic dipole), (iii) attachment probability (Sect. 3.1.3 and see Sect. 2.1.3 for theoretical detail), and (iv) an analysis of ionic motion in cluster deposition to illustrate the role of polarisation dynamics in QM/MM modelling (Sect. 3.1.4). Details on the computations of these observables can be found in [28, 54, 60].

2.2 *Quantum Modelling of End-Point Effects*

In order to study end-point effects such as LEE localization (attachment) and the subsequent chemical reactions like bond cleavage in a sufficiently realistic situation, the model systems must be subject to the following conditions: (1) they should be large enough to include DNA components and the environment, (2) it should be possible to accumulate sufficient statistics to compute thermodynamics quantities like free energy profiles, and (3) the capability of making and breaking chemical bonds should be retained.

Item (3) points to the use of electronic structure-based methods like DFT as opposed to empirical force fields. The latter are given explicitly in terms of the coordinates of the atoms and are thus much more efficient computationally, hence their use in QM/MM methodologies for the MM region. However, in general they do not allow for chemical modifications and for changes in the properties of the individual atoms according to their environment. For example, an oxygen atom in carbonyl group has to be described differently from an oxygen radical or a carboxyl group. The majority of the force fields, which we can group into the category of non-reactive, describe stretching coordinates via harmonic potentials that do not allow for dissociation, and also maintain the identity and properties of all the atoms. Clearly, these limitations can be relaxed and, indeed, they have been. For instance, it is possible to replace the harmonic bonds by Morse potentials that allow for dissociation. It is also possible to design a strategy that, by looking at the coordination and environment of an atom, assigns a type of bonding that evolves during the simulation. To a good extent, this is what reactive force fields like REAXFF [61] do, or a method recently proposed by Sushko et al. that was implemented in MBN explorer [62]. These methods are generally more expensive than non-reactive force fields, but still orders of magnitude cheaper than first-principles calculations. The onerous (in human time) parameterisation required for both types of force fields comes at a price of accuracy and transferability. Therefore, a thorough validation is required. Here, by choosing to describe interactions (potential and forces) at the first-principles level, via electronic structure calculations, we minimize accuracy and transferability issues and we bypass the validation step, compromising on the size of the systems that can be studied (a few thousand atoms vs millions) and the length of the simulations (tens of ps vs hundreds of ns). This fully quantum approach limits the class of systems and phenomena that can be studied to a maximum of a few base stacks of the solvated DNA double helix, e.g. an hexamer.

2.2.1 Electronic Structure Methods and Codes

In order to fulfil requirements (1) and (2), we need to use an efficient, still accurate approach to describe the electronic structure of large molecules and periodic systems, containing of the order of 1000 atoms. This excludes high-level quantum chemistry methodologies based on Hartree-Fock (HF) theory like MP2 or coupled clusters, and points towards DFT approaches. Semi-empirical quantum-mechanical methods like tight-binding are also a possibility [63, 64], but this reinstates the onerous parameterisation and validation steps. Within the DFT landscape, there are two choices to make. First is the theory level, i.e. the specific approximation to the exchange and correlation functional. This has been discussed briefly in Sects. 2.1.1 and 2.1.2 above, and will be expanded below. Second comes the mathematical representation of the electronic orbitals and density, i.e. the basis set. The TDDFT simulations, as mentioned in Sect. 2.1.5, were carried out using a grid representation with absorbing boundary conditions.

For the DFT simulations of end-point effects we have used the Quickstep module of the CP2K package [65]. The advantages of this choice are that it has implemented both DFT and HF (and hybrid) approaches including a good variety of functionals. It also permits to carry out efficient calculations for gas-phase and periodic systems, it offers both all-electron and pseudo-potential implementations of the nuclear-electron interaction, and the electronic orbitals are expanded in efficient Gaussian basis sets. For all-electron calculations, these are the standard Gaussian-Type Orbitals (GTO) used in quantum chemistry packages, and they are provided with the code distribution. For pseudo-potentials, the code uses the Goedeker-Teter-Hutter (GTH) version [66]. Pseudo-potentials and optimized associated Gaussian basis functions (MOLOPT) are also provided with the distribution [67]. The electrostatic potential is calculated by fast Fourier transforming (FFT) the density, which is expanded in an auxiliary plane-wave (PW) basis set. In all-electron calculations, the core orbitals require a large number of PW. To reduce the cost of the FFT, the density in the core region is augmented with Gaussians. This approach is called Gaussian and Augmented Plane Waves method (GAPW). For pseudo-potential calculations the augmentation is not needed, so the Gaussian and Plane Wave method (GPW) is used.

2.2.2 On the Choice of Functionals and Limitations Thereof

In this work we have used the Perdew-Burke-Ernzerhof (PBE) GGA to the exchange and correlation functional [68]. The GGA is necessary to describe hydrogen bonds that in the LDA are overbound and too short. GGA-PBE still lacks in a couple of important respects, which however are not particularly relevant in the calculations presented here. One of the limitations is the poor representation of non-bonded van der Waals (VDW) interactions. There are several methods introduced in past decades to overcome this problem. The most efficient one, which is implemented in CP2K, is the semi-empirical pair potential correction due to Grimme [69]. One of the effects of the VDW correction is to reduce the equilibrium volume of water and ice, thus increasing their density [70]. It also softens the microscopic structure of water by counteracting the highly-directional hydrogen bonds with the isotropic VDW interactions [71]. For most of the present investigations, namely solvated nucleobases and nucleotides, these effects are secondary and we have not considered VDW corrections. However, VDW interactions are crucial to account for the stacking of the bases. Therefore, they have to be used for any DNA fragment involving multiple bases. Here we have used Grimme's correction for the trinucleotide simulations.

The other limitation of the GGA-PBE is the incomplete cancellation of the electronic self-interaction, which was already mentioned in Sect. 2.1.2. Part of the calculations presented here concern nucleobases and nucleotides. For these systems, we have not observed signatures of an artificial electronic delocalization, and hence we have stayed within the GGA-PBE level of theory. For larger systems with several bases like polynucleotides or double-stranded DNA fragments we do observe delocalization amongst the bases, which is absent at the HF level. Therefore, in these cases we have used the SIC described in the last paragraph of Sect. 2.1.2 [47, 48]. A similar problem arises in fragmentation events (dissociation), where GGA will produce fractionally charged fragments, thus modifying the picture qualitatively.

2.2.3 Some Practical Considerations

In terms of basis sets, we conducted an initial validation [72] and decided to stay at a triple-zeta plus polarisation, namely TZVP-GTH for pseudo-potentials, and 6–311G* for all-electron calculations. We checked the GGA-PBE calculations against the DFT-HF hybrid PBE0, and found only small differences, mostly at the level of bond cleavage barrier heights, which were underestimated by about a 20 % in PBE [73]. This combination, GGA-PBE functional, GTH pseudo-potentials and TZVP-GTH basis sets, turned out to be sufficiently accurate for the present purposes, yet efficient enough to allow for long simulations (several tens of ps) on sufficiently large systems (around 1000 atoms). It is to be remarked that this is not the absolute computational limit. Presently we are running simulations for even larger systems, containing around 3000 atoms.

Free energy barriers were calculated using thermodynamic integration of the potential or mean force. This was done by identifying appropriate reaction coordinates ξ like the C–O distance in the phosphodiester bond, the C–N distance in the glycosidic bond, and the O–H distance in the hydrogen-bond for protonation reactions. We then ran first-principles molecular dynamics (FPMD) simulations with the reaction coordinate constrained to a sequence of values that interpolate between reactants and products. The value of the Lagrange multiplier λ_ξ associated with this coordinate is precisely the force required to maintain the constraint. When thermally averaged over the simulation, $\langle\lambda\rangle_\xi$ becomes the potential of mean force. The free energy is then computed using thermodynamic integration as $F_{A\to B} = -\int_A^B \langle\lambda\rangle_\xi d\xi$.

3 Illustrative Results

3.1 Electron Dynamics in Model Systems

3.1.1 Collision of CH_4 with a Proton

We first begin with a collision process between a methane molecule and a proton projectile [74], because ion-molecule collisions are considered to play an important role in many fields of science [75], such as chemistry, plasma physics, material science, astrophysics, and radiation therapy. In particular, when a projectile with several eV collision energy hits a target, some featured phenomena can occur, such as vibrational excitation, dissociation, charge transfer, and nuclear exchange. The complexity of collision processes in this energy range requires a highly elaborate, but still applicable numerically speaking, theoretical approach.

The $H^+ + CH_4$ collisions have been calculated in the framework of TDDFT-MD, presented in Sect. 2.1.1 [74]. To compare with existing experimental data [76], The proton is given an initial energy of 30 eV. We have simulated 9 different incident orientations in a $72 \times 72 \times 72$ simulation box with a grid spacing of 0.412 a_0. The collision dynamics is carried out as follows. The proton is placed at 25 a_0 distant from the methane (positioned at the origin of the numerical box and at rest) and it is given an initial velocity corresponding to the collision energy. The impact parameter b varies from 0 to 5 a_0 in a step of $\Delta b = 0.1$ a_0 and from 5 to 10 a_0 in a step of $\Delta b = 0.5$ a_0. All in all, 549 collision geometries have been computed to allow a statistical analysis of all possible reaction channels. The total simulation time is 40 fs with $\Delta t = 0.6$ as. The small time step ensures the stability of time propagation in the calculations. All the dynamical collision processes are considered in the laboratory system.

To analyze the collision process, we have specifically extracted ionic trajectory and the energy loss of proton for each event, as presented in [77]. In all collision events considered here, 8 different types of reactions are identified: non-charged transfer, charge transfer, proton exchange, collision-induced dissociation, proton exchange with dissociation, protonation, hydrogen molecular formation, and hydrogen molecular formation with dissociation.

We obtained in our calculations the same fragments, that is, CH_4^+, CH_3^+, and CH_2^+, as those found in experiments [76] and other calculations [78]. The estimated fragment intensities, shown in Fig. 3 are 79.6, 20.1, and 0.2 % for CH_4^+, CH_3^+, and CH_2^+ respectively, which is in agreement with experimental values (CH_4^+:CH_3^+:CH_2^+ = 74 %:22 %:4 %).

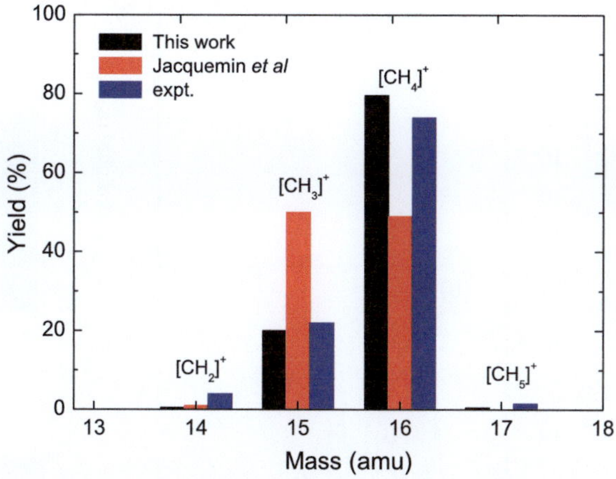

Fig. 3 Fragment yield in the collision process of H^++CH_4 at an impact energy of 30 eV for H^+: TDLDA-MD (black), experimental data (blue) [76] and other theoretical results (*red*) [78]. Adapted from [74]. See text for details on the TDLDA-MD calculations

Results from previous calculations [78] based on the extended Lagrangian molecular dynamics (single electronic potential energy surface), gave CH_4^+:CH_3^+:CH_2^+ = 49 %: 50 %:1 %, which drastically deviated from the experimental ones. This indicates that the intensity of CH_4^+ was underestimated by employing the adiabatic potential energy surface. More precisely, that of CH_3^+ was overestimated. Moreover, there is a small intensity peak at 17 amu in the experimental measurements, which is also found by our simulations but not in the other theoretical calculations. However, the origin of this small peak might be different. It was considered as an isotope effect ($^{13}CH_4/^{12}CH_4$) in the experiments, while we rather conclude that the CH_5^+ fragment may also contribute to this small intensity peak.

3.1.2 Self-interaction Error in Metal-Covalent Complexes

The second example explores the importance (or not) of SIC in mixed complexes, namely a Na atom in contact with a H_2O molecule [79]. The interest for such a system is the mixture of a metal atom with a covalent molecule. This raises specific issues at the side of the SIC treatment. We first compare the optical response of NaH_2O with that of the Na atom and of the H_2O molecule. Results calculated in LDA, ADSIC, see Eq. (8), and 2set-SIC, see Eq. (7), are compared in logarithmic scale in the left column of Fig. 4.

For each system, the spectrum is generally similar for the various SIC, especially below 5 eV with a prominent peak (a kind of plasmon peak) from the Na atom when present. This dominant response of the metal atom, even in contact with H_2O, demonstrates that this atom can be viewed as a chromophore: if one shines a laser with a frequency close to the plasmon peak, one can excite resonantly the complex. Furthermore, in the presence of a water molecule, a series of peaks, suppressed by at least two orders of magnitude with respect to the plasmon peak, are observed above 7 eV. Naturally, the Na atom does not contribute in this range. These peaks are covered

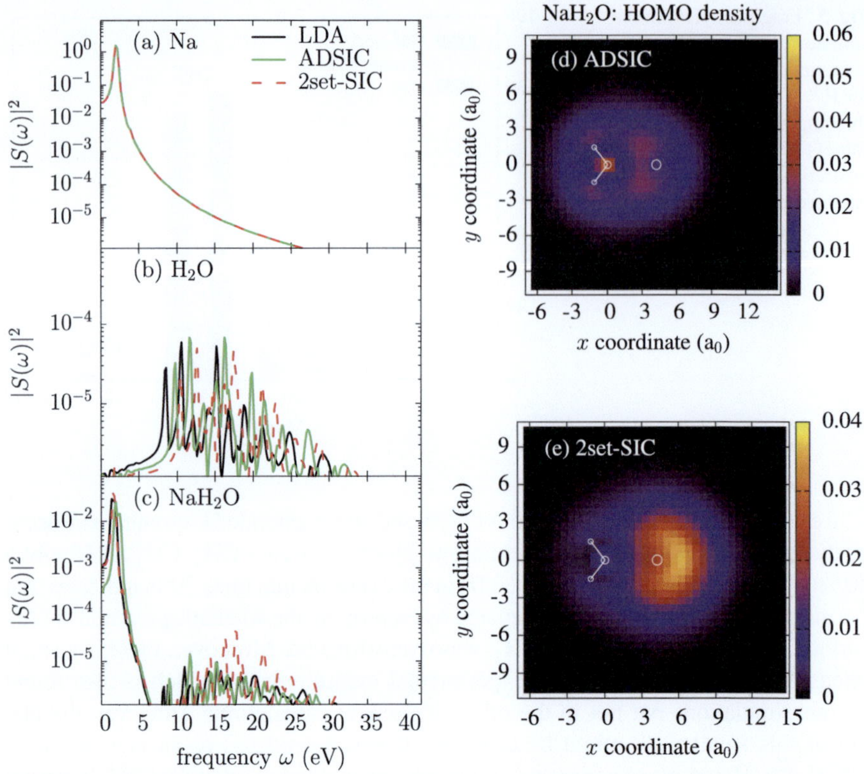

Fig. 4 *Left column*: optical response calculated in LDA (*black curves*), ADSIC (*light green curves*) and 2set-SIC (*red dashes*), for the Na atom, the H_2O molecule and the NaH_2O complex. *Right column*: electronic density of the HOMO in NaH_2O calculated in ADSIC (*top*) and 2set-SIC (*bottom*). Atoms are depicted by *open circles* with the water molecule on the left. Adapted from [79]

by signals "from" the H_2O molecule and the complex gives similar gross results as those for the water molecule alone, but there are differences in the remaining quantum fluctuations. Note that these spectra lie in the continuum where there are many unbound states (the ionization threshold is at 2.6, 5.6 and 4.3 eV in LDA, ADSIC and 2set-SIC respectively). Transitions from deeper lying shells to these states in the continuum are most probably the sources for the quantum fluctuations.

While the effect of SIC is not very large in the calculation of the optical response, other electronic properties can be strongly influenced by the level of SIC. To discuss this point, we show in the right column of Fig. 4 the electronic density of the HOMO in NaH_2O, integrated over the z coordinate and in the xy plane. The HOMO in the complex resembles the s state of a single Na atom, that is an orbital which is localized on the Na ion. However, ADSIC (top right panel) rather produces a localization of the HOMO density around the O atom. This is far from the expected result and demonstrates the failure of ADSIC for this system. A 2set-SIC calculation

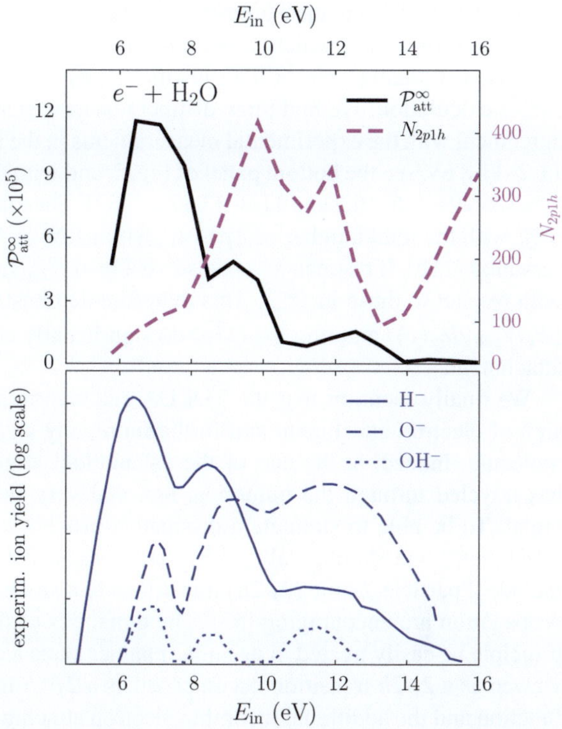

Fig. 5 *Top*: Number of $2p1h$ excitations N_{2p1h} fulfilling energy criterion (*dashed line* and vertical scale on the right) and asymptotic value of the probability of electron attachment $\mathscr{P}_{att}^{\infty}$, as functions of the energy of incoming electron E_{in} colliding on H_2O (adapted from [54]). *Bottom*: Experimental measurements in logarithmic scale of dissociative electron attachment of H_2O yielding H^- (*solid curve*) O^- (*dashes*), and OH^- (*dots*) from [80]

(bottom right panel) delivers a more reliable HOMO density which is mainly located around the Na atom and even repelled by the water molecule. In this complex, the bond is neither metallic nor covalent, as confirmed by the calculation of the static polarizability (not shown here). This excludes the use of ADSIC which tends to over-delocalize the electron cloud and therefore calls for a specific treatment at the side of the electronic description.

3.1.3 Dissipative Effects Explored in the Electron Attachment

We now discuss the case of a calculation exploring beyond mean-field effects. We briefly presented the theoretical framework in Sect. 2.1.3. We have explored the case of an electron of incident energy E_{in} between 4 and 16 eV, colliding a water molecule. The incoming electron is modeled by a Gaussian wavepacket and we propagate in TDLDA-ADSIC the wave functions of the water target and the colliding electron. The probability of electron attachment, see Eq. (11a), is accumulated along the time propagation. Details on the numerics can be found elsewhere [54]. In the top panel of Fig. 5 is plotted as a function of E_{in} the value of the probability of attachment at the end of the calculation, $\mathscr{P}_{att}^{\infty}$.

As expected from measurements of the cross section of dissociative electron attachment on water, which is less than 10^{-17} cm^2 [81], the probability of attachment is very small ($<10^{-4}$). This justifies a posteriori our perturbative approach for such a calculation. We find three distinct resonances at 6.8, 9.2 and 12.6 eV, well in agreement with the experimental measurements in the ranges of 6.4–6.9, 8.4–8.9 and 11.2–11.8 eV, see the bottom panel of Fig. 5, and with detailed R-matrix calculations at 6.5–6.994, 8.6–10.2 and 11.8–12.97 eV [82]. The top panel of Fig. 5 also compares $\mathscr{P}_{\text{att}}^{\infty}$ with the total number of $2p1h$ transitions N_{2p1h} which satisfy the energy conservation (11b). If resonances are also visible in N_{2p1h}, they are strongly blue-shifted with respect to those in $\mathscr{P}_{\text{att}}^{\infty}$. This therefore demonstrates that the matrix elements $\langle \Psi_n | \hat{V}_{\text{coll}} | \Phi_{\text{in}}(\tau) \rangle$ entering Eq. (11a) do significantly contribute to the probability of attachment.

We finally mention that the TDLDA calculation as such does not provide any sign of electron attachment and furthermore, any sign of dissociation of the water molecule. Indeed, at the end of the calculation, almost all the incoming electron has traveled through the numerical box and very little has attached to the water target. To be able to simulate a dissipative process such as a dissociative electron attachment, one should truly go beyond a mean-field approach. First tests which include 2 particle-2 hole ($2p2h$) transitions in a stochastic way on top of a TDLDA propagation are encouraging [53]. The dynamics of the electron attachment can in principle be easily treated in a similar manner since a core-excited resonance can be viewed as a $2p2h$ transition decomposed as a $2p1h$ transition from the water wave function and the additional incoming electron attaching an excited state of the water target. Work in this direction is under progress.

3.1.4 Dynamical Hierarchical Approach

We end up with an example of an extension of Quantum Mechanical/Molecular Mechanical (QM/MM) approach we developed some years ago (see Sect. 2.1.4) [60]. We here discuss the case of a dynamical deposition of a Na_6 cluster on an Ar(001) surface modeled by 6 layers of 8×8 squares which are copied periodically in both horizontal directions to simulate an infinite surface. In addition, the lowest two layers are frozen at bulk position to stabilize the underlying (supposed infinite) crystal structure. The Na_6 cluster is composed of a pentagon ring as base and an extra ion on the symmetry axis of the pentagon. Initially, the pentagon is positioned parallel to the Ar surface with the top Na ion above a hollow site of the surface and the center of mass of Na_6 15 a_0 above the uppermost Ar layer. Na_6 is initially given a kinetic energy $E_{\text{kin}}^0 = 0.82$ eV and a velocity normal (=z axis) to the surface. We compare in Fig. 6 the deposition process by looking at the time evolutions of the z coordinates of the Na ions and the Ar cores, using three different treatments of the Ar substrate: mobile Ar cores but no Ar dipoles (i.e. Ar dipoles frozen at static, initial value, see upper left panel), fixed Ar cores but dynamical dipoles (see upper right panel), and the full model (see lower panel).

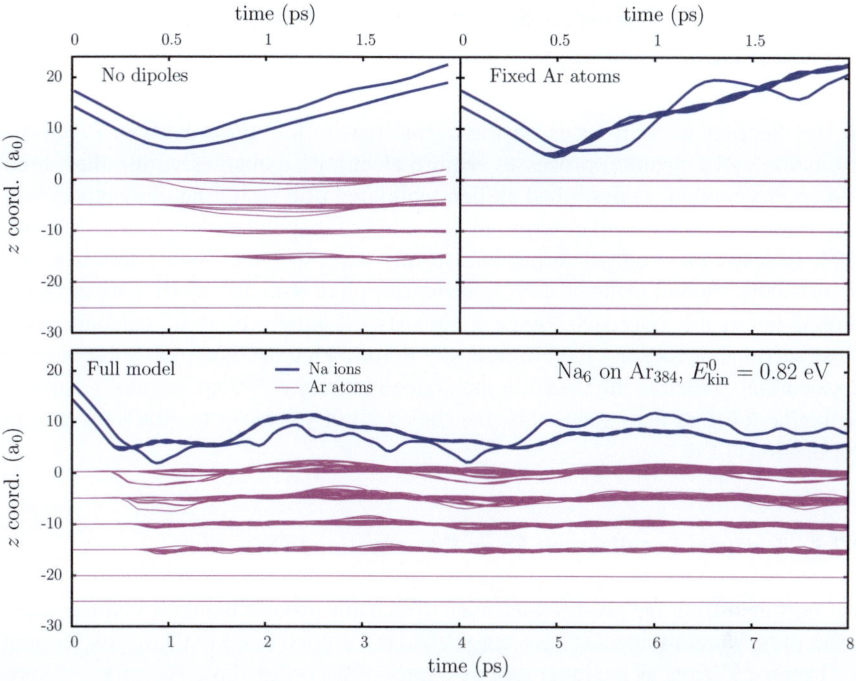

Fig. 6 Na_6 with an initial kinetic energy of $E^0_{\rm kin} = 0.82\,\rm eV$ and an initial velocity along the z direction (that is normal to the Ar surface), deposited on Ar(001). Time evolution of z coordinates of Na ions (*blue thick lines*) and Ar atoms (*pink thin lines*) for 3 levels of matrix treatment: fixed Ar cores and mobile dipoles (*top right*), moving Ar cores but no dynamical dipoles (*top left*), and fully active matrix (*bottom*). Adapted from [83]

Before 0.5 ps, that is before the impact time, the three cases exhibit a similar dynamics of the impinging Na_6. The evolutions after the collision are however qualitatively different. When the Ar cores are fixed (top right panel), the surface being rather rigid, the cluster bounces with strong internal excitations where the top ion oscillates forth and back through the pentagon. When one switches on the Ar cores dynamics while keeping the Ar dipoles at their static position (top left panel), a reflection of Na_6 is still observed. However, since dissipation through the substrate layers is now possible, much less excitation of internal motion of the cluster is produced. Finally, when all model components are fully active (bottom panel), more energy absorption at the side of the Ar substrate prevents Na_6 to be reflected. While the top ion still keeps going through the pentagon, the cluster finally sticks to the surface. This comparison highlights the importance of a full dynamical treatment of the Ar surface, especially at the side of the dynamics of their polarizabilities through dynamical dipoles.

3.2 *Excess Electrons in the Condensed Phase: Localization and Bond Cleavage*

In this Section we will focus on the events that follow pre-solvation. Electronic excitation and ionization processes require electronic dynamics within the Ehrenfest approximation, as described earlier, while inelastic scattering of electrons with water needs a higher level description that allows for spontaneous phonon emission [37]. Inelastic electron scattering is also important to DEA. While this has been traditionally studied using scattering techniques like R-matrix [84], there are some limitations in this approach. Since it consists essentially of static calculations of cross sections, it does not lend itself easily to following the dynamics subsequent to dissociation, which is important in the condensed phase. We are currently working on methodologies that enable this, but they are not yet ready to attack the present applications [38].

3.2.1 Electron Localization: From Pre-solvation to Solvation

We start then from the pre-solvated state. Since this process happens in times of the order of fs, water molecules have not yet had the time to reorient themselves around the excess electron, as the latter requires times of the order of ps. Therefore, the pre-solvated electron is found in a fairly delocalized state, but far from being the plane wave that characterises a particle propagating with high energy. This kind of state is partly localized on the unoccupied water orbitals, which become accessible at low energies. It appears clearly in our first-principles simulations of an excess electron in solvated nucleobases [72], but also in other solvated species and, more importantly, also in pure water [85]. In the left panel of Fig. 7 we show the spin density of such a pre-solvated state for a solvated thymine molecule. The spin density is a well-defined, measurable quantity that is accessible to DFT calculations, and tells about the spatial distribution of the unpaired (excess) electron.

Assuming that the excess electron is in a pre-solvated state, further dynamical evolution on the adiabatic surface, i.e. with the electronic density in the ground state corresponding to the instantaneous nuclear configuration, leads to fast geometric rearrangements that localize the electron. Notice the difference between this *adiabatic* dynamics with the *non-adiabatic* dynamics embodied in the TDDFT simulations in Sec. 3.1. In times of the order of 50 fs, the nucleobase evolves towards a geometry very reminiscent of the gas-phase anion, with the electron sitting in the formerly lowest unoccupied molecular orbital (LUMO), as shown in the right panel of Fig. 7. The same behaviour is observed for larger fragments such as nucleosides and nucleotides, as shown in Fig. 8 (left panel), with the excess electron always localizing on the base, which is the region with highest electron affinity. This is one form of the *solvated* electron, although this terminology is more commonly used to describe the electron localized in cavities in water.

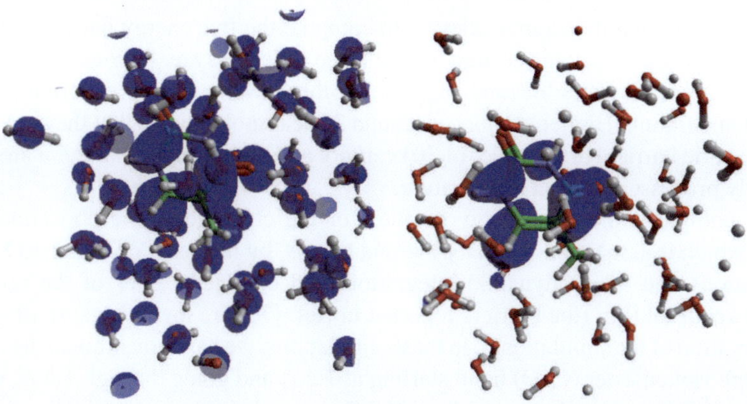

Fig. 7 Spin density characterising an excess electron in water containing a thymine molecule. *Left panel*: After vertical attachment of the electron to a neutral system at room temperature. This arrangement mimics the pre-solvated state in which the electron is moving at low velocity so that its wave function spreads over the water and thymine unoccupied (anti-bonding) orbitals. *Right panel*: After 50 fs of dynamics. The excess electron has localized in the LUMO of the thymine

Fig. 8 Spin density characterising an excess electron in water containing a deoxy-thymidine monophosphate (dTMPH) nucleotide. Left panel: After 50 fs of dynamics following vertical attachment of the electron to a neutral system at room temperature. The excess electron is localized in the LUMO of the nucleotide, which is located on the nucleobase end (thymine). Right panel: Upon stretching the phosphodiester C–O bond between sugar and phosphate groups. The electron moves towards the sugar-phosphate end of the nucleotide, after going through a transition state characterised by the crossing of electronic energy levels. At the transition state, the spin density is completely delocalized across the nucleotide [73]

3.2.2 Bond Cleavage

Further dynamical evolution of the solvated nucleotide anions does not lead to the spontaneous cleavage of the phosphodiester or the glycosidic bond [73]. This means that, under these circumstances, the bond cleavage process must be thermally activated, and then it becomes relevant to compute the free energy barriers for such processes, as these will tell us how viable these processes are. Barriers of the order of 10 kcal/mol are quite low and suggest that these events occur very frequently at room temperature. Higher barriers of around 20 kcal/mol indicate that they are infrequent, while barriers of the order of 30 kcal/mol and above imply that these are quite unlikely processes at room temperature.

We computed the free energy profiles for the cleavage of the phosphodiester bond, which is associated with DNA strand breaks, by means of constrained FPMD simulations and thermodynamic integration, and obtained values of the order of 15–20 kcal/mol [86] (the barriers reported in Ref. [73] are lower because of a poor equilibration of the simulations). In these simulations, we have constrained the length of the phosphodiester (C–O) bond starting at 1.2 Å and going through 1.4 Å, which is the equilibrium length, up to 2.2 Å. The free energy increases until the transition state (TS) at 1.8 Å and then it starts decreasing, as shown in the left panel of Fig. 9. It is interesting to observe the behaviour of the spin density upon stretching. It starts localized in the base, but when approaching the TS, the spin density delocalizes over the whole nucleotide, to eventually re-localize on the sugar-phosphate end, as shown in the right panel of Fig. 8. This has been rationalised in terms of the crossing of occupied and empty energy levels upon stretching of the C–O bond.

3.2.3 Protonation of Nucleotides

In our earlier work [73] we noticed that all the nucleotides Exhibited similar barriers except for the adenine-based dAMPH. Upon further inspection, we observed that in that case the base had spontaneously protonated. A proton had been transferred from a neighbouring, hydrogen-bonded water molecule. This did not happen in the other nucleotides for a variety of reasons. For dCMPH and dTMPH, the equilibration time was not sufficiently long to allow for the hydrogen bonds to form. However, once they were formed, protonation was found to be favourable and the barriers for protonation turned out to be very low (less than 1 kcal/mol). For dGMPH, instead, we found that protonation was not favourable at all (see Fig. 2 in Ref. [86]). We therefore decided to compute the free energy barriers by enforcing the nucleotide to be either non-protonated or protonated. The results are shown in Fig. 9. When the nucleotides are forced to be non-protonated at the base (left panel), all free energy barriers are similar (about 15 kcal/mol) except for dAMPH, for which it is higher. It is important to notice that, since protonation of adenine is spontaneous and the hydrogen-bonded situation is not a local minimum, this profile is somewhat artificial and has to be taken with care. This enforcing of the de-protonation is probably the reason why the barrier for dAMPH turns out to be higher.

The right panel of Fig. 9 shows similar profiles but now enforcing protonation of the base. Protonation is favourable for all nucleotides except for dGMPH. As a general trend we see that the barriers increase upon protonation. The exception is again dAMPH, for the reasons discussed above; the non-protonated state is not stable and the barrier is artificially raised, while the protonated state, is in line with

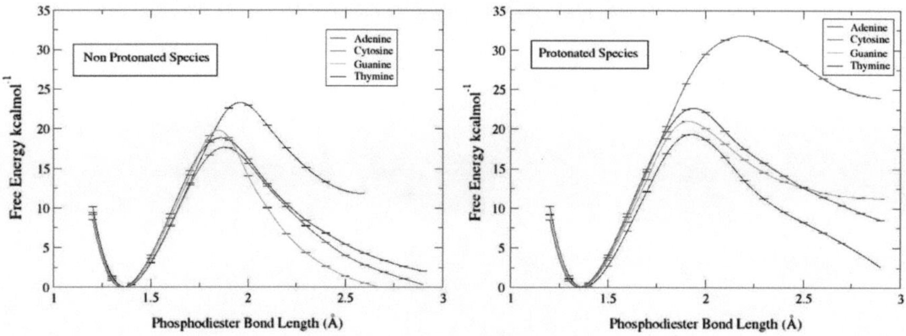

Fig. 9 Free energy profiles for the cleavage of the phosphodiester bond between sugar and phosphate in the four solvated nucleotides dAMPH, dCMPH, dGMPH, and dTMPH. *Left panel*: in the absence of protonation of the base. *Right panel*: after protonation at the most favourable sites on the base, i.e. O for thymine and N for the other bases

dTMPH and dGMPH, which exhibit barriers of the order of 20 kcal/mol. A second protonation of the adenine is possible and, interestingly, it restores the free energy profile to a situation similar to the non-protonated case (see Fig. 3 in Ref. [86]). The case of cytosine (dCMPH) falls outside the norm in the sense that the barrier is more than doubled, peaking at 30 kcal/mol. This is intriguing and, for that reason, we have looked at signatures in the spin density that could explain this behaviour. This is shown in Fig. 10. In the upper three panels we see the spin density for the ground state (GS, left), transition state (TS, middle) and product state (PS, right) in the non-protonated case. In the lower panels, we show the spin density at the same distances, but for the protonated dCMPH. In the right lower panel, we see the spin density delocalizes over the whole nucleotide at the distance of 2.2 Å. This delocalization pattern is characteristic of the TS, and not of the PS. In the other protonated nucleotides, the TS occurs at a shorter distance of 1.9 Å, so that breaking the bond requires less stretching than for dCMPH. Why this is different in cytosine is not completely clear yet, but it must be related to a modification of the energy levels upon protonation, such that the crossing is displaced towards at a longer distance, thus requiring more energy to keep stretching the bond for longer.

We have conducted similar studies also for the cleavage of the glycosidic C–N bond between base and sugar that leads to base excision. The resulting free energy profiles can be found in Fig. 4 of Ref. [86]. The main results are that base protonation increases the barriers for all nucleotides except for dGMPH, but one has to keep in mind that protonated dGMPH is unstable.

To summarise these results, we can say that the solvent has a significant effect that cannot be easily incorporated into continuum solvation models. Surprisingly we have found that there are marked differences between the various bases. The purine nucleotides (dAMPH and dGMPH) have the lowest barriers to strand breaking reactions and will break at the phosphodiester bond. Pyrimidine nucleotides (dTMPH and dCMPH), meanwhile, have higher barriers to reaction but the barriers for breaking the phosphodiester and the glycosidic bond are similar. The question now is

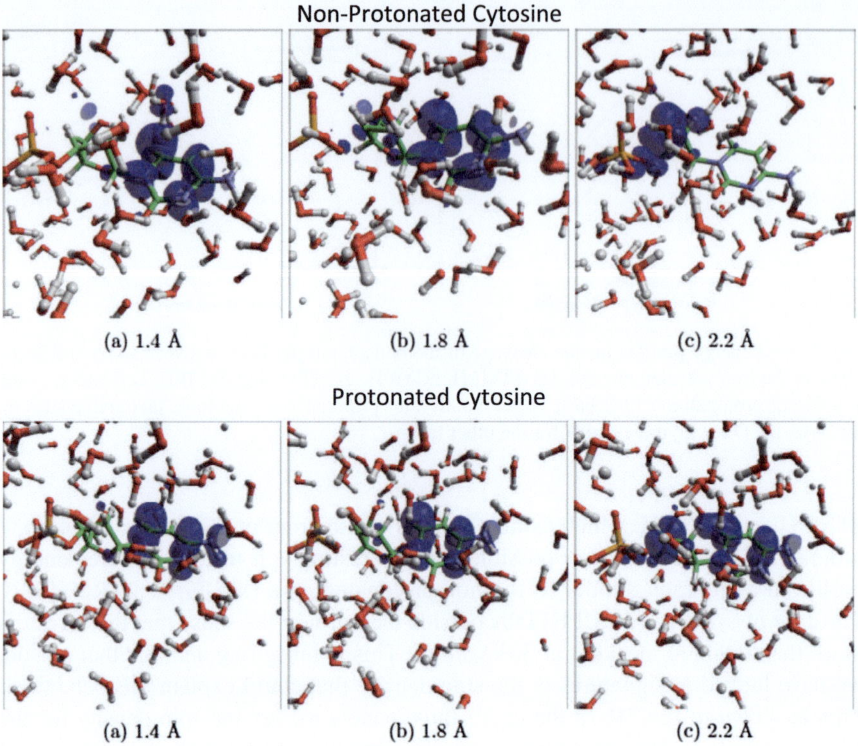

Fig. 10 Spin density characterising an excess electron in solvated dCMPH for a sequence of C–O distances when stretching the phosphodiester bond, in the case of non-protonated (*top panels*) and protonated (*bottom panels*) cytosine

how these differences will manifest themselves in larger DNA fragments such as polynucleotides or double-stranded DNA pieces, where the bases are less solvent accessible.

3.2.4 Trinucleotides

As a first step in this direction we conducted a series of simulations of solvated trinucleotides, where the three bases are stacked at a distance of approximately 3.4 Å. Preliminary work using the GGA-PBE functional showed some alarming features. The spin density of the excess electron was distributed amongst the three bases, and it fluctuated between them, driven by the thermal motion of the medium. This suggested the possibility that the poor cancellation of the self-interaction in PBE could lead to a spurious delocalization of the spin density. We therefore set out to test various functionals for the model case of two stacked thymines in the presence of an excess

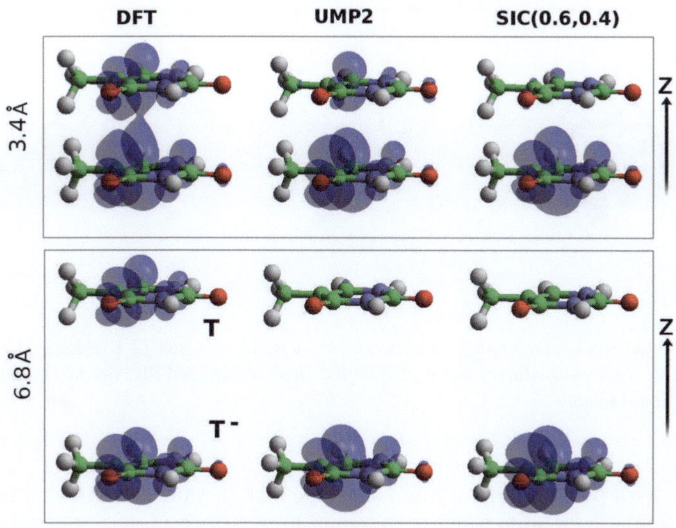

Fig. 11 Spin density characterising an excess electron in the gas-phase thymine dimer anion, Thy_2^-, for various theory levels: DFT-PBE (*left*), unrestricted MP2 (*middle*), and an optimized self-interaction corrected functional SIC (*right*)

electron, i.e. Thy_2^-. This was done for two different stacking distances, 3.4 and 6.8 Å. We took as a reference an MP2 quantum chemical calculation.

In the left and middle panels of Fig. 11, we show the PBE and unrestricted MP2 spin densities, respectively. It can be seen that the PBE calculation suffers from this very common delocalization error, not only at the shorter distance, but also at longer distances.

By splitting the density in this way, the Coulomb energy due to self-interaction of the unpaired electron is halved. The MP2 calculation, which treats exchange exactly, shows that this delocalization is spurious, and the excess electron should be localized on one side or the other. In these calculations, we have used different geometries for the two thymines, neutral (top) and anion (bottom). This is the reason why the electron localizes in the bottom one. It is then quite remarkable that PBE still prefers to split the charge shifting some of it to a molecule that is in an unfavourable geometry. Clearly, the self-interaction error is large enough to overcome this. The right panel shows the spin density for the SIC functional described in Sect. 2.1.2. Details of the functional, optimization and validation tests will be presented elsewhere.

Figure 12 shows the spin density for a trinucleotide with all thymine bases (TTT). The left panel shows the suspicious PBE result that motivated the study of SIC functionals (middle panel). The latter eliminates the delocalization error and localizes the excess electron on the central thymine. The hybrid DFT-HF functional PBE0 produces a similarly localized spin density, but at a much higher computational cost in the condensed phase.

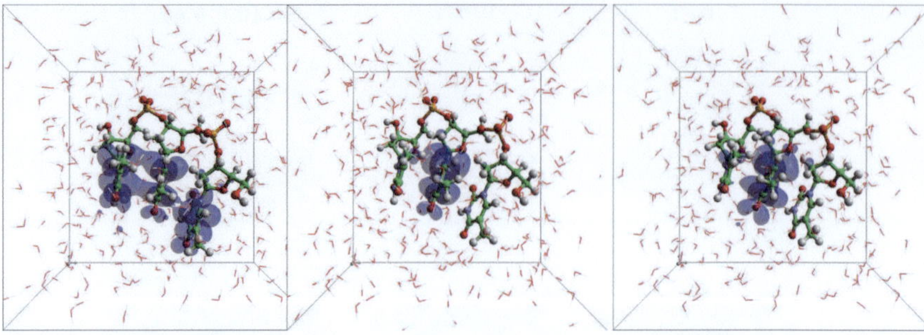

Fig. 12 Spin density characterising an excess electron in the solvated TTT trinucleotide (all bases are thymines), for various theory levels: DFT-PBE (*left*), optimized SIC (*middle*), and the PBE0 hybrid functional (*right*)

The study of phosphodiester bond cleavage here is more complicated than in the case of a single nucleotide because there are four different bonds that can break. A schematic view is presented in the inset to Fig. 13. Experimental data [87] suggests that the bonds that are more prone to break are the $C_{3'}$-$O_{3'}$, as opposed to the $C_{5'}$-$O_{5'}$. These correspond to the bonds indicated with the numbers 1 and 3 in the inset of Fig. 13. For these two bonds we have run FPMD simulations by constraining the corresponding bond lengths as reaction coordinates, both for the PBE and the SIC functionals. The free energy curves obtained by thermodynamic integration are shown in Fig. 13.

A first observation is that the barriers are of the order of 12–17 kcal/mol, consistently with the barriers obtained for single nucleotides. Bond 1 turns out to be weaker than bond 3 and hence should be more prone to break. This is consistent with experimental data, where a ratio of 2.5 to 1 has been measured [87]. A second observation is that the barrier heights appear insensitive to the functional, whether PBE or

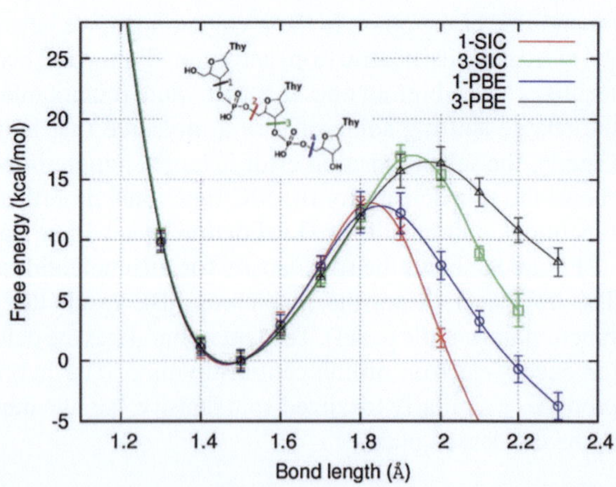

Fig. 13 Free energy profiles for the cleavage of phosphodiester bonds 1 and 3, defined in the inset. The system is a solvated TTT trinucleotide, where all bases are thymines. For each bond we report the profile using the PBE and the SIC functionals

SIC. The shape of the free energy curve is somewhat different, with the SIC TS (the maximum) slightly displaced towards shorter distances, and exhibits a faster decay at larger distances, but the height is remarkably similar. A possible explanation for this is that upon stretching the phosphodiester bonds, the excess electron quits the base and localizes on the specific sugar-phosphate region around the stretched bond. Since this spin density is quite localized, precisely because it is the only stretched bond, the SIC does not appear to have a major influence. The shape of the SIC free energy curve is somewhat worrying however, because it decays too rapidly upon breaking the bond. A possible explanation is that this functional has not been designed with the energetics in mind. It was done to reproduce the electronic spin density, but there is no guarantee that energetically it will behave correctly in every situation. A possible way out of this situation consists of recalculating the energy at the PBE level, but using the SIC density. This approach has proved successful for other systems and theory levels other than SIC [88], and may represent a simple and inexpensive alternative to higher level calculations such as DFT-HF hybrids.

4 Conclusions and Discussion

Connecting space and time scales in processes where microscopic ultrafast phenomena lead to chemical modifications of the materials, and thus to end-point macroscopic effects, is never a simple task. In our case, we have to think of irradiation producing secondary electrons and holes as well as radicals. These reactive species propagate through the biological medium, i.e. water containing a variety of biomolecular entities like membranes and proteins, to eventually interact with DNA and produce damage in the form of strand breaks, base excision, or cross-linking. The accumulation of such damage triggers signalling pathways that terminate the cell cycle when this becomes not viable or dangerous. Perhaps the latter is the most difficult part to model, as it involves a complex network of biochemical events triggered by enzymatic recognition of the damage. Therefore, at the present stage, it seems more reasonable to focus on the physicochemical processes, still multi-scale, that begin at the irradiation stage and conclude with DNA damage in its natural environment. We have presented here two types of such processes, considering the case of simple model systems at the most detailed level of description, and complementing these with calculations at a larger scale, accounting for environmental effects, but still within a fully quantum description. In both cases, we have been especially concerned with electron dynamics, which is a key issue in the field.

In the case of model systems, we have proposed a development of standard DFT-based approaches in order to address the key question of electron attachment. We have seen that such an extension provides realistic values of attachment probabilities. The strategy developed for this extension of standard TDLDA is of perturbative nature, which is well justified for electron attachment (as compared to elastic scattering) but it does not provide a state in which the electron is actually attached to the target molecule. Such a step, which would open the door to a direct analysis of the impact of the attached electron onto the dynamics of the target molecule (and thus directly lead to a microscopic analysis of DEA) has not been taken yet and would require some further developments, both methodological and computational.

Another important aspect concerns the actual description of ionization and the importance of the self-interaction error. Simple strategies can be used in many cases to solve the SIC problem at low cost. Still, there remain cases which require an elaborate SIC correction strategy. We have illustrated this problem in the very demanding case of a complex associating a metal atom to an organic molecule. In such systems, the "conflict" between localized electron orbitals associated to the organic compound and the delocalized orbitals characteristics of metals make simple SIC approaches inefficient. One then has to use a most elaborate approach to recover realistic ionization properties. This elaborate approach nevertheless does not yet solve all pending issues in SIC. One should in particular mention here the open question concerning scaling properties of the SIC corrections when considering larger and larger systems. This is certainly an issue to be explored further in the future, e.g. by trying the SIC approach used here to study solvated trinucleotides, or the OEP approach.

In the case of extended systems we have seen, using first-principles molecular dynamics simulations of solvated DNA fragments, that the role of the environment in relation to DNA damage by low-energy electrons (LEE) can be quite dramatic in many ways. Computation of free energy barriers for all solvated nucleotides produced values of the order of 20 kcal/mol. Interestingly, protonation from the aqueous medium plays an important role here. It is spontaneous for adenosine monophosphate (dAMP), activated with a very low barrier for thymidine and cytidine (dTMP and dCMP), and it does not occur for guanosine monophosphate (DGMP). The most dramatic effect is on dCMP, where the free energy barrier for phosphodiester bond cleavage, which is reminiscent of a strand break in DNA, increases twofold, to 30 kcal/mol. Therefore, strand breaks at cytidine sites seem less favourable than for the other nucleotides, whose barriers remain at a lower value of 20 kcal/mol. In summary, protonation appears to protect DNA from strand breaks.

We have then studied larger fragments, starting from solvated trinucleotides, but here the choice of functional becomes quite critical. Using standard DFT-GGA we see an artificial delocalization of the excess electron over several nucleobases, e.g. over the three thymines in TTT. Preliminary results show that this is an artefact created by the poor cancellation of the electronic self-interaction in GGA. So, we see that, as in the case of ionization, the self-interaction error crops up again. We have endeavoured to improve the functional by means of an approximate SIC that, however, is different from the one used in ionization studies. The use of the SIC functional changed completely the picture with respect to PBE by localizing the excess electron in one of the three thymines, generally the central one, although the barrier heights remained unaltered. Some issues remain open, however. In particular, the shape of the SIC free energy curve looks unrealistic at long distances, and this could be a consequence of focusing the parameterization only on the spin density while ignoring the energy.

Another important element in a realistic environment is the presence of amino-acids from the histones around which DNA is wrapped. We have looked into this matter by considering a thymine solvated in glycine instead of water [89]. There we saw that glycine (and potentially other amino-acids) also protect single-stranded DNA by protonating the base. However, a second mechanism also plays an important role. The amino-acids have electron affinities similar to nucleic acids so, effectively, they compete for the excess electron, sometimes scavenging it and avoiding bond cleavage. Ultimately, one would like to simulate solvated double-stranded DNA in the presence of amino-acids and ions, but this is in the future. There is still a lot to learn from simple systems by building gradually from those discussed here.

Acknowledgements LBB was supported by the COST action MP1002 (Nano-IBCT) through a STSM to QUB. MMcA and MS were funded by the Department of Employment and Learning of Northern Ireland. BG was partially supported by the National Natural Science Foundation of China (Grant No.11105075), and the Jiangsu Government Scholarship for Overseas Studies (Grant No. JS2012-105). CZG has received a financial support from China Scholarship Council (CSC) (No. [2013]3009), and LL from the CNRS and the Midi-Pyrénées region (doctoral allocation number 13050239). This work was also supported by the Institut Universitaire de France. Access to the HPC resources of IDRIS was granted under the allocation 2014-095115 made by GENCI (Grand Equipment National de Calcul Intensif), and of CalMiP (Calcul en Midi-Pyrénées) under the allocation P1238. QUB researchers are grateful for computational support from the UK national high performance computing service, ARCHER, for which access was obtained via the UKCP consortium and funded by EPSRC grant EP/K013564/1.

References

1. Fennel T, Meiwes-Broer KH, Tiggesbäumker J, Dinh PM, Reinhard PG, Suraud E (2010) Rev Mod Phys 82:1793
2. Dinh PM, Reinhard PG, Suraud E, Wopperer P (2015) Eur Phys J D 69:48
3. Campbell EEB, Hansen K, Hoffmann K, Korn G, Tchaplyguine M, Wittmann M, Hertel IV (2000) Phys Rev Lett 84:2128
4. Bartels C, Hok C, Huwer J, Kuhnen R, Schwöbel J, von Issendorff B (2009) Science 323:132333
5. Tribedi LC, Agnihotri AN, Galassi ME, Rivarola RD, Champion C (2012) Eur Phys J D 66:303
6. Agnihotri AN, Nandi S, Kasthurirangan S, Kumar A, Galassi ME, Rivarola RD, Champion C, Tribedi LC (2013) Phys Rev A 87:032716
7. Boudaïffa B, Cloutier P, Hunting D, Huels MA, Sanche L (2000) Science 287:1658
8. Sanche L (2009) Low-energy electron interaction with DNA: bond dissociation and formation of transient anions, radicals and radical anions, radicals in nucleic acids. Wiley, Hoboken, NJ
9. Arumainayagam CR, Lee H, Nelson RB, Haines DR, Gunawardane R (2010) Surf Sci Rep 65:1
10. Baccarelli I, Bald I, Gianturco FA, Illenberger E, Kopyra J (2011) Phys Rep 508:1
11. Alizadeh E, Sanche L (2012) Chem Rev 112(11):5578
12. Alizadeh E, Orlando T, Sanche L (2015) Annu Rev Phys Chem 66:379
13. Kumar A, Sevilla M (2012) Handbook of computational chemistry, vol 3, Applications–Biomolecules. Springer, Dodrecht, Netherlands
14. Caron L, Sanche L (2012) In: Čársky P, Čurik R (eds) Low-energy electron scattering from molecules, biomolecules and surfaces. Taylor & Francis, Boca Raton, FL
15. Gu J, Leszczynski J, Schaefer I (2012) H. Chem Rev 112:5603
16. Zheng Y, Hunting DJ, Ayotte P, Sanche L (2008) Phys Rev Lett 100:198101
17. Xiao F, Zheng Y, Cloutier P, He Y, Hunting DJ, Sanche L (2011) Nanotechnology 22:465101
18. Tippayamontri T, Kotb R, Paquette B, Sanche L (2013) Anticancer Res 33:3005

19. Rezaee M, Hunting DJ, Sanche L (2013) Int J Radiat Oncol Biol Phys 87:847
20. Zheng Y, Sanche L (2013) Therapy. Rev Nanosci Nanotechnol 2:1
21. Rezaee M, Alizadeh E, Cloutier P, Hunting DJ, Sanche L (2014) Chem Med Chem 9:1145
22. Sanche L (2015) Nat Mater 14:861
23. van der Burgt P, Finnegan S, Eden S (2015) Eur Phys J D 69(7):173
24. Schlatholter T, Alvarado F, Bari S, Lecointre A, Hoekstra R, Bernigaud V, Manil B, Rangama, Huber B (2006) Chem Phys Chem 7(11)
25. Behmand B, Marignier JL, Mostafavi M, Wagner JR, Hunting DJ, Sanche L (2015) J Phys Chem B 119
26. Liu B, Nielsen SB, Hvelplund P, Zettergren H, Cederquist H, Manil B, Huber BA (2006) Phys Rev Lett 97:133401
27. Neustetter M, Aysina J, da Silva FF, Denifl S (2015) Ang Chem Intern Ed 54(31):9124
28. Calvayrac F, Reinhard PG, Suraud E, Ullrich CA (2000) Phys Rep 337:493
29. Reinhard PG, Suraud E (2004) Introduction to cluster dynamics. Wiley, New York
30. Calvayrac F, Reinhard PG, Suraud E (1995) Phys Rev B 52:R17056
31. Dreizler RM, Gross EKU (1990) Density functional theory: an approach to the quantum many-body problem. Springer-Verlag, Berlin
32. Marques M, Gross E (2004) Annu Rev Phys Chem 55:427
33. Perdew JP, Burke K, Ernzerhof M (1996) Phys Rev Lett 77:3865
34. Perdew JP, Zunger A (1981) Phys Rev B 23:5048
35. Tapavicza E, Tavernelli I, Rothlisberger U (2007) Phys Rev Lett 98:023001
36. Escartin JM, Romaniello P, Stella L, Reinhard PG, Suraud E (2012) J Chem Phys 137:234113
37. McEniry EJ, Wang Y, Dundas D, Todorov TN, Stella L, Miranda RP, Fisher AJ, Horsfield AP, Race CP, Mason DR, Foulkes WMC, Sutton AP (2010) Eur Phys J B 77(3):305
38. Rizzi V, Todorov TN, Kohanoff J, Correa AA (2016) Phys Rev B 93:024316
39. Kohn W, Sham LJ (1965) Phys Rev 140:1133
40. Gross EKU, Kohn W (1990) Adv Quant Chem 21:255
41. Gross EKU, Dobson JF, Petersilka M (1996) Top Curr Chem 181:81
42. Hofmann D, Klüpfel S, Klüpfel P, Kümmel S (2012) Phys Rev A 85:062514
43. Messud J, Dinh PM, Reinhard PG, Suraud E (2008) Phys Rev Lett 101:096404
44. Messud J, Dinh PM, Reinhard PG, Suraud E (2008) Ann Phys f324:955
45. Legrand C, Suraud E, Reinhard PG (2002) J Phys B 35:1115
46. D'Avezac M, Calandra M (2005) M. F. Phys Rev B 71:205210
47. VandeVondele J, Sprik M (2005) Phys Chem Chem Phys 7:1363
48. Gu B Kohanoff J. Unpublished
49. Körzdörfer T, Kümmel S, Mundt M (2008) J Chem Phys 129(1):014110
50. Hofmann D, Körzdörfer T, Kümmel S (2012) Phys Rev Lett 108:146401
51. Beck MH, Jäckle A, Worth GA, Meyer HD (2000) Phys Rep 324:1
52. Reinhard PG, Suraud E (1992) Ann Phys 216:98
53. Suraud E, Reinhard PG (2014) New J Phys 16:063066
54. Lacombe L, Dinh PM, Reinhard PG, Suraud E, Sanche L (2015) Eur J Phys D 69:195
55. Warshel A, Levitt M (1976) J Mol Biol 103:227
56. Moret ME, Tapavicza E, Guidoni L, Röhrig U, Sulpizi M, Tavernelli I, Rothlisberger U (2005) Chimia 59:493
57. Matveev AV, Neyman KM, Pacchioni G, Rösch N (1999) Chem Phys Lett 299:603
58. Sokol AA, Bromley ST, French SA, Catlow CRA, Sherwood P (2004) Int J Quant Chem 99:695
59. Gresh N, Parisel O, Giessner-Prettre C (1999) J Mol Struct THEOCHEM 458:27
60. Dinh PM, Reinhard PG, Suraud E (2010) Phys Rep 485:43
61. van Duin A, Dasgupta S, Lorant F, Goddard W (2001) J Phys Chem A 105(41):9396
62. Sushko GB, Solov'yov IA, Verkhovtsev AV, Volkov SN, Solov'yov AV (2016) Eur Phys J D 70:12
63. Sheppard TJ, Lozovoi AY, Pashov DL, Kohanoff JJ, Paxton AT (2014) J Chem Phys 141(4)
64. Elstner M, Porezag D, Jungnickel G, Elsner J, Haugk M, Frauenheim T, Suhai S, Seifert G (1998) Phys Rev B 58(11):7260

65. VandeVondele J, Krack M, Mohamed F, Parrinello M, Chassaing T, Hutter J (2005) Comp Phys Comm 167:103
66. Goedecker S, Teter M, Hutter J (1996) Phys Rev B 54(3):1703
67. VandeVondele J, Hutter J (2007) J Chem Phys 127:114105
68. Perdew JP, Burke K, Ernzerhof M (1996) Phys Rev Lett 77(18):3865
69. Grimme S (2006) J Comp Chem 27(15):1787
70. Wang J, Roman-Perez G, Soler JM, Artacho E, Fernandez-Serra MV J Chem Phys 134(5)
71. Mogelhoj A, Kelkkanen AK, Wikfedt KT, Schiotz J, Mortensen JJ, Pettersson LGM, Lundqvist BI, Jacobsen KW, Nilsson A, Norskov JK (2011) J Phys Chem B 115(48):14149
72. Smyth M, Kohanoff J (2011) Phys Rev Lett 106:238108
73. Smyth M, Kohanoff J (2012) J Am Chem Soc 134:9122
74. Gao CZ, Wang J, Wang F, Zhang FS (2014) J Chem Phys 140(5):054308
75. Barkett C, Ray J, Ricci E, Wilker H, Mcdaniel E, Thohss H, Gilbody V (1977) Oak Ridge National Laboratory Report ORNL-5206
76. Chiu YN, Friedrich B, Maring W, Niedner G, Noll M, Toennies JP (1988) J Chem Phys 88(11):6814
77. Gao CZ, Wang J, Zhang FS (2013) Chem Phys 410:9
78. Jacquemin D, Morales J, Deumens E, Öhrn Y (1997) J Chem Phys 107(16):6146
79. Dinh PM, Gao CZ, Kluepfel P, Reinhard PG, Suraud E, Vincendon M, Wang J, Zhang FS (2014) Eur J Phys D 68:239
80. Fedor J, Cicman P, Coupier B, Feil S, Winkler M, Guch K, Husarik J, Jaksch D, Farizon B, Mason NJ, Scheier P, Mrk TD (2006) J Phys B 39(18):3935
81. Itikawa Y, Mason N (2005) J Phys Chem Ref Data 34(1):1
82. Gorfinkiel JD, Morgan LA, Tennyson J (2002) J Phys B 35(3):543
83. Dinh PM, Fehrer F, Reinhard PG, Suraud E (2007) Eur Phys J D 45:415
84. Tennyson J (2010) Phys Rep 491:29
85. Uhlig F, Marsalek O, Jungwirth P (2012) J Phys Chem Lett 3(20):3071
86. McAllister M, Smyth M, Gu B, Tribello G (2015) K. J J Phys Chem Lett 6:3091
87. Li ZJ, Cloutier P, Sanche L, Wagner JR (2010) J Am Chem Soc 132:5422
88. Kim MC, Park H, Son S, Sim E, Burke K (2015) J Phys Chem Lett 6(19):3802
89. Bin G, Smyth M, Kohanoff J (2014) Phys Chem Chem Phys 16:24350

Multiscale Modelling of Molecular Processes for Biomedical and Nanotechnology Applications with MBN Explorer

Andrey V. Solov'yov

Abstract This chapter introduces MesoBioNano Explorer (MBN EXPLORER) (Solov'yov et al. J Comput Chem 33:2412–2439, (2012), [1]), a software package for the advanced multiscale simulations of complex molecular structure and dynamics and highlights some of its biomedical and nanotechnology applications. MBN EXPLORER has many unique features, a wide range of applications in Physics, Chemistry, Biology, Material Science, and in related Industries. It is suitable for classical molecular dynamics, Monte Carlo and relativistic dynamics simulations of a large range of molecular systems of different kind, such as nano- and biological systems, nanostructured materials, composite/hybrid materials, gases, liquids, solids and various interfaces, with the sizes ranging from atomic to mesoscopic. MBN EXPLORER can be exploited together with MBN Studio (Solov'yov et al. MBN Studio, (2015), [2]), a specially developed graphical user interface.

1 Introduction

The Meso-Bio-Nano (MBN) Science is the interdisciplinary field of research studying structure-formation and dynamics of animate and inanimate matter on the nano- and the mesoscales. It bundles up several traditional topics in theoretical physics under a common theme. The range of open challenging scientific problems in this field is rather broad. They may include: structure and dynamics of clusters, nanoparticles and biomolecules; clustering, self-organization, growth and structure-formation processes, their multiscale nature and scaling laws; assemblies of clusters/nanoparticles and bio-macromolecules, hybrid bio-nano systems, nanostructured materials; surface phenomena; nanoscale phase transitions; thermal, optical and

A.V. Solov'yov (✉)
MBN Research Center at Frankfurter Innovationszentrum Biotechnologie,
Altenhöferallee 3, 60438 Frankfurt am Main, Germany
e-mail: solovyov@mbnresearch.com

A.V. Solov'yov
A.F. Ioffe Physical-Technical Institute,
Polytekhnicheskaya Ul. 26, 194021 Saint Petersburg, Russia

A.V. Solov'yov (ed.), *Nanoscale Insights into Ion-Beam Cancer Therapy*,
DOI 10.1007/978-3-319-43030-0_9

magnetic properties; collective phenomena; electron/spin transport and molecular electronics; nuclear magnetic resonance; collision, fusion, fission and fragmentation processes; channeling effects; radiation effects; radiobiological effects and many more. There are many important applications closely linked to the field. One of such important applications is the Ion-Beam Cancer Therapy (IBCT). Understanding the processes behind IBCT on the molecular, nano- and meso- scales is an open area of intensive current research, which is widely discussed in this book. The list of topical areas in the field grows rather rapidly facilitating also the development of the relevant theoretical and computational methods.

In many areas the future of many industrial products is associated with the creation of an integrated environment for numerical design and modeling. This encompasses a wide range of end-products and applications in nanoelectronics, nanomaterials and their adoption within transportation, avionics, polymer technologies, medicine, etc. In most of these areas simulations need to operate over a wide range of scales, ranging from the molecular and the nanoscale to the micro and sometimes even to macro-dimensions. Such multiscale modeling usually integrates different physical and chemical phenomena and is currently one of the hot topics of theoretical and computational research. Multiscale modeling may save crucial time and money in product development processes, and hence play a key role in industrial competitiveness. The development of multiscale modeling tools is necessarily parallel with the development and widening of modern methods of high-performance computing. The implementation and success of the versatile numerical design and modeling requires a close and wide cooperation of industrial and academic players.

Thus, the multiscale modeling is one of the most topical research fields nowadays. In order to fully exploit its potential in the field of IBCT, as well as for other biomedical and nanotechnological applications, one needs to be well familiar with the wide range of interdisciplinary topics, including

- Physics: providing the fundamental theories for the delivery of radiation and its interactions with biological targets, or for instance, explaining the fundamentals of variety of processes occurring during deposition of materials on surfaces and the formation of nanostructures;
- Chemistry: describing the chemical processes induced at specific physical conditions and providing tools for tailoring of nanoscale species to specific functions;
- Materials Science: searching for advanced materials with the unique properties or functionalization of the materials on the nanoscale;
- Life Sciences: elucidating effects on the cellular level and integrating this knowledge into clinical practices;
- Software Engineering and High Performance Computing: providing the basis for advanced computational/virtual modeling of a large variety of systems and phenomena on the scales ranging from atomic to macroscopic.

Any form of inanimate condensed matter, including biological, consists of many different components linked by numerous, different interactions. Important efforts in deepening of the molecular level understanding of different forms of condensed matter and their dynamical behaviour concern the origin, nature and evolution of various complex molecular systems, as well as the emergence of new features, properties, processes and functions involving the systems with increasing their complexity. On the meso- and nanoscales the physics and chemistry of biological and biomolecular systems, nanosystems and materials typically deal with such behaviour. Many examples of emergence of qualitatively new features can be quoted, e.g. the development of new collective properties when going from small molecules to large clusters or the nanoparticle aggregation on surfaces or inside cells leading to the appearance of fractally shaped morphologies. The fractal morphologies, being emerged in dynamical systems on the nanoscale, remain characteristic for many systems, including biological ones, at practically all larger scales, and are present in practically all living systems.

This book contains many examples of the molecular systems and molecular processes, like DNA strand breaks, which play the important role in IBCT. Most of these processes can be modelled and simulated using modern theoretical and computational techniques. These simulations provide a basis and necessary molecular level quantitative details for the construction of the inclusive Multi-Scale Approach [3, 4], which is systematically introduced and widely discussed in several chapters of this book. The best suited software package to set up the computer simulations of various processes occurring in the irradiated molecular system on the atomic, nano- and meso-scales is MBN Explorer [1]. MBN Explorer is equipped with an advanced graphical user interface, the so called MBN Studio [2], enabling the construction of input files, simple start of simulations, as well as visualisation and analysis of the results obtained.

Below in this Chapter MBN Explorer and some of its important features are introduced and illustrated this with a number of exemplar simulations and case studies.

2 MesoBioNano Explorer and MBN Studio

MesoBioNano Explorer (MBN EXPLORER) is a software package for the advanced multiscale simulations of complex molecular structure and dynamics. It has many unique features, a wide range of applications in Physics, Chemistry, Biology, Material Science, and in related Industries, see Fig. 1. It is suitable for classical molecular dynamics (MD), irradiation driven molecular dynamics (IDMD), Monte Carlo (MC) and relativistic dynamics simulations of a large range of molecular systems of different kind, such as nano- and biological systems, nanostructured materials, composite/hybrid materials, gases, liquids, solids and various interfaces, with the sizes ranging from atomic to mesoscopic. MBN EXPLORER permits computer simulations with the sizes of molecular systems ranging from the atomic to the mesoscopic scales.

Such knowledge is required in an enormous number of applications, e.g. in avionics and automobile industry for designing of nanostructured materials, functionalized surface coatings, stronger and lighter materials for aircrafts and cars suitable for high-performance at extreme conditions, in mechanical engineering for virtual design of superhard nanostructured materials, in medical applications for nanostruc-

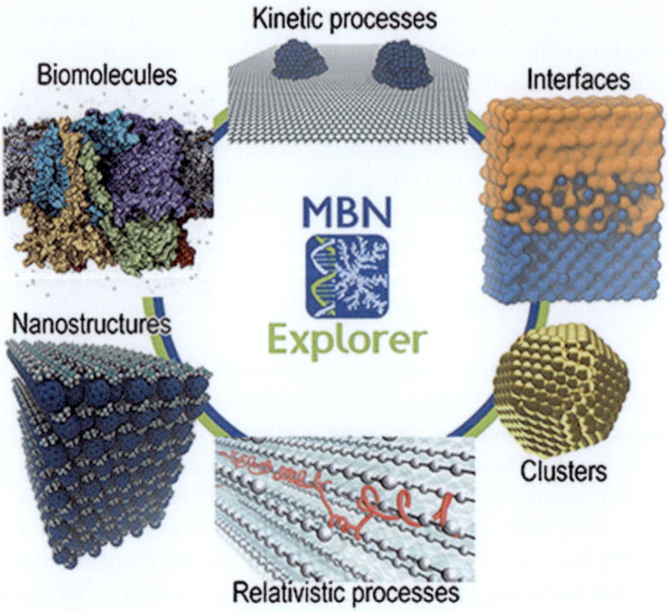

Fig. 1 Illustration of different application areas of MBN Explorer. Adapted from [5]

tured implants, in cement industry for the design of superplasticizers allowing the production of a concrete with higher compressive strength, in electronic and chemical industry for construction of highly efficient batteries, catalyzers, in pharm industry for drug design, etc. In many of these applications it is necessary to identify and/or design specific properties of the system determined by its molecular structure on the nanoscale and to ensure their transfer to the macroscopic scale in order to make them functional and usable. Such a transition implies a multiscale modeling supported in MBN Explorer through a combination of MD and MC simulations, algorithmic, coarse graining and phenomenon based approaches.

The ultimate goal of MBN Explorer is to expand the understanding of structure and dynamics of complex molecular systems, mechanisms of their stability, self-organization and growth, as well as the ways of their manipulation and control aiming a broad spectrum of application of this knowledge in nanotechnology, microelectronics, material science and medicine.

MBN Explorer version 2.0 [1] can be exploited together with MBN Studio [2], a specially developed graphical user interface. MBN Studio helps to set up and start MBN Explorer calculations, to monitor their progress and to examine the calculation results. It is supported by the graphical utility enabling to visualize selected inputs and outputs.

A number of built-in tools allow for the calculation and analysis of specific characteristics that are determined by the output of MD simulations. Examples include diffusion coefficients of various molecular species, heat capacities and melting temperatures, radial distribution function, etc. A special modeling plug-in allows one to construct a large variety of molecular systems quickly and efficiently. By means of this plug-in one can easily construct molecular systems of different geometry built of various elements of the periodic table.

Figure 1 highlights a variety of molecular systems, which can be simulated using MBN EXPLORER. In particular, MBN EXPLORER is suited to compute the system's energy, to optimize molecular structures, as well as to explore the molecular dynamics (classical, irradiation driven, Euler, relativistic) and random walk dynamics. MBN EXPLORER supports using a large library of interatomic potentials, allowing to model a large number of very different molecular systems, which are introduced briefly below in Sect. 4.

The first release of MBN EXPLORER has been the heritage of more than a decade development. The code has been thoughtfully tested and proved to be efficient and reliable in calculations. The structure of MBN EXPLORER, its main features and capabilities are described in detail in the reference article [1] published by the Journal of Computational Chemistry. The figures, see Fig. 2, highlighting MBN EXPLORER and its 3D kinetic Monte Carlo module, were chosen for the cover page of the two JCC issues.

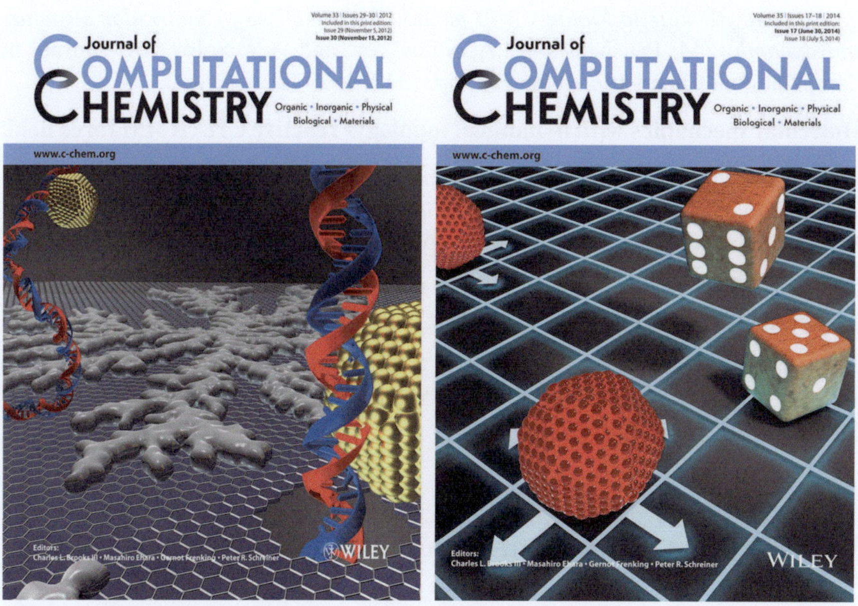

Fig. 2 Figures highlighting MBN EXPLORER for the cover page of volume 33, issue 30 (*left*) and volume 35, issue 17 (*right*) of the Journal of Computational Chemistry, in which the reference article about MBN EXPLORER [1] and its 3D kinetic Monte Carlo module [6] were published

The code is under continuous development conducted by the joined participation of world-class scientists and IT developers affiliated with MBN Research Center gGmbH, see website http://www.mbnresearch.com/.

Citing MBNEXPLORER. The authors request that all published work which utilizes MBN EXPLORER include the primary citation:

[1] *Meso Bio Nano Explorer—a universal program for multiscale computer simulations of complex molecular structure and dynamics*, I.A. Solov'yov, A.V. Yakubovich, P.V. Nikolaev, I. Volkovets, and A.V. Solov'yov, Journal of Computational Chemistry, volume **33**, pp. 2412–2439 (2012).

For specific algorithms the authors are requested to include additionally the following citations in their publications:

Relativistic Integrator:

[2] *Simulation of ultra-relativistic electrons and positrons channeling in crystals with* MBN EXPLORER, G.B. Sushko, V.G. Bezchastnov, I.A. Solov'yov, A.V. Korol, W. Greiner, and A.V. Solov'yov, Journal of Computational Physics, volume **252**, pp. 404–418 (2013).

Kinetic Monte-Carlo approach:

[3] *Efficient 3D kinetic Monte Carlo method for modeling of molecular structure and dynamics*, M. Panshenskov, I.A. Solov'yov, and A.V. Solov'yov, Journal of Computational Chemistry, volume **35**, pp. 1317–1329 (2014).

Molecular mechanics with dynamical topology- reactive CHARMM force field:

[4] *Studying chemical reactions in biological systems with MBN Explorer: implementation of molecular mechanics with dynamical topology*, G.B. Sushko, I.A. Solovyov, A.V. Verkhovtsev, S.N. Volkov, A.V. Solov'yov, European Physical Journal D, volume **70**, p. 12 (10pp) (2016).

Irradiation driven molecular dynamics:

[5] *Molecular dynamics for irradiation driven chemistry: application to the FEBID process*, G.B. Sushko, I.A. Solov'yov, A.V. Solov'yov, European Physical Journal D, volume **70**, 217 (15pp) (2016).

3 MBN EXPLORER Main Features

Universality

MBN EXPLORER is designed for studying a broad range of physical, chemical and biological systems and materials by computing their energies, optimizing molecular structures, as well as through molecular dynamics and random walk dynamics (kinetic Monte Carlo) simulations. Universality is an important feature of MBN EXPLORER, which allows modeling of a large number of molecular systems and processes (e.g. atomic clusters, fullerenes, nanotubes, polypeptides, proteins, DNA, nanostructured materials, nanofractals, etc., composite systems like a metallic nanoparticles interacting with a biomolecule, or a DNA penetrating through a nanopore) exploiting a broad variety of interatomic potentials of different kind.

Tunable Force Fields

MBN EXPLORER includes a large variety of interatomic potentials. A distinctive feature of the program is the possibility to combine various interatomic potentials from a large library of interatomic potentials available in MBN EXPLORER. The potentials implemented in MBN EXPLORER include pairwise, many-body, and molecular mechanics potentials which are widely accepted for studying bio- and nanosystems.

The file format of molecular mechanics force field used by MBN EXPLORER is the same as that used by the programs http://www.charmm.org/ CHARMM, http://cns-online.org/v1.3/ XPLOR and http://www.ks.uiuc.edu/Research/namd/ NAMD. This compatibility allows using MBN EXPLORER for calculations of a broad range of biological molecules with minimal efforts. The results of MBN EXPLORER calculations are made compatible with standard visualization programs http://www.ks.uiuc.edu/Research/vmd/ VMD and http://www.chemcraftprog.com Chemcraft.

Unique Algorithms

Apart from many standard algorithms MBN EXPLORER contains also unique algorithmic implementations, being useful in particular application areas. For instance, it allows flexible coarse graining, i.e. grouping of particles into rigid fragments, thereby significantly reducing the number of dynamical degrees of freedom, in the system. This algorithm is especially useful for molecular dynamics simulations of large molecular systems, having well defined interacting constituent parts, which could be treated as frozen. Note that most of other molecular dynamics codes do not allow grouping of atoms in rigid bodies. Another examples include the unique algorithm for simulations of relativistic particles channeling through oriented crystals, simulation of radiation damage processes and irradiation driven molecular dynamics.

Multiscale Approach

MBN EXPLORER allows one to perform stochastic Monte Carlo dynamics of molecular systems on the time scales significantly exceeding those of the conventional atomistic molecular dynamics simulations. Such multiscale dynamics approach is ideal for the systems, in which the details of the system dynamics on the atomic scale are not so important, and can be parameterized through the kinetic rates for the dominating transformations occurring in the system. This important feature of MBN EXPLORER expands significantly its application areas and goes beyond the limits of other molecular dynamics codes usually unable to deal with the multiscale modeling.

Computational Efficiency

Despite the universality, the computational efficiency of MBN EXPLORER is comparable to and often even higher then the computational efficiency of other software packages, making MBN EXPLORER often a favorable choice.

Object-Oriented Design

The primary design objective for MBN EXPLORER is extensibility and maintainability of the code. In order to achieve this goal, MBN EXPLORER code implements an object-oriented approach with C++. The modular design of the code allows an easy integration of new algorithms and techniques for molecular dynamics simulations.

4 Applications of MBN EXPLORER

There are many different areas of application of MBN EXPLORER some of which are briefly introduced below.

Crystals, Liquids, and Gases

With MBN EXPLORER one can simulate crystals, liquids, glasses and gases, and study numerous physical and chemical phenomena involving different phase states of matter [7, 8]. For each condensed matter state there are many examples of simulations which are collected in the library of MBN EXPLORER tests and examples [9]. These examples include simulations of metallic, carbon and silicon, atomic and molecular crystals, oxides, thin films and surface coatings, liquids and their interfaces with metals and biocompatible materials, their various properties and processes with their involvement. MBN EXPLORER also provides tools for multiscale modeling of various MBN systems. These tools allow one to model kinetic behaviour of such systems far beyond the time and spatial limits of the conventional molecular dynamics simulations [6, 10] (Fig. 3).

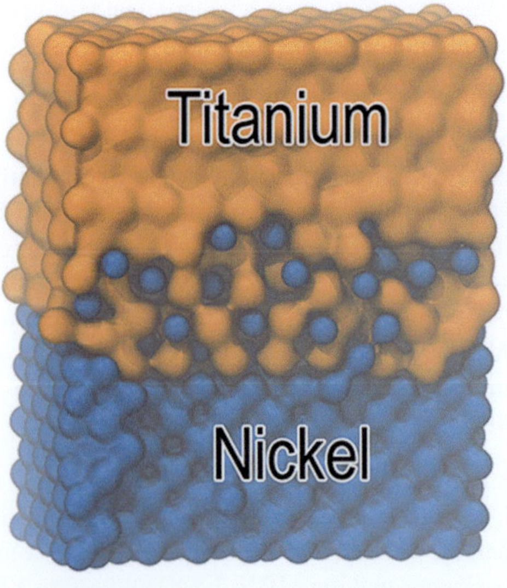

Fig. 3 Nickel-titanium interface. Adapted from [5]

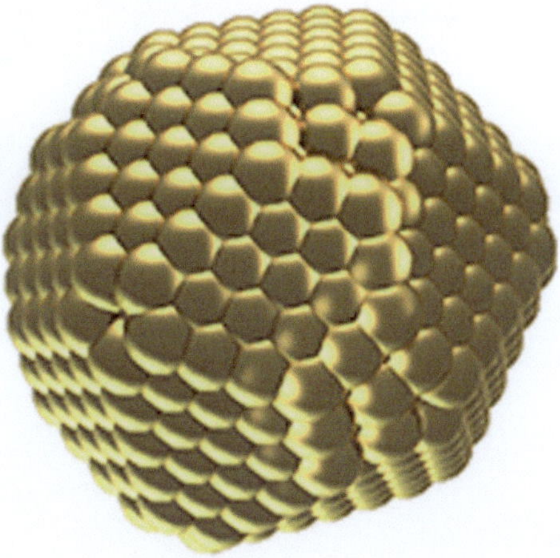

Fig. 4 Icosahedral atomic cluster. Adapted from [5]

Atomic Clusters and Nanoparticles

MBN EXPLORER is suitable for computer simulations of structure and dynamics of free, deposited and embedded atomic and molecular clusters, nanoparticles (NPs) of different types, e.g. metals, noble gases, semiconductor clusters, fullerenes, carbon nanotubes, graphene, as well as all other allotropic forms of nanocarbon materials, composite and functionalized NPs, nanoalloys etc., [11–16]. The sizes of these molecular systems could be varied from a few atoms up to a few million atoms. Possible simulations include the tasks on the structure analysis and optimization, various thermal effects, mechanical properties, nanoscale phase transitions, diffusion and a broad range of other dynamical and collision processes involving clusters and NPs [17–19] (Fig. 4).

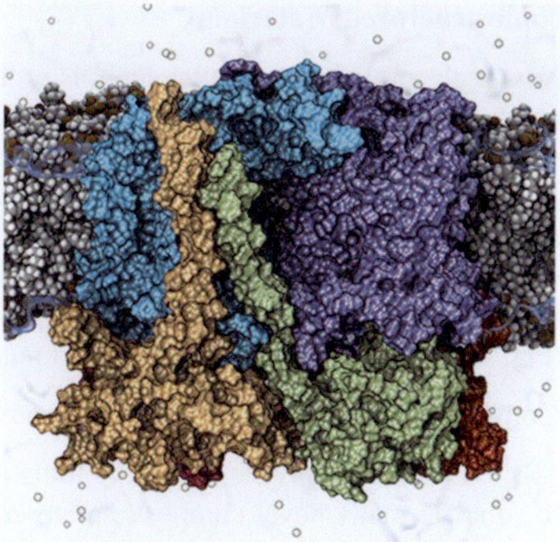

Fig. 5 Protein complex. Adapted from [5]

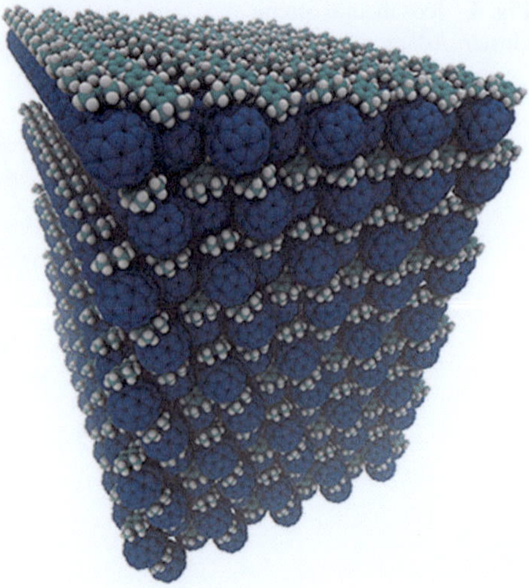

Fig. 6 C_{60}-based nanowire. Adapted from [5]

Biomolecular Systems

MBN Explorer allows one to simulate a large variety of biomolecules, biomolecular, hybrid bio-nano systems with various interfaces [20]. Transformations of these systems at different thermal and biologically relevant conditions, at various external stresses can be explored. Numerous possible case studies include proteins, DNA, lipid bilayers, interaction of these systems with NPs, external environments and many more. MBN Explorer allows one to simulate structure and dynamics of proteins, DNA, RNA and other biomolecules in ubiquitous environments [21]. Protein folding [22], antigen-antibody bounding [23], DNA unzipping [24], radiation damage phenomena [25] and many other processes involving biomolecules can be studied (Fig. 5).

Nanostructured Materials

Nanoscale molecular objects, such as atomic clusters, NPs, proteins, DNA fragments, etc., provide a possibility to construct new types of materials, the so-called nanostructured materials, thin films, surface coatings with the structure and properties determined by the molecular constituent building blocks. MBN Explorer allows one to simulate a wide spectrum of nanostructured materials and to study their properties [26]. Examples of such materials include: metals (e.g. Ni or Ti [27]), metal NPs crystals, nanocarbon (nanosilicon) based nanostructured materials (e.g. TMB-C_{60} nanowires [28–31]) and many more (Fig. 6).

Composite Materials and Material Interfaces

MBN Explorer has the necessary tools (appropriate force fields and algorithms) to simulate many novel composite materials consisting of components of different nature, ordered or disordered. Examples of such hybrid systems include nanoalloys,

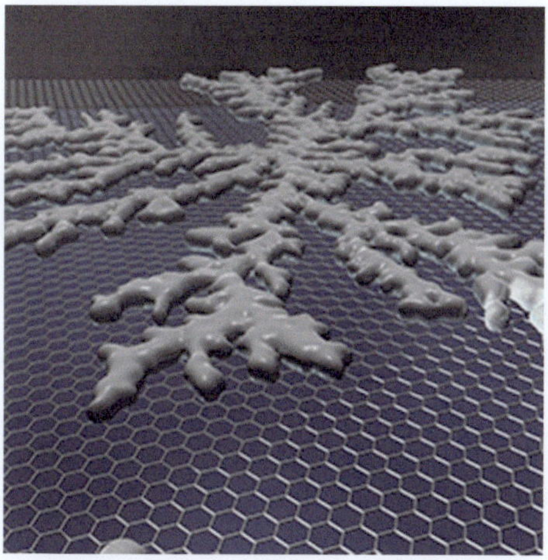

Fig. 7 Silver fractal on graphite surface. Adapted from [5]

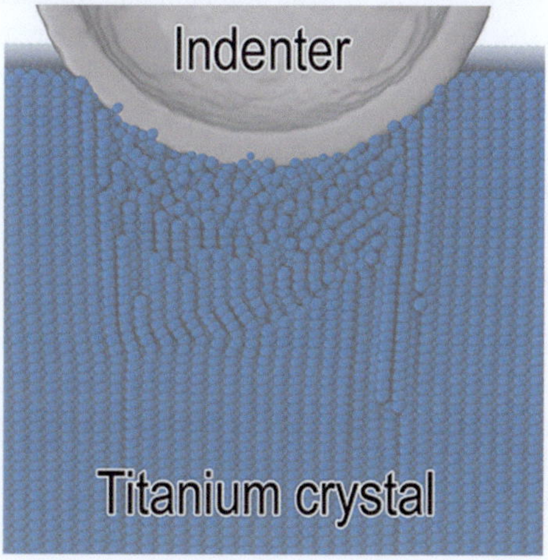

Fig. 8 Nanoindentation of Ti crystal. Adapted from [5]

nanofractals [10, 32, 33], crystalline superlattices of metal NPs linked by different organic or biological molecules, or NPs placed into the biological environments (e.g. attached to DNA, protein, or cell membrane). The latter systems appear to be of significant interest and importance in connection with the analysis of toxicity of nanomaterials and the development of advanced radiotherapies exploiting nanoprocesses and technologies [34]. With MBN EXPLORER one can simulate and investigate a variety of complex multiscale dynamical processes, for instance, diffusion and surface pattern formation (e.g. nanofractals, droplets etc.) in the course of NP, atomic or molecular deposition, morphological transitions and many more [35–38] (Fig. 7).

Thermo-Mechanical Properties of Materials

MBN EXPLORER can be utilized for simulations and investigation of the mechanical properties and thermal effects of a broad variety of the materials mentioned above. This includes analysis of elastic and plastic deformations [39] (e.g. Young's modulus, Poisson's ratio, hardness, etc.), dynamics of dislocations, nanoindentation [8], phase transitions [40], thermo-mechanical damage [41] and many more. For most of these processes and phenomena the thermal dependence of various characteristics of materials is of significant importance and interest (Fig. 8).

Collision Processes and Related Phenomena

MBN EXPLORER supports the most advanced molecular dynamics simulations for a large variety of complex molecular systems. With these methods one can study many different dynamical processes, including collisions, that occur in molecular systems (Fig. 9).

These studies include collision and fragmentation processes involving atomic clusters, NPs and biomolecules, molecular association and dissociation, nano- and microscale conformational, morphological, and phase transitions, proteins folding, DNA unzipping, NP and molecular diffusion, propagation of particles through a medium (channeling, multiple scattering, track structure analysis), collision induced thermo-mechanical medium effects, and many more. Some of these processes are discussed in detail in several chapters of this book (Fig. 10).

Novel Technologies

MBN EXPLORER is a very useful and powerful tool for the exploration of the challenging problems arising in connection with the development of new technologies [42]. There are several research areas, in which simulations performed with the use of MBN EXPLORER play an important role. One of such areas concerns the construction of novel light sources based on charged particles channeling in crystalline undulators. Another example deals with simulations of the nanoscopic molecular processes playing the key role in the ion-beam cancer therapy [3, 43, 44]. Combined with

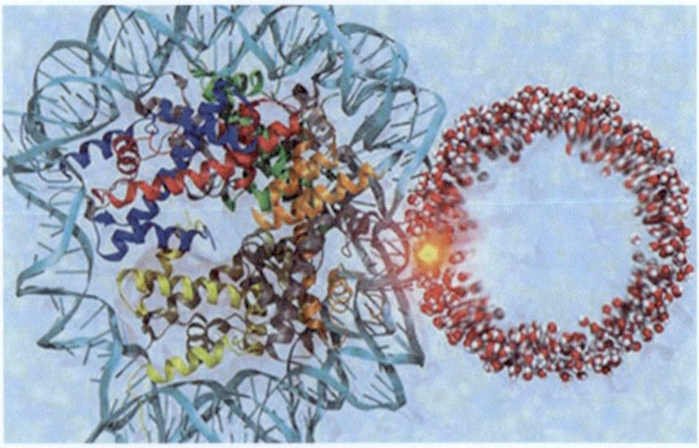

Fig. 9 Ion induced shock wave interacting with nucleosome. Adapted from [5]

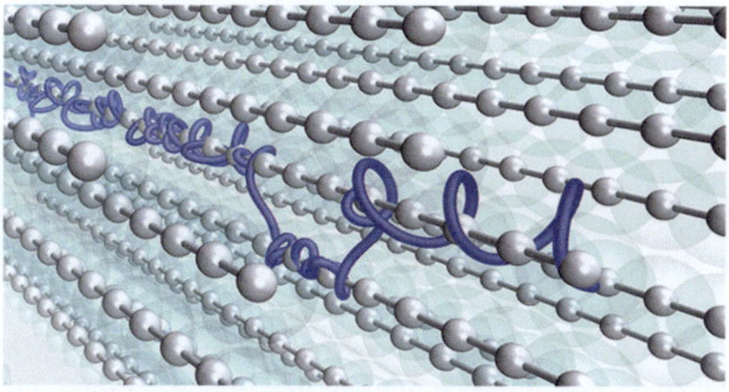

Fig. 10 Axial channeling of an ultrarelativistic electron along crystal axis. Adapted from [5]

the visualisation interface through MBN Studio or other similar visualisation tools, MBN EXPLORER in many cases can help to optimise or even substitute expensive laboratory experiments by computational modeling. Predictive power, the possibility to visualise structure and dynamics of complex molecular systems allow to percept the MBN EXPLORER based computational approach as a kind of 'computational nano- and microscope'.

It is suitable for relativistic dynamics simulations [42, 45–47]. Among other applications, MBN EXPLORER can be used to simulate thermo-mechanical damage of a biological medium, e.g. a DNA nucleosome, which is caused by the propagation of a shock wave initiated by irradiation with fast ions [48]. The results of such simulations are used then to evaluate the efficiency of radiation with different projectiles [49] within the framework of the multiscale approach to the physics of radiation damage [3] and can be applied in the field of ion-beam cancer therapy [3, 49–51].

5 Exemplar Case Studies

It is impossible in one Chapter to overview all the case studies supported by MBN EXPLORER. This deserves a whole book. Some of them are already introduced in other chapters of this book, e.g. simulation of ion induced shock waves and their damaging effects on DNA, structure and dynamics of sensitising nanoparticles, etc. Here, other important case studies relevant to the topics of this book are briefly introduced.

5.1 Reactive CHARMM Force Fields

Nowadays, it has become feasible to study structure and dynamics of molecular systems that constitute of millions of atoms [52, 53] and evolve on time scales up to hundreds of nanoseconds [54] by employing the classical molecular dynamics approach, often also called molecular mechanics (MM). In this approach, a molecular system is treated classically, so that constituent atoms interact with each other through a parametric phenomenological potential that is governed by the type of individual atoms and by the network of chemical bonds between them. This network defines a so-called molecular topology, that is a set of rules that impose constraints in the system and permit maintaining its natural shape, mechanical, and thermodynamical properties. The MM method has been widely used throughout the last decades [53, 55–57] and implemented in the well-established computational packages, such as CHARMM [58], AMBER [59], GROMACS [60], NAMD [61] and MBN EXPLORER (www.mbnexplorer.com, www.mbnresearch.com) [1].

Despite numerous successes, the conventional MM method is primarily capable of studying processes where chemical reactions do not take place. This leads to significant limitations of the method and makes it practically unsuitable for studying highly non-equilibrium processes in biomolecular systems, e.g. thermo-mechanical biodamage. This particular example involves rupture and formation of covalent bonds that cannot be simulated by the conventional MM method due to a fixed topology of the system.

Simulation of the rupture and formation of covalent bonds can be performed by using Quantum Mechanical/Molecular Mechanical (QM/MM) methods or *ab initio* MD simulations [62–64]. Both methods are computationally rather demanding and, thus, the *ab initio* approach is used typically for studying fragmentation of small biomolecules, such as DNA nucleobases or nucleotides [65, 66]. The size of such systems is far from the typical sizes of systems of biological relevance, consisting of hundred thousands of atoms, and more. This problem is addressed to some extent in QM/MM methods where a core part of a large biomolecular system is described quantum mechanically while all the surroundings are described classically using, for example, the conventional MM method [67, 68]. Thus, the rupture or formation of covalent bonds can be simulated only in a small part of the system, which is treated quantum mechanically.

In recent work [69] an extended version of the conventional MM method was implemented in MBN EXPLORER. It is based on the newly introduced reactive CHARMM force field, being an important extension of the standard CHARMM force field. It was demonstrated that this extension describes correctly the dynamically changing molecular topology of a system within the classical MD framework. The presented modification takes into account additional parameters of the system, such as dissociation energy of bonds, bonds multiplicity and the valence of atoms. The functional form of the interatomic interactions is also adjusted to account for the finite dissociation energy of the chemical bonds.

To illustrate these modifications here let us now consider the two examples that go beyond the standard MM methodology. The first example illustrates the rupture of a single C–N bond in an alanine dipeptide molecule, being one of the simplest building blocks of larger biomolecular systems like polypeptides or proteins. The second example shows the reverse process of the new bond formation.

Fragmentation of Alanine Dipeptide

To illustrate the bond breakage, following [69], let us consider the dynamics of alanine dipeptide consisting of 20 atoms, solvated in a simulation box with 95 water molecules. The alanine dipeptide molecule was considered with neutral terminals.

In order to clearly illustrate the difference between the standard CHARMM force field, utilizing the harmonic interatomic potential, and the dissociative CHARMM potential implemented in MBN EXPLORER, two simulations were carried out [69]. In these simulations, the rupture of the central C–N bond in the dipeptide, leading to the formation of two isolated alanine molecules, was monitored.

To facilitate the process, let us set the initial velocity of the C and N atoms high, which corresponds to an energy fluctuation sufficient for the bond rupture. In the simulation performed with the standard force field, the peptide bond is modeled through the harmonic potential, therefore, the bond cannot break. The behavior of the C–N bond in the harmonic approximation is illustrated in the left part of Fig. 11, and the corresponding atoms are marked with red circles. In this case, the distance between the atoms oscillates around the equilibrium value as the atoms always return to their equilibrium positions.

In the second simulation, the Morse potential was used for the description of the peptide bond. In this case the C and N atoms do not oscillate around an equilibrium

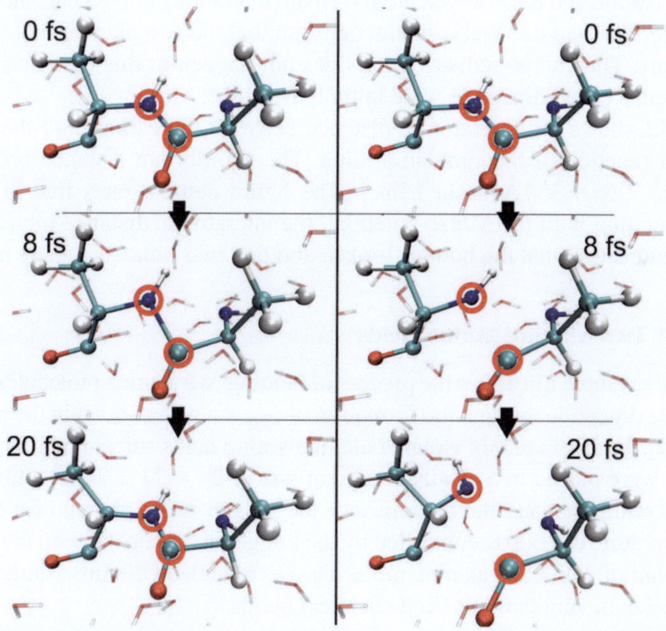

Fig. 11 Snapshots illustrating dynamics of alanine dipeptide and the C–N bond rupture simulated with the harmonic (*left*) and Morse (*right*) potentials at 0 fs (*top*), 8 fs (*middle*) and 20 fs (*bottom*). Adopted from [69]

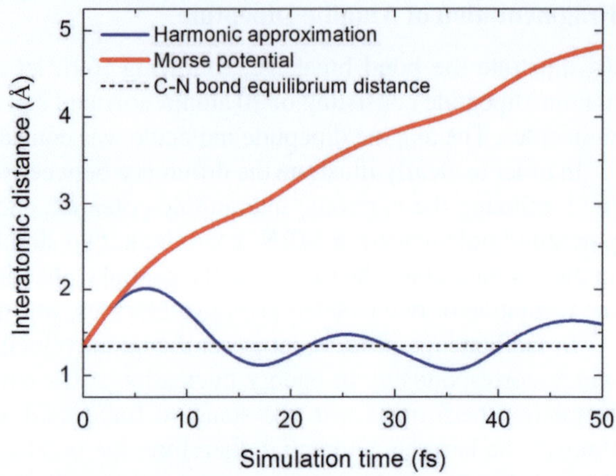

Fig. 12 Dependence of the C–N interatomic distance in alanine dipeptide as a function of the simulation time. Adopted from [69]

position, and the structure of the system after 20 fs of simulation changes significantly from the one considered above (see the right part of Fig. 11). It is evident from the snapshots that the distance between the atoms increases already after 8 fs.

When the distance between the atoms exceeds a given cutoff radius (which is equal to 2.5 Å in this example), the bond is considered as broken. Once this has happened, the carbon and the nitrogen atoms remain interacting only via the electrostatic potential and the van der Waals interactions, so that the two alanine molecules can diffuse apart. The charge redistribution does not happen in this case because both new fragments of the dipeptide were initially neutral.

Figure 12 shows the interatomic distance between the carbon and the nitrogen atoms as a function of the simulation time. The equilibrium distance between the atoms is $r_0^{C-N} = 1.354$ Å (dashed line). The figure demonstrates that in the case of the simulation with the Morse potential, the interatomic distance monotonically increases indicating that the bond is broken and that two isolated alanine molecules drift apart.

Binding of Two Alanine Amino Acids

The second example illustrates the process of binding two alanine molecules together into a single dipeptide through the formation of a new covalent bond in the molecular system [69]. In this case, six isolated alanine amino acids surrounded by 54 water molecules were placed in a small simulation box of $24 \times 24 \times 24$ Å^3 with periodic boundary conditions, and the dynamics of the system was simulated for 80 ps at a fixed temperature of 1000 K controlled by the Langevin thermostat with the damping time constant of 1 fs. Each alanine molecule was modeled with unsaturated N– and C–termini, i.e. having two unpaired chemical bonds.

In the course of the simulation, all distances between the different termini of alanines are monitored. When the distance between a pair of terminal atoms became smaller than the predefined cutoff radius (equal to 3 Å in this example), a new covalent bond is considered to be formed. Figure 13a illustrates a spacial conformation of two

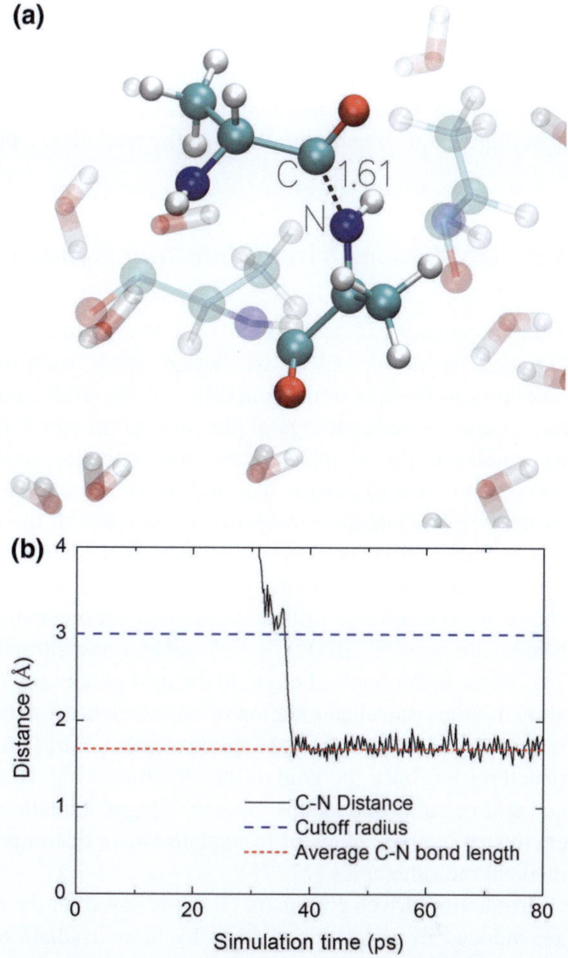

Fig. 13 **a** Two alanine molecules approaching each other to form a new C–N bond. **b** Dependence of the distance between C and N atoms for the two alanine molecules. Adopted from [69]

amino acids in the simulation leading to the formation of a new bond. Figure 13b gives the dependence of the distance between C and N atoms for the two molecules shown in the upper part. At some point, this distance becomes smaller than the cutoff radius (blue dashed line), and the two molecules become connected. Note that after 40 ps the distance between the C and N atoms oscillates around a constant value corresponding to the C–N bond equilibrium length. Since six alanines are considered in this simulation, more of them could self-assemble in a polypeptide chain but this would require longer simulation. In this system, each initial alanine molecule has a total charge equal to zero. Therefore, after the formation of a new molecule the charge redistribution step was not necessary.

Having proven the force field to work on a simple molecules one can generalize the framework towards macromolecules. This allows for studying the systems of biologically relevant sizes, on the time scales which are not accessible by means of *ab initio* methods. In [69] the example illustrating the process of water splitting and the issue of chemical equilibrium were analysed using the reactive CHARMM force field. It was demonstrated that the results of the simulation are in a reasonable quantitative agreement with those of the analytical calculations.

5.2 *Irradiation Driven Molecular Dynamics*

There are many examples in which chemical transformations of complex molecular systems are driven by irradiation. Often such modifications carry important outcomes to the functional properties of the irradiated molecular systems. Enough to mention the radiobiological phenomena, in which living cells can be inactivated by irradiation due to the induced DNA complex strand breaks [3], the formation and composition of cosmic ices and dusts in the interstellar medium and planetary atmospheres is largely a result of the interplay of the molecular surface adsorption and surface irradiation [70], the formation of biologically relevant molecules under extreme conditions involving irradiation [71],[1] and many more. Irradiation driven chemistry is nowadays utilized in modern nanotechnology, such as focused electron beam deposition (FEBID) [72, 73] and extreme ultraviolet lithography (EUVL) [74, 75]. These technologies belong to the next generation of nanofabrication techniques allowing the controlled creation of nanostructures with nanometer resolution which is attractive in both, basic and applied research. The fabrication of smaller and smaller structures has been the goal of the electronics industry for more than three decades and still remains one of this industry's biggest challenges. Furthermore, irradiation chemistry is a key element in nuclear waste decomposition technologies [76] and medical radiotherapies [77, 78].

Irradiation driven chemistry (IDC) is based on the quantum transformations that are induced in molecular systems by their irradiation by external fields of different modality (X-rays, lasers, electrons/positrons, ions, etc.) and the dynamics of molecular system which can be also influenced by external factors like temperature, pressure, external fields, etc. Highly perturbed dynamical molecular systems can only be described from first principles within the time dependent density functional theory (TDTDFT), or any of its equivalents, if the size of the molecular system is sufficiently small, typically hundreds of atoms [79–82]. This strong limitation makes TDDFT of limited use for the description of the IDC of complex molecular systems.

Classical molecular dynamics (MD) could be considered as an alternative for the theoretical description of IDC. However, in spite of the manyfold advantages, classical MD is often inapplicable for simulations of chemical reactions and IDC processes, because it does neither account for coupling of the molecular system to radiation, nor does it describe quantum transformations in the molecular system

[1]See COST Action "The Chemical Cosmos: Understanding Chemistry in Astronomical Environments" www.cost.eu/COST_Actions/cmst/CM0805.

induced by the irradiation. In the recent work [83] these deficiencies of MD were overcome and a new methodology for simulation of irradiation driven chemical transformations of complex molecular systems was suggested. There it was suggested to model irradiation induced quantum transformations in a molecular system as random, fast and local processes involving small molecular fragments (typically on sub-nanometer scale) of the entire system. The modeled transformations include molecular bond breakages, saturation of dangling bonds, chemical reactions in the system, and changes in the molecular topology of the system. These transformations are introduced according to the specific rates that are coupled to the irradiation field and the probabilities of the corresponding quantum processes are established through *ab initio* quantum approaches, such as many-body theory, DFT, collision theory, or taken from experiments. The fundamental basis for such an approach relies on the Born-Oppenheimer theory justifying uncoupling of the fast electronic motion in molecular systems from the slow motion of the ionic subsystem and the fact that the characteristic time scale for the fast quantum transformations in the system is typically within the femtosecond range, i.e., about the duration of one time step in MD simulations. Furthermore, the spatial dimension of the region where an irradiation induced quantum transformation takes place is much smaller than the size of the molecular systems under consideration. Therefore, if the outcomes of the quantum transformations are properly accounted for on the basis of quantum mechanics or simply taken from experiment and correctly embedded as random and local modifications of the classical force fields, it becomes feasible to model structure and dynamics of large molecular systems under irradiation through the irradiation driven molecular dynamics (IDMD), as demonstrated in [83]. This methodology is designed for the molecular level description of the irradiation driven chemistry of complex molecular systems arising in various circumstances introduced above.

Classical molecular dynamics (MD) has been introduced for the description of quantum molecular systems with the use of the classical Newtonian equations [84]. The justification of this approach is based on the Born-Oppenheimer theorem, which separates the light electronic and heavy ionic subsystems and elucidates the quasi-classical motion of the nuclei in the system. In this approach all the information about the quantum-mechanical properties of the system is included in the parameters of the classical force fields guiding the motion of the nuclei. Within the classical MD the trajectories of atoms and molecules are determined through the numerical solution of Newton's equations, where forces between the particles and their potential energies are calculated using interatomic potentials and force fields. Such a simplification of the description of motion of a quantum system provides significant advantages for computer simulations as already discussed above. The method was originally developed within the field of theoretical physics in the late 1950s [85] but is applied today mostly in chemical physics, materials science and the modelling of biomolecules.

The classical MD approach does not describe the electron dynamics and, therefore, most of the quantum transformations that may occur during the system dynamics. These transformations are often induced in the system through exposure to external perturbations such as external fields or irradiation by charged particles (electrons, protons, ions, etc.) or photons. The resulting effects may have a global character (electric current, spin ordering, polarisation, magnetisation, etc.) or be local (atomic or molecular excitation, ionisation, dissociation, charge transfer, etc.).

Irradiation induced local quantum perturbations of a molecular system typically occur on the sub-femtosecond time scale and involve only those atoms that are directly affected by the irradiation. This results in the creation of secondary electrons, ions, reactive species (radicals), and excited molecules, which can further interact with the molecular system and cause further chemical transformations. This complex local dynamics typically involves the nearest environment of the targeted molecular site, being a small part of the entire molecular system, and is completed within femtoseconds. During this time some of the initial perturbations of the system, such as quasi-free electrons, electron holes, ionic charges, relax and vanish, due to the high electronic mobility and the Coulomb attraction. The femtosecond time scale is, however, still significantly shorter than the characteristic timescales responsible for the motion of the entire molecular system. Indeed, in classical MD a typical integration time step is 1–2 fs, corresponding to the oscillation period of a hydrogen atom at room temperature.

The notable outcome of the process described above will be the emergence of bond breaks in the system. These events are most significant as they affect the dynamical behaviour and chemical transformations in the molecular system on the larger time scales, up to nanoseconds and beyond, being the typical time frame for the classical MD. The bond breaks arise in those parts of the molecular system which are targeted by the irradiation. They occur randomly with a probability depending on the intensity and the modality of irradiation. The probabilities of these events are related to the cross sections of the involved irradiation induced processes (elastic and inelastic scattering, electronic and vibrational excitation, dissociative electron attachment, collision dissociation, etc.) occurring in the system on the femtosecond time scale and can be elaborated from the collision theory or be taken from experiment.

Irradiation conditions of a molecular system can differ substantially and depend on the radiation modality, duration of the system exposure to irradiation and the system geometry. Irradiation can be a swift single event, like a single ion track crossing the molecular system, or it can last a certain period of time up to some nanoseconds and even longer. In the latter case the irradiation induced bond breaks and charge redistribution in the system occurs during the entire irradiation period. Irradiation can be homogeneous within a certain volume or strongly inhomogeneous. The choice of the irradiation conditions corresponds to each particular case study. In the follow up sections we consider these for the FEBID process.

The above described scenario defines the irradiation driven molecular dynamics (IDMD). The IDMD can be introduced as classical MD with the superimposed random process of molecular bond breakage related to the irradiation conditions. The bond breakage is defined as the local alteration of the system force fields, which involves (i) creation of reactive atomic species (radicals) with dangling bonds, (ii) the possibility of dangling bonds' closure and creation of new molecular bonds or molecules, (iii) accounting for molecular topology changes (in the cases when it is defined, e.g. molecular mechanics force fields). The characterisation of these modifications of the classical MD force fields can be elaborated on the basis of quantum chemistry methods. Examples of such characterisation for the FEBID process are given below.

The IDMD methodology aims to account for the major dissociative transformations of the molecular system induced by its irradiation and possible paths of further reactive transformations. The latter are sensitive to statistical mechanical factors, like the concentration of the reactive spices, their mobility (diffusion), the temperature of the medium, etc. All these factors are automatically accounted for in a correct way through the Langevin MD describing the molecular system as a NVT statistical mechanics ensemble. The local deviations from the statistical mechanics equilibrium arising in the vicinity of the breaking bonds caused by the local deposition of energy into the system leads to minor perturbations of the large molecular system and can be incorporated into IDMD as a perturbation.

The concept of IDMD introduced recently in [83] is general. It is applicable to any kind of molecular system treated with any type of classical force field. The method is implemented in the MBN EXPLORER software package and can operate with the large library of classical potentials, many-body force fields (including the recently implemented reactive CHARMM force field [69]) and their combinations. The limited number of parameters that determine molecular force fields, and their irradiation driven perturbations, results in a countable number of modifications that could occur in a molecular system upon irradiation and makes the method efficient and accurate.

This implementation opens a broad range of possibilities for modelling of irradiation driven modifications and chemistry of complex molecular systems. In order to highlight these possibilities let us present an example of the FEBID process simulated by means of IDMD and reported in [83]. These simulations have been thoroughly examined and compared with experiment.

A snapshot of MD simulation of the initial irradiation phase in the FEBID process is shown in Fig. 14. The interaction of $W(CO)_6$ precursor molecules with the cylindrical electron beam depicted in green leads to the precursor fragmentation and the formation of W clusters on the surface, illustrated in Fig. 14 in blue.

Following [83], let us now demonstrate that the morphology, the type of emerging surface nanostructure and its composition depend strongly on the irradiation driven chemistry of precursors. For this purpose let us consider the models A and B. In both models, for the atoms with open bonds the algorithm of searching for the atomic open bond neighbours of the suitable type is implemented. In the case of model A the searching for the reactive neighbours is performed only among the atoms located beyond the the molecular structure to which a chosen atom belongs. In the model B the searching is performed over all open valence atoms in the system including the molecular structure to which a chosen atom belongs.

The nanostructures presented in Figs. 15 and 16 emerge after 150 ns of simulated irradiation (15 rounds of irradiation 10 ns each at the conditions corresponding to the e-beam current 1.2 μA). These figures show that the chained structures of W (blue dots) with the C-O fragments attached to the most of W atoms are formed within model A, while model B results in the formation of more compact and dense molecular structure with the larger W content. The relative content of tungsten in

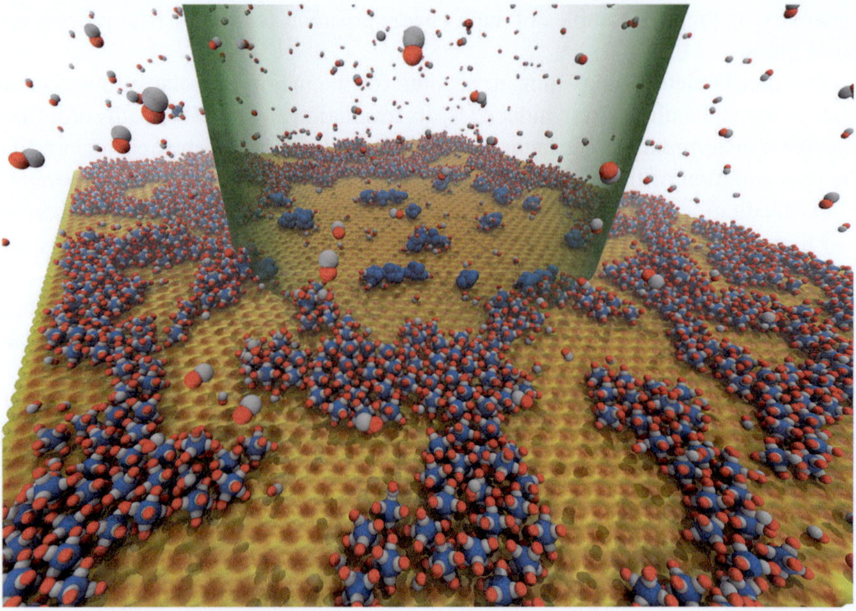

Fig. 14 Snapshot of MD simulation of adsorption of $W(CO)_6$ precursor molecules on a SiO_2 surface in the case of model A (see text) experiencing the early stage of irradiation by the electrons beam (shown as transparent green cylinder). The interaction of precursor molecules with the beam leads to fragmentation of molecules and formation of W clusters, shown in blue. Adopted from [83]

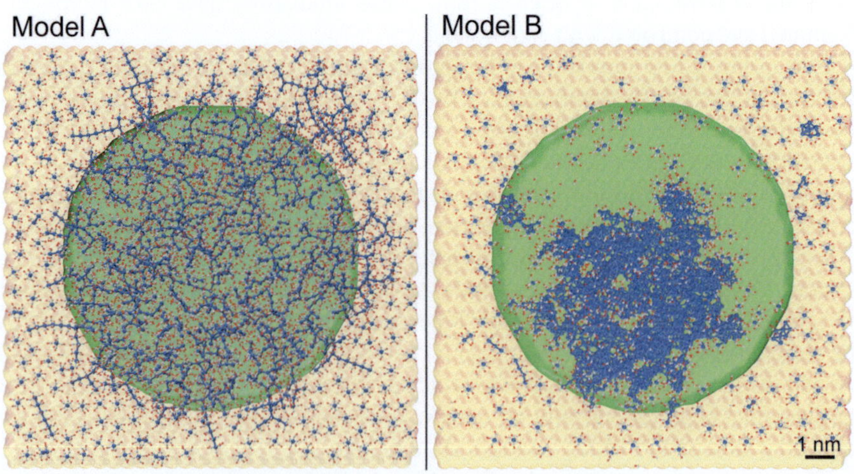

Fig. 15 Top view of morphologies of W enriched nanostructures atop the hydroxylated SiO_2 surface simulated within models A and B after 15 irradiation/adsorption cycles. Adopted from [83]

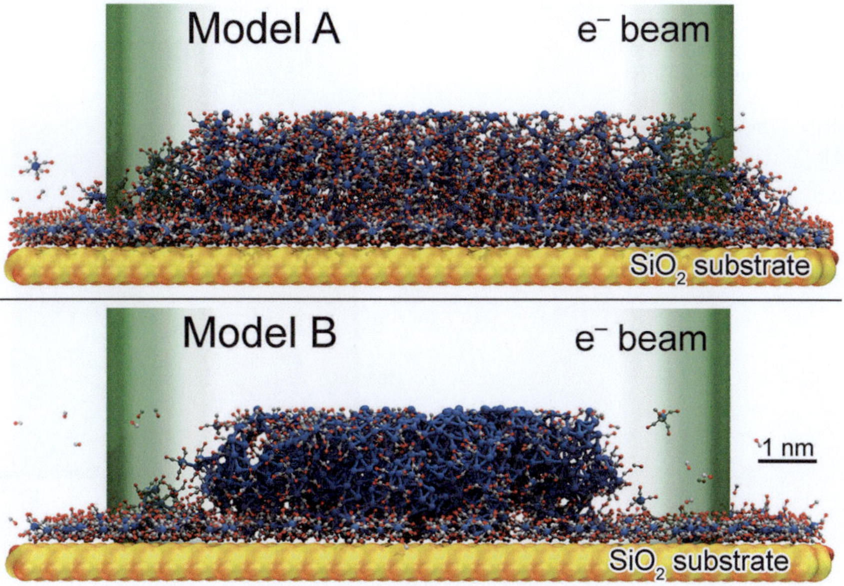

Fig. 16 Side view of morphologies of W enriched nanostructures shown in Fig. 15. Adopted from [83]

these nanostructures is ∼15 % (model A) and ∼46 % (model B). These simulations also indicate that formation of chemical bonds within the growing nanostructure is essential for the emergence of the crystalline-like molecular structure with higher content of tungsten atoms. The increase of the number of bonds between W atoms leads to the decrease of the CO content and the total number of atoms in the growing nanostructure, i.e. its growth rate.

These simulations demonstrate that the IDMD approach provides a powerful computational tool to model the growth process of W granular metal structures emerging in the FEBID process at the atomistic level of detail. The morphology of the simulated structures, their composition and growth characteristics are consistent with the available experimental data as demonstrated in [83]. Moreover, the observed dependencies like increasing of the volume growth rate per incident electron with decreasing the e-beam current, or the growth of the W content with increase of the e-beam current are reproduced correctly by within the developed models.

The performed analysis indicates also the need of further wide exploitation of the IDMD methodology in FEBID and many other processes in which the irradiation of molecular systems and irradiation driven chemistry play the key role.

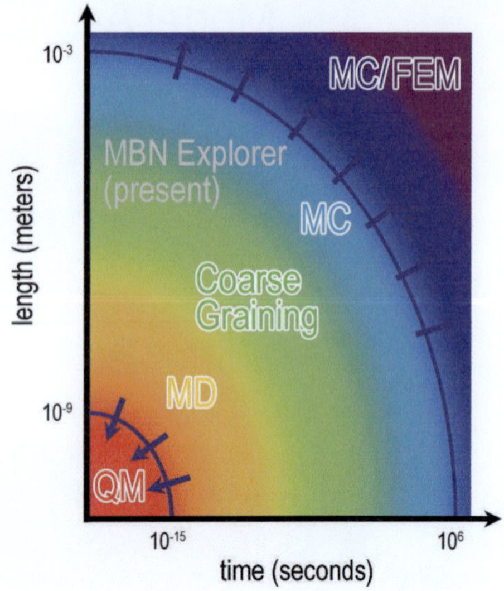

Fig. 17 Temporal and spatial system scales and the corresponding simulation methods. Adopted from MBN EXPLORER Tutorials [5]

6 Conclusions and Outlook

The further technical development of MBN EXPLORER, MBN Studio and their biomedical and nanotechnology applications will involve the creation of new modules allowing various types of multiscale modeling by linking different modeling methodologies (Quantum Mechanics (QM), Molecular Dynamics (MD), Coarse Graining, Monte Carlo (MC), Finite Element Method (FEM) and others) well-established for different temporal and spatial scales as illustrated in Fig. 17. In the figure, lines indicate the limits of the current version of MBN EXPLORER and arrows show the directions for the further development.

One of the important directions for the further development concerns the multiscale approach to the ion-beam cancer therapy allowing quantitative understanding of the medical treatments on the molecular level. The recent advances in this direction that are reported in this book will be integrated in a form of special module of MBN EXPLORER. This module could be utilised for further studies of the molecular processes behind the IBCT and the optimisation of the existing treatment planning protocols. The similar modules have already been implemented in other areas of application of MBN EXPLORER, for details see [83].

The realization of these plans should allow increasing the number of application areas, case studies and the universality of the software package much beyond of its current limits. Some of the above mentioned multiscale methodologies have already been successfully implemented in the latest release of MBN EXPLORER [5]. This process will be continued in the future. The complete realization of this programme means a long term development aiming at a large number of customers and wide exploitation of this universal and powerful software package in numerous areas of its application.

7 How to Get MBN Explorer and MBN Sudio?

MBN Explorer and MBN Studio are the software products developed by MBN Research Center gGmbH, http://www.mbnresearch.com/. Different types of licences for MBN Explorer and MBN Studio can be acquired from MBN Research Center via its website http://www.mbnresearch.com/, and the access to the software, the user's guide and a library of representative examples can be obtained. Inquiries about the more detailed information on the types of licences and the prices should be sent to info@mbnexplorer.com.

The further details about the current and future releases of MBN Explorer and MBN Studio can be found on the website: http://www.mbnexplorer.com/.

References

1. Solov'yov IA, Yakubovich AV, Nikolaev PV, Volkovets I, Solov'yov AV (2012) MesoBio-Nano Explorer–a universal program for multiscale computer simulations of complex molecular structure and dynamics. J Comput Chem 33:2412–2439. http://mbnresearch.com/, http://mbnexplorer.com/
2. Solov'yov, IA, Sushko G, Solov'yov AV (2015) MBN Studio. http://mbnresearch.com/mbn-studio
3. Surdutovich E, Solov'yov AV (2014) Multiscale approach to the physics of radiation damage with ions. Eur Phys J D 68:353
4. Verkhovtsev AV, Surdutovich E, Solov'yov AV (2016) Multiscale approach predictions for biological outcomes in ion-beam cancer therapy. Sci Rep 6:27654
5. Solovyov IA, Sushko GB, Verkhovtsev AV, Korol AV, Solovyov AV (2015) MBN Explorer: simulations of nanomaterials structure and dynamics. Polytechnic University Publishing House, St. Petersburg State Polytechnic University, St. Petersburg, pp 1–102. ISBN 978-5-7422-4893-4. MBN Explorer: dynamics of biomolecular systems and self-organization. Polytechnic University Publishing House, St. Petersburg State Polytechnic University, St. Petersburg (2015), pp 1–112. ISBN 978-5-7422-4894-1
6. Panshenskov MA, Solov'yov IA, Solov'yov AV (2014) Efficient 3D kinetic Monte Carlo method for modeling of molecular structure and dynamics. J Comput Chem 35:1317–1329
7. Sushko GB, Verkhovtsev AV, Yakubovich AV, Schramm S, Solov'yov AV (2014) J Phys Chem A 118:6685
8. Verkhovtsev AV, Yakubovich AV, Sushko GB, Hanauske M, Solov'yov AV (2013) Molecular dynamics simulations of the nanoindentation process of titanium crystal. Comput Mater Sci 76:20–26
9. http://mbnexplorer.com/
10. Solov'yov IA, Solov'yov AV, Kébaili N, Masson A, Bréchignac C (2013) Thermally induced morphological transition of silver fractals. Phys Status Solidi B 251:609–622
11. Solov'yov IA, Solov'yov AV, Greiner W (2003) Int J Mod Phys E 13:697–736
12. Obolensky OI, Solov'yov IA, Solov'yov AV, Greiner W (2005) Comput Lett 1:313–318
13. Koshelev A, Shutovich A, Solov'yov IA, Solov'yov AV, Greiner W (2003) In: Proceedings of international workshop from atomic to nano-scale. Old Dominion University, pp 184–194
14. Solov'yov IA, Solov'yov AV, Greiner W, Koshelev A, Shutovich A (2003) Phys. Rev. Lett. 90(1–4):053401
15. Solov'yov IA, Mathew M, Solov'yov AV, Greiner W (2008) Phys Rev E 78(1–13):051601
16. Verkhovtsev AV, Schramm S, Solov'yov AV (2014) Eur Phys J D 68:246

17. Solov'yov AV, Surdutovich E (eds) (2009) Atomic Cluster Collisions: structure and dynamics from the nuclear to the MesoBioNano scale. AIP Conf. Proc. 1197:1–236
18. Verkhovtsev AV, Hanauske M, Yakubovich AV, Solov'yov AV (2013) Comput Mater Sci 76:80
19. Sushko GB, Verkhovtsev AV, Solov'yov AV (2014) J Phys Chem A 118:8426
20. Solov'yov AV, Broyer M (eds) (2009) Topical Issue on Bio macromolecules. Eur Phys J D 51:1–172
21. Solov'yov IA, Yakubovich AV, Solov'yov AV, Greiner W (2007) Phys Rev E 75(1–9):051912
22. Solov'yov AV (2011) Supervisor's foreword. In: Yakubovich AV (ed) Theory of phase transitions in polypeptides and proteins. Springer Ph.D. Thesis series, Springer, Heidelberg, London, New York
23. Henriques E, Solov'yov AV (2008) A rational method for probing macromolecules dissociation: the antibody-hapten system. Eur Phys J D 46:471
24. Volkov SN, Paramonova EV, Yakubovich AV, Solov'yov AV (2012) Micromechanics of base pair unzipping in the DNA duplex. J Phys: Condens Matter 24:035104
25. Huber BA, Malot C, Domaracka A, Gauduel YA, Solov'yov AV (eds) (2012) 1st Nano-IBCT conference 2011, radiation damage of biomolecular systems: nanoscale insights into Ion-Beam Cancer Therapy. J Phys: Conf Ser 373:1–210
26. Mishnaevsky L, Levashov E, Solov'yov AV (eds) (2013) Book of abstracts, international conference on computational modelling of nanostructured materials (ICCMNM), Frankfurt am Main, Germany, 4–6 Sept 2013
27. Verkhovtsev AV, Sushko GB, Yakubovich AV, Solov'yov AV (2013) Benchmarking of classical force fields by ab initio calculations of atomic clusters: Ti and NiTi case. Comput Theor Chem 1021:101–108
28. Geng J, Solov'yov IA, Reid DG, Skelton P, Wheatley AEH, Solov'yov AV, Johnson BFG (2010) Fullerene-based one-dimensional crystalline nanopolymer formed through topochemical transformation of the parent nanowire. Phys Rev B 81:214114
29. Geng J, Solov'yov IA, Zhou W, Solov'yov AV, Johnson BFG (2009) J Phys Chem C 113:6390–6397
30. Solov'yov IA, Geng J, Solov'yov AV, Johnson BFG (2009) AIP Conf Proc 1197:89–102
31. Solov'yov IA, Geng J, Solov'yov AV, Johnson BFG (2009) Chem Phys Lett 472:166–170
32. Panshenskov M, Solov'yov IA, Solov'yov AV (2014) J Comput Chem 35:1317
33. Solov'yov IA, Solov'yov AV (2013) J Phys: Conf Ser 438:012006
34. Solov'yov AV (ed) (2013) Dynamics of systems on the nanoscale. J Phys: Conf Ser 438
35. Moskovkin P, Panshenskov M, Lucas S, Solov'yov AV (2014) Phys Status Solidi B 251:1456
36. Dick VV, Solov'yov IA, Solov'yov AV (2009) AIP Conf Proc 1197:76–88
37. Dick VV, Solov'yov IA, Solov'yov AV (2010) J Phys: Conf Ser 248(1–11):012025
38. Dick VV, Solov'yov IA, Solov'yov AV (2011) Phys Rev B 84(1–14):115408
39. Yakubovich AV, Solov'yov IA, Solov'yov AV (2013) Molecular dynamics simulations of bionano systems with MBN Explorer. Phys Proc 40:93–99
40. Yakubovich AV, Sushko GB, Schramm S, Solov'yov AV (2013) Kinetics of liquid-solid phase transition in large nickel clusters. Phys Rev B 88:035438
41. Surdutovich E, Yakubovich AV, Solov'yov AV (2013) DNA damage due to thermomechanical effects caused by heavy ions propagating in tissue. Nucl Instrum Meth B 314:63–65
42. Sushko GB, Bezchastnov VG, Solov'yov IA, Korol AV, Greiner W, Solov'yov AV (2013) Simulation of ultra-relativistic electrons and positrons channeling in crystals with MBN Explorer. J Comput Phys 252:404–418
43. http://www.mbnresearch.com/project-nanoibct
44. http://www.cost.eu/domains_actions/mpns/Actions/MP1002
45. Sushko GB, Korol AV, Greiner W, Solov'yov AV (2013) J Phys: Conf Ser 438:012018
46. Sushko GB, Bezchastnov VG, Korol AV, Greiner W, Solov'yov AV, Polozkov RG, Ivanov VK (2013) J Phys: Conf Ser 438:012019
47. Sushko GB, Korol AV, Solov'yov AV (2015) Nucl Instrum Meth B 355:39
48. Yakubovich AV, Surdutovich E, Solov'yov AV (2012) J Phys: Conf Ser 373:012014
49. Surdutovich E, Yakubovich AV, Solov'yov AV (2013) Sci Rep 3:1289

50. Solov'yov AV, Surdutovich E, Scifoni E, Mishustin I, Greiner W (2009) Phys Rev E 79:011909
51. Surdutovich E, Yakubovich AV, Solov'yov AV (2010) Eur Phys J D 60:101
52. Sanbonmatsu KY, Tung CS (2007) J Struct Biol 157:470
53. Zhao G, Perilla JR, Yufenyuy EL, Meng X, Chen B, Ning J, Ahn J, Gronenborn AM, Schulten K, Aiken C et al (2013) Nature 497:643
54. Salomon-Ferrer R, Götz AW, Poole D, Le Grand S, Walker RC (2013) J Chem Theory Comput 9:3878
55. Rappé AK, Casewit CJ (1997) Molecular mechanics across chemistry. University Science Books
56. Gumbart J, Schreiner E, Wilson DN, Beckmann R, Schulten K (2012) Biophys J 103:331
57. Shim J, Humphreys GI, Venkatesan BM, Munz JM, Zou X, Sathe C, Schulten K, Kosari F, Nardulli AM, Vasmatzis G et al (2013) Sci Rep 3:1389
58. Brooks B, Bruccoleri R, Olafson B, States D, Swaminathan S, Karplus M (1983) J Comput Chem 4:187
59. Case DA, Cheatham TE, Darden T, Gohlke H, Luo R, Merz KM, Onufriev A, Simmerling C, Wang B, Woods RJ (2005) J Comput Chem 26:1668
60. Spoel DVD, Lindahl E, Hess B, Groenhof G, Mark A, Berendsen H (2005) J Comput Chem 26:1701
61. Phillips JC, Braun R, Wang W et al (2005) J Comput Chem 26:1781
62. Senn HM, Thiel W (2009) Angew Chem Int Ed 48:1198
63. Car R, Parrinello M (1985) Phys Rev Lett 55:2471
64. Marx D, Hutter J (2000) Ab initio molecular dynamics: theory and implementation. Modern methods and algorithms of quantum chemistry (John von Neumann Institute for Computing). Jülich, Germany, p 301
65. Smyth M, Kohanoff J (2012) J Am Chem Soc 134:9122
66. McAllister M, Smyth M, Gu B, Tribello GA, Kohanoff J (2015) J Phys Chem Lett 6:3091
67. Lüdemann G, Solov'yov IA, Kubař T, Elstner M (2015) J Am Chem Soc 137:1147
68. Sjulstok E, Haugaard Olsen JM, Solov'yov IA (2015) Sci Rep 5:18446
69. Sushko GB, Solovyov IA, Verkhovtsev AV, Volkov SN, Solovyov AV (2016) Eur Phys J D 70:12
70. Tielens AGGM (2013) Rev Mod Phys 85:1021
71. Horneck, G Baumstark-Khan C (eds) (2002) Astrobiology: the quest for the conditions of life. Springer, Berlin. ISBN 978-3-642-59381-9
72. Huth M, Porrati F, Schwalb C, Winhold M, Sachser R, Dukic M, Adams J, Fantner G (2012) Beilstein J Nanotechnol 3(1):597
73. Utke I, Hoffmann P, Melngailis J (2008) J Vac Sci Technol B 26(4)
74. Wu B, Kumar A (2007) J Vac Sci Technol B 25(6)
75. Hawryluk AM, Seppala LG (1988) J Vac Sci Technol B 6(6):2162
76. Heidet F, Brown NR, Tahar MH (2015) Rev Accel Sci Technol 8:99
77. Schardt D, Elsässer T, Schulz-Ertner D (2010) Rev Mod Phys 82(1):383
78. Surdutovich E, Solovyov AV (2012) Eur Phys J D 66:1
79. Dinh P, Reinhard PG, Suraud E (2010) Phys Rep 485:43
80. Maruhn J, Reinhard P, Suraud E (2010) Simple models of many-fermion systems. Springer, Berlin
81. Jacquemin D, Wathelet V, Perpéte EA, Adamo C (2009) J ChemTheory Comput 5(9):2420
82. Bochevarov AD, Harder E, Hughes TF, Greenwood JR, Braden DA, Philipp DM, Rinaldo D, Halls MD, Zhang J, Friesner RA (2013) Int J Quantum Chem 113(18):2110
83. Sushko GB, Solov'yov IA, Solov'yov AV (2016) Eur Phys J D 70:217
84. Brooks B, Bruccoleri R, Olafson B, States D, Swaminathan S, Karplus M (1983) J Comp Chem 4:187
85. Fermi E, Pasta J, Ulam S (1955) Studies of nonlinear problems. Los Alamos report LA-1940

Thermo-Mechanical Damage of Biomolecules Under Ion-Beam Radiation

Pablo de Vera, Nigel J. Mason, Eugene Surdutovich and Andrey V. Solov'yov

Abstract The prediction of the relative biological effectiveness of ion beams requires the quantification of all the biomolecular damage processes involved in the interaction of energetic ions with biological media. Traditionally, the damage pathways have been classified as direct or indirect, the former being related to the direct action of the secondary electrons produced along the ion path with DNA molecules, while the latter are referred to the damage produced by the other chemical species generated, mainly free radicals. However, the development over the last years of the multiscale approach to ion beam cancer therapy has revealed the contribution of a new damage mechanism, not present in conventional therapy with photons or electrons: the thermo-mechanical DNA damage arising from the development of shock waves on the nanometer scale around the swift ion path. The present chapter explains the theoretical framework in which this effect is predicted and reviews the work performed over the last years to try to understand the role of this damage pathway in the mechanisms of ion beam cancer therapy.

P. de Vera · N.J. Mason
Department of Physical Sciences, The Open University,
Milton Keynes, England MK7 6AA, UK
e-mail: nigel.mason@open.ac.uk

P. de Vera
MBN Research Center, 60438 Frankfurt am Main, Germany

P. de Vera (✉)
School of Mathematics and Physics, Queen's University Belfast, BT7 1NN, Belfast
Northern Ireland, UK
e-mail: p.devera@qub.ac.uk

E. Surdutovich
Department of Physics, Oakland University, Rochester, MI 48309, USA

A.V. Solov'yov
MBN Research Center at Frankfurter Innovationszentrum Biotechnologie,
Altenhöferallee 3, 60438 Frankfurt am Main, Germany
e-mail: solovyov@mbnresearch.com

A.V. Solov'yov
A.F. Ioffe Physical-Technical Institute,
Polytekhnicheskaya Ul. 26, 194021 Saint Petersburg, Russia

A.V. Solov'yov (ed.), *Nanoscale Insights into Ion-Beam Cancer Therapy*,
DOI 10.1007/978-3-319-43030-0_10

1 Introduction

In order to achieve a better control of ion beam cancer therapy (IBCT) an accurate picture of its underlying physico-chemical mechanisms is needed, which includes all the series of events that occur from the initial propagation of the energetic ions in tissue to the final irreparable DNA damage. This mechanistic understanding of IBCT should rely on a multiscale description, taking into account the very different space, time and energy scales involved in the problem [1]. The events considered start with the initial propagation of the ion beam in tissue on the macroscopic scale, although in very short times, of the order of 10^{-18}–10^{-17} s. After that, a large number of secondary electrons is ejected, which propagate mostly on the nanometer scale, and in times $\sim 10^{-16}$–10^{-15} s, and can damage DNA molecules by ionization or electronic excitation. Diffusion of the generated free radicals follows, also on the nanometer scale and in times of the order of nanoseconds. The interaction of all these secondary species (electrons and radicals) with the biological molecules results in their damage, which if the target is nuclear DNA might be irreparable in case the lesions are clustered together causing cell death or mutation.

However this picture may be oversimplified and there are other events that contribute to the damage of the biomolecules. For example, the work performed by Sanche and co-workers [2] showed that even very-low energy electrons (with energies even below ionization threshold) can produce dissociation of DNA components, a discovery that has changed our view of cellular damage which traditionally considered ionization to be the major damage mechanism together with free radical induced damage.

More recently a multiscale physics approach to ion beam cancer therapy has been developed over the last years [1] in order to try to include and identify all the relevant mechanisms involved, based on physical considerations. Such a multiscale approach models all of the above phenomena, but in addition revealed a new effect related to the induction of thermo-mechanical damage as the result of the generation of a shock waves around the ion track [3].

These shock waves appear due to the large energies deposited by ion beams on the nanometer scale, as a result of the transport of secondary electrons. Since most of these secondary electrons are ejected with low energies (below 50 eV), they are thermalized in liquid water within lengths of the order of one nanometer. Also, as discussed above, the transport of the secondary electrons occurs on a time scale of the order of several tens of femtoseconds. These times are very short in comparison with those characterizing the mechanisms capable of dissipating such energies, such as the electron-phonon coupling, which occurs on the sub-picosecond scale, or molecular diffusion which occurs on much longer times. Therefore, this means that the fast and very spatially restricted deposition of energy produces high local temperatures and pressures around the ion path starting an hydrodynamics process in a form of strong cylindrical explosion of the excited medium or, in other words, a shock wave.

Temperatures and pressures in the shock wave can be very high and sometimes may be sufficient to break chemical bonds. However such shock waves act on very short time scales ($\sim$10 ps) and also over very short distances ($\sim$10 nm). Therefore a careful analysis of the evolution of these high pressures and temperatures in time and distance is needed, in order to assess to which extent such effects can produce biomolecular damage and also whether this mechanism is comparable to the better well known electron and radical effects. The present chapter reviews the research performed on thermo-mechanical effects over the last years in order to establish the current status of the topic. Section 2 gives an overview of the physical origin of the shock waves. A hydrodynamics description of the process is explained in Sect. 3, what gives the main characteristics of the shock waves. Further evaluation of the effects of shock waves in biomolecules is reviewed in Sect. 4, by means of molecular dynamics simulations. The biological implications of shock waves are analyzed in Sect. 5. Then, open questions on the topic are set out in Sect. 6, and the final conclusions and outlook are given in Sect. 7.

2 Electron Track Structure Around Ion Tracks, and the Onset of the Thermo-Mechanical Effects

A detailed analysis of the track structure produced by the secondary electrons ejected around an ion path has been performed using a multiscale physics approach to ion beam cancer therapy [1, 4]. First of all, the characteristic energies of the electrons ejected by ion impact were evaluated, and found to be $\leq$50 eV in the Bragg peak region, irrespective of the type of incident ion and the irradiated biomaterial [5, 6]. The propagation of these electrons can be described analytically within the random walk approximation [1, 4]. In this approximation, the electrons are assumed to propagate in random directions due to the elastic scattering with the target nuclei, until they lose their energy due to inelastic collisions. Within the simplest version of this model, the probability of finding an electron at a given distance r from its point of ejection at a time t is simply given by:

$$P(t,r) = \left(\frac{3}{2\pi \overline{v} t l}\right)^{3/2} \exp\left(-\frac{3r^2}{2\overline{v} t l}\right), \tag{1}$$

which is determined by the average velocity of the electrons $\overline{v}$, and their elastic mean free path l. Note that r is distance from the point of ejection, while we will refer to the radial distance from the track as ρ. Both quantities are related by $r^2 = \rho^2 + z^2$, where z is the longitudinal coordinate. From this expression, it is possible to estimate the average radius at which the secondary electrons travel, which is given by:

$$\overline{\rho} = \int \rho P(t,r) \mathrm{d}r^3 = l\sqrt{\frac{\pi k}{6}}, \tag{2}$$

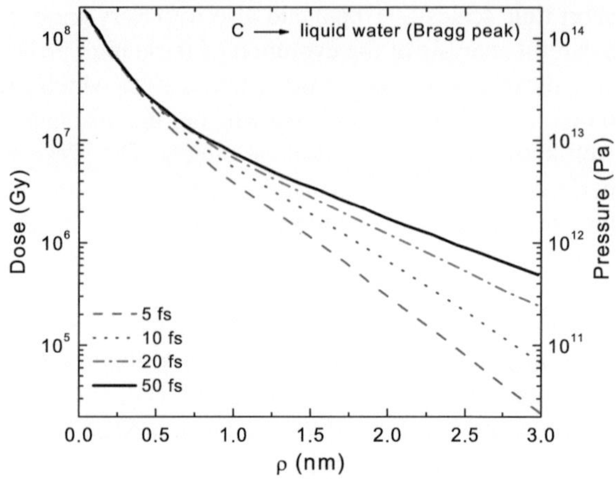

Fig. 1 Build up of the radial dose around a carbon ion track in the Bragg peak region, due to the transport of the first and second generation of secondary electrons, as calculated by the random walk approximation [4]. The right axis shows the correspondence with the pressure developed along the radial coordinate, as explained in the text

where $k = \overline{v}t/l$ is the number of elastic collisions experienced by time t. By comparing the ratio of the inelastic to the elastic mean free paths of low energy electrons, a good estimation of the number of elastic collisions until the electron loses its energy by inelastic collision is $k \simeq 20$ [1]. The average radius of propagation of electrons is then given by Eq. (2) with $k = 20$, what gives $\overline{\rho} \simeq 1$ nm. Further on, we will refer to this radius as the radius of the "hot cylinder", since most of the energy transported by secondary electrons is deposited within this cylinder.

This estimation is based on the propagation of the first generation of electrons. A more complete analysis of the whole cascade of electrons was performed recently [4]. In this work, the random walk approximation was extended to account for the slowing down of the first generation of electrons, together with the production and transport of a second generation. Some results of this approach are shown in Fig. 1, where the build up of the radial dose generated around a carbon ion in the Bragg peak region is depicted. As it can be seen, the radial dose is built up within 50 fs, and it is quite steep, with most of the energy deposited within 1 nm, as estimated above. This correspondence arises from the fact that most of the energy is deposited by the first generation of electrons.

The next question to be answered is what happens when this energy is deposited quickly within a small cylinder around the track. Such a question was already raised during the early times of the development of the multiscale approach, and the first answers based, at the beginning, on simple thermodynamics estimates [5]. It was observed that if the heat capacity of water is used to compute the increase in temperature of the water around the track due to these energy transfers, the temperatures might increase by more than 100 °C, enough to melt DNA.

Better estimates of the temperatures reached within the hot cylinder were obtained using the inelastic thermal spike model [7]. Within this approximation the energy deposited by the secondary electrons (i.e., the radial dose) is coupled to the electron-phonon interaction. Two heat transfer equations are used to describe the system, one for the electronic system and another one for the nuclei system. By using this approach it was estimated that $\sim 10^{-13}$ s after ion passage the temperature of liquid water around a carbon track in the Bragg peak region can increase up to 700–1200 K, producing pressures differences of 25–50 MPa/nm at small distances from the track.

A similar analysis of the pressures generated by the energy deposition was done by means of the random walk approximation in Ref. [4]. The radial dose around a carbon ion track in the Bragg peak region has been already shown in Fig. 1. In this reference it was demonstrated that, indeed, this energy distribution profile determines as well the pressure profile along the radial coordinate. If we define $\varepsilon(\rho)$ as the energy per unit volume deposited at times long enough for all the electrons being thermalized (i.e., $t \geq 50$ fs), then the energy deposited within a cylinder of radius ρ and length L is given by:

$$U(\rho) = \int_0^\rho \varepsilon(\rho')2\pi\rho' \mathrm{d}\rho' L \,. \tag{3}$$

The pressure at the surface of a cylinder of radius ρ is then given by the ratio of the force $F(\rho)$ normal to the surface applied on the cylinder surface $2\pi\rho L$. The force is simply calculated as $F(\rho) = -\mathrm{d}U(\rho)/\mathrm{d}\rho = 2\pi\rho L\varepsilon(r)$, which gives:

$$P(\rho) = \frac{F(\rho)}{2\pi\rho L} = \varepsilon(\rho). \tag{4}$$

This is a very interesting result, demonstrating that the radial dose profile gives the profile of the pressure along the cylinder radius. It can be seen from Fig. 1 that the pressure differences around the track go up to 700 MPa/nm, even larger than the values estimated with the inelastic spike model.

It is quite obvious that such large pressure differences will prompt a hydrodynamic response in the liquid medium. This feature can not be accounted for in the inelastic thermal spike model, where the medium is considered to be static. Therefore the problem has to be studied solving hydrodynamics equations, what is analyzed in the next section.

3 Hydrodynamics Description: The Shock Wave

The pressures estimated around the ion path can be used to determine whether the medium in the "hot cylinder" will produce a violent explosion or not. Water molecules with all their degrees of freedom available have $\gamma = C_\mathrm{P}/C_\mathrm{V} = 1.222$ (where C_P and C_V are the molar heat capacity coefficients of water at constant pressure and volume, respectively). This gives the ratio $\frac{\gamma+1}{\gamma-1} = 10$. The ratio of the pressure inside the cylinder to the pressure outside it is much larger than 10, according to the estimates given in the previous section. This satisfies the conditions for a "cylindrical strong explosion" [8].

A very good description of the solution of the hydrodynamic problem describing the strong explosion regime of the shock wave, as well as its mechanical features and limitations, can be found in Refs. [8–10]. This solution was adapted to the case of the cylindrical geometry that characterizes the present problem in Ref. [3]. The main results of this work are reproduced in the following paragraphs.

The flow of water and heat transfer is self similar, depending on a single variable ξ, a dimensionless combination of the radial distance ρ from the axis (i.e., the ion's path), the time t after the ions passage, the energy dissipated per unit length along the axis (i.e., the stopping power S), and the density of undisturbed water, $\delta = 1$ g/cm^3, given by:

$$\xi = \frac{\rho}{\beta t}\left[\frac{\delta}{S}\right]^{1/4} . \tag{5}$$

In this equation, β is a dimensionless parameter depending on C_P and C_V, and which value was determined to be 0.86 in Ref. [3].

The radial position of the wave front at time t is given by:

$$R(t) = \frac{\rho}{\xi} = \beta\sqrt{t}\left[\frac{S}{\delta}\right]^{1/4} . \tag{6}$$

The rest of physical characteristics of the shock wave can be obtained by solving the equations of the cylindrically symmetric adiabatic flow, i.e., the continuity equation, the Euler equation and the energy conservation equation, subject to a series of boundary conditions along the wave front [3]. One of these boundary conditions establishes that the pressure at the wave front should be given by:

$$P_{\text{front}} = \frac{2}{\gamma + 1}\delta u^2 , \tag{7}$$

where u is the wave front speed. The latter can be straightforwardly obtained from the derivative of Eq. (6), what gives the pressure of the wave front as a function of its position R:

$$P_{\text{front}}(R) = \frac{\beta^4}{2\,(\gamma + 1)}\frac{S}{R^2} . \tag{8}$$

Both Eqs. (6) and (8) define the main space and time characteristics of the ion-induced shock waves. Typical values for the propagation of the wave front R with time and of the pressure at the wave front position are given in Fig. 2, where lines show the results predicted by this hydrodynamics model for a carbon ion and an iron ion in the Bragg peak region. The stopping powers of these ions are, respectively, 900 eV/nm and 7195 eV/nm. The shock wave produced by the iron ion propagates 1.68 times faster than for carbon, while the pressure at the front is 8 times larger, according to the dependences shown by Eqs. (6) and (8).

Such large velocities of propagation and pressures might have several important consequences for DNA damage. First of all a collective mass flow is carried by the

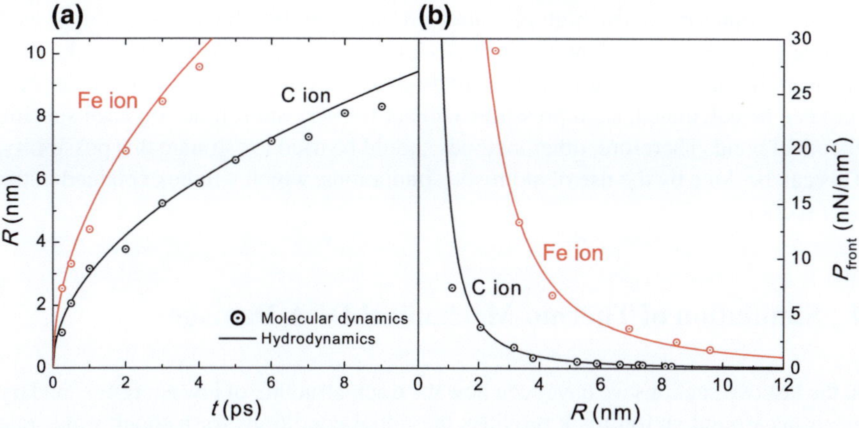

Fig. 2 Evolution of the shock wave front for shock waves produced by a carbon and an iron ion in the Bragg peak region. **a** Position of the wave front as a function of time. **b** Pressure of the wave front as a function of its position. *Lines* are the predictions of the analytical hydrodynamics model, while symbols represent the results from molecular dynamics simulations [17] (see Sect. 4)

wave with its maximum at the wave front. The mass flow produced by an ion along a path length L, at the wave front, is given by [11]:

$$J = \delta_{\text{front}}\, u = \frac{\gamma+1}{\gamma-1}\pi\beta^2\sqrt{S\delta}L\ , \tag{9}$$

where $\delta_{\text{front}} = \frac{\gamma+1}{\gamma-1}\delta$ [3] and u is derived from Eq. (6). This equation shows that the amount of matter transported by the shock wave is proportional to $\sqrt{S}$, so it can be quite substantial for ions with large stopping powers. Since this mass transport is expected to follow the wave front we can obtain the time needed by the front to travel a distance $\rho = R$, which is, from Eq. (6), $t = \frac{\rho^2}{\beta^2}\sqrt{\frac{\delta}{S}}$. We can compare this with the time needed for a species to reach the same distance by diffusion, which is ρ^2/D, with D being the diffusion coefficient. The ratio of both quantities is given by:

$$\frac{t_{\text{front}}}{t_{\text{diffusion}}} = \sqrt{\frac{\delta}{S}}\frac{D}{\beta^2}\ . \tag{10}$$

The diffusion coefficients for the relevant reactive species are always $<10^{-4}$ cm^2/s [12], what means that the ratio of Eq. (10) is always inferior to $10^{-3}/\sqrt{S(\text{keV/nm})}$. This quantity is much lower than 1, even for low stopping power ions such as protons, what may indicate that the shock wave is a means of transport of reactive species that is much more efficient than diffusion. The biological implication of this point will be discussed in Sect. 5.

On the other hand, the high pressures produced by the shock waves might well be able to break the covalent bonds in DNA or other biomolecules. This evaluation, however, lies beyond the limits of the hydrodynamics model. Even though the pressure can be calculated, such pressures will act for very short times on each specific chemical bond. Therefore, other methods should be used to estimate this possibility. This can be done by the use of atomistic simulations, which will be explained in the next section.

4 Simulation of Thermo-Mechanical DNA Damage

In the last two sections we have seen how the track-structure of low energy secondary electrons around an ion track provides the initial conditions for a shock wave, due to the high pressure differences developed on the nanometer scale (Sect. 2), and how the development of the shock wave propagates high pressures at high speeds around the ion track (Sect. 3). However, the hydrodynamics model does not tell us if these high pressures act for times long enough on the biomolecules to effectively produce bond breaking. Such question has to be answered by means of molecular dynamics simulations in which the detailed effect of the shock wave on the biomolecule bonds can be studied. Molecular dynamic approaches have been explored by several authors in recent years [3, 13–18].

The general procedure for setting up a molecular dynamics simulation of a shock wave in liquid water consists in selecting the water molecules which have been excited by the secondary electron shower and giving them velocities according to the energy lost by the ion. Typically, the most convenient way of doing so is by defining a cylinder around the ion track (the "hot cylinder"), and assuming that all the water molecules inside it have been electronically excited. Then, after some time, the electronic excitations will decay to the vibrational and translational degrees of freedom, due to the electron-phonon coupling mechanism. From estimates given by the inelastic thermal spike model [7] we know that this time for liquid water is of the order of 0.1 ps, which is the time at which the temperatures predicted by this model reached the maximum. This time can be considered as the starting point for the molecular dynamics simulations. Another question is the choice of the radius of this hot cylinder. As an estimate, it can be calculated from Eq. (2), so 1 nm is a reasonable value for this radius. Then, the velocities of the atoms inside the hot cylinder can be scaled by a factor α, which is calculated by the following equation [14]:

$$\sum_{i}^{N} \frac{1}{2} m_i (\alpha \cdot v_i)^2 = \frac{3 N k_{\mathrm{B}} T_0}{2} + S \cdot L \,, \tag{11}$$

where the sum is done over all the atoms N inside the cylinder, of mass m_i and velocity v_i (at body equilibrium temperature T_0). k_{B} is the Boltmann's constant, and L is the distance traveled by the ion, i.e., the length of the simulation box. The left hand side of Eq. (11) is the total kinetic energy of the atoms inside the hot cylinder after excitation. The first term at the right hand side is their total kinetic energy before excitation, and the product $S \cdot L$ corresponds to the energy deposited by the ion inside the hot cylinder, i.e., the stopping power S (energy loss per unit path length) times the length of the simulation box L.

It should be noted that, according to Eq. (11), all the energy lost by the ion is transferred to the translational and rotational degrees of freedom. This might not be actually the case. The energy initially transferred to secondary electrons is mainly spent in producing electronic excitations of the water molecules, which should decay to the nuclear system by $\sim$0.1 ps. However, a part of the energy lost by the ion is spent in promoting the electrons to the continuum and in creating holes in the system. This energy might be relaxed in different time scales, maybe longer than the time needed to develop the shock wave. Therefore, it is possible that part of the energy lost by the projectile is not effectively transferred to the shock wave. In order to account for this fact, in Ref. [16] a 75 % of the energy loss was used to scale the hot cylinder atom velocities, as a conservative estimate. This amount was estimated from the energy typically spent in overcoming the electronic binding energy: for a typical secondary electron of 45 eV, the binding energy of water being $\sim$15 eV represents $\sim$25 % of the energy lost in the ionizing collision. Even though, this conservative estimate was not used in Ref. [17], where all the energy loss was transferred to the motion of the hot cylinder atoms. The effect of these different approximations on the simulation results will be analyzed later on.

Before starting the discussion on the effects of shock waves in DNA, some important results can be extracted just from the simulations in water. The shock wave induced by carbon ions in pure liquid water was studied in Ref. [13], and some results about the shock wave in water can also be extracted from Ref. [17]. The results from both works demonstrate that the molecular dynamics simulations can reproduce the properties of the shock waves in accordance with the predictions of the hydrodynamics model reviewed in Sect. 3. The work of Ref. [13] showed how the radial velocity distributions of the water molecules in time at different distances from the ion track follow the front propagation predicted by Eq. (6). The evolution of the pressure waves produced by carbon and iron ions in liquid water was studied in Ref. [17]. The results for a carbon ion in the Bragg peak are shown in Fig. 3. The characteristics of the wave front can be obtained from the maxima of these pressure profiles. The results for carbon and iron ions in the Bragg peak region are shown by symbols in Fig. 2, and compared by the predictions of the hydrodynamics model, Eqs. (6) and (8). As it can be seen, the calculated position of the front and its pressure agree with the analytical model. The molecular dynamics results from these two works, then, reinforce the theoretical prediction of the shock wave effect.

A biological molecule can now be placed inside the simulation box to study the effect of the shock wave on its chemical bonds. The most important molecule is DNA since any damage to it has biological consequences. Several works have focused on DNA in its native form in cells, wrapped around histone proteins to form the nucleosome [14–16], while in others simpler DNA strands in water have been used for the evaluation [17, 18]. As a visual example of the most obviouseffects of shock waves in DNA, Fig. 4 shows some snapshots of the evolution of the geometry

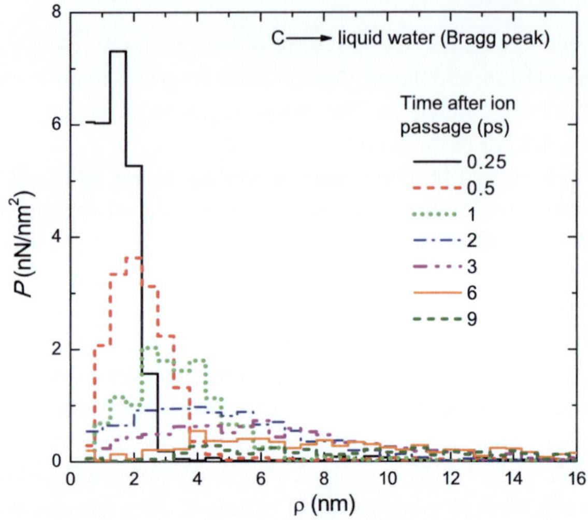

Fig. 3 Time evolution of the pressure wave produced by a carbon ion in liquid water in the Bragg peak region, as obtained from molecular dynamics simulation [17]

of a short DNA strand impacted by shock waves induced by carbon and iron ions in the Bragg peak region ($S = 900$ eV/nm and 7195 eV/nm, respectively) passing by at 1 nm distance from the DNA surface [17]. The atoms represented as spheres are the ones initially inside the hot cylinder. As time passes, they produce an explosion that expels matter in the radial direction, creating nano-cavities. This cavity is larger for iron than for carbon, since the stopping power of the former is eight times larger than the latter. At some point, the shock wave impacts the DNA molecule, distorting its structure, the more heavily the larger the stopping power of the ion. As another illustration, a movie showing the effect of a carbon ion induced shock wave on a nucleosome located at the same distance from the ion path can be watched in the supplementary material of Ref. [16]. The evolution of the system is quite similar, with the difference that the free DNA strand in water can be more stretched by the shock wave.

A visual inspection of Fig. 4 just reveals the rupture of the secondary structure of the DNA, what is a reversible process. Therefore it is interesting to analyze whether covalent bonds are broken or not, damage that might be permanent. For that purpose it is possible, during the molecular dynamics simulation, to monitor the potential energy stored in the covalent bonds of DNA. This is specially important for the backbone bonds, since their rupture may lead to a single strand break (SSB). The simulation of biomolecules such as DNA is generally done using the CHARMM forcefield [19]. This forcefield takes into account the geometrical constrains to which chemical bonds are held, due to the molecular orbital hybridization. The force acting on the atom i is obtained from the potential energy $U(\mathbf{R})$ as $\sum_{j\neq i} \mathbf{F}_{ij} = \mathrm{d}U(\mathbf{R})/\mathrm{d}\mathbf{r_i}$ which corresponds to a given set of atomic coordinates $\mathbf{R} = \{\mathbf{r_i}\}$ and is expressed as a combination of energies arising from the distances between pairs of bonded atoms, the angles formed between groups of three sequentially bonded atoms, the dihedral torsion angle formed by groups of sequentially four bonded atoms, the improper angles formed between groups of atoms that should form a plane, and the

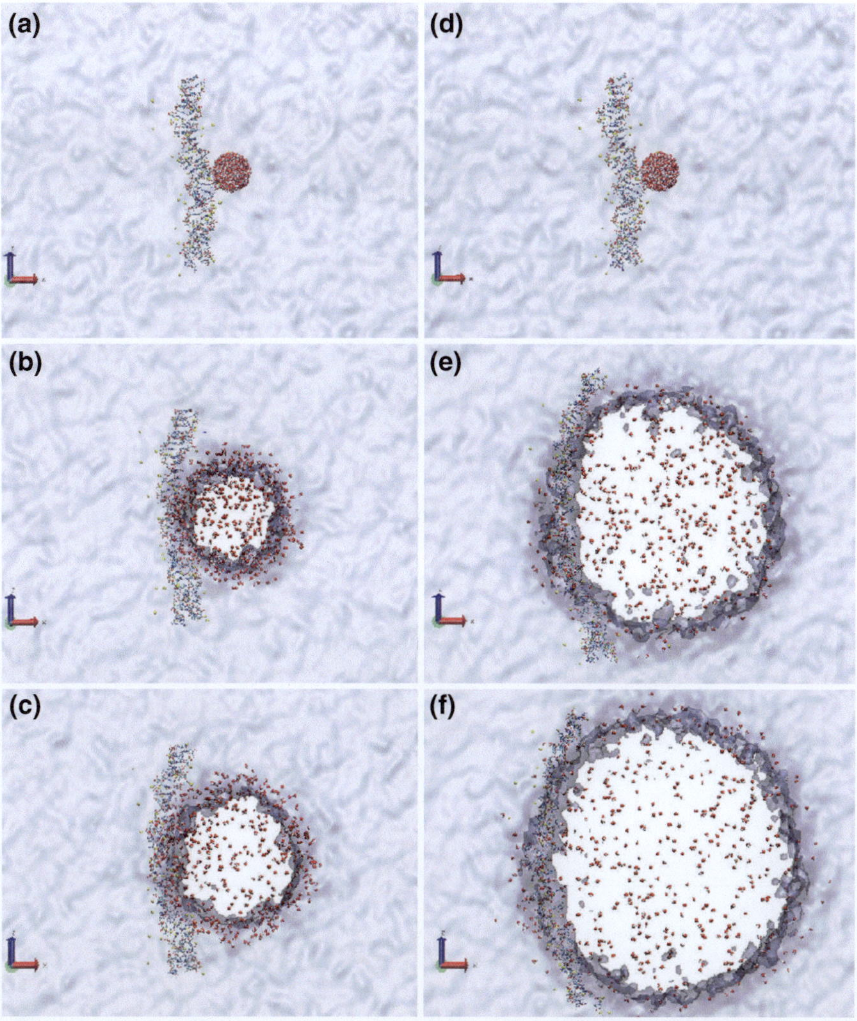

Fig. 4 Snapshots of the structure of a DNA strand in liquid water, impacted by shock waves induced by a carbon and an iron ion in the Bragg peak region. *Panels a–c* correspond to times 0, 5, and 10 ps, respectively, for carbon in water, while panels *d–f* correspond to the same times for iron in water [17]

nonbonded interactions represented by the pure Coulomb force and the van der Waals interaction between pairs of atoms:

$$U(\mathbf{R}) = \sum_{\text{bonds}} K_{\text{b}}(b - b_0)^2 + \sum_{\text{angles}} K_\theta(\theta - \theta_0)^2 + \sum_{\text{dihedr.}} K_\chi \left(1 + \cos(n\chi) - \Delta\right) \tag{12}$$

$$+ \sum_{\text{improp.}} K_\varphi(\varphi - \varphi_0)^2 + \sum_i \sum_{j \neq i} \frac{q_i q_j}{\varepsilon r_{ij}} + \left[\sigma_{ij}\left(\frac{R_{\text{min},ij}}{r_{ij}}\right)^{12} - \left(\frac{R_{\text{min},ij}}{r_{ij}}\right)^{6}\right].$$

In this equation b is the bond distance between two bonded atoms, θ is the bond angle between every triplet of sequentially bonded atoms, χ is the dihedral torsion angle formed by every four atoms connected via covalent bonds and φ is the improper torsion angle, used to maintain planarity between groups of sequentially bonded atoms; b_0, θ_0 and φ_0 correspond to the equilibrium quantities, while n and Δ determine the periodicity of the dihedral interaction. K_b, K_θ, K_χ and K_φ are the corresponding force constants. The Coulomb interaction is characterized by the atomic partial charges q_i, the interatomic distances r_{ij} and the effective dielectric constant ε. The van der Waals interaction is defined by a 6–12 Lennard-Jones potential with well depth σ and the minimum energy distance R_{min}. All of these parameters can be obtained for many biological molecules, including nucleic acids and proteins, from the CHARMM potential [19].

Thus, the potential energy of a particular covalent bond can be obtained from the first term in the right hand side of Eq. (12) as a function of time. Figure 5a shows, as an example, the bond distance of given bond in the DNA backbone, when exposed to a shock wave produced by a iron ion in the Bragg peak region, as depicted in Fig. 4. Figure 5b shows the corresponding potential energy of this bond. If this energy overcomes the given dissociation threshold then it can be assumed that this bond is broken. The dissociation energies for the bonds present in the DNA backbone range

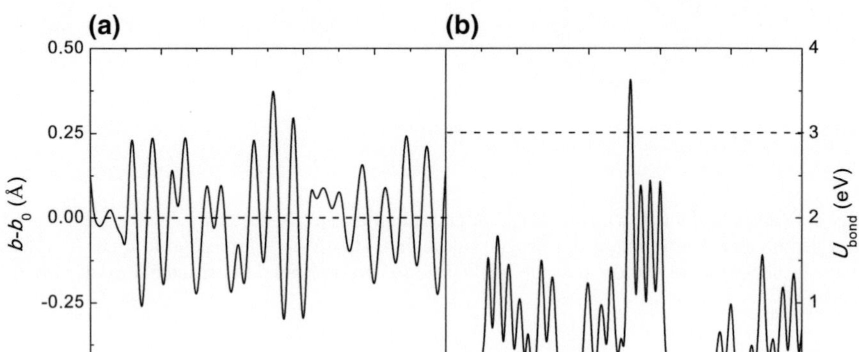

Fig. 5 Evolution of **a** the bond length and **b** the bond potential energy on time, for a DNA backbone bond during an iron-ion induced shock wave in the Bragg peak region. The *dashed line* in *panel a* represents the equilibrium length of the bond, while the *dashed line* in *panel b* marks a bond breaking threshold energy of 3 eV

between 3 and 6 eV [20]. However, these energies can be as low as 0.3 eV, due to the interaction of the reactive species present in the medium due to irradiation [21]. An example of bond breaking can be seen in Fig. 5, when the potential energy goes above the dashed line, representing a given bond breaking threshold of 3 eV.

The distribution of energy deposited in DNA backbone bonds can be obtained by monitoring all the energy deposition events, which we will define as all the local maxima in the bond potential energy profiles, as shown in Fig. 5b. Then, all these events can be represented in an histogram, where the number of events between $U_{\rm bond}$ and $U_{\rm bond} + dU_{\rm bond}$ is represented. Such histograms are shown in Fig. 6 for carbon ($S = 900$ eV/nm) and iron ($S = 7195$ eV/nm) ions induced shock waves in the Bragg peak region, both for the nucleosome [16] and the DNA strand [17] targets, both located at 1 nm from the ion path. As can be seen the systematics are quite similar for both targets, with an increase in the number of high energy events with the increase in the stopping power of the ion. For iron, a slightly larger probability of high energy events is observed for the case of the free DNA strand (full squares) as compared with the nucleosome (open squares). This difference comes mainly from the fact that, in Ref. [17], all the energy lost by the ion was used for the development of the shock wave, while only 75 % was used in Ref. [16], as explained when discussing Eq. (11). Even though, it can be checked that, when only 75 % of the energy is used in the case of the free DNA strand, the distribution is much closer to the one of the nucleosome, as it can be seen by the diamond symbols presented in the figure. This reveals that the geometry of the system is not very relevant for the evaluation of the high energy deposition events. Also, it is clear that the amount of energy considered for the development of the shock wave can affect the results. This is why, in the

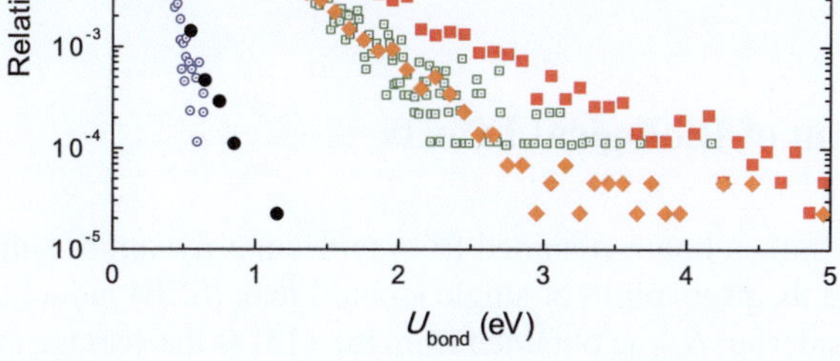

Fig. 6 Distribution of energy deposition events in the DNA backbone bonds, for shock waves produced in the Bragg peak region by a carbon and an iron ion passing by at 1 nm from DNA. *Filled symbols* are the results for a short DNA duplex [17], while *open symbols* are results for nucleosome [16]. *Orange diamonds* are calculations for the DNA duplex, assuming that only 75 % of the energy loss by the ion is used for the shock wave. See the text for the details

following, the conservative estimate of 75 % will be used, although it is clear that the effects can be even larger.

A systematic analysis of these trends for the case of nucleosome was performed in Ref. [16] for a series of ions in the Bragg peak region having different stopping powers: carbon ($S = 900$ eV/nm), neon ($S = 1700$ eV/nm), argon ($S = 4700$ eV/nm) and iron ($S = 7195$ eV/nm). The bond energy distributions for neon and argon lies in between of those shown for carbon and iron in Fig. 6. All of them are characterized by a linear behavior in the linear-log scale, evidencing the fact that they correspond to Boltzmann distributions of the type:

$$\frac{1}{N_0}\frac{\mathrm{d}N_{\mathrm{events}}}{\mathrm{d}U_{\mathrm{bond}}} = \frac{1}{k_{\mathrm{B}}T}\exp\left[-\frac{U_{\mathrm{bond}}}{k_{\mathrm{B}}T}\right], \tag{13}$$

corresponding to some characteristic temperature T. N_0 is the total number of recorded energy deposition events used to normalize the distributions. These two parameters can be obtained by fitting to the molecular dynamics results. Having taken into account that the equilibrium temperature is $T_0 = 310$ K (body temperature), it was found from the analysis of the data in Ref. [16] that the increase in temperature of the system follows a linear behavior with the stopping power S:

$$T - T_0 = \mu S\,, \tag{14}$$

where $\mu = 494$ K nm keV^{-1}.

An advantage of the identification of the energy deposition profiles with Boltzmann distributions is that this allows us to determine the number of above-threshold events as:

$$N_{\mathrm{breaks}} = \int_{U_0}^{\infty}\frac{\mathrm{d}N_{\mathrm{events}}}{\mathrm{d}U_{\mathrm{bond}}}\mathrm{d}U_{\mathrm{bond}} = N_0\exp\left[\frac{-U_0}{k_{\mathrm{B}}T}\right], \tag{15}$$

where the parameters T and N_0 have been already obtained from simulation results, so we can readily obtain N_{breaks} as a function of the stopping power S. U_0 is the threshold energy considered for bond dissociation.

5 Estimation of Biological Effects

The number of broken bonds obtained from molecular dynamics simulation can be used to estimate the probability of single strand break (SSB) induction in the DNA backbone. Considering N_{breaks} obtained from Eq. (15) as the average number of SSB, we can use Poisson statistics to obtain the actual probability of producing ν SSB as:

$$P(\nu) = \frac{\exp\left(-N_{\mathrm{breaks}}\right)N_{\mathrm{breaks}}^{\nu}}{\nu!}\,. \tag{16}$$

Fig. 7 Probability of SSB induction by shock waves produced by ions with different stopping powers, assuming three different threshold energies for bond breaking U_0. The results are compared with the probability of SSB induction by chemical effects [16]

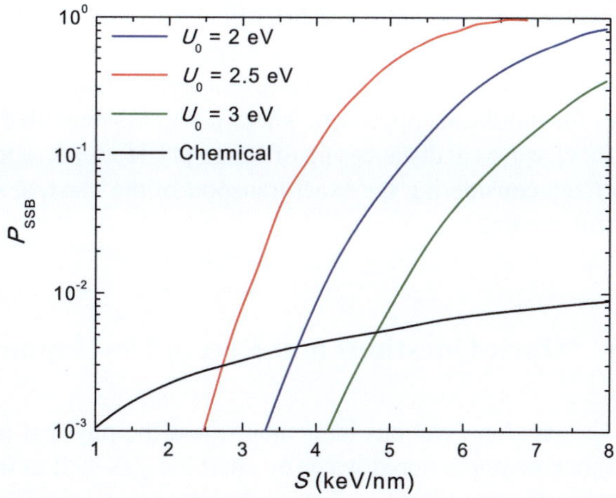

Then, the probability of having at least one SSB in a given segment of DNA molecule is:

$$P_{\text{SSB}} = 1 - P(0) = 1 - \exp\left(-N_{\text{breaks}}\right) . \tag{17}$$

This probability is represented as a function of the ion stopping power in Fig. 7 for three values of the threshold energy for breaking bonds U_0, 2, 2.5, and 3 eV. Of course, the choice of U_0 influences this calculation. However, a better picture of the relevance of shock waves on the induction of SSB is got by comparing these values with the probability of inducing SSB by chemical effects (i.e., secondary electrons and free radicals) in a similar DNA segment located at the same distance. Such an estimate can be obtained from the multiscale approach as [1, 16]:

$$P_{\text{SSB}}^{\text{chem}} = \kappa \frac{S}{S_0} , \tag{18}$$

where S is the ion's stopping power, $S_0 = 0.9$ keV/nm is the value for carbon, and $\kappa = (1.1 \pm 0.5) \times 10^{-3}$. The result from this expression is also shown in Fig. 7. For the most conservative estimate of $U_0 = 3$ eV, the shock wave starts to produce SSB at $S \geq 4$ keV/nm, and it actually overcomes other effects by 5 keV/nm. Therefore, these results mean that, for ions heavier than Ar, passing by at distances shorter than 2 nm from the DNA, the shock wave is the dominant mechanism of SSB induction [16].

Even though, as already discussed in Sect. 3, the biological role of the shock wave does not only consist on the direct production of SSB by thermo-mechanical stress. Even for lower stopping power ions, such as protons or carbons, the radial transport of chemical species by the collective flow is much faster than diffusion. Indeed, owing to the diffusion coefficient of the relevant secondary species ($<10^{-4}$ cm^2/s [12]), it would take about 9 ns for them to reach distances of 5 nm from the track, while their lifetimes are shorter than 5 ns [12, 22, 23]. Therefore, the shock wave might be the only means of transporting these species away from the track. This idea, indeed, has been used to estimate the effect of radicals on the biological effects in the multiscale approach, where it was assumed that they were propagated by the shock waves at distances up to 10 nm [1]. However, a more refined estimation of this effect, considering the exact transport of the reactive species by the shock wave, is still pending.

6 Open Questions and Recent Developments

The previous sections have manifested the physical origin for the development of shock waves in liquid water by swift ions, as well as their main physical properties, their effects on DNA and, thus, their biological significance. However, all the works reviewed do not cover the whole problem, and some open questions remain to be clarified.

First of all, the simulations of DNA damage described in the previous section correspond to one specific impact parameter of the ion with respect to the DNA molecule, either it is the nucleosome or the DNA strand. In both studies, the impact parameter was set to 2 nm from the center of the DNA to the center of the track [16, 17]. In Ref. [16], some estimations about the radius of predominance of the shock wave effect over chemical contributions were made, based on the predicted properties of the shock wave front by the hydrodynamics model. However, it has to be taken into account that, as it has been shown in Fig. 3, the shock wave profile widens with time, so the front is less defined as the time increases. This fact might have some consequences on the dependence of the shock wave effect with distance. Also, the case of direct ion traversal through DNA was studied in Ref. [18]. From the analysis shown in this work, the probability of inducing SSB seems to be quite larger, even having into account that these authors consider lower stopping powers that the considered in Sects. 4 and 5. Moreover, different orientations between DNA and the ion track can also produce different effects. Therefore, all these aspects have to be taken into account in more systematic studies.

There are two more different aspects between the work performed in Ref. [18] and the results shown in Sects. 4 and 5. One of them is the definition of the "hot cylinder" radius. While we have shown in Sect. 2 that it is possible to estimate it in 1 nm, based on physical considerations, the authors of Ref. [18] use a radius of 2 Å. It is still a question how this choice may affect the results. Even though, future work could even disregard the choice of this radius: the results from the random walk approximation shown in Sect. 2 [4] very clearly establish the initial conditions of the shock wave, in terms of the radial pressure distribution arising from the energy deposition by the electrons (see Fig. 1). As shown in Sect. 2, this pressure profile is built up within 50 fs after ion's passage. According to the estimations from the inelastic thermal spike model [7], the transfer of this energy from the electronic excitations to the vibrational and translational degrees of freedom does not occur until $\sim$0.1 ps. Therefore, the pressure profile around the ion track at the initial time of the shock wave development is the one given by the track structure calculation (Fig. 1), so it can be directly used to set up the molecular dynamics simulation, avoiding the problem of choosing a given hot cylinder radius.

The second difference from the work reported in Ref. [18] and the results shown previously in this chapter is the forcefield used. In all the simulations previously reported, the CHARMM forcefield [19] was used to describe the biomolecules. This forcefield is developed to reproduce the properties of biomolecules in equilibrium, so it uses harmonic potentials to describe the bond, angle, and improper angle energies. This means that such a potential does not allow bond dissociation, for what a potential which becomes less powerful as the bond length increases is needed. In turn, the reactive forcefield ReaxFF [24] is used in Ref. [18]. This forcefield is especially designed to allow bond breaking and reformation through chemical reactions. Therefore, the differences between the forcefields used might contribute to the different results reported by these two groups. It is worth to note, though, that a new reactive CHARMM forcefield has been recently implemented [25] in the simulation package MBN Explorer [26], program that has been used to perform some of the simulations reported in Sect. 4 [17]. This reactive CHARMM forcefield has demonstrated, for example, to be able of reproducing the known reactions of water splitting occurring at high temperatures [25]. Therefore, the use of this forcefield within the MBN Explorer software will allow to better study the bond breaking in DNA within the CHARMM forcefield, what may compare better with the results reported in Ref. [18], and will represent an improvement on the estimations of the biological relevance of shock waves. Also, the water dissociation itself can be studied as a consequence of the shock wave, what might contribute to the total number of chemical species present in solution. Finally, molecular dynamics simulations can also be used to study in detail the propagation of all these secondary species by the collective flow.

7 Summary, Conclusions and Outlook

In this chapter, we have reviewed the research developed during the last years on thermo-mechanical damage of biomolecules under ion beam irradiation, what is a newly proposed mechanism of biodamage in ion beam cancer therapy. First of all, the physical origin of the thermo-mechanical effects has been explained in Sect. 2, where it has been shown how the space and time scales characteristic of the secondary electron propagation around the ion track give place to the large temperatures and pressures that can induce the hydrodynamic response of the liquid medium.

The solution to this hydrodynamics problem has been reviewed in Sect. 3, where it has been demonstrated how the pressure differences built up around the ion track satisfy the conditions for a strong cylindrical explosion, i.e., to develop a shock wave in the nanoscale. The main physical properties of these shock waves, namely their speed of propagation and pressure, have been also determined on the basis of this hydrodynamics model. These results show that ion tracks induce an important collective flow of matter, that can transport reactive species faster than diffusion, and that the high pressures developed might be able of breaking chemical bonds in biomolecules.

This aspect has been analyzed in detail in Sect. 4, where results from molecular dynamics simulations have been used to determine the number of bond breaks happening in DNA molecules located within nm distance of the ion track as a function of the stopping power of the ion. In Sect. 5, the previous results have been used to estimate the biological significance of the shock waves. From the number of broken bonds calculated by molecular dynamics, Poisson statistics has been used to obtain the probability of single strand break (SSB) induction by the shock waves. It has been compared with the probability for SSB induction owing to chemical effects (i.e., secondary electrons and free radicals), and it has been determined that, for the short distances analyzed ($\leq$2 nm), the shock wave effect, in terms of SSB induction by thermo-mechanical stress, overcomes that of chemical effects for ions having stopping powers larger than 5 keV/nm, i.e., ions heavier than Ar in the Bragg peak region.

Finally, some open questions have been set out Sect. 6. It turns out that a more systematic analysis of the dependence of shock waves effects on the DNA-track distance is required, including the possibility of direct ion hit, what might produce larger numbers of SSB. Also, better estimations of the biological effects may be got by the use of reactive forcefields, such as ReaxFF, or the recently implemented reactive CHARMM forcefield in the MBN Explorer simulation package. The use of these reactive forcefields will yield better predictions of bond breaking, as well as the contribution from the reactive species generated around the ion track, which transport by the shock wave and its interaction with DNA should be better understood. Once a better understanding of all the properties of ion induced shock waves has been got, it might be possible to design experiments in which some of these properties can be observed, what will be needed for the final confirmation of this newly proposed damage mechanism.

Acknowledgements Part of the research reviewed in this chapter was developed with the support of the European Union's COST Action MP1002, Nanoscale insights into Ion Beam Cancer Therapy. The most recent work has been done within the European Union's (FP7-People Program, Marie Curie Actions) Initial Training Network No. 608163 "ARGENT", Advanced Radiotherapy, Generated by Exploiting Nanoprocesses and Technologies, to which PdV acknowledges financial support.

References

1. Surdutovich E, Solov'yov AV (2014) Multiscale approach to the physics of radiation damage with ions. Europ Phys J D 68:353-1–353-30
2. Boudaïffa B, Cloutier P, Hunting D, Huels MA, Sanche L (2000) Resonant formation of DNA strand breaks by low-energy (3 to 20 eV) electrons. Science 287:1658–1660
3. Surdutovich E, Solov'yov AV (2010) Shock wave initiated by an ion passing through liquid water. Phys Rev E 82:051915-1–051915-5

4. Surdutovich E, Solov'yov AV (2015) Transport of secondary electrons and reactive species in ion tracks. Europ Phys J D 69:193-1–193-7
5. Surdutovich E, Obolensky OI, Scifoni E, Pshenichnov I, Mishustin I, Solov'yov AV, Greiner W (2009) Ion-induced electron production in tissue-like media and DNA damage mechanisms. Europ Phys J D 51:63–71
6. de Vera P, Abril I, Garcia-Molina R, Solov'yov AV (2013) Ionization of biomolecular targets by ion impact: input data for radiobiological applications. J Phys: Conf Ser 438:012015-1–012015-10
7. Toulemonde M, Surdutovich E, Solov'yov AV (2009) Temperature and pressure spikes in ion-beam cancer therapy. Phys Rev E 80:031913-1–031913-9
8. Landau L, Lifshitz E (1987) Fluid Mechanics, vol 6 of Course of Theoretical Physics, 2nd edn. Reed-Elsevier, Oxford, Boston, Johannesburg
9. Zeldovich Y, Raiser Y (1966) Physics of shock waves and high-temperature hydrodynamic phenomena, vol 1. Dover Publications, Oxford, New York
10. Chernyj G (1994) Gas dynamics. Nauka, Moscow
11. Surdutovich E, Yakubovich AV, Solov'yov AV (2013) DNA damage due to thermomechanical effects caused by heavy ions propagating in tissue. Nucl Instr Meth B 314:63–65
12. LaVerne J (1989) Radical and molecular yields in the radiolysis of water with carbon ions. Radiat Phys Chem 34:135–143
13. Yakubovich AV, Surdutovich E, Solov'yov AV (2011) Atomic and molecular data needs for radiation damage modeling: multiscale approach. AIP Conf Proc 1344:230–238
14. Yakubovich AV, Surdutovich E, Solov'yov AV (2012) Thermomechanical damage of nucleosome by the shock wave initiated by ion passing through liquid water. Nucl Instr Meth B 279:135–139
15. Yakubovich AV, Surdutovich E, Solov'yov AV (2012) Damage of DNA backbone by nanoscale shock waves. J Phys: Conf Ser 373:012014-1–012014-7
16. Surdutovich E, Yakubovich AV, Solov'yov AV (2013) Biodamage via shock waves initiated by irradiation with ions. Sci Rep 3:1289-1–1289-6
17. de Vera P, Mason NJ, Currell FJ, Solov'yov AV (2016) Molecular dynamics study of accelerated ion-induced shock waves in biological media. Europ Phys J D 70:183
18. Bottländer D, Mücksch C, Urbassek HM (2015) Effect of swift-ion irradiation on DNA molecules: a molecular dynamics study using the REAX force field. Nucl Instr Meth B 365:622–625
19. MacKerell AD Jr, Bashford D, Bellott M, Dunbrack RL Jr, Evanseck JD et al (1998) All-atom empirical potential for molecular modeling and dynamics studies of proteins. J Phys Chem B 102:3586–3616
20. Range K, McGrath MJ, Lopez X, York DM (2004) The structure and stability of biological metaphosphate, phosphate, and phosphorane compounds in the gas phase and in solution. J Am Chem Soc 126:1654–1665
21. Smyth M, Kohanoff J (2012) Excess electron interactions with solvated DNA nucleotides: strand breaks possible at room temperature. J Am Chem Soc 134:9122–9125
22. von Sonntag C (1987) The chemical basis of radiation biology. Taylor & Francis, London
23. Alpen EL (1998) Radiation biophysics. Academic Press
24. van Duin ACT, Dasgupta S, Lorant F, Goddard WA III (2001) ReaxFF: a reactive force field for hydrocarbons. J Phys Chem A 105:9396–9409
25. Sushko GB, Solov'yov IA, Verkhovtsev AV, Volkov SN, Solov'yov AV (2016) Studying chemical reactions in biological systems with MBN Explorer: implementation of molecular mechanics with dynamical topology. Eur Phys J D 70:12-1–12-10
26. Solov'yov IA, Yakubovich AV, Nikolaev PV, Volkovets I, Solov'yov AV (2012) MesoBioNano Explorer—a universal program for multiscale computer simulations of complex molecular structure and dynamics. J Comput Chem 33:2412–2439. http://www.mbnexplorer.com

Predictive Assessment of Biological Damage Due to Ion Beams

Alexey Verkhovtsev, Eugene Surdutovich and Andrey V. Solov'yov

Abstract This chapter presents recent achievements in validation of the Multiscale Approach (MSA) to the physics of radiation damage with ions. An analytical recipe for the assessment of biological damage, developed using the phenomenon-based MSA, has been applied to numerous experiments, where survival curves were obtained for different cells and irradiation conditions. Contrary to other, in essence empirical methods for evaluation of macroscopic effects of ionizing radiation, the MSA predicts the biodamage based on the physical effects related to ionization of the medium, transport of secondary particles, chemical interactions, thermo-mechanical pathways of biodamage, and heuristic biological criteria for cell survival. An extensive comparison with experimental data for cell survival probability demonstrates the validity of the MSA to predict the macroscopic effects of ionizing radiation through an understanding of biological damage on the nanoscale. The analysis performed allows us to conclude that the biodamage can be accurately predicted in both aerobic and hypoxic conditions. Therefore, we anticipate this method to give great impetus to the practical improvement of ion-beam cancer therapy and the development of more efficient treatment protocols.

A. Verkhovtsev (✉)
Instituto de Fisica Fundamental, Consejo Superior de Investigaciones Cientificas (CSIC),
Serrano 113-bis, 28006 Madrid, Spain
e-mail: verkhovtsev@iff.csic.es

A. Verkhovtsev · E. Surdutovich
MBN Research Center,
60438 Frankfurt am Main, Germany
e-mail: surdutov@oakland.edu

A.V. Solov'yov
MBN Research Center at Frankfurter Innovationszentrum Biotechnologie,
Altenhöferallee 3, 60438 Frankfurt am Main, Germany
e-mail: solovyov@mbnresearch.com

A.V. Solov'yov
A.F. Ioffe Physical-Technical Institute,
Polytekhnicheskaya Ul. 26, 194021 Saint Petersburg, Russia

E. Surdutovich
Department of Physics, Oakland University, Rochester, MI 48309, USA

A.V. Solov'yov (ed.), *Nanoscale Insights into Ion-Beam Cancer Therapy*,
DOI 10.1007/978-3-319-43030-0_11

1 Introduction

The damaging effects of ionizing radiation have been studied for many years. Regardless whether the goal is to kill cancerous cells using radiotherapy or to protect cells exposed to radiation, the relation of physical conditions to the biological effects has always been the major challenge for radiobiology [1]. Historically, this relation is presented as the dependence of the clonogenic cell survival probability on the deposited dose. These dependencies are known as survival curves and they are the starting point for any treatment plan. Having an objective of inactivating certain fraction of cells in a given region, radiation oncologists determine the dose that has to be delivered in this region. Photons have been the most common radiation used for therapy and a vast experience has been gained for different cells in different conditions. The biological diversity of survival curves is such that there was no hope to predict their shape theoretically.

When ion-beam modality [2, 3] became a reality in 1990s [4], the survival curves for ions were obtained by and large either experimentally or semi-empirically, based on the photon experience. However, the MultiScale Approach (MSA) to the assessment of radiation damage with ions suggested a possibility to predict the survival curves theoretically based on major physical effects [5]. This possibility arises because in the case of ions the physical and chemical actions may be more definitive for the biological response. The MSA considers these effects on different scales in space, time, and energy, and explores their relation to biological damage. The realization of the goal of the MSA was described in the form of a recipe for calculating survival curves [5].

In both x-ray and ion-beam irradiations, secondary electrons, free radicals, and other reactive species do the major damage to the cells, but the qualitative features such as the dose dependence of the probability of cell survival are different for these radiation modalities. When tissue is exposed to x-rays, the dose distribution in the plane perpendicular to the beam axis on the cellular ($\sim$10 μm) scale is uniform. Therefore, the model for determining the probability of cell survival can be built on a single physical quantity such as dose [1]. With ions, the above picture changes since the radial distribution of dose varies on the nanometer scale and the dose on the cellular and sub-cellular scales becomes a probabilistic function of a number of ions traversing a given target [5]. Physical parameters, such as number density of reacting species or their fluence, describe radiation damage in a more straightforward way. This is why all semi-empirical approaches [6, 7] (incorporated in existing treatment protocols) based on the linear-quadratic (LQ) model,

$$-\ln F = \alpha d + \beta d^2, \tag{1}$$

where F is a surviving fraction of cells, d is the dose, with the coefficients α and β being determined empirically, become artificial. However, the full potential of ion-beam therapy can only be realized if the fundamental mechanisms [8–10] leading to lethal damage of cells under ion irradiation are well understood. This understanding is fostered by means of the MSA. Having been developed specifically for ions [11, 12], this approach joins the knowledge about ion stopping in the medium, the production of secondary electrons and other reactive species in the vicinity of ion's path, the transport of these species, the interaction of secondary particles with biomolecules, relaxation of energy in the medium that leads to thermo-mechanical damage due to the formation of nanoscale shock waves [13–15], and the interaction of secondary species with DNA molecules to calculate the probability of important lesions, such as double and single strand breaks (DSBs and SSBs) per unit length of ion's path [5]. On this basis, a criterion for lethality of damage is established, and, finally, the probability of production of lethal lesions is obtained. All these features make the MSA unique because it can predict the macroscopic effects of ionizing radiation, based on the inclusive scenario and fundamental science.

In this chapter, we present the results of evaluation of radiobiological effects due to ion beams by means of the MSA. An analytical recipe for the assessment of biological damage has been applied to numerous experiments, where survival curves were obtained for different cells and irradiation conditions. Capability and predictive power of the MSA is demonstrated by an extensive comparison with experimental data for numerous mammalian cancerous and normal cell lines, such as A549, AG1522, A172, CHO, V79, HeLa, and NB1RGB. The analysis performed allows us to conclude that the macroscopic effects of ionizing radiation can be accurately predicted through an understanding of biological damage on the nanoscale. The understanding of the phenomena at play on a solid physical basis is crucial for technological advances of new treatment techniques.

2 Calculation of Cell Survival Probability Within the MSA

The MSA is a phenomenon-based approach that aims at obtaining knowledge about biological damage on the nanoscale and finding the relation between the characteristics of incident particles and the resultant biodamage [5]. Contrary to other methods which are mostly based on a number of empirical parameters, the MSA elucidates physical, chemical, and biological effects that happen on a variety of temporal, spatial, and energy scales, and thus allows one to assess relative biological effectiveness (RBE) of ionizing radiation. A comprehensive description of different aspects of the MSA is given in Ref. [5] and Chap. 1 of this book. In this section, we briefly outline the formalism used for the estimation of radiobiological effects within this approach.

The calculation of a survival curve starts with establishing the relation between physical effects and the lethality of radiation damage. In regard to irradiation with heavy charged particles, the key assumption adopted in the MSA following Refs. [16–18] is that the leading cause of cell inactivation is the complexity of nuclear DNA damage. However, this may change in different conditions, e.g., when biodamage takes place in presence of sensitizing nanoparticles [19, 20]. It is currently established that radiosensitizing nanoparticles delivered to the cells are preferentially localized outside the cell nucleus [21, 22], so that the damage of other organelles may become more important.

In this work, we have focused our attention on cell damage brought about by pathways that involve only nuclear DNA damage. The criterion for lethality of damage suggested in Ref. [5] is based on the well-established hypothesis that among different DNA lesions caused by interaction with reactive species (e.g., secondary electrons, free radicals, and solvated electrons), the multiply damaged sites with a sufficient complexity may not be repaired [16, 18, 23]. In the formulated recipe for the assessment of biodamage [5], it was postulated that a complex lesion combined of a DSB and at least two other simple lesions such as SSB within two DNA twists is lethal for a cell. This criterion required further justification that has been carried out in the work summarized in this chapter.

The multiple damage sites containing clustered damage are brought about by several independent agents [24, 25]. In this analysis, the number of simple lesions is equated to the number of agents (secondary electrons, free radicals, or other reactive species) attacking a DNA segment. The average number of lesions per target produced at a distance r from the ion's path is defined as:

$$\mathcal{N}(r) = \mathcal{N}_{\rm e}(r) + \mathcal{N}_{\rm r}(r) = \Gamma_{\rm e}\mathcal{F}_{\rm e}(r) + \Gamma_{\rm r}\mathcal{F}_{\rm r}(r) \, , \tag{2}$$

where the functions $\mathcal{N}_{\rm e}(r)$ and $\mathcal{N}_{\rm r}(r)$ define the average number of lesions like SSBs, base damages, etc., done by secondary electrons and other reactive species, respectively.

The calculation of the probability for the clustered damage to occur starts with the calculation of the number of secondary electrons incident on a given DNA segment [24]. This number is averaged over all angles and as a result the number of hits with secondary electrons, $\mathcal{F}_{\rm e}(r)$, is obtained as a function of distance from the ion's path. The function $\mathcal{N}_{\rm e}(r)$, is then calculated by multiplying $\mathcal{F}_{\rm e}(r)$ by the probability of inducing a lesion per hit, $\Gamma_{\rm e}$. The same is done for free radicals and other reactive species. Function $\mathcal{F}_{\rm r}(r)$ includes the physics pertinent to transport of reactive species, such as the relaxation of ionization energy in the medium and the (predicted) cylindrical shock wave around the ion's path [5, 13].

Then, the criterion for lethality is introduced as

$$P_l(r) = \lambda \sum_{\nu=3}^{\infty} \frac{[\mathcal{N}(r)]^{\nu}}{\nu!} \exp\left[-\mathcal{N}(r)\right] , \tag{3}$$

where ν is the number of simple lesions per cluster. The DSBs consequent to SSBs are more probable than those due to independent nearby SSB events [26–28]. This is accounted for by introducing the factor λ that is a probability of conversion of an SSB into a DSB. The introduction of this factor relies on experimental findings [26, 27] that the DSBs caused by the electrons with energies higher than about 5 eV happen in one hit. In this case, the subsequent break in the second strand is due to the action

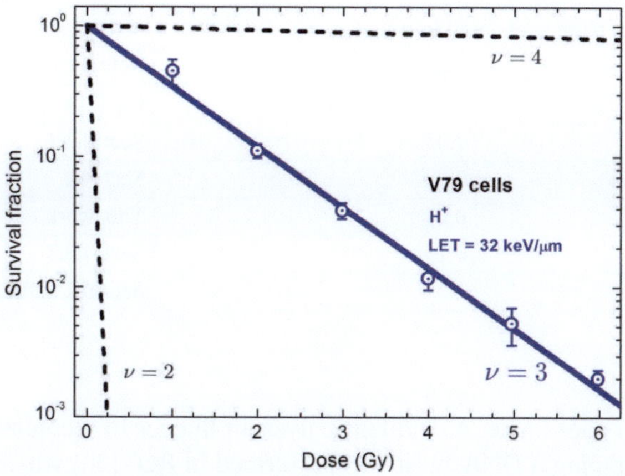

Fig. 1 Dependence of cell survival after irradiation with ions on the criterion for lethality of radiation damage. All the calculations performed in this work were carried out assuming that at least three simple damages within two DNA twists are required in order for damage to be lethal (the sum in Eq. (3) starts with $\nu = 3$). The variation of this criterion, i.e. when it is supposed that lethal damage is done by at least two ($\nu = 2$) or by four ($\nu = 4$) simple damages, leads to systematically incorrect survival curves (*dashed lines*). Symbols denote the experimental data for the Chinese hamster V79 cell line [29]

of debris generated by the first SSB. In the cited works, it was shown that if a single electron causes an SSB, the same electron causes a DSB with a probability of about 0.1–0.2 of that to create an SSB. The value $\lambda = 0.15$ was suggested earlier [5] and has been left unchanged in the presented analysis. It was utilized in all the calculations presented.

The sum in Eq. (3) starts with $\nu = 3$, which means that at least three simple damages (one of which should convert into a DSB) within a certain region are required in order for damage to be lethal. This criterion, introduced in Ref. [5] heuristically, is fully applicable for quantitatively correct prediction of cell survival, while the assumption that lethal damage is done either by a smaller ($\nu = 2$) or by a larger ($\nu = 4$) number of simple damages, yields systematically incorrect results (see Fig. 1). Function $P_l(r)$, Eq. (3), represents the radial distribution of lethal lesions. Finally, it has to be integrated over the area perpendicular to the ion's path and multiplied by the number density of sites on chromatin, n_s, (assumed to be uniform) to obtain the number of lethal lesions per ion's path $\mathrm{d}x$:

$$\frac{\mathrm{d}N_l}{\mathrm{d}x} = n_\mathrm{s} \int_0^\infty P_l(r)\, 2\pi r\, \mathrm{d}r = n_\mathrm{s}\, \sigma \,, \tag{4}$$

where $\sigma = \int_0^\infty P_l(r)\, 2\pi r\, \mathrm{d}r$ is the effective cross section of the complex damage site, which depends on linear energy transfer (LET) [5]. The latter quantity is approximately equal to the stopping power, $\mathrm{d}E/\mathrm{d}x$, with E being the ion's energy and x

Table 1 Values of parameters used in calculations

Parameter	Value	Reference (in Ref. [5])
$\mathscr{F}_e(r)$		Eq. (20)
Γ_e	0.03	Sect. 4.3.1
$\mathscr{N}_r$ (normal)	0.08	Sect. 4.4
$\mathscr{N}_r$ (hypoxic)	0.04	This work
λ	0.15	Sect. 4.3.1
$P_l(r)$		Sect. 6.1 and Fig. 13

the longitudinal coordinate. A detailed analysis of fluence of secondary electrons on a cylinder enwraping a DNA twist was performed in Ref. [30] where the results of the analytical approach were compared to those of Monte Carlo simulations. Values of the parameters entering Eqs. (2)–(4) as well as the references for more detailed explanation of these numbers are summarized in Table 1.

The number density of targets, n_s, is proportional to the ratio of base pairs accommodated in the cell nucleus to the nuclear volume, $n_s \sim N_{bp}/V_n$. The coefficient of proportionality takes into account that a target represents a double DNA twist comprising 20 bp [5]:

$$n_s = \frac{N_{bp}}{20\,V_n} = \frac{3\pi}{8} \times \frac{N_{bp}}{20\,A_n\,\bar{z}}\,, \tag{5}$$

where the cross sectional area, A_n, of the cell nucleus, its diameter, D_n, and an average length of ions' traverse through a nucleus, $\bar{z} = \pi D_n/4$, are listed in Table 2. Taking into account the chromatin dynamics during the cell cycle and that diploid cells contain a double set of chromosomes, one gets the final expression for n_s:

$$n_s = \frac{1.67 \times 2}{20} \times \frac{3\pi}{8} \times \frac{N_g}{A_n\,\bar{z}} = \frac{\pi}{16} \times \frac{N_g}{A_n\,\bar{z}}\,, \tag{6}$$

where N_g is genome size, equal to 3.2 Gbp for human cells [31] and to 2.7 Gbp for Chinese hamster cells [32], which we have considered in this study. The factor 1.67 arises because of dependence of N_{bp} on the phase of the cell cycle. During interphase, the number density n_s remains constant during G_1 phase, which takes about 1/3 of the total cell cycle duration (T_c) in human cells [33], but becomes doubled in the S and G_2 phase after DNA replication has taken place. Averaging the number density of DNA over the different phases, one gets

$$N_{bp} = \frac{2\,N_g}{T_c}\int_0^{T_c} f(t)\,dt = \frac{2\,N_g}{T_c}\left(\frac{1}{3}T_c + 2 \times \frac{2}{3}T_c\right) = \frac{2\,N_g}{T_c} \times 1.67T_c = 3.33\,N_g\,. \tag{7}$$

The obtained number density of targets n_s for all cell lines considered in this study is listed in Table 2.

Table 2 Characteristics of the cells studied. Cross sectional area, A_n, and diameter, D_n, of the cell nucleus, as well as an average length of ions' traverse through a nucleus, $\bar{z}$, and the number density of complex damage sites on chromatin, n_s, for different cell lines considered in this study. Experimentally measured values of A_n and D_n are taken from the indicated references

Cell line	A_n (μm^2)	D_n (μm)	References	$\bar{z}$ (μm)	n_s (nm^{-3})
A549		9.6	[34]	7.5	1.2×10^{-3}
AG1522	144 ± 45	13.4	[35]	10.6	4.2×10^{-4}
HeLa	219 ± 3.5	16.7	[36]	13.0	2.2×10^{-4}
NB1RGB	172 ± 2.2	14.8	[37]	11.6	3.1×10^{-4}
A172	209 ± 3.2	16.3	[38]	12.7	2.4×10^{-4}
V79	88	10.6	[39]	8.2	7.2×10^{-4}
CHO	127 ± 1.2	12.7	[36]	9.9	4.2×10^{-4}

The probability Γ_e that a single electron hitting a DNA molecule induces an SSB, was taken to be equal to 0.03 in all the calculations. This value was estimated in Ref. [5] by fitting the experimental probability for SSBs induced in plasmid DNA by secondary electrons as a function of dose [40].

The effect of free radicals was also considered in Ref. [5]. Since the exact radial distribution of the number of SSBs induced by radicals is not known, it is assumed to be uniform within a certain distance from the ion's path:

$$\mathcal{N}_r(r) = \begin{cases} \mathcal{N}_r \, , \, r \le R_r \\ 0 \quad , \, r > R_r \, , \end{cases} \tag{8}$$

where R_r is the effective distance for free radicals propagation which depends on the projectile's velocity and charge. In the below presented analysis, we considered this value in the range between 5 and 10 nm (see Fig. 2). A uniform distribution of radicals within a certain distance from the ion's path implies that the reactive species, formed in the nearest proximity to the path, are transported by a shock wave (SW) and their number density is nearly uniform inside the cylinder that enwraps the decayed SW [5]. It was estimated that a SW produced by a single carbon ion at the Bragg peak (LET $\approx$ 900 keV/μm) propagates free radicals to the distances of about 10 nm from the ion's track [13], and this value gradually decreases with decreasing projectile's velocity and charge. In the plateau region of the depth-dose distribution (LET $\sim$ 10–20 keV/μm), the shock wave is much weaker, if at all significant, and the reactive species may produce damage to the DNA in a narrower region around the ion's path [41]. For the low-LET irradiations considered in this work, an effective distance of free radicals distribution is taken equal to 5 nm (see Fig. 2). This value corresponds to a characteristic diffusion range for radicals in mammalian cells. Beyond this distance, the probability of DNA damage induced by $OH^{\bullet}$ radicals rapidly decreases [42, 43].

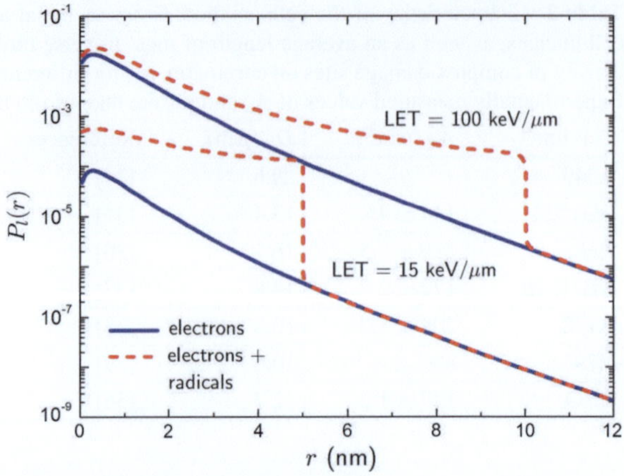

Fig. 2 Dependence of the probability for complex damage, $P_l(r)$, to be induced by secondary electrons (*solid lines*) and electrons together with other reactive species (*dashed lines*) on the radial distance from the ion's path. See the text for further explanation

The average number $\mathcal{N}_{\rm r}$ of SSBs caused by free radicals to take place was estimated as 0.08 from the comparison of the experimental results [40] for plasmid DNA dissolved in pure water or in a scavenger-rich solution. However, this value is affected by environmental conditions of an irradiated target. In the case of hypoxic conditions, the value $\mathcal{N}_{\rm r}$ is reduced because the radical-induced damage may be repaired if oxygen is not present. The quantitatively correct description of the experimental survival probabilities of cells irradiated under hypoxic conditions was achieved by utilizing the value $\mathcal{N}_{\rm r} = 0.04$ which is two times smaller than that in the aerobic environment; this corresponds to experimental data on the induction of DSBs and non-DSB clustered DNA lesions in mammalian cells at normal concentration of oxygen and at deep hypoxia [43]. Further work, however, is needed to explore, in more detail, how the probability of lethal lesion production by free radicals depends on the environmental conditions, e.g. at intermediate concentrations of oxygen [44].

According to the analysis of Ref. [5], the effect of each ion can be treated independently from others, since the average distance between the paths is considerably larger than the radii of tracks. Then, the number of lethal lesions per ion, traversing distance z through a cell nucleus is given by $\frac{{\rm d}N_l}{{\rm d}x}\, z$ and the average number of lethal lesions per cell nucleus is given by [5]

$$Y_l = \frac{{\rm d}N_l}{{\rm d}x} \sum_{i=1}^{\infty} z_i \, i \, \mathscr{P}_i(d) \,, \tag{9}$$

where the sum $\sum_{i=1}^{\infty} z_i \, i \, \mathscr{P}_i(d) = \bar{z} \sum_{i=1}^{\infty} i \, \mathscr{P}_i(d)$ yields an average length of traverse of all ions passing through a cell nucleus for a given dose. The probability

$$\mathscr{P}_i(d) = \frac{N_{\rm ion}^i}{i!} \, e^{-N_{\rm ion}} \tag{10}$$

that exactly i ions traverse the cell nucleus depends on the average number of ions traversing it, $N_{\rm ion}$. The latter in its turn depends on dose, LET, and the size of cell nucleus: $N_{\rm ion} = A_{\rm n}\, d/S_{\rm e}$, where $A_{\rm n}$ is the cross sectional area of the cell nucleus and $S_{\rm e} = |{\rm d}E/{\rm d}x|$ is a part of LET spent on ionization of tissue. At large values of $N_{\rm ion}$, $\bar{z}$ becomes dose-dependent. For values of $N_{\rm ion}$ relevant for this study, $N_{\rm ion} \lesssim 10^2$, $\bar{z}$ is nearly constant and substitution of Eq. (10) into (9) yields a linear dependence of the number of lethal lesions per cell nucleus on dose:

$$Y_l = \frac{{\rm d}N_l}{{\rm d}x}\,\bar{z}\, N_{\rm ion}(d) = \frac{\pi}{16}\sigma\, N_{\rm g}\,\frac{1}{S_e}\,d\,. \tag{11}$$

Equations (9) and (11) give the number of lethal damage sites per cell nucleus, therefore the probability Π_l of producing damage lethal to the cell is given by,

$$\Pi_l = 1 - e^{-Y_l}\,, \tag{12}$$

since a single lethal lesion is sufficient for the cell inactivation. Then, the probability of cell's survival as a function of absorbed dose is given by unity less the above probability,

$$\Pi_{\rm surv} = e^{-Y_l}\,. \tag{13}$$

3 Results and Discussion

3.1 Survival Probability for Different Cells Lines

In this section, the above presented formalism is utilized to evaluate survival probability for various cell lines irradiated with ions. Figure 3a, b shows the survival curves for human adenocarcinomic A549 cells and normal fibroblasts AG1522, irradiated with protons and alpha-particles at different values of LET. The calculated curves (lines) are compared to the experimental data (symbols) on survival of the same cells in the same conditions. Different cell lines have different cross sectional area of their nuclei, and, thus, the average distance $\bar{z}$ of the ion's traverse through the nucleus (see Table 2). This results in different slopes of the survival curves calculated for A549 and AG1522 cell lines at comparable values of LET. More comparisons of calculated survival curves for other human cell lines, such as glioblastoma A172 cells and normal skin fibroblasts NB1RGB, with experiments are presented in Fig. 3c, d.

For a more complete picture, we also analyzed the widely studied Chinese hamster V79 cells irradiated with protons and alpha-particles (see Fig. 4), thus confirming the capability of the MSA to reproduce a large number of experimental results, based on the understanding of fundamental molecular and nanoscale mechanisms of radiation damage. With this understanding, it becomes possible to evaluate the probability of cell survival under different environmental conditions of irradiated targets. This issue

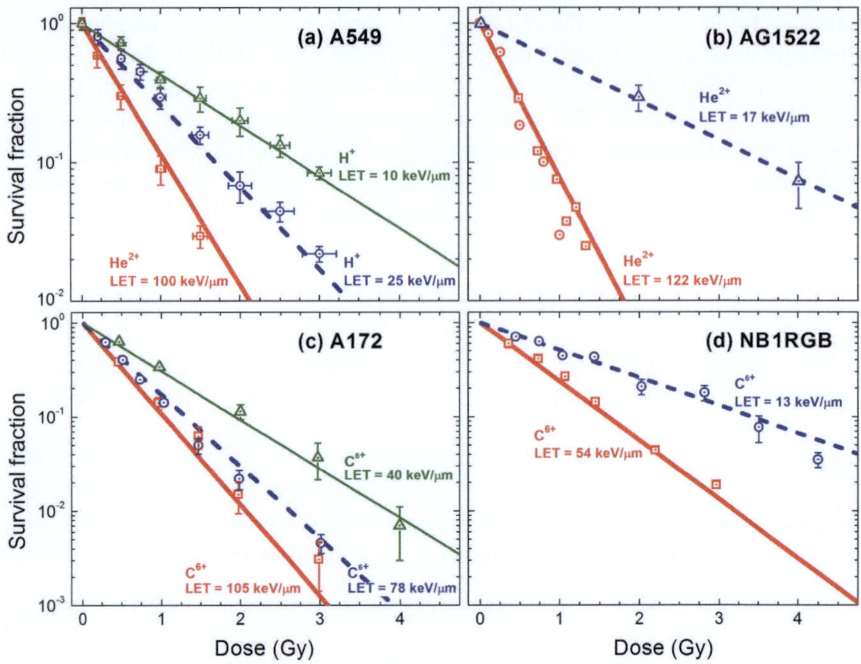

Fig. 3 Survival curves for different human cell lines: adenocarcinomic A549 cells (**a**), normal fibroblasts AG1522 (**b**), glioblastoma A172 cell line (**c**), and normal skin fibroblasts NB1RGB (**d**). The calculated survival probabilities are shown with lines and experimental data from Refs. [45, 46] (A549), Refs. [35, 47, 48] (AG1522), Refs. [38, 49] (A172) and Refs. [37, 38] (NB1RGB) are shown by symbols

Fig. 4 Survival curves for Chinese hamster V79 cell line. The calculated survival probabilities are shown by lines and experimental data from Refs. [29, 35, 50, 51] are shown by symbols. Experiments performed under normal and hypoxic conditions are depicted by open and closed symbols, respectively

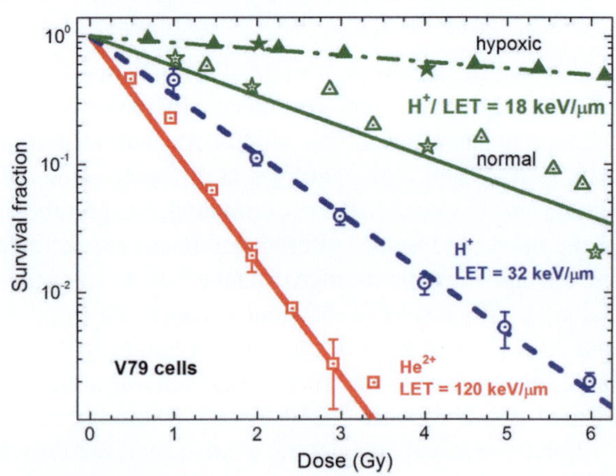

is crucial for medical applications because in many clinical cases, especially in the center of large tumors, one can find regions with reduced oxygen concentration [44]. It is established that the presence of molecular oxygen substantially changes chemical interactions with biological molecules as it affects both the content of reactive species and the possibility of damage fixation. The survival curves calculated for the V79 cells irradiated under aerobic and hypoxic conditions are presented in Fig. 4 alongside with the corresponding experimental data [29, 35, 50, 51]. Under hypoxic conditions, the experimental studies (closed symbols) were performed at high level of hypoxia, since they were carried out in the atmosphere of nitrogen with no addition of pure oxygen.

3.2 *Evaluation of the Oxygen Effect*

Evaluation of cell survival under different environmental conditions allows one to analyze the oxygen enhancement ratio (OER). It is defined as the ratio of the dose delivered under hypoxic conditions to that under normal aerobic conditions, leading to the same biological effect, such as the probability of an irradiated cell inactivation [1]. The OER is about 3 for low-LET radiation and gradually approaches unity as the LET of the radiation increases. In Fig. 5, we present the OER at the 10 % survival level calculated for Chinese hamster CHO and V79 cells irradiated with carbon ions. The calculated curves cover a broad range of LET and are compared to existing experimental results for carbon and heavier ions. The MSA adequately describes the main features of the OER as a function of LET, namely it predicts the decrease of the OER with increasing the LET and its asymptotical value equal to unity at high LET. It also provides good quantitative agreement with experimental data [52] in a broad range of LET. At the LET ranging from approximately 100 to 150 keV/μm, where the RBE for carbon ion beams reaches its maximal value [52], the OER is within the range from 1.5 to 2.0 and nicely agrees with different experimental measurements [44, 53, 54].

The effect of reacting species formed near ion paths strongly depends on their transport. If the latter were driven exclusively by diffusion, free radicals would not be able to propagate further than a few nanometers from the ion's path. Their high reactivity in the region of their high concentration would result in their annihilation [41]. Shock waves predicted in Ref. [13] significantly change this picture as they are capable of effectively propagating the reactive species to much larger distances. As mentioned above, a shock wave produced by a single carbon ion at the Bragg peak (LET $\approx$ 900 keV/μm) propagates free radicals to the distances of about 10 nm from the ion's track [13]. In the plateau region of the depth-dose distribution (LET $\sim 10 - 20$ keV/μm), the shock wave is much weaker and the reactive species may produce damage to the DNA in a narrower region around the ion's path [41]. The low-LET (less than 20 keV/μm) survival curves, shown in Figs. 3 and 4 were calculated with an effective distance of free radicals distribution equal to 5 nm, which corresponds to a characteristic diffusion range for radicals in mammalian cells [42, 43].

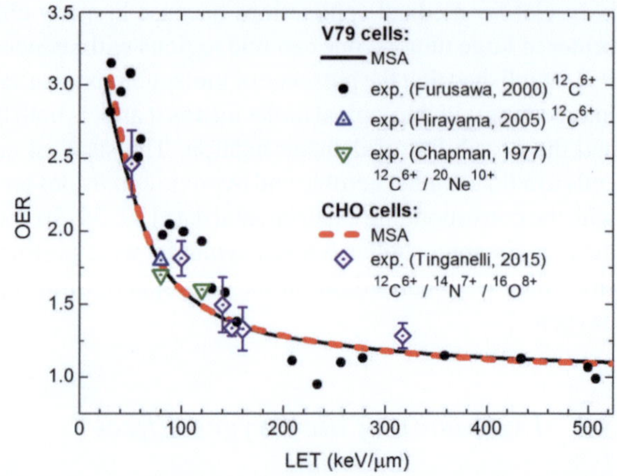

Fig. 5 Oxygen enhancement ratio at the 10 % survival level for V79 and CHO cells irradiated with carbon ions. Symbols denote the experimental data taken from Refs. [44, 52–54]

The probability for lesion production by free radicals is also sensitive to environmental conditions of irradiated targets. At the early stages of the radiation-matter interaction, a decrease of the concentration of diluted oxygen in the cell environment can modify the water radiolysis process that results in modification of primary DNA damage yields [55]. On the other hand, it has been discussed that the effect of oxygen can be explained mainly by chemical repair or oxygen fixation of primary DNA damages, which come into play at later stages of the radiation-matter interaction depending on the oxygen concentration [43, 55]. In the case of hypoxic conditions, the damage induced by secondary species may be repaired chemically through reduction of DNA radicals by endogenous thiols such as glutathione or other sulfur-containing cellular constituents [56], thus decreasing the number of individual and clustered DNA lesions processed by enzymatic repair mechanisms. All these mechanisms suggest that in hypoxic conditions, the average probability for radical-induced lesion production at a given distance from the ion's path should be smaller than that in the aerobic environment. Experimental survival probabilities of cells irradiated under hypoxic conditions (Fig. 4) are nicely described with the probability which is two times smaller than that used to describe aerobic conditions; this corresponds to experimental data on the induction of DSBs and non-DSB clustered DNA lesions in mammalian cells at normal concentration of oxygen and at deep hypoxia [43]. Reduction of the oxygen concentration under hypoxia results in a decrease in the rate of formation of free radicals and, thus, in a decrease in the effectiveness of free radicals to produce DNA damage.

3.3 Analysis of Survival of Repair-Efficient Cells

As described above, the probability of cell survival, Π_{surv}, decreases exponentially with respect to the yield of clustered damage events, Y_l, and the latter linearly increases with dose if the probability of at least one ion to traverse a cell nucleus is sufficiently large (see Eqs. (9)–(11)). This results in an exponential dependence of cell survival on dose (see Fig. 3), which is a common feature for cells irradiated with ions [5].

However, in some cases, there is an evidence that survival curves as a result of irradiation with ions can be "shouldered" consequent to successful damage repair; i.e., in the language of LQ model, Eq. (1), the coefficient β may be noticeably large [57]. In the framework of the MSA, this means that in these cases, some complex lesions can still be repaired. The appearance of shoulders in survival curves have been observed in experiments for specific cells [39, 58]. An example for such curves for the more repair-efficient cervical cancer HeLa cells and Chinese hamster CHO cell line is shown in Fig. 6. In what follows, we present an extension of the MSA formalism which allows one to account for an enhanced repair capability.

The deviation from a purely exponential behavior of survival probability can be explained by a possibility of repair of complex lesions. In these cases, a biological parameter, the probability of a successful repair of a complex lesion, χ, is introduced and Eq. (13) transforms into

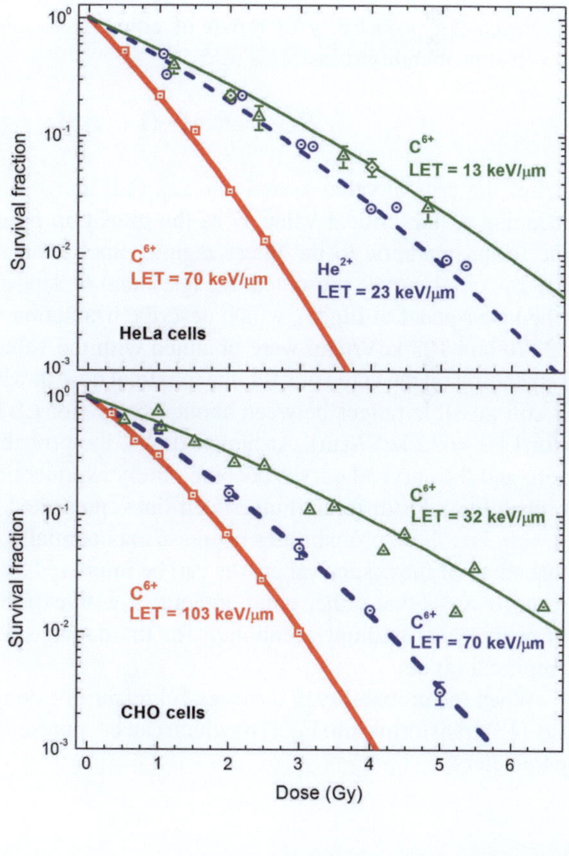

Fig. 6 Survival curves for repair-efficient HeLa (*upper panel*) and CHO (*lower panel*) cell lines. The calculated survival probabilities are shown with lines and experimental data from Ref. [59–61] (HeLa) and [39, 58] (CHO) are shown by symbols. The survival curves are calculated using Eq. (16) with the probability (15), where $\chi_0 = 0.08$ and $\chi_1 = 0.07$ for HeLa cells and $\chi_0 = 0.4$ and $\chi_1 = 0.045$ for CHO cells

$$\Pi_{\text{surv}} = e^{-Y_l} + \sum_{\mu=1}^{\infty} \chi^{\mu} \frac{Y_l^{\mu}}{\mu!} e^{-Y_l} = e^{-(1-\chi)Y_l} \,, \tag{14}$$

where each term in the sum represents the probability of exactly μ complex lesions to be induced multiplied by χ^{μ}, since all of these lesions must be repaired.

The probability of repair of a complex damage may depend on the cell's response to radiation, which involves specific biological mechanisms of damage repair [62]. Although the exact form of this dependence is unknown, the simplest function of probability, χ, can be introduced as a linear function of Y_l,

$$\chi = (\chi_0 - \chi_1 Y_l)\, \Theta(\chi_0 - \chi_1 Y_l) \tag{15}$$

where the positive parameters χ_0 and χ_1 of the function of probability are likely to depend on a cell line, cell phase, and irradiation conditions, and $\Theta(x)$ is the Heaviside step function. A more detailed study of these dependencies as well as biological reasons for such a functional dependence requires a significant effort; this analysis will be presented in future works on this topic. The probability χ gradually approaches zero with increasing the number of lesions until it becomes equal to zero at a critical value, $\tilde{Y}_l = \chi_0/\chi_1$, which depends, in particular, on dose and LET.

When the probability of repair of complex lesions is taken into account, the survival probability transforms into

$$\Pi_{\text{surv}} = \exp\left[-(1-\chi_0)Y_l - \chi_1 Y_l^2\right] \tag{16}$$

below the critical value $\tilde{Y}_l$ and into Eq. (13) above it. These equations explain the meaning of the critical value $\tilde{Y}_l$ as the transition point in the survival curve from the linear-quadratic to the linear regime. Such a behavior can be observed in the experimental curves presented in Figs. 3 and 6. The survival curves for CHO cells (the lower panel of Fig. 6), which describe irradiation with carbon ions with LET = 32, 70 and 103 keV/μm, were obtained with the values $\chi_0 = 0.4$ and $\chi_1 = 0.045$. Depending on the value of LET, the maximal dose at which repair of complex lesions is still possible ranges between about 4.5 Gy (for LET = 103 keV/μm) and 11 Gy (for LET = 32 keV/μm). At higher doses, the probability of repair, χ, is equal to zero and the survival curves become purely exponential functions of dose. Survival probabilities for different human cell lines, presented in Fig. 3, are calculated with $\chi = 0$, i.e., these probabilities decrease exponentially even at small doses. We note that some of these survival curves can be improved by introducing χ. However, we want to stress that rather good agreement with experimental data can be achieved in many cases without accounting for the damage repair and thus the associated empirical inputs.

When the probability of a successful repair of a complex lesion, χ, is introduced, Eq. (13) transforms into Eq. (16) which can be represented as a function of the system parameters,

$$-\ln \Pi_{\rm surv} = (1-\chi)Y_l = Y_l - \Theta(\chi_0 - \chi_1 Y_l)\,(\chi_0 - \chi_1 Y_l)\,Y_l\,. \tag{17}$$

At $Y_l < \chi_0/\chi_1$, the survival probability in virtue of Eq. (11) is as follows,

$$-\ln \Pi_{\rm surv} = (1-\chi_0)\,\frac{\pi}{16}\sigma\,N_{\rm g}\,\frac{d}{S_e} + \chi_1\left(\frac{\pi}{16}\sigma\,N_{\rm g}\right)^2\frac{d^2}{S_e^2}\,. \tag{18}$$

This result provides the molecular-level justification of the empirical LQ parameters α and β for doses $d \leq \frac{16}{\pi}\frac{S_e}{\sigma N_g}\frac{\chi_0}{\chi_1}$:

$$\alpha = (1-\chi_0)\,\frac{\pi}{16}\sigma\,N_{\rm g}\,\frac{1}{S_e}\,, \qquad \beta = \chi_1\left(\frac{\pi}{16}\sigma\,N_{\rm g}\right)^2\frac{1}{S_e^2}\,. \tag{19}$$

At $Y_l > \chi_0/\chi_1$, i.e. at $d > \frac{16}{\pi}\frac{S_e}{\sigma N_g}\frac{\chi_0}{\chi_1}$, one derives the linear regime,

$$-\ln \Pi_{\rm surv} = \frac{\pi}{16}\sigma\,N_{\rm g}\,\frac{d}{S_e}\,, \tag{20}$$

and the parameter α then transforms into

$$\alpha = \frac{\pi}{16}\sigma\,N_{\rm g}\,\frac{1}{S_e}\,. \tag{21}$$

To justify the applicability of the above relations, we have analyzed the dependence of the parameter α on LET for the CHO cells. Figure 7 compares the values

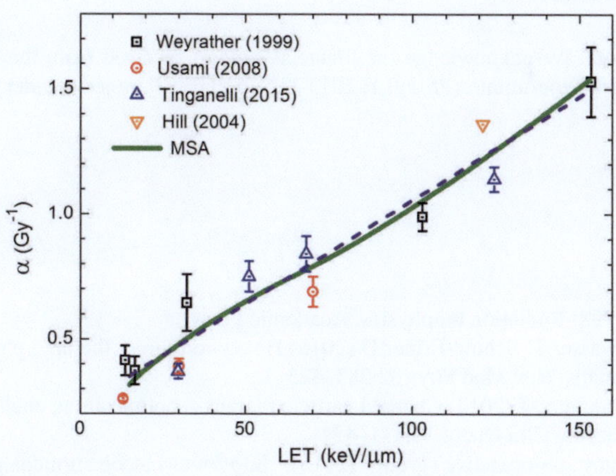

Fig. 7 Dependence of the parameter α on LET for the CHO cells. *Solid green line* shows the values of α calculated using Eq. (19) with $\chi_0 = 0.4$. Open symbols represent the values obtained by fitting experimental data in Refs. [39, 44, 58, 63] with the LQ model (open symbols). *Dashed line* presents the best fit of the array of data taken from the four indicated experimental papers

of α, calculated using Eq. (19) (solid green line) with $\chi_0 = 0.4$, with the values obtained by fitting experimental data in Refs. [39, 44, 58, 63] with the LQ model (open symbols). The figure demonstrates that thus calculated values of α are in very good agreement with the best fit for the array of data taken from the four indicated experimental papers (dashed line).

4 Conclusions and Outlook

In conclusion, novel techniques of radiation therapy, such as ion-beam therapy, can be fully exploited only after the complete scenario of biological damage consequent to irradiation with ions is well understood. This understanding is fostered by means of the MultiScale Approach to the physics of radiation damage with ions—an analytic approach that constructs the scenario of biodamage accounting for the key physical, chemical, and biological effects that take place on different spatial, time, and energy scales. Our extensive comparison with experimental data on survival probability of a broad range of cell lines, irradiated with protons and heavier ions at different values of linear energy transfer and under aerobic and hypoxic conditions, demonstrates the capability of this method to accurately predict the probability of cell survival and related phenomena such as oxygen enhancement ratio. The advantages of the method allow one to extend it to many other cell lines, including radiosensitive and radioresistive cells, different cell phases, irradiation conditions (e.g. in the presence of sensitizers) and make predictive evaluation of radiobiological effects. This analysis will be continued as the predictions are experimentally verified in the future. Then, judgements on practical implementation of the new methodology in treatment planning can be made.

Acknowledgements We acknowledge the financial support received from the European Union Seventh Framework Programme (PEOPLE-2013-ITN-ARGENT project) under grant agreement no. 608163.

References

1. Alpen, EL (1998) Radiation biophysics. Academic Press
2. Schardt D, Elsässer T, Schulz-Ertner D (2010) Heavy-ion tumor therapy: physical and radiobiological benefits. Rev Mod Phys 82:383–425
3. Loeffler JS, Durante M (2013) Charged particle therapy—optimization, challenges and future directions. Nat Rev Clin Oncol 10:411–424
4. Particle therapy co-operative group (2015). http://www.ptcog.ch/index.php/facilities-in-operation. Accessed on 10/2015
5. Surdutovich E, Solov'yov AV (2014) Multiscale approach to the physics of radiation damage with ions. Eur Phys J D 68:353

6. Elsässer T, Krämer M, Scholz M (2008) Accuracy of the local effect model for the prediction of biologic effects of carbon ion beams in vitro and in vivo. Int J Radiat Oncol Biol Phys 71:866–872
7. Hawkins RB (1996) A microdosimetric-kinetic model of cell death from exposure to ionizing radiation of any LET, with experimental and clinical applications. Int J Radiat Biol 69:739–755
8. Boudaïffa B, Cloutier P, Hunting D, Huels MA, Sanche L (2000) Resonant formation of DNA strand breaks by low-energy (3 to 20 eV) electrons. Science 287:1658–1660
9. Baccarelli I, Gianturco FA, Scifoni E, Solov'yov AV, Surdutovich E (2010) Molecular level assessments of radiation biodamage. Eur Phys J D 60:1–10
10. Surdutovich E, Garcia G, Mason N, Solov'yov AV (2016) Nano-scale processes behind ion-beam cancer therapy. Eur Phys J D 70:86
11. Solov'yov AV, Surdutovich E, Scifoni E, Mishustin I, Greiner W (2009) Physics of ion beam cancer therapy: a multiscale approach. Phys Rev E 79:011909
12. Surdutovich E, Solov'yov AV (2014) Multiscale physics of ion-induced radiation damage. Appl Radiat Isot 83:100–104
13. Surdutovich E, Solov'yov AV (2010) Shock wave initiated by an ion passing through liquid water. Phys Rev E 82:051915
14. Surdutovich E, Yakubovich AV, Solov'yov AV (2013) Biodamage via shock waves initiated by irradiation with ions. Sci Rep 3:1289
15. de Vera P, Currell FJ, Mason NJ, Solov'yov, AV Molecular dynamics study of accelerated ion-induced shock waves in biological media. Eur Phys J D 70:183
16. Ward JF (1995) Radiation mutagenesis: the initial DNA lesions responsible. Radiat Res 142:362–368
17. Amaldi U, Kraft G (2005) Radiotherapy with beams of carbon ions. Rep Prog Phys 68:1861–1882
18. Malyarchuk S, Castore R, Harrison L (2009) Apex1 can cleave complex clustered DNA lesions in cells. DNA Repair 8:1343–1354
19. McMahon SJ et al (2011) Biological consequences of nanoscale energy deposition near irradiated heavy atom nanoparticles. Sci Rep 1:18
20. Zhang X-D et al (2015) Ultrasmall glutathione-protected gold nanoclusters as next generation radiotherapy sensitizers with high tumor uptake and high renal clearance. Sci Rep 5:8669
21. Porcel E et al (2014) Gadolinium-based nanoparticles to improve the hadrontherapy performances. Nanomed Nanotechnol 10:1601–1608
22. McQuaid HN et al (2016) Imaging and radiation effects of gold nanoparticles in tumour cells. Sci Rep 6:19442
23. Sage E, Harrison L (2011) Clustered DNA lesion repair in eukaryotes: relevance to mutagenesis and cell survival. Mutat Res 711:123–133
24. Surdutovich E, Gallagher DC, Solov'yov AV (2011) Calculation of complex DNA damage induced by ions. Phys Rev E 84:051918
25. Surdutovich E, Yakubovich AV, Solov'yov AV (2010) Multiscale approach to radiation damage induced by ion beams: complex DNA damage and effects of thermal spikes. Eur Phys J D 60:101–108
26. Sanche L (2005) Low energy electron-driven damage in biomolecules. Eur Phys J D 35:367–390
27. Huels MA, Boudaïffa B, Cloutier P, Hunting D, Sanche L (2003) Single, double, and multiple double strand breaks induced in DNA by 3–100 eV electrons. J Am Chem Soc 125:4467–4477
28. Surdutovich E, Solov'yov AV (2012) Double strand breaks in DNA resulting from double ionization events. Eur Phys J D 66:206
29. Folkard M, Prise KM, Vojnovic B, Davies S, Roper MJ, Michael BD (1989) The irradiation of V79 mammalian cells by protons with energies below 2 MeV. Part I: experimental arrangement and measurements of cell survival. Int J Radiat Biol 56:221–237
30. Bug MU, Surdutovich E, Rabus H, Rosenfeld AB, Solov'yov AV (2012) Nanoscale characterization of ion tracks: MC simulations versus analytical approach. Eur Phys J D 66:291

31. Alberts B, Johnson A, Lewis J, Raff M, Roberts K, Walter P (2007) Molecular biology of the cell, 5th edn. Garland Science, Hamden, CT
32. Lewis NE et al (2013) Genomic landscapes of Chinese hamster ovary cell lines as revealed by the Cricetulus griseus draft genome. Nat Biotechnol 31:759–765
33. Morgan DO (2006) The cell cycle: principles of control. New Science Press
34. Jiang RD, Shen H, Piao YJ (2010) The morphometrical analysis on the ultrastructure of A549 cells. Rom J Morphol Embryol 51:663–667
35. Raju MR, Eisen Y, Carpenter S, Inkret WC (1991) Radiobiology of α particles III. Cell inactivation by α-particle traversals of the cell nucleus. Radiat Res 128:204–209
36. Konishi T et al (2005) Number of Fe ion traversals through a cell nucleus for mammalian cell inactivation near the Bragg peak. J Radiat Res 46:415–424
37. Tsuruoka C, Suzuki M, Kanai T, Fujitaka K (2005) LET and ion species dependence for cell killing in normal human skin fibroblasts. Radiat Res 163:494–500
38. Suzuki M, Kase Y, Yamaguchi H, Kanai T, Ando K (2000) Relative biological effectiveness for cell-killing effect on various human cell lines irradiated with heavy-ion medical accelerator in Chiba (HIMAC) carbon-ion beams. Int J Radiat Oncol Biol Phys 48:241–250
39. Weyrather WK, Ritter S, Scholz M, Kraft G (1999) RBE for carbon track-segment irradiation in cell lines of differing repair capacity. Int J Radiat Biol 75:1357–1364
40. Dang HM, van Goethem MJ, van der Graaf ER, Brandenburg S, Hoekstra R, Schlathölter T (2011) Heavy ion induced damage to plasmid DNA: plateau region vs. spread out Bragg-peak. Eur Phys J D 63:359–367
41. Surdutovich E, Solov'yov AV (2015) Transport of secondary electrons and reactive species in ion tracks. Eur Phys J D 69:193
42. Nikjoo H, O'Neill P, Goodhead DT, Terrissol M (1997) Computational modelling of low-energy electron-induced DNA damage by early physical and chemical events. Int J Radiat Biol 71:467–483
43. Stewart RD, Yu VK, Georgakilas AG, Koumenis C, Park JH, Carlson DJ (2011) Effects of radiation quality and oxygen on clustered DNA lesions and cell death. Radiat Res 176:587–602
44. Tinganelli W et al (2015) Kill-painting of hypoxic tumours in charged particle therapy. Sci Rep 5:17016
45. Wéra A-C, Riquier H, Heuskin A-C, Michiels C, Lucas S (2011) In vitro irradiation station for broad beam radiobiological experiments. Nucl Instrum Meth B 269:3120–3124
46. Wéra A-C, Heuskin A-C, Riquier H, Michiels C, Lucas S (2013) Low-LET proton irradiation of A549 non-small cell lung adenocarcinoma cells: dose response and RBE determination. Radiat Res 179:273–281
47. Autsavapromporn N, de Toledo SM, Little JB, Jay-Gerin J-P, Harris AL, Azzam EI (2011) The role of gap junction communication and oxidative stress in the propagation of toxic effects among high-dose α-particle-irradiated human cells. Radiat Res 175:347–357
48. Hamada N, Funayama T, Wada S, Sakashita T, Kakizaki T, Ni M, Kobayashi Y (2006) LET-dependent survival of irradiated normal human fibroblasts and their descendents. Radiat Res 166:24–30
49. Tsuboi K, Tsuchida Y, Nose T, Ando K (1998) Cytotoxic effect of accelerated carbon beams on glioblastoma cell lines with p53 mutation: clonogenic survival and cell-cycle analysis. Int J Radiat Biol 74:71–79
50. Prise KM, Folkard M, Davies S, Michael BD (1990) The irradiation of V79 mammalian cells by protons with energies below 2 MeV. Part II. Measurement of oxygen enhancement ratios and DNA damage. Int J Radiat Biol 58:261–277
51. Staab A, Zukowski D, Walenta S, Scholz M, Mueller-Klieser W (2004) Response of chinese hamster. Radiat Res 161:219–227
52. Furusawa Y et al (2000) Inactivation of aerobic and hypoxic cells from three different cell lines by accelerated ^{3}He-, ^{12}C- and ^{20}Ne-ion beams. Radiat Res 154:485–496
53. Hirayama R, Furusawa Y, Fukawa T, Ando K (2005) Repair kinetics of DNA-DSB induced by x-rays or carbon ions under oxic and hypoxic conditions. J Radiat Res 46:325–332

54. Chapman JD, Blakely EA, Smith KC, Urtasun RC (1977) Radiobiological characterization of the inactivating events produced in mammalian cells by helium and heavy ions. Int J Radiat Oncol Biol Phys 3:97–102
55. Štěpán V, Davídková M (2008) Impact of oxygen concentration on yields of DNA damages caused by ionizing radiation. J Phys Conf Ser 101:012015
56. von Sonntag C (2006) Free-radical-induced DNA damage and its repair. Springer-Verlag, Berlin, Heidelberg
57. Krämer M, Scifoni E, Wälzlein C, Durante M (2012) Ion beams in radiotherapy—from tracks to treatment planning. J Phys Conf Ser 373:012017
58. Usami N et al (2008) Mammalian cells loaded with platinum-containing molecules are sensitized to fast atomic ions. Int J Radiat Biol 84:603–611
59. Zhao J et al (2013) The potential value of the neutral comet assay and γH2AX foci assay in assessing the radiosensitivity of carbon beam in human tumor cell lines. Radiol Oncol 47:247–257
60. Usami N, Kobayashi K, Furusawa Y, Le Sech C (2016) Hadrontherapy enhanced by combination with heavy atoms: role of Auger effect in nanoparticles. In: Grumezescu AM (ed) Nanobiomaterials in cancer therapy: applications of nanobiomaterials, Ch. 14. Elsevier, Oxford, UK, pp 471–503
61. Goodhead DT et al (1992) Direct comparison between protons and alpha-particles of the same LET: I. Irradiation methods and inactivation of asynchronous V79, HeLa and C3H 10T2 cells. Int J Radiat Biol 61:611–624
62. Falk M, Lukasova E, Kozubek S (2012) Repair of DNA double strand breaks. In: García Gómez-Tejedor G, Fuss MC (eds) Radiation damage in biomolecular systems, Ch. 20. Springer, pp 329–357
63. Hill MA, Herdman MT, Stevens DL, Jones NJ, Thacker J, Goodhead DT (2004) Relative sensitivities of repair-deficient mammalian cells for clonogenic survival after α-particle irradiation. Radiat Res 162:667–676

New Research in Ionizing Radiation and Nanoparticles: The ARGENT Project

Marta Bolsa Ferruz, Vladimir Ivošev, Kaspar Haume, Lilian Ellis-Gibbings, Ali Traore, Vivek Thakare, Soraia Rosa, Pablo de Vera, Vu-Long Tran, Arkadiusz Mika, Daria Boscolo, Sophie Grellet, Alexey Verkhovtsev, Bernd A. Huber, Karl T. Butterworth, Kevin M. Prise, Frederick J. Currell, Nigel J. Mason, Jon Golding, Emanuele Scifoni, Gustavo García, Frédéric Boschetti, F. Lux, O. Tillement, Cédric Louis, Kurt Stokbro, Andrey V. Solov'yov and Sandrine Lacombe

Abstract This chapter gives an overview of "ARGENT" ("Advanced Radiotherapy, Generated by Exploiting Nanoprocesses and Technologies"), an ongoing international Initial Training Network project, supported by the European Commission. The project, bringing together world-leading researchers in physics, medical physics, chemistry, and biology, aims to train 13 Early Stage Researchers (ESRs) whose research activities are linked to understanding and exploiting the nanoscale processes that drive physical, chemical, and biological effects induced by ionizing radiation in the presence of radiosensitizing nanoparticles. This research is at the forefront of current practices and involves many experts from the respective scientific disciplines. In this chapter, we overview research topics covered by ARGENT and briefly describe the research projects of each ESR.

S. Lacombe—On behalf of the ARGENT consortium, see http://www.itn-argent.eu.

M. Bolsa Ferruz (✉) · V. Ivošev · S. Lacombe
Institut des Sciences Moléculaires d'Orsay (ISMO), CNRS,
Univ. Paris-Sud, Université Paris-Saclay, 91405 Orsay Cedex, France
e-mail: marta.bolsa-ferruz@u-psud.fr

V. Ivošev
e-mail: vladimir.ivosev@u-psud.fr

S. Lacombe
e-mail: sandrine.lacombe@u-psud.fr

K. Haume · P. de Vera · N.J. Mason
Department of Physical Sciences, The Open University,
Milton Keynes MK7 6AA, UK
e-mail: Kaspar.Haume@open.ac.uk

P. de Vera
e-mail: p.devera@qub.ac.uk

N.J. Mason
e-mail: nigel.mason@open.ac.uk

K. Haume · P. de Vera · A. Verkhovtsev
MBN Research Center, 60438 Frankfurt am Main, Germany
e-mail: verkhovtsev@iff.csic.es

A.V. Solov'yov (ed.), *Nanoscale Insights into Ion-Beam Cancer Therapy*,
DOI 10.1007/978-3-319-43030-0_12

1 Introduction

Cancer remains a major health concern. Around 50 % of patients receive radiotherapy as part of their cancer treatment. The main limitation of this treatment is the lack of tumor selectivity, which causes severe side effects, and radioresistance. The most promising developments to improve the performances of radiation-based therapies are the use of fast ion-beam radiation (proton and carbon therapy) and nanoparticle-enhanced therapies.

The new FP7 European multi-ITN (Marie Curie Action Initial Training Network) project, named "Advanced Radiotherapy, Generated by Exploiting Nanoprocesses and Technologies" (ARGENT), has been in progress since March 2014. This project was built upon the strong foundations of the Nano-IBCT community and the "Nano-IBCT" COST Action [1]. The main objective of the intersectoral and multidisciplinary ARGENT ITN is to create a new generation of researchers and experts able to develop and propose to society new tools and concepts for the improvement of cancer therapy treatments.

ARGENT brings together world-leading researchers of different disciplines, namely, physicists and medical physicists, chemists, biologists, medical doctors, and small or medium-sized enterprises (SMEs), with the aim of understanding and exploiting the nanoscale processes induced by ionizing radiation. The consortium aims at training 13 Early Stage Researchers (ESRs) whose research activities are split into three work packages, entitled "Nanodosimetry", "Therapeutic Nanoagents", and

A.V. Solov'yov
MBN Research Center at Frankfurter Innovationszentrum Biotechnologie, Altenhöferallee 3, 60438 Frankfurt am Main, Germany
e-mail: solovyov@mbnresearch.com

A.V. Solov'yov
A.F. Ioffe Physical-Technical Institute, Polytekhnicheskaya Ul. 26, 194021 Saint Petersburg, Russia

L. Ellis-Gibbings · A. Traore · A. Verkhovtsev · G. García
Instituto de Física Fundamental, Consejo Superior de Investigaciones Científicas, Serrano 113-bis, 28006 Madrid, Spain
e-mail: l.ellisgibbings@csic.es

A. Traore
e-mail: traoredubuis@iff.csic.es

G. García
e-mail: g.garcia@iff.csic.es

V. Thakare · F. Boschetti
CheMatech, 9 Avenue Alain Savary, 21000 Dijon, France
e-mail: vivekthakare22@gmail.com

F. Boschetti
e-mail: fboschetti@chematech-mdt.com

V. Thakare
Institute of Molecular Chemistry, University of Bourgogne, Dijon, France

"Preclinical Evaluation". The ARGENT scientific objectives, which are the main concerns of these work packages, are the following:

- to advance understanding of the physicochemical processes initiated by the interaction of various forms of radiation with biological matter in the perspective to predict and control the effects of new treatments;
- to develop new nanodrugs able to direct and improve the application of these nanoscale phenomena for best patient benefit;
- to further the understanding of how the effects of these nanoscale phenomena can change clinical practice, and to evaluate the use of the new methods and tools developed in this project for better patient outcomes.

The "Nanodosimetry" unit combines experimental and computational techniques to answer the most fundamental questions regarding the mechanisms present in radiation-induced damage in cells. A group of ESRs study biomolecules, nanoagents and their mutual interaction when activated by slow and fast incident charged particles. This approach is crucial for the optimization of the interactions of potential nanodrugs with radiotherapy protocols.

The "Therapeutic Nanoagents" unit is composed of ESRs with a background in chemistry, pharmacy, biology and medical physics. This multi-disciplinary approach aims at synthesizing, characterizing and testing the properties and effects of potential new generation nanodrugs able to amplify radiation effects and to improve diagnostic performance. Cell uptake and localization of the nanoagents are also included to complete the investigation. The combination of nanoparticles with medical radiation

S. Rosa · K.T. Butterworth · K.M. Prise
Centre for Cancer Research and Cell Biology,
Queen's University Belfast, Belfast BT9 7AE, UK
e-mail: s.rosa@qub.ac.uk

K.T. Butterworth
e-mail: k.butterworth@qub.ac.uk

K.M. Prise
e-mail: k.prise@qub.ac.uk

P. de Vera · F.J. Currell
School of Mathematics and Physics, Queen's University Belfast, Belfast BT7 1NN, UK
e-mail: f.j.currell@qub.ac.uk

V.-L. Tran · F. Lux · O. Tillement
Team FENNEC, Institut Lumière Matière, UMR5306, Université Claude
Bernard Lyon1-CNRS, Université de Lyon, 69622 Villeurbanne Cedex, France
e-mail: vu-long.tran@univ-lyon1.fr

F. Lux
e-mail: francois.lux@univ-lyon1.fr

O. Tillement
e-mail: olivier.tillement@univ-lyon1.fr

V.-L. Tran · C. Louis
Nano-H S.A.S, 2 Place de l' Europe, 38070 Saint-Quentin-Fallavier, France
e-mail: c.louis@nano-h.com

(X-rays and fast ions used in radiation therapy) is studied from molecular to cellular scales, up to in vivo.

The "Preclinical Evaluation" unit combines their efforts to establish a link between nanoscale interactions and clinical effects, through investigating how nanoscale processes initiated by the interaction of radiation with living matter affect biological responses. Combining advanced experimental, theoretical and modeling tools, this team investigates nanoscale interactions for preclinical testing in cell-based models and explores their clinical applicability. The major goal of this team in ARGENT is to evaluate the use of the new methods and tools developed in the project for better patient outcomes.

This chapter provides an overview and background on activities that are being carried out within the ARGENT project. The chapter is organized as follows.

Sections 2–9 are devoted to experimental and theoretical studies of the properties of nanoparticles (NPs) proposed for radiation therapy applications and their interaction with cells. Section 2 introduces the idea of using NPs in radiation therapy and outlines different types of the systems used in ARGENT. Section 3 gives an overview of different techniques for the NP synthesis and characterization. Section 4 is devoted to functionalization of NPs by different ligands aimed for a better localization of NPs in tumors. Section 5 is devoted to the computational modeling of NPs and the investigation of their properties. Section 6 presents an overview of the NP toxicity studies. This problem is of crucial importance because the biological response induced by NPs is governed by physical and chemical properties that impact important cellular processes. Section 7 describes recent studies of the structure and stability of blood

A. Mika · B.A. Huber
Centre de Recherche sur les Ions, les Matériaux et la Photonique (CIMAP),
Unité Mixte CEA-CNRS-EnsiCaen-Université de Caen Basse-Normandie,
UMR 6252, 6 Boulevard Maréchal Juin, 14050 Caen, France
e-mail: mika@ganil.fr

B.A. Huber
e-mail: huber@ganil.fr

D. Boscolo · E. Scifoni
Biophysics Department, GSI Helmholtzzentrum für Schwerionenforschung GmbH,
Planckstraße 1, 64291 Darmstadt, Germany
e-mail: d.boscolo@gsi.de

E. Scifoni
e-mail: e.scifoni@gsi.de

S. Grellet · J. Golding
Department of Life, Health and Chemical Sciences,
The Open University, Milton Keynes MK7 6AA, UK
e-mail: sophie.grellet@open.ac.uk

J. Golding
e-mail: jon.golding@open.ac.uk

K. Stokbro
QuantumWise A/S, Fruebjergvej 3, 2100 Copenhagen, Denmark
e-mail: kurt.stokbro@quantumwise.com

proteins upon interaction with NPs. Section 8 provides a description of the cell lines which are used within ARGENT to perform experimental studies with different *in vitro* models. Section 9 describes the methods for measuring the NP uptake into different cells and makes an overview of preliminary results of the corresponding experiments.

The physical and chemical phenomena appearing due to interaction of biomolecular systems with ionizing radiation are the main topic of the research described in Sects. 10–13. Section 10 gives a background for particle track simulations along with a deeper discussion of the computer codes used in the project. Section 11 is devoted to the computational simulation of nanoscale shock waves induced by fast heavy ions traversing biological media. This mechanism of radiation damage may affect the distribution of free radicals and other reactive species formed after irradiation. The production of free radicals and the oxygen effect are described in Sect. 12. Section 13 describes experimental and theoretical methods for analyzing the effects of secondary species formed due to radiation on biomolecules and water.

Sections 14–16 are devoted to experimental and theoretical studies of the interaction of NPs with ionizing radiation and the impact of nanoscale phenomena on the resulting biological damage. Section 14 describes experimental studies of electron emission from ion-irradiated metallic NPs; the emission of low-energy secondary electrons induced by ionizing radiation is currently considered as one of the main mechanisms underlying the radiosensitizing properties of NPs. The impact of nanoscale processes on biodamage complexity is covered in Sect. 15. It describes some results obtained up to date towards the understanding of the physicochemical processes initiated by the interaction of various forms of ionizing radiation with biological matter. Section 16 describes the various modes of experimentation and pre-clinical trials to be undertaken by the ARGENT group—using both photon and ion radiation *in vitro* and in vivo. These studies should allow us to define optimal treatment protocols that are able to improve tumor treatment whilst decreasing the side effects on healthy tissue.

Finally, Sect. 17 briefly summarizes the different aspects covered in the chapter.

2 Exploration of Nanoparticles for Radiation Therapy Applications

Nanoparticles (NPs) represent the diverse set of colloidal structures that encompasses metals, inorganic materials (e.g., carbon or silica), peptides, proteins, bio- or synthetic polymers or hybrid compounds in conjugated or unconjugated forms. Although many different shapes have been reported, the spherical model has been widely studied and often discussed in this context with the expected size range of 10–100 nm. Nanoparticles having this size can be expected to be preferentially accumulated in the cancerous tissue owing to the widely cited phenomenon "Enhanced Permeation and Retention (EPR) Effect" [2].

Different aspects need to be taken into account when developing new NPs, such as the composition, size and shape, as well as surface coating and charge. These parameters can influence the uptake by cells, their biological response and interaction with radiation.

Metallic NPs composed of elements with high atomic number (high-Z elements) have been widely investigated because of strong electron emission after interaction with ionizing radiation, which is more pronounced as compared to small metal-containing molecular agents, good biocompability, an easy surface functionalization by attachment of ligands, and the possibility of synthesis in a wide range of sizes [3–9]. The use of NPs for radiosensitization was first demonstrated by Hainfeld et al. [10] using 1.9 nm gold nanoparticles (AuNPs) delivered systemically prior to irradiation in mice exhibiting mammary carcinomas. Other metal-based NPs made of platinum and silver have also been successfully used to radiosensitize cells although they are not as easily functionalized as gold, and silver has less biocompatibility and more toxicity [3, 6, 11–14]. Another choice is gadolinium which is used mainly for its contrast properties for magnetic resonance imaging (MRI). Gadolinium is toxic for cells and therefore must be chelated onto a core of another material such as polysiloxane or gold, thus reducing its cytotoxicity [15, 16] (see Sect. 3).

Some of the physical properties of NPs heavily influence their biological compatibility, effect and range. In this section, we detail some of the main properties that can be controlled in the synthesis process and the choice of NPs used within the ARGENT project.

2.1 Effects of Nanoparticle Size on the Uptake

The size of NPs used for radiosensitization not only affects how they interact with the biological system, but also how they interact with the radiation source. Moreover, biodistribution and the route of elimination from the body are also dependent on the NPs size.

Avoiding accumulation of NPs in non-targeted organs such as the heart and liver is a major concern as it may potentially cause long-term side effects; therefore, non-biodegradable NPs should be designed to be rapidly eliminated from the body. Elimination is mainly achieved through renal clearance, which has been demonstrated for NPs smaller than 6 nm [17–19].

Although experiments have a tendency to point to a maximum cell uptake of NPs between 20 and 60 nm [20–23], smaller NPs can accumulate in cancer tissues due to porous blood vessels near tumors [19]. Smaller NPs can diffuse further into tumor tissue and, therefore, present a more even distribution in larger tumors than larger NPs. This may counteract the fact that the actual cell uptake is less than that of larger NPs [20, 23]. Moreover, if the NPs are small enough, they can enter a cell directly by diffusion through the cell membrane [24].

In regards to the interaction between NPs and radiation, the latter interacts mostly with the interior of the NPs; therefore, if the NP size is increasing, the dose deposition of radiation in the medium from this interaction decreases [4]. Carter et al. [8] found that the production of low-energy electrons was larger for 3 nm NPs than for 6 nm NPs exposed to X-rays, and Lin et al. [25] found improved cell killing in their theoretical study of the X-ray and proton irradiation for 2 nm AuNPs compared to sizes up to 50 nm.

2.2 *Effects of Nanoparticle Charge on the Uptake*

The bi-lipid membrane of a cell is negatively charged, which has led to the suggestion that positively charged NPs might exhibit improved uptake due to electrostatic forces [20, 26–29]. However, the optimal charge on NPs for cell uptake is still unclear [26].

da Rocha et al. [30] simulated the uptake of differently charged NPs into cells and found that passive uptake (diffusion) was favored for neutral or slightly negatively charged NPs while for positively charged NPs endocytosis mediated uptake was dominant. Positively charged NPs have also been shown to interfere with certain cell functions, such as ion transport, and to perturb the membrane potential due to the stronger interaction between the positive NPs and the negative membrane [20, 26]. Furthermore, in vivo studies show that a positive charge of the NPs is associated with opsonization and therefore quicker elimination from the bloodstream [18]. However, this can be circumvented by properly coating the NP, as discussed below.

2.3 *Active and Passive Targeting*

To specifically target tumor tissue, the coating of NPs is an indispensable tool and can be used in mainly two ways: passive targeting and active targeting [31], which are discussed below.

Passive targeting relies on EPR and the fact that tumor tissue tends to see higher uptake of macromolecules (e.g., NPs) from the bloodstream due to the presence of leaky vasculature and reduced lymphatic clearance from the tumors compared to healthy tissue [17]. The abnormal blood vessel growth near tumors, induced by these cells' growth mechanisms, create pores in nearby healthy blood vessels [32]. The size of the pores can be as large as 200–1000 nm which allows NPs to extravasate the blood vessels and accumulate in the tumor. In addition, tumor tissue has subnormal lymphatic clearance, which means that anything that is absorbed tends to be retained for longer [32].

In the body, specialized opsonin proteins found in blood serum tend to adsorb onto the surface of any foreign substance, labeling it for clearance from the body [7]. To take advantage of the random uptake during passive targeting, a protracted bloodstream circulation time is required. It has been shown that NPs coated with the molecule poly(ethylene glycol) (PEG) reduces opsonization, thereby increasing the NP circulation time by effectively giving it a "stealthy" nature [17, 33]. The mechanism for the reduced opsonization has been linked to a repelling force from the PEG molecules when opsonins attempt to adsorb and cover the NP [34, 35]. The uptake dynamics of NPs changes when they become covered by opsonins [36]. The reduced interaction granted by a PEG coating thus ensures that uptake properties are more controllable [34, 37].

Active targeting involves the attachment of molecules on the NPs for which suitable receptors exist on the surface of cancer cells. One example is the growth factor Her2, which is overexpressed in a large portion of human breast cancers. By attaching the antibody anti-Her2, NPs can be specifically targeted towards these cells [38].

Another, more general, kind of targeting exploits the higher and faster proliferation of cancer cells compared to healthy cells. The increased growth of cancer cells requires more glucose for their metabolism, therefore glucose-coated NPs tend to have a higher interaction with cancer cells [39–41].

It is also possible to combine passive and active targeting by coating NPs with a combination of PEG and a targeting molecule. However, this requires carefully balancing the two coating molecules: too much PEG will reduce the effect of the targeting molecule, whereas too much targeting molecule will reduce the circulation time [38]. Furthermore it has been shown that the length of the PEG molecules should not exceed the length of the targeting molecules to avoid hindering their interaction with their targeted receptors [42].

2.4 Nanoparticles Used in ARGENT

ARGENT industrial partners, Nano-H and Chematech, have developed 3 different NPs, named Au@DTDTPA, Au@DTDTPA-Gd and AGuIX (see Fig. 1). Au@DTDTPA are gold-based NPs with a ∼2.4 nm core coated by dithiolated diethylenetriamine pentaacetic acid (DTDTPA), presenting ∼6.6 nm of total diameter (Fig. 1a). These NPs have shown encouraging experimental results in combination with ionizing radiation (see Sect. 16). To improve the imaging and possibly the radiosensitizing potential, the Au@DTDTPA NPs were chelated with gadolinium ions (Fig. 1b). The Au@DTDTPA-Gd50 NPs (50 ions of gadolinium per NP) promote a contrast enhancement in MRI and the gadolinium ions do not change the size of the particles since there is no change in the spectrum of the gold NPs colloid [18]. These NPs have shown promising results with low toxicity, theranostic (therapeutic + diagnostic) properties, and easy excretion from the body [18, 43].

AGuIX (Activation and Guidance for X-ray Irradiation) are ultrasmall (under 5 nm) NPs that are made up of a polysiloxane core grafted with Gd^{3+} DOTA (1, 4, 7, 10-tetra-azacyclododecane-1-glutaric anhydride-4, 7, 10-triacetic acid) chelates and primary amines (Fig. 1c). The AGuIX NPs can also be used for theranostics as they can be imaged by both MRI and CT scan [19]. Experimental studies performed by

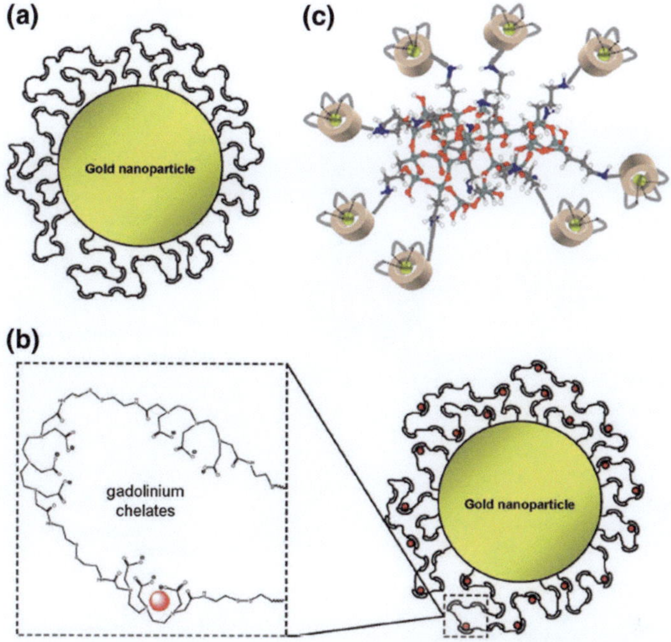

Fig. 1 Au@DTDTPA (**a**), Au@DTDTPA-Gd (**b**) and AGuIX (**c**) nanoparticles which are studied experimentally within the ARGENT project. Figures are adapted from Refs. [16, 19]

ESRs within the ARGENT project are mostly carried out with these three types of NPs. Small platinum NPs with the size of less than 5 nm are also tested; a radioenhancement effect produced by these NPs was demonstrated earlier [3].

3 Synthesis and Characterization of Nanoparticles for Cancer Treatments

3.1 Synthesis Methods

"Metal based nanoparticles for biomedical applications" is a very wide field of research including many different structures published in literature. Therefore, a very wide range of synthesis methods have been introduced to date, as summarized in Fig. 2. Generally, the first step is the synthesis of a metal or a metal oxide core. The second step is usually the formation of a coating around the core for biocompatibility.

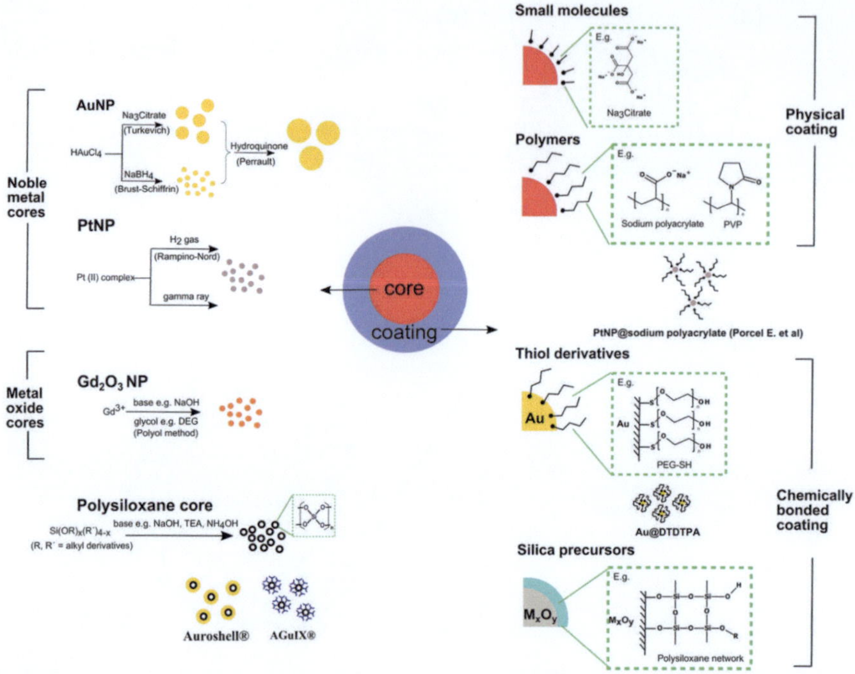

Fig. 2 Illustration of various methods for synthesizing and coating some inorganic nanoparticles used for radiosensitization in cancer therapy

3.1.1 Methods for Synthesizing Metal Cores

For noble metals, such as gold (Au) or platinum (Pt), the most common methods are the reduction of respective salts by different reductants to create metal particles [3, 44]. Gold nanoparticles (AuNPs) can be synthesized from $HAuCl_4$ mainly by three methods depending on the targeted size [44]. The first method to produce AuNPs was proposed by Turkevich et al. [45, 46]; this produces NPs in the range from 15 to 150 nm using citrate as a reducing agent. Ultrasmall AuNPs of less than 10 nm can be achieved from the method described by Brust-Schiffrin [47] with $NaBH_4$ as reductant and alkylthiols as stabilizers. The AuNP coated by DTDTPA, one of the nanoparticles commonly used in the ARGENT project (see Fig. 1a), is an example of this method [48]. Bigger NPs might be achieved with the Perrault method which utilizes hydroquinone to slowly reduce Au(III) onto small AuNP seeds [49]. Pt nanoparticles (PtNPs) are classically synthesized from K_2PtCl_4 using hydrogen gas as the reducing agent [50]. Different kinds of polymers such as sodium polyacrylate, polyvinylpyrrolidone (PVP) or ethylene glycol are used to stabilize the solution [51]. Porcel et al. [3] have applied a radiolytic reduction method using γ-rays to synthesize ultrasmall PtNPs of ca. 3 nm; these NPs are studied in ARGENT as radiosensitizers. In the cited work, radiolytic reduction of platinum complexes was performed in an aqueous solution, containing or not different polymers used as a coating [3].

3.1.2 Methods for Synthesizing Metal Oxide Cores

For other heavy metals, e.g., for gadolinium (Gd), oxide cores are mostly created from their salts [52]. Gd_2O_3 particles are usually produced by reacting $GdCl_3$ or $Gd(NO_3)_3$ with sodium hydroxide in polyethylene glycol (PEG) or diethylene glycol (DEG). This method called "polyol approach" takes advantage of the high viscosity and high boiling temperature of polyols to prevent the aggregation of particles and induce thermolysis of hydroxide into oxide [53, 54].

3.1.3 Methods for Coating Metal Based Nanoparticles

It is almost impossible to use metal NPs in biological medium without a biocompatible coating, due to the fact that metal oxides like Gd_2O_3 can be easily dissolved in water to produce toxic Gd^{3+} ions. Even with inert and biocompatible materials such as Au or Pt, a coating layer might offer higher colloidal stability which leads to longer shelf life and stealth effect which prevents protein adsorption, macrophage clearance and finally liver accumulation [55]. Normally, AuNPs are naturally coated with a layer of citrate or alkylthiols after the synthesis based on the Turkevich or Brust method, respectively. However, these layers are either quite weakly bonded (citrate) or hydrophobic (alkylthiols) and therefore not biocompatible. The most commonly used material for coating is PEG [56]. On AuNP, mercaptopolyethyleneglycol (PEG-SH) can be readily used [57]. A wide range of macromolecules, i.e., natural and synthetic polymers or dendrimers, have also been used as thiol derivatives to coat AuNP [56]. In the case of metal oxides, silane precursors can be used to form stable chemically bonded silica coating layers on their surfaces [54, 58]. Other polymers, e.g., oleic acid or PVP have also been reported to form stable coatings on Gd_2O_3 surface [59].

In some studies, the coating layer may provide other functionalities. For example, in Au@DTDTPA NPs, the gold core is coated by a layer of dithiol DTPA, a strong chelator which can be used later to incorporate Gd^{3+} or radioisotopes such as ^{111}In or ^{99}Tc in order to turn the particle into an efficient MRI contrast agent and gamma emitters for scintigraphy [18].

3.1.4 Methods for Synthesizing Silica Cores

Instead of metal and metal oxide cores, other authors have created silica nanoparticles (SNPs) and coated them with metals, for example gold as in Auroshell® [60], or functionalized them with different chelators which strongly coordinate metals [61].

There are two main methods used to synthesize SNPs, namely the Stober method and water-in-oil microemulsion [62]. The former is based on the hydrolysis-condensation reaction of a silane precursor catalysed by bases, e.g., ammonia in mixtures of water and alcohol to form polysiloxane 3D networks [63]. This technique is simple and easy to scale up but unable to create homogeneous SNPs under 10 nm. However, recently, different terminating approaches have been introduced to achieve ultrasmall SNPs [64, 65]. The microemulsion approach is basically the Stober process carried out in tiny water droplets separated by oil and surfactants. Although this method is believed to be able to produce homogeneous small SNPs, it suffers from a complicated and time-consuming purification procedure to get rid of the oil and surfactants [66].

Gadolinium-based AGuIX NPs (Fig. 1c) are produced from an original top-down approach starting from a conventional Gd_2O_3 NP coated with a silica layer functionalized by DOTA, a macrocycle chelator which is very stable and commonly used for Gd^{3+}. The presence of DOTA, a thin silica coating layer and an acidic synthesis solution accelerate the dissolution of Gd_2O_3 core, break down the silica coating layer into pieces and allow Gd^{3+} ions to be chelated by DOTA finally leading to the stable AGuIX NP [61]. A description of the AGuIX structure and properties by different analytical tools as well as an exploration of the synthesis and functionalization of these NPs are the main topics of the ESR research project "Development of lanthanides based nanosensitizers for theranostics", carried out by Vu-Long Tran in Lyon, France. His project is supervised by François Lux and Olivier Tillement (University Lyon I) and Cédric Louis (Nano-H).

3.2 Characterization Techniques

As with synthesis methods, a very wide range of characterization techniques have also been developed and adapted in order to control the quality of the NPs. Some of these techniques are classified and briefly described below according to the characteristics of the particles they are able to characterize.

3.2.1 Determination of Particle Size and Shape

The size of NPs can be determined by several techniques. The most straightforward method is scanning/transmission electron microscopy (SEM/TEM) which offers direct observation of the size of particles. However, this technique requires complicated equipment and sample preparation [67]. Meanwhile, the most commonly used method is dynamic light scattering (DLS) which deduces the diffusion coefficient of the particles in the solution from the correlation function of scattering light intensity over time. Then, the hydrodynamic radius of the particle, which is the radius of a hypothetical hard sphere that has the same diffusion coefficient as the particle, can be calculated from the Stokes-Einstein equation [68]. Another optical spectroscopy technique that is used to measure particle size is small angle X-ray scattering (SAXS).

This method allows determination of the radius of gyration which is defined as the root-mean-square distance of all elemental scattering volumes from their center of mass weighted by their scattering densities [67]. Diffusion ordered nuclear magnetic resonance spectroscopy (NMR DOSY) can also be described as a method for size determination. This technique utilizes an encoding/decoding process of the nuclear magnetization by spin-echo pulses to probe translational diffusion so that the result will be a 2D plot of diffusion coefficients and chemical shifts of studied nuclei e.g. ^{1}H, ^{13}C or ^{29}S [69, 70]. Through diffusion coefficients, hydrodynamic diameter can be inferred again from the Stokes-Einstein equation. However, this method cannot be used to study paramagnetic materials such as Gd because they can dramatically shorten the relaxation time and broaden the width of chemical shift peaks so that it makes the encoding process impossible to achieve. In reality, several methods need to be exploited in order to have a precise idea about the size of particles [71]. Some of the above techniques might be used also to determine their shape. SEM/TEM can undoubtedly reveal the morphology of particles. Also, the ratio between radius of gyration and hydrodynamic radius can give an idea of the particle morphology [67].

3.2.2 Determination of Overall Elemental Composition, Mass and Morphology

Another level of control highly demanded in the case of NPs is the knowledge of elemental composition and crystallinity. The former can be investigated with standard methods such as inductively coupled plasma mass spectrometry (ICP-MS), inductively coupled plasma optical emission spectroscopy (ICP-OES) or X-ray photoelectron spectroscopy (XPS). Meanwhile, the average mass of NPs can be determined by electrospray ionization (ESI) or matrix assisted laser desorption ionization (MALDI) mass spectrometry. These data may help to establish the average molecular formula of the particles. Interestingly, in the case of AGuIX, STEM was used with high-angle annular dark-field imaging (HAADF) and electron energy loss spectroscopy (EELS) imaging approaches to visualize different elements within individual particles [61]. The crystallinity of the particles, especially their core in core-shell structures, might be answered by high resolution transmission electron microscopy (HRTEM) or X-ray diffraction [72].

3.2.3 Determination of Surface Charge

Surface charge of nanoparticles can be determined by measuring the zeta potential existing between the first ion layer surrounding a particle and the ions in solution [68].

3.2.4 Detection and Quantification of Functionalized Ligands

Existing NPs in the literature are highly varied in terms of functionalized ligands. Hence, numerous techniques have been utilized to quantify the ligands of interest including thermogravimetric analysis (TGA), chemical assays, infra-red spectroscopy (IR), NMR and MS. In functionalized inorganic NPs, TGA can be used to determine the presence and quantity of organic content by weight loss as a function of increasing temperature. This method has been applied for Au@DTDTPA [48]. However, this technique cannot give precise structural information about the grafted functions. Chemical assays such as the quantification of free DOTA by Eu^{3+} used in AGuIX development can be a very useful approach [61]. Nevertheless, this technique requires handling many samples. Therefore, it might be time and material consuming. IR is a handy and well developed technique that allows detection of chemical changes by the shift of the infrared absorption band of material. This approach is very frequently used in literature for detecting functionalized ligands and has also been used for AGuIX and Au@DTDTPA but is hardly quantitative [48, 73, 74]. Similarly, MS can hardly be used as a quantitative method due to its complicated procedure of nebulization, ionization, ion selection and detection although it is a powerful qualitative technique for determining the grafting of ligands [73, 74]. Finally, NMR is a technique with great potential for both detecting and quantifying the grafted organic functions on particles as long as the spectra are not too complicated to have good peak resolution. Interestingly, with the NMR-DOSY technique, the presence of organic ligands can be directly correlated to the integrity and the size of the particle. This allows a straightforward approach for evaluating the stability of degradable NPs [61]. However, as previously mentioned, it is limited to non-magnetic materials.

3.2.5 Evaluation of Stability, Purity and Degradation

Since NPs applied in the biomedical field are strictly regulated, their stability, purity and degradation are of great importance and can be evaluated by high performance liquid chromatography (HPLC) in combination with optical detectors (UV-Vis, Fluorescence) or MS, as shown in the work of Truillet et al. [75].

3.2.6 Characterization of AGuIX and Au@DTDTPA Nanoparticles

Specific NPs might be evaluated very differently in terms of efficacy depending on their applications. In the case of radiosensitizers, no ready method has been developed to test the radiosensitization effect before in vitro and in vivo studies. However, with theranostic NPs such as AGuIX and Au@DTDTPA, their efficiency as contrast agents might be rapidly tested with relaxometry. This gives the value of longitudinal relaxivity, r_1, and transverse relaxivity, r_2. Higher r_1 indicates better efficacy as positive contrast agents [77].

Table 1 Summary of AGuIX and Au@DTDTPA characterization

No.	Characteristics	Nanoparticle			
		AGuIX	References	Au@DTDTPA	References
1	Hydrodynamic radius (R_h) by DLS	1.4 ± 0.3 nm	Unpubl.	Core: 1.2 ± 0.3 nm whole: 3.3 ± 0.9 nm	Alric C et al. [18]
2	Diameter of gyration (R_g) by SAXS	1.4 ± 0.1 nm	Unpubl.	N/a	
3	Ratio R_g/R_h	1		N/a	
4	Hydrodynamic radius (R_h) of Y^{3+} alternative by DOSY	~2.3 nm	Unpubl.	N/a	
5	SEM/TEM image	Yes	Mignot et al. [61]	Yes	Alric et al. [18], Debouttiere et al. [48]
6	X-ray Diffraction	N/a		Yes	Debouttiere et al. [48]
7	Elemental analysis by ICP-MS	%Gd: 14.0; %Si:11.5; %C: 29.0; %N: 8.1	Unpubl.		
8	Elemental analysis by XPS	N/a		Ratio of elements	Debouttiere et al. [48]
9	Zeta potential at pH 7.4	9.03 ± 5.5 mV	Kotb et al. [76]	−30 mV	Debouttiere et al. [48]
10	Mass	~8.5 ±1 kDa	Kotb et al. [76]	N/a	
11	Chemical composition	$Gd_{10}Si_{40}C_{200}N_{50}O_{150}Hx$	Kotb et al. [76]	N/a	
12	Thermogravimetric analysis	N/a		150 DTDTPA ligands per particle	Debouttiere et al. [48]
13	Purity by HPLC UV detector	>95 %	Unpubl.	N/a	
14	Lifetime at pH 7.4 by HPLC-MS	3 h	Truillet et al. [75]	N/a	
15	Longitudial relaxivity r_1	9.4 $s^{-1}mM^{-1}$ at 60 MHz	Sancey et al. [19]	9.4 $s^{-1}mM^{-1}$ at 60 MHz	Miladi et al. [52]
16	Ratio r_2/r_1	1.14 at 60 MHz	Sancey et al. [19]	1.3 at 60 MHz	Miladi et al. [52]

Many of the techniques described above have been applied to AGuIX and Au@DTDTPA NPs that are used in the ARGENT project. The results of characterization of these NPs are summarized in Table 1.

4 Functionalization of Nanoparticles for Tumor Targeting and Biocompatibility

Active targeting of nanosized structures is an exciting avenue in terms of adding the exceptional ability of localization into tumor tissue. Active targeting harnesses the capability of affinity ligands to selectively bind to the receptors predominantly overexpressed on the cancer cells. The approach based on the premise of the "Magic Bullet" is aimed towards maximizing the interactions between NPs and cells, thus augmenting the internalization of NPs containing drugs without altering its overall biodistribution [78]. This section details the importance and options for targeted ligand attachment, the various methods for conjugation of ligands, and the status and future outcomes of using nanomedicines for targeted therapeutic treatments.

At present, active targeting has been a widely pursued strategy to complement the enhanced permeation and retention (EPR) effect and further enhance the efficacy of nanomedicine. The current research in the context of nanomedicine is mainly dedicated towards the delivery of chemotherapeutic drugs in the oncology space. The ARGENT program aims to widen the scope of nanomedicine through the development of nanotechnological tools to improve the outcome of the radiotherapy, which so far has not been rigorously investigated. The ESR research project "Nanoagent functionalization aiming at tumor targeting and biocompatibility" is carried out by Vivek Thakare at Chematech and University of Bourgogne (Dijon, France) under the supervision of Frederic Boschetti (Chematech) and Franck Denat (University of Bourgogne). The project is focused on the development of novel strategies to synthesize NPs for the imaging and therapy of cancer. The project also aims to conjugate these NPs with suitable bio recognized ligands so as to selectively dart the cancerous tissue. Such an approach demands substantial manipulation and investigations at the chemical/biophysical level.

4.1 Targeting Ligands

Active targeting encompasses anchoring the affinity ligands onto the surface of the NPs. These ligands could be small molecules such as sugars, lipids, peptides, proteins, antibodies or antibody fragments. The design of actively-targeted NP drug carriers is an intricate affair owing to the NP's supramolecular architecture, the ligand conjugation chemistry and the several types of ligands at scientists' disposal. The key determinants for the development of ligand based nanotherapeutics include: the target/receptor on the cancer cells of interest, surface chemistry of the nanoparticles, ligand structure and the associated linker chemistry [79].

Several factors determine the success of active targeting, however the basic premise is that the NP needs to reach sufficient concentration at the cancerous site so as to affect the advantage of ligand conjugation. In most of the instances, particularly with smaller ligands, the pharmacokinetics and biodistribution of the NPs may not be affected and hence they form the best choice. However, off target accumulation of the targeted NPs may occur in cases where the receptors for the ligands are also expressed in the healthy tissue (e.g., folic acid). To some extent this can be resolved through the use of highly specific ligands like monoclonal antibodies.

From the viewpoint of ARGENT, it is interesting to integrate different modalities to sensitize cancer cells to the ionizing radiation. For instance, chemotherapeutic platinum based drugs like cisplatin have been shown to improve radiotherapy through their synergistic effects on the DNA damage [80]. Monoclonal antibodies also exhibit their ability to sensitize cells to ionizing radiation through their effects on the subcellular machinery [81]. The project delves to integrate the above modalities into nanoparticulate platform so as to harness and maximize their synergy.

4.2 Conjugation Strategies

Bioconjugate chemistry used for the grafting of ligands on the NP surface is typically based on aqueous reactions owing to instability of the NPs/ligands in organic solvents. One widely explored strategy is the so-called "EDC-NHS" chemistry, which involves activation of the carboxylate groups on the NP surface with a zero length linker 1-Ethyl-3-(3-dimethylaminopropyl) carbodiimide hydrochloride (EDC) and ester/intermediate stabilizer N-Hydroxy Succinimide (NHS) followed by reacting it with the amine group present on the ligand (Fig. 3a). The reaction can be performed vice-versa depending on the functional groups present on the NP or ligands. The efficiency of the reaction depends on the pH of activation and coupling, ratio of NP/ligand concentration and concentration of activating agents [82].

Another strategy that has found widespread mention is the maleimide chemistry (Fig. 3b). The focus on this chemistry has gained relevance owing to the increased number of approved biologics/protein based drugs which could also be used as ligands (e.g., monoclonal antibodies). The reaction involves the maleimide group (typically present on NPs) and the thiol group (typically present on protein/peptide ligands) resulting in the formation of the 3-thiosuccinimidyl ether bond. If free thiol groups are not present on the proteins they could be generated through partial reduction of proteins or through thiolation using Traut's reagent or other thiolation strategies [84].

"Click chemistry" (Fig. 3c) also deserves a mention here, although it is not as popular as the above two methods owing to the synthetic modifications required for both the ligands and the NP. The copper(I)-catalyzed azide-alkyne cycloaddition (CuAAC) occurs in the presence of copper(II) (e.g., copper(II)sulfate) and a

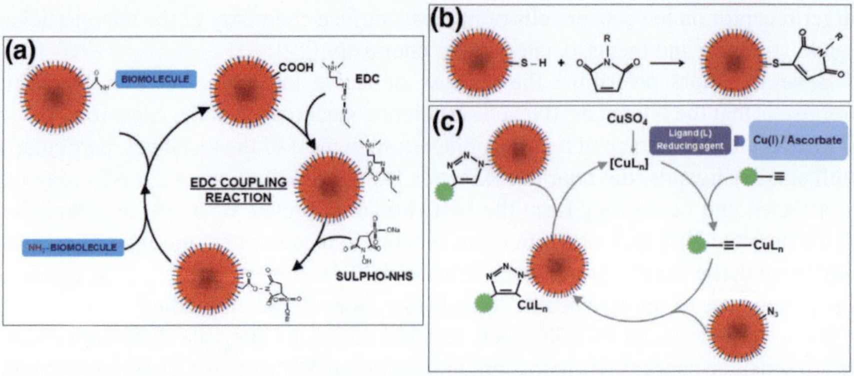

Fig. 3 Conjugation strategies: EDC-NHS coupling chemistry (**a**), maleimide-thiol chemistry (**b**), and click chemistry (**c**). The figure is adapted from Ref. [83]

reducing agent (e.g., sodiumascorbate) to produce Cu(I) in situ and is regarded as a bioorthogonal chemistry for site specific bioconjugation. Azides and alkynes are highly energetic functional groups with selective reactivity resulting in the formation of triazole providing good yields hence referred to as "click chemistry". The reaction occurs at room temperature, showing a high degree of solvent and pH insensitivity and high chemoselectivity [85].

4.3 Future Challenges in Nanomedicine Development

Nanomedicine is a burgeoning field with a diverse set of applications across several disease areas, but so far the progress witnessed in the oncology space has been overwhelming. More than half a dozen products are already approved while many more are undergoing clinical trials. These products mainly address the drug delivery issues with the chemotherapeutic drugs, however it should not be surprising to see the applications in imaging or theranosis making their headway. With the advances in the immunological tools for cancer therapy, the biologics/protein based drugs have assumed a prominent role in adding value to nanomedicine and represent an integrated approach towards developing targeted therapeutics. In the future, nanomedicine has the potential to improve the outcome of radiotherapy through its complex interactions with biological systems, which remain to be fully elucidated and explored. The objective of ARGENT is to steer in this direction, so as to contribute to maximizing the outcome of radiotherapeutic strategies. However, development of nanomedicines is bridled with several challenges in terms of the manufacturing, the biological efficacy and the lack of an established approval pathway unlike other conventional drugs. At the chemistry, manufacturing and control (CMC) level, the challenge is to produce nanomaterials that are rigorously characterized and reproducible across all scales. Moreover, this needs to be strongly justified through the stringent safety and efficacy data from clinical trials that validate the potential of such nanometric tools over the existing standard therapeutics. Nonetheless, the pace at which the advances in this field have been made certainly forecast nanomedicine to be a bright spot in the future of cancer therapy and diagnosis.

5 Computational Modeling of Nanoparticle Coatings and Shapes

The coating and the shape of nanoparticles (NPs) is a crucial tool in manipulating their properties and behavior [6]. Due to the vast number of parameters involved in optimizing the shape and the composition of NPs (e.g., which coating molecules to use, how many, should they have a specific charge), it is a daunting task to systematically go through each of them experimentally. Modeling the different parameters with computer simulations can provide a significant speedup in this process.

5.1 Modeling of Coatings

The ESR project "Development of new modules for ATK code for modeling radiosensitizing nanoagents" is carried out by Kaspar Haume at the Open University (Milton Keynes, UK) in close collaboration with MBN Research Center (Frankfurt am Main, Germany) and QuantumWise (Copenhagen, Denmark). This work is coordinated by Nigel Mason (The Open University), Andrey Solov'yov (MBN Research Center) and Kurt Stokbro (QuantumWise). The research project aims to study the structure and behavior of the coating on NPs intended for use in cancer treatment. One of the tasks is to determine the optimal number of coating molecules to obtain a realistic system, one suitable for further simulations regarding the interaction of these NPs with radiation and the intracellular environment. Currently, the focus is on 1.5 nm gold nanoparticles (AuNPs) coated with polyethylene glycol (PEG).

In the simulations, between 32 and 60 PEG molecules are attached to the AuNP, and the system is then annealed from 1000 K down to 0 K. In this way, we can make sure that the coating molecules have had enough energy to overcome local barriers, and have been free to attain their optimal structure. Additionally, by doing this, we can extract information about the behavior of the coating at intermediate temperatures not just related to body temperature (310 K).

In this project, classical molecular dynamics (MD) simulations are employed to model the structure and behavior of the coating of NPs for different combinations of NP material, shape and size, and coating molecules [86].

Classical MD is basically the solution of Newton's equations of motion for a set of atoms. From an initial starting point, forces are calculated on all atoms, which leads to an acceleration on all atoms as given by Newton's law, $\mathbf{F} = m\mathbf{a}$. By integration of the acceleration with respect to time (for given infinitesimal time steps $\mathrm{d}t$), new positions are given to all atoms and the process starts again. Temperature effects are included by perturbing the equations in a suitable way, given by the used thermostat. More specifically, the positions of atoms are updated by solving the following equation, known as the Langevin equation [87]:

$$m_i \frac{\mathrm{d}^2\mathbf{r}_i}{\mathrm{d}t^2} = \sum_{j\neq i} \mathbf{F}_{ij} - \frac{1}{\tau_\mathrm{d}} m_i \mathbf{v}_i + \mathbf{f}_i, \tag{1}$$

where $\sum_{j\neq i} \mathbf{F}_{ij}$ is the total force acting on atom i as a consequence of its interaction with all the rest of atoms j (i.e., second Newton's law). The second and third terms in the right-hand side of Eq. (1) correspond to the thermostat, used to keep the temperature of the system nearly constant to T, when coupled to a thermal bath. In the simulations performed within the project of Kaspar Haume, the Langevin thermostat is used, which acts as a viscous force on each particle of velocity v_i. τ_d is the damping time of the thermostat, while f_i is a random force with zero mean and dispersion $\sigma_i^2 = 2m_i k_\mathrm{B} T/\tau_\mathrm{d}$, with k_B being the Boltzmann's constant.

The force in Eq. (1) is given by the applied force field, which takes into account various parameters, such as bond distance, bond angle, and others. Utilizing a suitable force field for the given task is a crucial part of doing good MD simulations.

For biomolecular systems, which are the topics of interest in ARGENT, the structure of the molecules is not only dominated by interatomic distances, but also by the geometric configuration of groups of atoms, due to the molecular orbital hybridization. In this case, it is common to use specialized force fields describing such interactions, such as the CHARMM force field [88], which is one of the most commonly used force fields for describing the interactions of biomolecules. The force acting on the atom i is obtained from the potential energy $U(\mathbf{R})$ as $\sum_{j\neq i} \mathbf{F}_{ij} = \mathrm{d}U(\mathbf{R})/\mathrm{d}\mathbf{r}_i$, which corresponds to a given set of atomic coordinates R, and is expressed as a combination of energies arising from the distances of bonds between pairs of atoms, the angles formed between groups of three sequentially bonded atoms, the dihedral and improper angles formed by groups of four bonded atoms, and the non-bonded interactions, represented by the pure Coulomb force and the van der Waals interaction between pairs of atoms.

One of the parameters we monitor is the radius of the system versus the number of coating molecules, see Fig. 4. It is clear from the figure that the radius of the coated NP increases with increasing number of coating molecules and for increasing temperature of the system. The increase with number of coating molecules is related to the mutual repulsion of the molecules. The PEG molecule is composed of a C–C–O backbone which makes it almost linear in its native state. By increasing the coating concentration, the PEG molecules will repel each other and enter an increasingly linear structure, to increase their mutual distance, which leads to a larger radius of the system.

Fig. 4 The radius of the system as a function of the number of coating PEG molecules

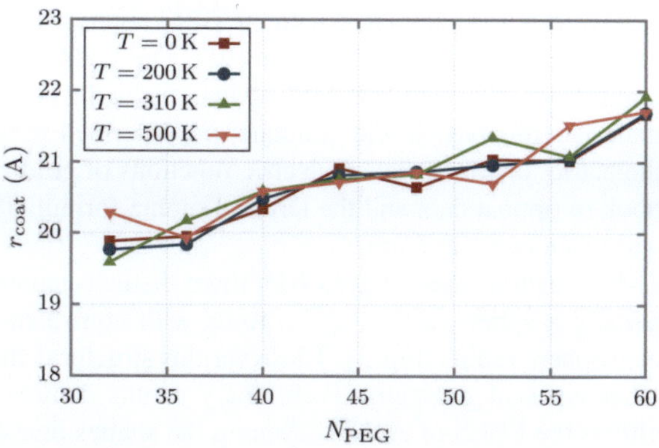

The increase in radius for increasing temperature is likely related to the mutual interactions as well. Increasing the kinetic energy of the PEG molecules will increase the interaction between them, and this is again reduced by attaining a more linear structure.

Other results from simulations include the density distribution of the coating, the penetration of water molecules into the coating, the energetic stability of the coating, and more. All of this will aid in enabling future calculations of electron production under irradiation, propagation of electrons through the coating, and interaction with biological compounds of the cell such as proteins and DNA.

5.2 Impact of the Nanoparticle Shape on Its Electronic Properties

Within ARGENT, one of the goals is to investigate how shape and size influence the yield of electrons emitted from radiosensitizing NPs. In the ESR project "Validation of low energy scattering model in medical radiation planning", carried out by Ali Traore at CSIC (Madrid, Spain) under the supervision of Gustavo Garcia, this problem is partially covered by performing the simulations of the electromagnetic properties of irradiated AuNPs. Using the Metallic Nanoparticle Boundary Element Method (MNPBEM) toolbox of Matlab, it is possible to calculate the electron energy loss spectra (EELS) which give information about the electron yield; the latter is related to the radiosensitization efficiency of NPs [89, 90]. The goal of this work is to determine the size and the shape which provides the highest efficiency in terms of electron yield.

Up to date, this study has considered a 60 keV electron beam, primarily for validation with existing experimental data available at that energy [89]. The study of electromagnetic properties is performed by solving the classical Maxwell equations around the NPs. EELS of AuNPs were computed from the surface charges and current distribution calculated with the BEM approach at the NP boundaries embedded in a water dielectric environment [91]. The BEM method is based upon a rigorous numerical solution of Maxwell's equations in the Fourier domain, assuming that the materials involved in the structure are described in terms of frequency-dependent dielectric functions. The dielectric functions of gold were extracted from the handbook of optical data and the Drude-Lorentz formula [92] and the impact parameter was taken equal to 5 nm.

For several sizes of gold NPs three distinct geometries have been investigated, namely a sphere, a rod, and a torus with equivalent volumes determined by the nanosphere radius (Fig. 5). These various structural shapes should give insights into other types of geometry. Preliminary results demonstrate that the NP's geometry affects the EELS of AuNPs. Among the shapes investigated, the metallic structure in the form of a rod yields the highest EEL spectrum. Future work will be devoted to the analysis of the coating effects on electron production following the methodology schematically illustrated in Fig. 6. Such investigations should provide a better understanding of nanoparticle radiation physics applied to biomedical applications.

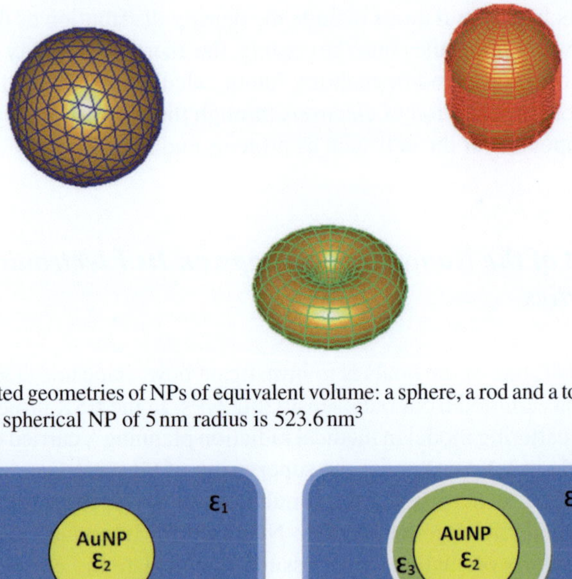

Fig. 5 Investigated geometries of NPs of equivalent volume: a sphere, a rod and a torus. The volume determined for a spherical NP of 5 nm radius is 523.6 nm^3

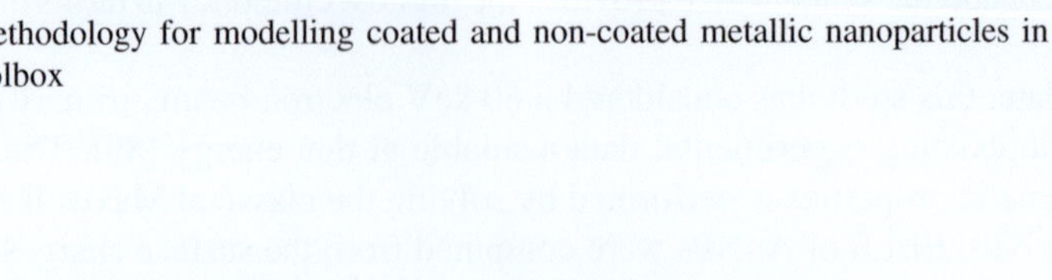

Fig. 6 Methodology for modelling coated and non-coated metallic nanoparticles in MNPBEM Matlab toolbox

6 Nanoparticle Toxicity

The range of biological responses induced by NPs is governed by chemical and physical properties that impact a number of important cellular processes including biocompatibility, biodistribution, toxicity and cellular uptake. Gold has gained increased attention as it exhibits good biological compatibility, low cytotoxicity and is relatively easy to synthesize. Elemental gold has been considered to be biologically inert, however, on the nanoscale, it has been shown that the surfaces of gold NPs are chemically reactive [93]. This occurs for NPs below 5 nm as they react differently with the environment and generate an electromagnetic field due to surface plasmon resonance effects [94]. This phenomenon is due to a coherent oscillation of the free electrons which generates the electromagnetic field on the surface of the gold NPs. Other metal NPs such as platinum, silver and gadolinium have also been investigated. However, they are generally more toxic to cells [3, 6, 11, 12, 14].

The biodistribution and elimination of NPs from the body is an important consideration in the context of NP toxicity. It is desirable for non-biodegradable NPs to be rapidly eliminated from the body to avoid long-term side effects. This is achieved through renal clearance which prevents accumulation in organs such as the heart and liver [18]. If the NPs are smaller than 5–6 nm in size, they are usually eliminated by renal clearance independently of their charge. These small sized NPs are excreted faster [17, 19] and are taken up less by cells [20, 23] but they present a more even distribution as they diffuse further within the tumor [20, 23]. It has been reported that NPs with a size range of 6–10 nm are cleared by the liver, particularly when positively charged. In contrast, NPs greater than 10 nm are retained by the liver [18].

Biodistribution studies of gold NPs with a DTDTPA coating (Au@DTDTPA) have revealed accumulation in the heart and liver, which has been shown to gradually decrease as the signal increases in the bladder and kidneys after 30 min. This indicates that the main elimination pathway was through renal clearance with non-specific uptake by the mononuclear phagocyte system, which is reduced due to the NPs negative charge. For Au@DTDTPA-Gd NPs, contrast enhancement for MRI was achieved in rats with low toxicity observed for a 72 h exposure, despite the known toxicity of gadolinium. These results may be attributed to chelation onto the surface coating of the AuNP [16]. These NPs start to accumulate in the kidneys 3 min after intravenous injection in mice and reach a maximum concentration in the tumor 3–7 min after injection. No accumulation in vital organs or aggregation was detected 6 weeks after injection. AGuIX NPs significantly improve image enhancement with MRI shortly after being injected into mice (1 min), and this remains constant for 24 h. Furthermore, 95 % of these NPs are excreted by renal clearance during a period of 18 days. AGuIX NPs also have very low toxicity. No visible accumulation in the healthy tissues, inflammatory processes or pathologies in the lungs were detected [95].

AuNPs toxicity can vary with the characteristics mentioned above but also differ in various cell types. For example, Pan et al. [96] showed that the toxicity of AuNPs on different types of cells, including mouse fibroblast, macrophages and HeLa cervical cancer cells, is size-dependent but does not depend on the type of coating or the cell type [97]. However, differences in toxicity were observed in HeLa cells using AuNPs with different coatings [96, 98] Moreover, a study presented in Ref. [99] explored the toxicity of 33 nm AuNPs on three types of cells, namely on the A549 carcinoma lung cell line, the BHK21 baby hamster kidney cell line, and HepG human hepatocellular liver carcinoma cells. It was demonstrated that these NPs were toxic only in the A549 cells [99].

To explore the potential toxic effect of NPs used within the ARGENT network, different NPs have been tested in vitro using a range of different exposure times, concentrations and cell assays such as the MTT, MTS short term viability assays and clonogenic assays. The clonogenic assay gives information about the long term proliferative capacity whereas the two colorimetric assays, the MTT (3-(4, 5-dimethylthiazol-2-yl)-2, 5-diphenyltetrazolium bromide) or the MTS (3-(4, 5-dimethylthiazol-2-yl)-5-(3-carboxymethoxyphenyl)-2-(4-sulfophenyl)2H-tetrazoli-um) give information about cell metabolism.

Two ESRs Sophie Grellet and Vladimir Ivošev are working specifically on "Uptake dynamics of nanoagents and effect on radio-enhancement" and "Exploring site specificity, structure and sequence dependence of radiation-induced damage", respectively. The project of Sophie Grellet, conducted at the Open University (Milton Keynes, UK), is supervised by Jon Golding, Nigel Mason and Malgorzata Smialek-Telega at a multidisciplinary level. The project of Vladimir Ivošev is conducted at the Molecular Sciences Institute of Orsay (ISMO, France) and is supervised by Sandrine Lacombe. Another research project devoted to the nanoscale understanding of cell signalling and biological response in the presence of NPs and radiation is conducted by the ESR Soraia Rosa at the Centre for Cancer Research and Cell Biology, Queen's University Belfast under the supervision of Karl Butterworth and Kevin Prise.

7 Blood Toxicity

When NPs interact with proteins, even non-covalently, they may cause structural and conformational variations in proteins thereby inducing unexpected biological reactions and ultimately, leading to toxicity. Recent clinical studies have indicated adverse health effects of exposure to NPs through breathing. It has been evidenced that the addition of NPs may induce changes in blood viscosity and blood clotting capacity [100]. Medical use of NPs requires administration by intravenous injection. In order to design the next generation of NPs, it is necessary to measure the structure and stability of blood proteins upon interaction with NPs (see Fig. 7).

Circular dichroism (CD) is an excellent and sensitive method for rapidly evaluating the secondary structure, folding and binding properties of proteins, and recently has also been used to detect structural changes of proteins interacting with NPs. The use of synchrotron radiation (SR) to perform CD experiments (SRCD) shows several advantages with respect to the conventional CD technique. The major advantage is the high flux provided by a SR source with respect to a conventional xenon lamp

Fig. 7 Illustration of nanoparticles passing through the blood stream

that allows CD data to be measured both with very low amounts of proteins and in the presence of highly absorbing chemicals, such as suspensions of high-Z (composed of high atomic number elements) NPs [101, 102]. The high photon flux in the far-UV region, small cross section size, and highly collimated synchrotron radiation beam from the DISCO beamline at SOLEIL Synchrotron (St. Aubin, France) have allowed the group of Sandrine Lacombe at the Molecular Sciences Institute of Orsay (ISMO, France) to measure the SRCD spectra using cells with low volume capacity. The effect of Gd-based AGuIX NPs on the structure of blood proteins has been studied within the ESR research project "Improvement of the hadrontherapy protocols using nanosensitizers" carried out by Marta Bolsa Ferruz at ISMO under the supervision of Sandrine Lacombe. Human serum albumin (HSA) has been chosen for this study since it is the most abundant protein in the circulatory system and a multifunctional transporter molecule. The inactivation of HSA would lead to life threatening problems. The goal of this study is not only to assess the toxicity of AGuIX under intravenous injection but to optimize a sensitive test of NPs toxicity in blood.

8 Cell Lines

The scope of the ARGENT project is to better understand the processes of radiotherapy at the nanoscale in order to develop a new generation of radiation-based therapies able to treat different types of cancer. Towards this aim, several ESRs are performing experimental studies using a wide range of different in vitro models. In addition, the potential selective effect of NPs is explored in cancer cell models and also in appropriate normal tissue cell models. Here we detail the cell lines under investigation and the rationale for their selection.

One of the cancer types that we are focused on is prostate cancer as it corresponds to 8 % of the total of newly diagnosed cases and 15 % of the diagnosed cancers in men [103]. We are using prostate cancer cell lines, such as PC-3 and DU-145, and comparing these with normal epithelial prostate cells such as PNT2-C2. As PC-3 cells are derived from bone metastasis and DU-145 from brain metastasis, this gives us the possibility of comparing different behaviors of metastatic cells that occupy distinct microenvironments.

Astrocytoma is also a cancer of interest. The national cancer institute estimates an incidence rate of two to three per 100,000 adults per year and it represents 17 % of all brain tumours. The life expectancy with standard treatment is about 2–3 years [104]. We are using the U87 cell line in our experiment, a model of glioblastoma multiform, which represents a high grade of astrocytoma and the most challenging brain tumor to treat [104].

Pancreas cancer follows the same tendency. The estimated incidence of pancreatic cancer is 4.1 per 100,000 adults per year. This cancer is almost always fatal. Approximately 5 % of adults in England survive 5 years after being diagnosed [104]. Indeed, it is the seventh most common cause of death from cancer. The cell line used as a model of pancreatic cancer is BxPC-3. It was derived from the pancreas of a 61 year old female patient. The morphology of this cell line is epithelial and the cells are adherent in cell culture.

Breast cancer is the leading killer among women aged 20–59 years worldwide [103] and was the cause of 521,000 deaths worldwide in 2012. For studying the effect of NPs on this cancer we used MCF-7 as a model of breast cancer and MCF-10A as its respective healthy tissue.

We are also interested in squamous carcinoma skin cancer. It is the second most common skin cancer after basal carcinoma and is also the most likely to metastasize. About 5.4 million of these two types of cancers are diagnosed each year [105]. The first line of treatment involves excision but it can also include radiotherapy [106]. In ARGENT, we use two different cell lines, the HSC-3 cancer line and the HaCaT normal skin keratinocyte line.

Cervical cancer (represented by HeLa cells) is explored in this network since it is the second most common cancer in women living in developing countries [103]. HeLa is a human adenocarcinoma cell line extracted from the cervix of 31 year old African female patient. Its morphology is epithelial and it is adherent in cell culture. It is also a good model of a radioresistant cancer. Since this cell line has been used for many decades it is well understood and can give key information on NPs response and radiosensitization.

In some experiments conducted by the ARGENT consortium, other cell lines have been utilized. For instance, the studies of NP uptake (see Sect. 9) considered also U-CH1 (human chordoma cancer) cells derived from the sacral bone of a 56 year old Caucasian male patient and primary dermal fibroblasts that are human fibroblasts derived from the foreskin of a male African new-born.

9 Gold Nanoparticle Uptake and Retention in Cancerous and Healthy Cells

Improving our knowledge about the dynamics of nanoparticles' (NPs) uptake by cancer cells and effects on radiosensitization is essential [107]. The goal of the ESR project "Uptake dynamics of nanoagents and effect on radio-enhancement" carried out by Vladimir Ivošev at the Molecular Sciences Institute of Orsay (ISMO, France) and supervised by Sandrine Lacombe is to better understand the uptake of gold nanoparticles (AuNPs) in different human cell lines. The methods detailed in this section can measure the uptake of NPs into cells, and techniques that give insight into the mechanisms of cellular NP uptake for different cell lines are also described along with preliminary results.

To measure the cellular uptake, inductively coupled plasma mass spectrometry (ICP-MS) is one of the best quantitative techniques. In ICP-MS, the sample is ionized and passed through a magnetic field to separate and quantify the ions produced, on the basis of their charge and size. This technique can give high specificity and excellent limits of detection [93].

The uptake of NPs by cells can also be explored by transmission electron microscopy (TEM). This technique takes advantage of the high electron density of AuNPs or other high atomic number NPs. It uses electromagnetic lenses and directs high intensity beams of electrons through the samples [108]. It can give an image resolution of 1 nm. In addition, dynamic light scattering can be performed on living cells to visualize the location of AuNPs by using their elastic light scattering properties. Fluorescence microscopy can also be used on living cells to study NPs localization in specific organelles by using fluorescent probes [109] but it can also follow the NPs inside cells if a fluorophore is attached to it. All these microscopy techniques allow one to gain special information and the precise localization of NPs in cells.

Small AuNPs (<10 nm) studied in this research project were synthetized by S. Roux and collaborators [110]. These NPs are advantageous as they amplify the effects in vivo of gamma rays and can be detected by imaging. In this study, the AuNPs are coated with dithiolated diethylenetriaminepentaacetic acid (DTDTPA). For the purpose of our experiments AuNPs were tagged with an organic dye, cyanine 5.

The following cell lines have been considered: U87-MG (human glioblastoma), HeLa (human adenocarcinoma), PC-3 (human prostate cancer), BxPC-3 (human pancreatic cancer), U-CH1 (human chordoma cancer) and primary dermal human fibroblasts (see the description of different cell lines in Sect. 8). All cell lines were bought from ATCC, Manassas, Virginia, USA.

Uptake of the AuNPs in the cells was measured by flow cytometry. Two-photon fluorescence-lifetime imaging microscopy (FLIM) was used to confirm the uptake of AuNPs in cells. Finally, ICP optical emission spectroscopy (ICP-OES) was used to quantify the number of AuNPs internalized.

The first results have demonstrated that the uptake of AuNPs varies significantly between the different cancer cell lines. The internalization of NPs is often attributed to endocytosis [111]. In this work, we investigated the role of other uptake mechanisms by using various chemical inhibitors [112, 113]. These preliminary results show that the route of uptake is also cell type dependent. First results demonstrate that cell lines with low uptake levels of AuNPs are less sensitive to the inhibitors of endocytosis, which indicates that the uptake takes place by other routes and it needs to be further analyzed. Lastly, the measurement of retention times of AuNPs in cells show that the NPs tend to be stored in cancer cells (tumors) at high concentrations for more than 48 h, whilst they are efficiently excreted from primary fibroblasts (healthy tissues).

This work highlights how diverse and complex the response of cancer cells to a perturbation, such as the presence of NPs, is, and how specific treatment must be in order to be effective. More importantly, the results have shown that the NPs used for radiotherapy tend to reside longer in cancer cells compared to healthy cells, even in in vitro conditions. This fact gives NPs a promising future as radio enhancers in radiotherapy and increases the likelihood of their use in clinical conditions. However, improvements of NPs in terms of targeted tissue specificity and harmless elimination from the organism after the tumor eradication are essential for their future application in radiotherapy.

10 Particle Tracks

Understanding how the physical interaction of primary ions, photons and electrons with biological media guides dose deposition is a fundamental issue in clinical treatment planning. Indeed, the different dose deposition patterns arising from photon and ion interaction with biomaterials are responsible for their different relative biological effectiveness (RBE). While photons mainly interact with the condensed phase target by Compton scattering, producing a large number of high energy electrons, ions mainly transfer their energy to the system by glancing collisions with the target electrons, producing a large number of low-energy electrons. This fact differentiates the track structure of both radiation qualities: photons deposit dose in a quite homogeneous manner (photoelectrons generally travel large distances and experience many collisions), while the low-energy electrons ejected by ion beams experience a lower number of elastic and inelastic collisions, creating a nanometer track structure around the ion's path. As a consequence, intense radial doses are deposited around the ion's path, which produces a larger number of clustered damage events in surrounding molecules, leading to an increased RBE. Experimental quantification of radiation damage at the molecular scale is considerably more difficult than performing a predictive numerical experiment covering all the physical channels leading to energy deposition. This section presents a description of the background of particle track simulations along with a deeper discussion of the two codes used in the ARGENT project.

In fact, an essential task of medical physicists is to fit the actual dose delivered by the ion beam (or any incident radiation modality) within the patient as close as possible to the dose calculated by the computerized treatment planning system [114, 115]. Since hadron therapy (as well as conventional radiotherapy with X-rays) involves several stochastic processes due to the enormous number of collisions, dose planning is performed by computer simulations. A direct consequence of this is the development of dose calculation algorithms mostly using the Monte Carlo approach, which can be divided into two areas: radiation transport and track structure codes. While the former are more suited for macroscopic dose calculations in the whole irradiated volume, the latter are focused on a small portion of the track in order to study the particular type of dose delivery on the nanoscale, e.g., its density, which is thought to correlate with the biological efficiency. Within the ARGENT project, we preferentially select the latter since it allows a complete understanding of the nanoscale radiation effects, including secondary species such as radicals and secondary electrons. However, it should be noted that other kinds of Monte Carlo codes, useful for specific purposes, have been discussed within this book, e.g., in the Chap. "Propagation of Swift Protons in Liquid Water and Generation of Secondary Electrons in Biomaterials" by P. de Vera, R. Garcia-Molina, and I. Abril. As examples of track structure codes, PARTRAC [116], KURBUC [117], NOTRE DAME [118], EPOTRAN [119], Geant4-DNA [120], LEPTS [121] and TRAX [122] have been designed for track structure simulations, among others [123, 124]. Within ARGENT, two track structures codes, namely LEPTS and TRAX, are being improved and undergoing validation studies for late biomedical applications. Therefore, in what follows we focus on these two modeling approaches.

LEPTS (Low Energy Particle Track Simulation) tracks the secondary electrons produced by the primary projectiles by sampling the appropriate cross section data and the corresponding energy loss spectra [121, 125]. A more detailed description of this code and its capabilities are provided in another chapter of this book (see the Chap. "Monte Carlo-Based Modeling of Secondary Particle Tracks Generated by Intermediateand Low-Energy Protons in Water" by A. Verkhovtsev, P. Arce, A. Muñoz, F. Blanco, and G. García). The total scattering cross sections are based on calculations and transmission beam experimental data and ultimately define the mean free path of the electrons, hence the spatial location of the interaction. Then the energy loss distribution, taken from experimental data, determines the amount of energy transferred. These are discussed in further detail in Sect. 13. Throughout the scattering process until thermalization the path is modified by the integral and differential elastic cross sections, both derived from measurements and calculations. Finally, Monte Carlo sampling according to the integral and differential inelastic interaction probability function drive the multiscale electron-driven processes responsible of the radiation damage [123].

The ultimate goal of the simulation procedure is to provide a full track of the primary ion (or photon or electron) and its secondaries, down to 1 eV, tracking biomolecular dissociations that occur. In order to build up a full energy deposition model, LEPTS is run jointly with classical Monte Carlo codes. For instance, Geant4 or Penelope can be used above 10 keV and LEPTS can take over below 10 keV to account for quantification of low energy physics based on reliable cross section data for ionization, electronic excitation, vibrational excitation, rotational excitation, electronic attachment and neutral dissociation [123]. These low energy scattering processes are difficult to take into account in a physical, clinical detector, and require technical expertise not yet present for medical physicists working in radiation oncology. The numerical experiment, using LEPTS, offers an excellent alternative to reach the molecular level of dosimetry for an accurate description of secondary species' (radicals and low energy electrons) interactions. The validation of the LEPTS methodology by a comparison with the outcomes of commercial treatment planning systems is the task of the ESR research project "Validation of low energy scattering model in medical radiation planning" carried out by Ali Traore at CSIC (Madrid, Spain) under the supervision of Gustavo García.

TRAX is a Monte Carlo particle track structure code developed during the last 20 years at GSI (Gesellschaft für Schwerionenforschung, Darmstadt, Germany). This code has been designed to simulate the passage of ion and electron radiation in dosimetric devices and radiation damage in biological systems. The software can, then, cover a range of radiation energies compatible with these applications: from a maximum energy of few hundred MeV/u for ions and few hundred keV for electrons to a lower threshold of between 10 and 1 eV. The lower threshold and then the accuracy of the simulations are given by the available cross sections. These cross sections can either be calculated from semi-empirical and analytical formulae or read from external cross section tables.

The ions and electron tracks are simulated considering ionization, excitation and elastic scattering interactions with the target material. The target material can be atomic or molecular or a mixture of atomic and molecular materials. Some complex materials like air or plastic are included as well. In more recent versions of TRAX, information regarding the yield of Auger electrons is included in the target characterization, and plasmon excitations for solid state material (volume plasmon) are also accounted for [126]. Thanks to the capability to follow the secondary electrons down to very low energies, the code allows the calculation of many relevant radiation quantities, such as the radial dose distribution, depth dose distribution, W-values, micro-dosimetric quantities such as the lineal and specific energy transfer, secondary electrons spectra and the ionization distribution. Recently the effect of radiosensitization of gold NPs has been also implemented in TRAX. This code, indeed, has proven to be very suitable for such applications since it can simulate the electron paths on a nanoscopic scale and can handle complex geometry, such as NPs in water solutions [127].

11 Computer Simulations of Ion-Induced Shock Waves

As explained before, one of the main characteristics of the interaction of energetic ion beams with tissue is that most of the energy is lost by ejection of low energy secondary electrons, mostly with energies around or below 50 eV around the Bragg peak region, irrespectively of the biological target considered [128–130]. The further propagation and interaction of these electrons with the biological environment can be studied by different techniques, including track-structure Monte Carlo simulation codes, as explained in Sect. 10, or analytical techniques such as the random walk approximation [128, 131].

A recent analysis of the track-structure performed within the random walk approximation revealed not only the well-known feature of the very intense and steep radial dose deposited around ion tracks, but also that this radial dose is built up very quickly, in around 50 fs [131]. As a result, a large amount of energy is deposited in very small volumes (cylinders with a radius of $\leq$1 nm), and in a time scale much shorter than the times in which the electron-phonon coupling operates, the mechanism capable of dissipating this energy. As a consequence, large amounts of energy will be delivered at once to the translational and vibrational degrees of freedom of liquid water surrounding ion tracks (what we will refer to as "the hot cylinder"), which posses the initial conditions for a violent explosion. Such observations are in agreement with previous estimates, based on the thermal spike model that suggested large increases in temperature around ion tracks in the sub-ps scale [132, 133]. Therefore, it is quite clear that the conditions for this violent explosion of "hot" nanocylinders surrounding ion tracks are plausible and may lead to the formation of cylindrical shock waves on the nanoscale.

Such ion-induced shock waves were first analyzed in terms of a classical hydrodynamics model [133]. Based on the initial conditions explained above, the hydrodynamics equations for the self-similar flow of water were solved for the cylindrical case of energy deposition around an ion track on the nanoscale. Several useful quantities characterizing the ion-induced shock waves were obtained, such as the velocity of propagation of the front and its pressure. The front velocity is proportional to the stopping power to one fourth, while the pressure of the front is directly proportional to the stopping power [133–135]. From these dependencies, it is possible to predict the main features of ion-induced shock waves, depending on the stopping power of the incident ion. The space and time scales in which shock waves operate can also be known: shock waves damp in times of about 10 ps, and they propagate in distances of a few tens of nanometers [133]. This time scale lies in between the initial propagation of secondary electrons and the further chemical effects produced by the diffusion and reaction of free radicals.

Although the main shock wave characteristics can be predicted by the hydrodynamics model, its effect on biological molecules cannot be obtained. For these purposes, it is possible to use atomistic simulations, such as the classical molecular dynamics described in Sect. 5, to predict whether the high pressures produced during the shock wave are enough to produce damage of biomolecules such as DNA. Within the ARGENT project, this analysis is being performed by the ESR Pablo de Vera under the supervision of Andrey Solov'yov (MBN Research Center, Frankfurt, Germany), Frederick Currell (Queen's University Belfast, UK) and Nigel Mason (The Open University, UK). The objective of the project is to theoretically characterize the features and effects of ion-induced shock waves, and also to provide data that can be checked experimentally, to demonstrate the existence of these shock waves.

Several works have been performed in this direction over the last few years [135–138], where, most significantly, it was shown that shock waves can deposit enough energy to produce single strand breaks (SSBs) in DNA, with probabilities, for high stopping power ions, even larger than the effect of free radicals [136].

More recently, within ARGENT, another molecular dynamics study has been performed by Pablo de Vera on the effects of ion-induced shock waves in short DNA duplexes [135]. This study showed that molecular dynamics simulations can perfectly reproduce the main wave front features predicted by the hydrodynamics model. Such results are relevant, since these characteristics are found in agreement from two different methodologies, this gives more support to the theoretical prediction of shock waves.

Final confirmation of the existence of shock waves should come from their experimental observation. Several quantities related to the effects of shock waves in DNA molecules might be observed experimentally, such as their geometrical distortion or the number of strand breaks produced. If we take into account that backbone bonds can be broken with energy deposition events $\geq$2.5 eV (as a conservative estimate), it is quite obvious that the probability of producing an SSB by carbon ions is not very large, according to the molecular dynamics simulations results. However, it is much more important for the case of heavier ions, such as iron [135, 136]. Such results for the short DNA duplex are compared with the previous results for the nucleosome. It has also been shown how heavier ions (iron) produce larger nano-channels in liquid water, as a result of the nano-cavitation effect induced by the shock waves [135]. Thus, the use of ions heavier than carbon would be more convenient for experimental evaluations.

We expect that any of the above described events (production of single strand breaks, production of nano-channels, bending/fragmentation of DNA) could be observed experimentally in the near future, and that the simulations described might help on the setting up and interpretation of potential experiments for shock wave detection. This, among other possibilities, such as the detection of the shock wave signature on the distribution of free radicals after irradiation (due to the velocities of the shock wave, much faster than diffusion) [136] are being considered within ARGENT, for a combined experimental and theoretical study of the biological effects of ion-induced shock waves. Further and more detailed explanations of all the topics discussed can be found in Chap. "Thermo-Mechanical Damage of Biomolecules Under Ion-Beam Radiation" of this book.

12 Radical Production

When describing the effect of radiation on biological media, it has to be considered that up to 70 % of the radiation damage is induced indirectly by the action of the water-derived free radicals. Water radiolysis produces highly reactive free radicals able to inflict severe damages on biological media, mostly $OH^{\cdot}$, $H^{\cdot}$, $^{\cdot}O^{\cdot}$ and e^-_{aq}. Among them, hydroxyl radicals ($OH^{\cdot}$) are supposed to be the most injurious as they are believed to be the main component responsible for cell killing via indirect damage [139]. Being very powerful oxidizers with high electron affinity, $OH^{\cdot}$ radicals can bond with free electrons produced during the irradiation preventing their recombination with the positive ion. In addition, two very toxic species (hydroperoxyl radical, $HO_2^{\cdot}$, and hydrogen peroxide, H_2O_2) are generated by the interaction of $OH^{\cdot}$ with water molecules from the medium and other free radicals produced during the radiolysis. In this section, the radical production, oxygen effect, influence of NPs on these and their respective timescales are articulated.

In radiation chemistry, the quantity used for describing the yield of free radical production is the G-value. This is the number of radicals produced per 100 eV of energy deposited by radiation in the medium. Experimentally this quantity can be established with high time resolution by means of pulsed radiolysis experiments. Useful information on the impact of the indirect damage can, in addition, be obtained by performing experiments with radical scavengers. In these experiments the direct and indirect damage are disentangled by the use of chemicals able to interact with reactive species, such as free radicals, thereby blocking the indirect effect of radiation. Dimethylsulfoxide (DMSO), for example, is a scavenger commonly used to inhibit the effect of $OH^{\cdot}$ radicals. With this approach, a significant decrease of the indirect effect has been observed for high linear energy transfer (LET) radiation [140, 141]. In particular, a drastic decrease of the yield of hydroxyl radicals has been measured with the increase of the LET. This kind of behavior, also predicted from several simulation tools [117, 142–144], is supposed to be related with the track density: for high LET the track is more dense and recombination processes between radicals are more likely to happen [145].

Another important parameter, which significantly affects the impact of the indirect damage, is the oxygenation level of the irradiated material: the increasing radioresistance of hypoxic tumors seems to be strongly related with the indirect effect of radiation [146] and in particular with the yield of OH radicals [147]. This effect is generally explained as a combination of two main phenomena. On one hand, the molecular oxygen in the solution reacts with the radical species generated during the irradiation and increases the yield of production of HO_2 and H_2O_2. On the other hand, the oxygen seems to be able to fix the DNA damage caused by radicals making it permanent. Reacting with a radical, the DNA can be ionized becoming $DNA^{\cdot}$. In absence of oxygen this damage can easily be repaired and come back to its original state of DNA-H; however, if there is a molecular oxygen in the proximity, the $DNA^{\cdot}$ can react with it leading to the formation of a hydroperoxyde radical, in this case the damage is fixed and irreparable. This process is described by the following reactions:

$$\mathrm{DNA}^{\cdot} + \mathrm{O}_2 \rightarrow \mathrm{DNA} - \mathrm{O} - \mathrm{O}^{\cdot} \quad \text{(DNA peroxide radical)}$$
$$\mathrm{DNA} - \mathrm{O} - \mathrm{O}^{\cdot}\mathrm{H}^{\cdot} \rightarrow \mathrm{DNA} - \mathrm{O} - \mathrm{OH} \quad \text{(DNA hydroperoxide).}$$

Recently a significant increase in radical production has been observed in the presence of gold and platinum NPs [148, 149]. This phenomenon seems to be related to surface effects such as a high concentration of low-energy electrons due to the surface plasmon effect [150], catalytic interaction between the metal surface and the water dipole [148], or geometrical factors like the surface/volume ratio for high-Z materials [149]. Even though this effect has still to be deeply investigated and a thorough theoretical description is not available, these observations are very promising in the context of using noble metal NPs as radiosensitizers, especially in conditions where the suppression of the indirect damage is the main cause of radioresistance, e.g., in the case of hypoxia.

Despite the high relevance of the indirect damage in the radiation effect, many theories are still not completely proven and a mechanistic description able to describe the impact of the indirect effect from different irradiation conditions is yet to be developed. At the moment only a few Monte Carlo particle track structure codes are able to describe the evolution of the track on the chemical level as well: PARTRAC [143],

CHEM-KURBUC [117], RITRACKS [142], and GEANT4-DNA [144]. In all these software the effect of radiation is represented as a three-stage process. These stages, characterized by different time scales, are identified, according to the standard paradigm of radiation damage, as: the physical stage, the pre-chemical stage and the chemical stage.

Physical stage: this lasts for the first 10^{-15} s after the irradiation. The primary particles and secondary electrons excite and ionize the water molecules. This phase corresponds to the direct effect of radiation.

Pre-chemical stage: in this stage the chemical species produced by the ionization and the dissociation of the ionized and excited water are generated. Ionized water molecules generate H_3O^+ radical ions and $OH^{\cdot}$ radicals. Excited water molecules instead have different channels of dissociation depending on the molecular excitation level. The species produced by the dissociation of excited water molecules are H_2O, $OH^{\cdot}$, $^{\cdot}O^{\cdot}$, $H^{\cdot}$, H_2, H_3O^+ and sub-excitation electrons. This stage ends with the thermalization of all the chemical species around 10^{-12} s after the irradiation.

Chemical stage: includes the Brownian diffusion and the reactions between all the chemical species generated during the pre-chemical stage. During this phase the track evolution is generally described by a step by step approach. During each time step, t, the radicals diffuse randomly with a step size dependent on their diffusion coefficient, D; the distance traveled, l, is sampled by a Gaussian distribution with a root mean square calculated according to the Smoluchowski diffusion equation: $\langle l \rangle = \sqrt{6Dt}$. The reactions between radicals are instead defined via a proximity parameter generally indicated as the reaction radius, a: two radicals interact if their separation is smaller than $2a$. The definition of the reaction radius changes depending on if the reactions are assumed to be diffusion controlled (as in CHEM-KURBUC), partially diffusion controlled (as in PARTRAC) or if more complex models are taken into account (as in RITRACKS). The chemical stage lasts up to 10^{-6} s; after that time the chemical development of the track is supposed to be over and the radical yields are assumed to be constant.

Within the ARGENT project, the action of water-derived free radicals is investigated, theoretically and experimentally, under different irradiation and oxygenation conditions. This is the main task of the ESR research project "OER prediction on the nanoscale for a target tissue in different conditions of irradiation and oxygenation" carried out by Daria Boscolo at GSI (Darmstadt, Germany) under the supervision of Emanuele Scifoni and Marco Durante. For this purpose a chemical extension of TRAX is currently under development and experiments in collaboration within the ESR project of Marta Bolsa Ferruz, "Improvement of the hadrontherapy protocols using nanosensitizers" are currently ongoing (see Sect. 16).

13 Collision-Induced Processes

The in-depth understanding and ongoing research described above for the production of secondary species, be they electrons or radicals, ions and neutrals, are vitally important for understanding the total impact of radiation on living tissue. The cascade of low-energy (<5000 eV) particles produced in photon and ion radiotherapy, with and without the presence of nanoparticles, is capable of inducing irreparable damage in living cells through a range of low-energy physical processes [151–154]. A full understanding and control of nanodosimetry, the effect of the radiation dose in nanovolumes in the cell, plays an important role in the effectiveness of radiotherapy modeling [155]. This is the major work of the ESR Lilian Ellis-Gibbings, a physicist working under Gustavo García in the Institute of Fundamental Physics, Consejo Superior de Investigaciones Científicas in Madrid, Spain. This section describes experimental and theoretical methods for the production of statistical data relating to the effects of secondary species of irradiation (electrons and anions) on biomolecules and water.

The channels these low energy secondary particles follow for their physical distribution and damage to important biological molecules include elastic scattering, rotational, vibrational and electronic excitations, as well as electron attachment and ionization. Appropriate modeling of these interactions within particle tracks is commonly and effectively undertaken using Monte Carlo codes [156], which require precise and accurate cross sections of interaction on both the medium and the molecules of interest [157]. This data, primarily in the form of total, differential (in angle and energy) and integral cross sections, along with the energy loss spectra, can be determined experimentally and theoretically. The theoretical methods have varied accuracy under different physical energy regimes [158, 159].

Electrons make up the bulk of secondary particles from photon and ion irradiation, and the majority of them are produced with less than 50 eV [160]—as such the interactions described above are possible before the electron thermalizes [151]. Since these electrons are clearly able to impact and damage biological material, part of this study is in the experimental and theoretical determination of the interaction types and cross sections for the collisions of low energy (0–5000 eV) electrons with molecules of biological importance.

Theoretical calculations of electron–molecule (and positron–molecule) impact collisional cross sections follow several main themes. Ab initio calculations are highly accurate and methods based on the close-coupling expansion, such as the R-matrix, the Kohn variational principle, and the Schwinger variational principle are popular amongst these [158, 161]. Unfortunately they are computationally expensive and inappropriate for complex molecules at intermediate to high energies [162]. In order to produce the cross sections necessary for radiotherapy particle track simulations (see Sect. 10), approximate methods have been developed. A popular starting point is the Independent Atom Model (IAM), which has proved fruitful due to its accuracy above approximately 100 eV and the ease of calculation. The IAM postulates that the scattering cross section of a molecule can be substituted with the scattering cross sections of each of the atoms in their respective positions within the molecule. It assumes each atom scatters independently, there is no multiple scattering within the molecule, and the molecular orbitals of electrons are ignored [163]. This simplification breaks down at low energies, where interference and multiple scattering begin to have an effect.

Equation (2) describes the basic additivity rule (AR) of the IAM for any molecule, derived through application of the optical theorem, where σ denotes a cross section. The first modification of this approximation is a correction to the additivity rule—based on the fact that at lower incident energies, scattering cross sections become much larger to the point of overlapping—so to sum them completely effectively counts some areas twice and hence overestimates the calculated molecular cross section [164, 165].

$$\sigma^{\text{molecule}} = \sum_{\text{atoms}} \sigma^{\text{atoms}}. \tag{2}$$

Equation (3) describes the Screening Corrected Additivity Rule, where s_i is the screening coefficient dependent upon the cross section of the other atoms in the molecule, and α_{ij} a term that gives the overlap of the cross sections within the molecule. This treatment managed to bring the accuracy of the IAM method down to 6 eV for C_3F_2 and has since been used in the >10 eV incident energy range, for total and differential elastic and inelastic cross sections. The differential cross section is treated with a redispersion factor described in detail elsewhere [165] which deals with the shielding of waves both into and out of the molecule. Multiple scattering events are included [159, 162] and are shown to be necessary between 1–300 eV incident electron energy.

$$\sigma^{\text{molecule}} = \sum_i s_i \sigma_i \qquad s_i = 1 - \frac{1}{2} \sum_{j \neq i} \frac{\sigma_j}{\alpha_{ij}}. \tag{3}$$

These improvements in the calculations for cross sections of low energy electron-molecule collisions are verified by experiment and ab initio calculations, and both experimental and calculated values are used as data tables in particle track simulations. Experiments also provide information on electron energy loss spectra and fragmentation pathways that are difficult to ascertain using approximations as simple as the IAM. Experimental criteria include low energy spread, simplistic experimental setups to enable appropriate analysis, and close approximation to the physical reality.

Within our group, total cross sections and electron energy loss spectra for collisions of electrons with various biomolecules in the gas phase can be collected for incident electron energies between 80 and 5000 eV. These are collected using a transmission experiment shown in Fig. 8. The total cross section is the probability of any interaction between the incoming particle and the target molecule—in the transmission experiment described this is defined using the Beer-Lambert law by measuring the drop in intensity of the incident beam of electrons as a function of the gas pressure in the scattering chamber. These values are used in simulations to define how many of the electrons at each energy will interact with their surroundings. Electron energy loss spectra are primarily used for identifying the excitation resonances and ionization energy of the target molecules (Fig. 9). They are recorded in the same transmission experiment as the total cross section, see Fig. 8. The characteristic energies and relative probabilities of a molecule to undergo electronic excitation, ion-

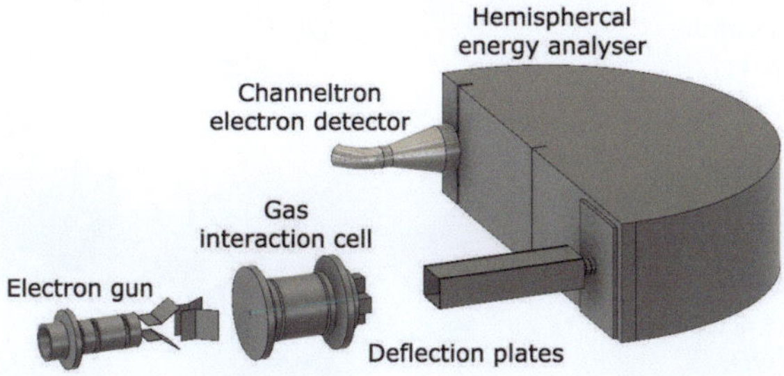

Fig. 8 Apparatus for both total cross sections and energy loss spectra, located in Madrid, Spain. Electrons from the electron gun cross biomolecules in the interaction chamber, and are analyzed by energy in a hemispherical analyzer and channeltron

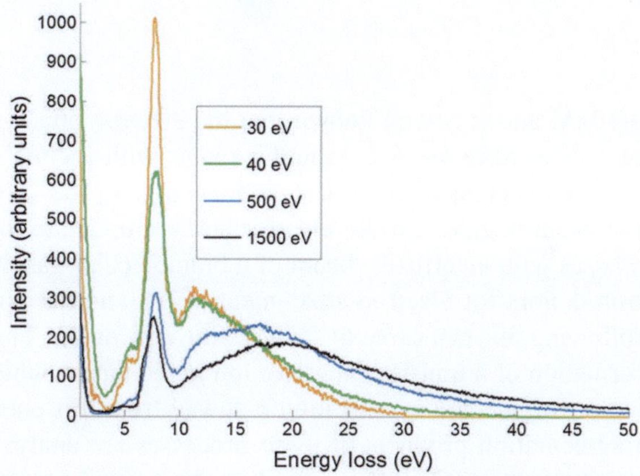

Fig. 9 Electron-molecule collisions result in various energy loss processes for the initial electron. As an example here you can see clearly the electronic excitation peak in pyrimidine of the ${}^1A_1$, ${}^1B_2$ $\pi \rightarrow \pi^*$ transitions at 7.6 eV. The initial energy of the impacting electron is shown in the legend. The figure is adapted from Ref. [166]

ization and autoionization following excitation are all clear in the spectra. With some deconvolution it is also possible to determine detail about the vibrational excitation and highest energy inner shell ionization.

Through their interactions with the medium, electrons, like the incident therapeutic radiation, are capable of ionizing molecules and creating radical species such as negative ions. As mentioned, these species are an immensely important part of radiation damage. Reactive oxygen species, such as the superoxide radical O_2^-, are particularly important in DNA damage and have been studied in biological experiments to prove their potential. As yet, no experimental cross sections of low energy anion–biomolecule collisions are available, despite their importance to radiotherapy. The following experimental setup is designed to bridge this gap.

A technique for investigating collisions of low-energy (<400 eV) negative ions with neutral biomolecules is being developed to provide details of fragmentation pathways through detection of positive, negative and neutral fragments [167]. The experiment shown schematically in Fig. 10 consists of a pulsed hollow cathode discharge, whereby the negative ions are formed in the plasma afterglow by electron attachment. The electron temperature in a hollow cathode discharge is typically

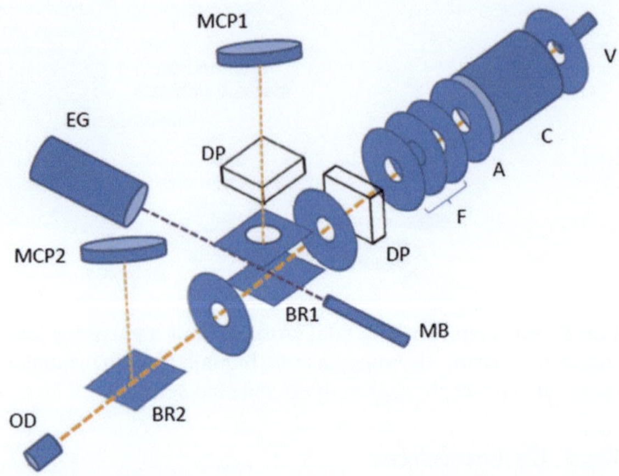

Fig. 10 Schematics of the experimental setup. *V*, pulsed supersonic valve; *C*, hollow cathode discharge; *A*, anode; *F*, focusing lens; *DP*, deflecting plates; *BR*, beam reflector; *MB*, effusive molecular beam; *MCP*, multi-channel plate detector; *EG*, electron gun; *OD*, optical detector. The figure is adapted from Ref. [167]

$<10\,\text{eV}$, and it is well known that the electron attachment cross sections for small molecules have resonances in this range, with excited species also having enhanced attachment cross sections at resonant energies [168–170]. Once formed, the negative ion beam is guided using electrostatic ion optics to the interaction region where it crosses with an effusive beam of a biomolecular gas. Negative ions can act as electron donors for electron attachment [171], and the presence of the neutral species following this can serve to stabilize the new anion. This charge transfer leads to the formation of a transient negative ion in the target, which can subsequently autoionize or fragment as a relaxation pathway from this energetically unstable state. The fragmentation products of these processes are analyzed using time of flight mass spectrometry (TOF-MS), including the neutral fragments, which are investigated with an electron gun.

All of these interactions contribute to the biological dose, and they occur at much lower energies than the incident radiation. Their contributions are slowly being uncovered and tabulated, implemented in particle track modeling, giving direction for radiosensitization, and slowly added to clinical radiotherapy software to increase the accuracy of life saving treatment. With strong data from experiment and calculations we are able to improve our modeling far beyond the initial particle in radiotherapy, to the end goal of modeling the survival of cell lines under different conditions with different radiosensitizers.

14 Experimental Study of Electron Emission from Ion-Irradiated Metallic Nanoparticles

As mentioned above, one highly promising idea in the field of nanotechnology enhanced radiation therapy is the administration of high-Z metal-based nanoparticles (NPs) composed, for instance, of gold, platinum, or gadolinium, as potential radiosensitizers during treatment. Their therapeutic effects are proven on the cellular level by irradiation of cell cultures and observation of the corresponding survival curves [172]. However, the full explanation of their radiosensitive properties on the fundamental level is yet to be determined. It has been associated with mechanisms that appear on different time scales, among which one of the most important is the emission of secondary low energy electrons induced by radiation and the catalytic (redox) properties due to the high surface/volume ratio that can enhance the production of the reactive oxygen species in the cellular environment [15].

In this context, it is important to determine the number of electrons which are emitted during the interaction of a specific ion with the NP. This value depends on the ion charge and velocity, the material of the NP, its size and its molecular coating as well as on the chemical environment in which the collision takes place. A theoretical approach is difficult to perform due to the large size of the radio-sensitizing system. As such the experimental approach is a promising alternative, while also representing a great challenge. Such an approach is realized in the framework of the activity of the AMA Group in the CIMAP Laboratory in Caen, France within the ARGENT project. This work is undertaken by Arkadiusz Mika under the supervision of Bernd Huber in the ESR project "Molecular efficiency of radiosensitizers in ion-induced radiation damage processes". The main goal is to construct and develop the experimental setup where under high vacuum conditions ($<10^{-8}$ mbar) a molecular beam of NPs will collide with highly charged ions and the resulting products (charged molecular fragments and electrons) will be detected. This experiment is under development, and preliminary experiments with ion irradiation of thin gold foils [173], as well as the irradiation of gold NPs by X-rays [174] have been performed.

14.1 Development of the Experimental Setup

14.1.1 Production of Nanoparticles

In order to perform a systematic study of the electron emission occurring in collisions of ions with various nanoscale targets, the first step is to create well-defined molecular beams containing biomolecules (e.g., nucleobases) in their natural cell environment (nano-solvated) with/without the presence of radiosensitizing agents or metallic clusters. This experimental approach is based on the use of a magnetron gas-aggregation cluster source [175]. Such a device is composed of 2 main elements—the condensation chamber cooled to liquid nitrogen temperatures and the magnetron head carrying the metallic target, for example a gold disk (see Fig. 11). A magnetic field-assisted discharge is ignited by a voltage of about 300 V in an argon atmosphere ($\sim$1 mbar). As the metallic target serves as the cathode, Ar^+ ions are accelerated in the $E \times B$ field towards the metal surface, forming a metal vapor via sputtering. The formation of clusters occurs outside of the discharge in the cooled condensation region.

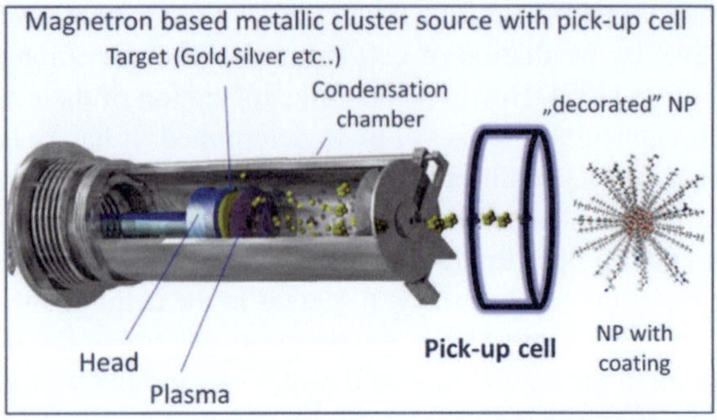

Fig. 11 A schematic view of the magnetron based gas aggregation cluster source with pick-up cell. The drawing of a NP was made by Kaspar Haume

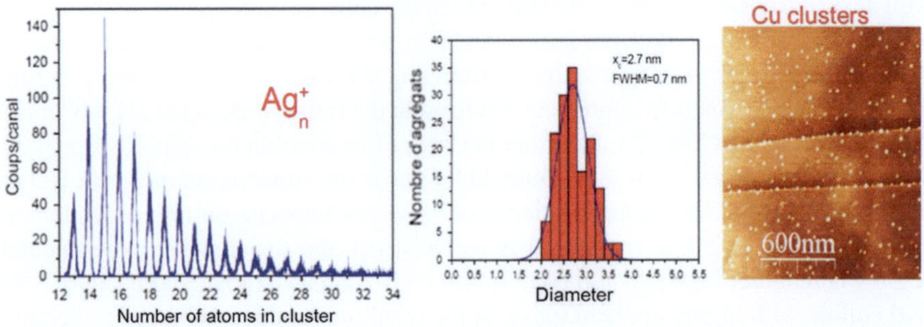

Fig. 12 Mass spectrum of small (a few atoms) silver clusters (*left panel*), the size distribution of size selected copper clusters (*middle panel*) and their deposition on a HOPG surface (*right panel*). The figure is adapted from Ref. [176]

The metallic clusters can be neutral or positively/negatively charged. After leaving the source they pass through a skimmer into a low pressure chamber where they traverse through a so called pick-up cell (Fig. 11). Here the molecular environment around the NP can be formed by the attachment of water molecules or evaporated (bio)molecules.

The thermodynamic conditions inside the source strongly influence the size and intensity of the produced NPs. These depend on the number of collisions in the aggregation source and, therefore, can be varied and controlled using the buffer gas pressure and the position of the magnetron head. For example, the formation of larger clusters is favored when the condensation length is longer. Some typical results [176] obtained with this source are given in Fig. 12. The left part shows the mass distribution of small positive, singly charged silver clusters containing between 10 and 30 atoms. In the middle part of Fig. 12, the size distribution of larger singly charged Cu^{n+} clusters is shown corresponding to particle diameters of $\sim$3 nm. The right part demonstrates the homogeneity of Cu clusters deposited on a clean highly oriented pyrolytic graphite (HOPG) surface, immobilized onto defects induced by prior ion bombardment.

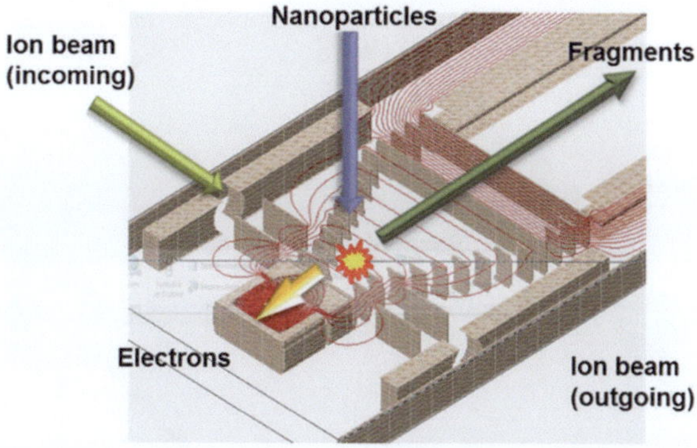

Fig. 13 The geometry of the extraction region with arrows showing the directions of the beams and products before and after the collision

14.1.2 Multicoincident Detection of Molecular Fragments and of the Number of Emitted Electrons

Extraction region: A technological challenge is represented by the coincident detection and separation of the collision products, either in the form of positive molecular fragments or the number of free electrons. For this purpose the ion extraction region (Fig. 13) was designed to be coupled with an electron statistics counter. Simulations using the SIMION software have been performed to optimize the electric fields and particle trajectories. The positively charged fragments are extracted along one direction (perpendicular to the axis of molecular beam and ion beam) into the Wiley McLaren Time-of-Flight Mass Spectrometer, whereas electrons are extracted in the opposite direction to a Passivated Implanted Planar Silicon (PIPS) detector that is kept at a high potential (more than +20 kV).

14.1.3 Electron Detection—PIPS Detector

One of the most challenging tasks is the detection of the signal coming from just one emitted electron (of the order of 0.4 mV) and a successful decoupling of this signal from the high voltage potential. As the PIPS detector and the connected electronics (pre-amplification and amplification stage) operate at a high potential, it is necessary to use an optical fiber data link to successfully decouple the energy and time signal for the multicoincidence hardware (FASTER developed at LPC Caen).

The collision process between low energy projectile ions (P) and the NP target (T) is described by the equation:

$$P^{n+} + T \rightarrow P^{(n-s)+} + T^{m+} + (m-s)e^{-}. \qquad (4)$$

Here P^{n+} is a projectile with charge $n+$, T^{m+} stands for the target with charge m, s is the number of electrons stabilized at the projectile and $(m-s)$ is the number of emitted electrons. The last remaining element to provide full information about the

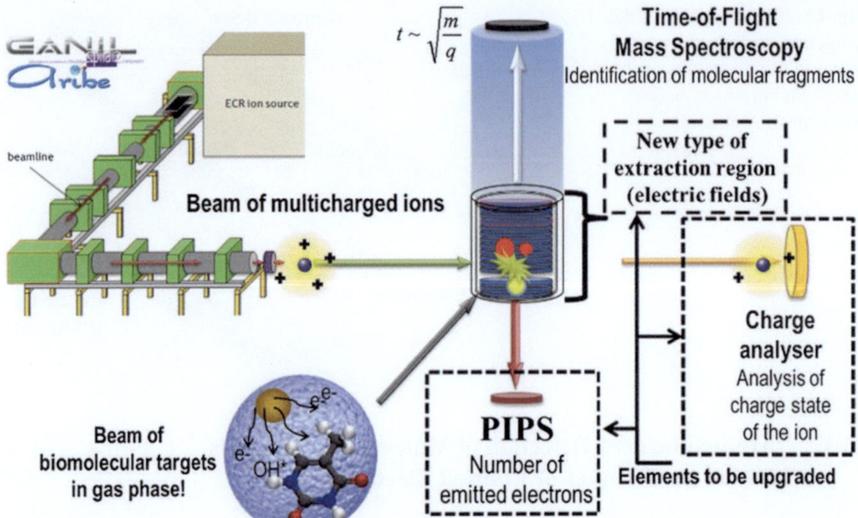

Fig. 14 Scheme of the experimental setup currently being developed for multicoincidence measurements between emitted electrons, molecular fragments and the projectile charge state after ion-nanoparticle collisions

collision process is the detection of the charge state of the projectile after collision. This will allow determination of the initial charge state of the mostly fragmented multiply charged target. In the experiment, this will be realized by the installation of a PSD type detector and the deflection of the charged projectiles after the collision with an electrostatic field.

The overall schematics presenting all elements of the experimental setup capable of characterizing the efficiency of radiosensitizing agents at the molecular level is shown on Fig. 14.

15 Impact of Nanoscale Processes and Nanoagents on Biodamage Complexity

Even though both x-rays and ion beams are ionizing radiation, and in both of them secondary electrons, free radicals, and other reactive species do the major damage to the cells, the qualitative features such as the dose dependence of the probability of cell survival are quite different for low- and high- linear energy transfer (LET) radiation [128, 177, 178]. The complexity of biodamage, which is an important distinction between the actions of high and low LET radiation, reduces the chances of damage repair and thus strongly affects the relative biological effectiveness (RBE) of ionizing radiation. This quantity is defined as the ratio of the dose delivered by photons to that delivered by different radiation modalities (e.g., electrons, protons, or heavier ions) leading to the same biological effect, such as the probability of an irradiated cell inactivation. Determining the impact of nanoscale processes and nanoagents on RBE is also one of the goals of the ESR research project carried out by Alexey Verkhovtsev at CSIC (Madrid, Spain) and MBN Research Center (Frankfurt am Main, Germany) under the supervision of Gustavo García (CSIC) and Andrey Solov'yov (MBN-RC). This project aims to improve the understanding of the physicochemical processes initiated by the interaction of various forms of ionizing radiation with biological matter.

A nanoscale understanding of radiotherapy requires evaluation of the effects of dose deposition on the nanoscale, i.e., nanodosimetry. The research conducted within this ESR project should give new physical and chemical insights necessary for developing a nanoscale model of radiation nanodosimetry. This concept aims at a detailed description of the interaction processes occurring in nanometer-size volumes of the medium and of implications of these processes in terms of radiation damage, such as the number of ionization or dissociative events, type of generated secondary species, etc. For that purpose, deep knowledge of numerous interactions induced by charged particles traversing living matter is strongly essential and can be obtained, for example, by means of Monte Carlo track-structure simulations (see Sect. 10). This approach is discussed in more details in the Chap. "Monte Carlo-Based Modeling of Secondary Particle Tracks Generated by Intermediateand Low-Energy Protons in Water" of this book.

This research project aims to quantify the effects of the presence of nanoagents on the complexity of biodamage—damage of biomolecules, such as DNA, leading to tumor cell death. Up to this date, particular attention has been paid to the assessment of direct and indirect damage of living cells as a function of LET and has been performed within the MultiScale Approach (MSA) to the physics of radiation damage with ions (see Ref. [128] and references therein). It has been demonstrated that the phenomenon-based MSA is capable of making quantitative predictions of macroscopic biological effects caused by ion radiation on the basis of physical and chemical effects related to the ion-medium interactions on a nanometer scale. Using this approach, evaluation of the survival probability for different cell lines has been performed. We applied the MSA to describe numerous experiments, where survival curves were obtained for different mammalian cell lines under various irradiation conditions. The details of this analysis are described in the Chap. "Predictive Assessment of Biological Damage Due to Ion Beams" of this book. The advantages of the MSA allow one to extend it to many other cell lines, including radiosensitive and radioresistive cells, different cell phases, irradiation conditions (e.g., in the presence of radiosensitizers) and make predictive evaluation of radiobiological effects.

Future research in this direction should include the effects due to radiosensitizing NPs. Recently we completed the first step in this direction by performing a theoretical and numerical analysis of secondary electron production by small metallic NP irradiated by fast ions [90, 179]. Nanometer-size NPs composed of gold, platinum, silver and gadolinium, which are of current interest for application in cancer treatments for reasons discussed above, have been considered. With this analysis, we provided a physical explanation for enhancement of the low-energy electron production by sensitizing metallic NPs irradiated with fast ions. It has been demonstrated that metal NPs significantly enhance the electron yield due to the collective response to an external electric field of a charged projectile. The role of collective electron excitations in the formation of secondary electron spectra in metal clusters and NPs is discussed in greater detail in the Chap. "Irradiation-Induced Processes with Atomic Clusters and Nanoparticles" of this book.

16 Radiation Experiments with Radiosensitizing Nanoparticles

Metallic nanoparticles (NPs) have been proposed as enhancers of radiation treatments in cancer therapy. Their high atomic number gives them higher mass energy absorption coefficients than soft tissue. Functionalized NPs can target cancer cells which allows more selectivity of the dose deposition within the tumor whilst sparing the surrounding healthy tissue. The use of NPs for radiosensitization was first demonstrated by Hainfeld et al. [10] using 1.9 nm gold nanoparticles (AuNPs) delivered systemically, prior to irradiation, in mice exhibiting EMT-6 mammary carcinomas. The authors reported a 1 year survival of 86 % in animals treated under these conditions compared with only 20 % in those irradiated without AuNPs injection. The functionalization of NPs provides further benefit in the clinical setting by increasing contrast properties in radiological imaging and MRI.

In vitro experiments have shown the potential of the gadolinium-based AGuIX NPs to amplify radiation-effects under X-ray irradiation in the keV and MeV range. The efficiency of AGuIX has also been demonstrated in vivo. The median survival time in mice was 112 days when combining AGuIX with radiation compared to 77 days for the radiation only treatment [19]. Gadolinium compounds can be used as theranostic agents due to their magnetic properties leading to improved MRI contrast. This would allow monitoring of the NPs within the tumor, providing a more precise definition of the uptake and biodistribution within the tumor. Another possibility is to attach gadolinium to metallic NPs, e.g., composed of gold. When using Au@DTDTPA-Gd NPs (see Sect. 2), it is possible to obtain the same biological effect in prostate cancer cells under gamma irradiation using half of the radiation dose [18, 43]. An improvement of the efficiency of heavy ion radiation using AGuIX NPs was observed for the first time by Porcel et al. [15]. This has opened the perspective of using theranostics approaches in hadrontherapy. An enhancement of 18.5 % and 11.3 % for carbon and helium ions respectively, was demonstrated in mammalian cells [15].

The efficacy of platinum NPs (PtNPs) to amplify ionizing radiation was quantified using plasmid DNA as a nanobiodamage probe [3]. When metallic compounds are added, additional ionizations take place due to the high ionization cross section of high-Z atoms. Incident ionizing particles and secondary electrons produced along the track may excite inner and outer electron energy shells of the metal atom. The ionizations in inner shells in particular are followed by Auger de-excitations processes, which result in an amplification of the electron emission by the metal. The presence of high-Z atoms in the medium locally amplifies the density of ionization and the dose deposition. When DNA is loaded with platinum, Auger electrons may either interact with the DNA directly and induce stand breaks or interact with surrounding water molecules to produce clusters of radicals that may further damage DNA. The role of free radicals and reactive oxygen species (ROS) has been investigated by adding dimethyl sulfoxide (DMSO) radical scavenger. In the case of pure DNA as well as in DNA loaded with platinum compounds, the induction of single and double strand breaks (SSBs and DSBs) was strongly reduced [3]. This result confirmed that the induction of DNA damage and the amplification of radiation effects due to the metal are mostly related to the production of water radicals close to the NP [180].

On a cellular level, the interaction between NPs and the ionizing radiation produces photons and electrons [181]. The interaction of these particles with water and oxygen yields ROS that will alter the activity of biomolecules in the vicinity of the radiosensitizers. In vitro experiments have shown an increase of the cell surviving fraction when adding DMSO [147, 182]. The different studies mentioned previously confirmed that the amplification of radiation effects due to metallic NPs are mostly related to the production of water radicals close to the metal. Furthermore, the level of oxygen in cells plays an important role in ROS production and therefore in cell killing (see Sect. 12).

The precise mechanism of how the NPs act as radio-enhancers and the crucial NP characteristics required to improve this effect remain unclear. Within the ARGENT project, a set of radiation experiments on cell and animal models is planned in order to further understand mechanisms to amplify radiation-induced cell killings. The radiochemical mechanisms involved in this effect will be studied as well as the increase of radio sensitivity of cells at low levels of oxygenation. To achieve this, different LET radiation and ions are being combined with diverse NPs, such as gold, gadolinium and platinum.

This section details the various necessary modes of experimentation and preclinical trials to be undertaken by the ARGENT group, using both photon and ion radiation in vitro and in vivo and the analysis techniques undertaken by the researchers involved.

16.1 In Vitro Studies Under Photon Radiation

Different types of AuNPs are investigated in combination with radiotherapy within ARGENT. The first types of AuNPs, with an average size of 5 nm, are used with different coatings to explore their toxicity and biocompatibility potential, in combination with radiotherapy. The second type of NPs tested, also with an average of 5 nm, are provided by the ARGENT industrial partners (see Sect. 2). These NPs are composed of gold and gadolinium (Au@DTDTPA, Au@DTDTPA-Gd and AGuIX), and their properties of amplifying the effects of photon and ion beam irradiation have been explored above. Given their theranostic properties, good biocompatility and low toxicity [18, 19, 43] they will be further characterized and tested regarding other cancer cell lines and other radiation sources.

Kilovoltage and Megavoltage X-ray radiation ranges have been shown different effects when interacting with NPs [183, 184]. The photoelectric effect is predominant at the kilovoltage range of energy and is responsible for a high photon absorption by the NPs [183]. Therefore, the radiosensitization effect of AuNPs has been mainly explored using kilovoltage photon energies. Although this energy is still used for brachytherapy treatment, Megavoltage X-rays are the main approach used to achieve an adequate dose deposition to deep tumors [185]. Despite the fact that theoretical models generally predict a lower radiosensitization effect of NPs in combination with MeV X-rays [181, 183], several groups have reported an enhancement with MeV irradiation when using gold NPs [4, 183, 184]. Therefore, a collaborative study between different laboratories within the network is exploring these two radiation energy ranges to better characterize the NPs in use. Experiments using kV and MV X-ray radiation are being performed at Northampton Hospital and Linford Wood Genesis Care Medical Centre (Milton Keynes, UK), conducted by the research team from the Open University, and also at the Centre for Cancer Research and Cell Biology in Queen's University and the Northern Ireland Cancer Centre (Belfast, UK), conducted by the research team from the Queen's University. In addition, a comparison within the ARGENT network between theoretical predictions and experimental results on AuNPs radiosensitization with X-rays of different energies will allow us a better understanding of this effect.

16.2 In Vivo Experiments Under Photon Radiation

Moving towards translational research, our studies need to focus in vitro initially and then in vivo. Our experimental in vitro results will allow us to determine the best radiosensitization conditions for each cell line and NP type. As these conditions are established, providing significant radiosensitizing results with minimal toxicity, in vivo experiments will be undertaken in the Centre for Cancer Research and Cell Biology (Belfast, UK) to validate these findings. These experiments will not only allow the evaluation of the radiosensitizing capacity of the NPs in a tumor microenvironment, but they will also assess the imaging potential to fully characterize the theranostic potential. Within this project it will be possible to make a comparative study involving gold, gadolinium and combinations of these regarding their theranostic properties.

16.3 In Vitro Experiments with Heavy Ion Radiation

As mentioned above, the low oxygen level in tumor cells is responsible for the decrease of radiosensitivity of solid tumors. Heavy ion radiation has been positioned as a possible solution to eradicate hypoxic tumors. When the therapeutic effects of ions are compared with those of photon radiation, the ions have the advantages of a large RBE [140, 186] and a small oxygen enhancement ratio (OER) [187, 188] due to the increasing ionization density and high LET. Clinical studies show that the large RBE and the small OER of ions are responsible for enhanced biological effects and improved response in overcoming hypoxia-induced radioresistance [189, 190]. A few studies have been reported on the LET dependence of OER [191]. The objective is to evaluate OER for different ions and LETs, including carbon, helium and oxygen, in the presence and the absence of diverse NPs such as gold and platinum. Since the OER may also depend on other characteristics of the beam quality, such as the dose rate, the existing techniques to shape beam distribution on the tumor will be tested: passive beam at the Heavy Ion Medical Accelerator in Chiba (HIMAC, Japan) and active beam at the Heidelberg Ion Therapy Center (HIT, Germany). These radiation experiments will be conducted by the research team from the University of Paris-Sud. The theoretical study of the OER will be done in collaboration with the GSI research team (see Sect. 12).

It has been demonstrated that the contribution of indirect damage to cell killing decreased with increasing LET. Nevertheless, indirect action still played a significant role in cell killing and contributed around 30 % to cell killing even at an LET of 2106 keV/μm [140]. A thorough study of the indirect contribution of ROS in hypoxia when NPs are present and how it varies regarding the nature and the quality of the beam is planned. NPs open the possibility of increasing cell killing in hypoxic radioresistance cells.

16.4 Analysis

For the data analysis of both photon and ion beam experiments, we mainly use clonogenic assays. These quantify reproductive cell death after treatment with ionizing radiation, but can also be used to determine the effectiveness of other cytotoxic agents. The survival fraction (SF) of cells is plotted versus the dose and the result can be modeled, for instance, by the linear-quadratic law, $\mathrm{SF}(D) = e^{-(\alpha D+\beta D^2)}$. The coefficient α corresponds to the contribution of lesions, which are directly lethal for the cell, whereas β is attributed to the contribution of additive sub-lethal lesions. The efficiency of NPs to amplify radiation-induced cell deaths can be quantified by calculating the enhancing factor, $\mathrm{EF} = (D^{10}_{\mathrm{control}} - D^{10}_{\mathrm{NPs}})/D^{10}_{\mathrm{control}}$. Here, D^{10} represents the radiation dose used to reach 10 % of cell survival in the control ($D^{10}_{\mathrm{control}}$) and in cells loaded with NPs (D^{10}_{NPs}). The dose at 50 % of cell survival is also used for comparison of the EF depending on the nature of the radiation.

Alongside clonogenic assays DNA damage is an important readout of biological efficacy. Radiation itself is known to induce DNA DSBs which can be imaged using immunofluorescent staining to quantify the number of foci present in each sample. An increased number of foci is associated with radiation exposure and directly related to cell death [192]. One of the potential radiosensitizing mechanisms of action of NPs could be due to an increase in DNA damage post-irradiation or interference with the repair mechanisms of the cell. An indication of such consequences can be identified by looking at early (1h post-irradiation) and late (24h post-irradiation) DNA damage. After only 1 h of being exposed to radiation, cells present increased foci numbers. However after 24 h these have been reduced due to repair mechanisms. If NPs are present, they can potentially increase the initial yields of DNA damage. Moreover, if there are differences after being exposed to radiation in the presence of NPs compared to the controls at 24 h this may indicate that there is interference in the repair mechanism. Thus, analyzing DNA damage could further lead to a better understanding of NPs' radiosensitizing mechanism by providing clues towards how they affect cells.

The comparison between different radiation sources, beam types, NPs and cell lines will allow us to define optimal treatment protocols that are able to improve tumor cure whilst decreasing the side effects on healthy tissue. Moreover, the theranostic potential of these NPs will be exploited to combine both diagnostic and treatment possibilities while maintaining the same therapeutic effect. These should improve the therapeutic outcome and benefit the patient's life after treatment.

17 Conclusions

As has been shown throughout this chapter, the current status of nanoparticle assisted medicine and in particular enhancement of radiation (photon and ion) for cancer therapy is improved in a big way by interdisciplinary studies and collaboration amongst research leaders in their respective fields. From the beginning, expertise is required in understanding the synthesis and effects of different nanoparticle properties on not just the cancer cells, but the entire organism, and then to go further and fundamentally control those effects. The nanoparticles show high cell specificity, complex uptake dynamics, and their effect is highly dependent on their surface coatings, shape, size and target cell type. Their interaction with radiation in biological material is being investigated along with a deepening understanding of the initial and secondary effects of ion and photon irradiation, along with the cellular and systematic effects of these combination therapies in high and low oxygen environments. All of the areas of research described here are in a perpetual feedback loop, generating direction and improvements back and forth from simulation and experiment to provide the best possible environment for a full understanding of the essentially human problem of advanced cancer treatment. The ARGENT ITN is dedicated to advancing this vast and growing field, and expects to see large scale introduction of nanomedicine to society over the coming decades.

Acknowledgements The authors acknowledge financial support from the European Union's FP7 People Program (Marie Curie Actions) within the Initial Training Network No. 608163 "ARGENT".

References

1. COST Action Nano-IBCT. http://mbnresearch.com/project-nanoibct
2. Peer D et al (2007) Nanocarriers as an emerging platform for cancer therapy. Nat Nanotechnol 2:751–760
3. Porcel E et al (2010) Platinum nanoparticles: a promising material for future cancer therapy? Nanotechnology 21:085103
4. McMahon SJ et al (2011) Biological consequences of nanoscale energy deposition near irradiated heavy atom nanoparticles. Sci Rep 1:18
5. Hainfeld JF, Dilmanian FA, Slatkin DN, Smilowitz HM (2008) Radiotherapy enhancement with gold nanoparticles. J Pharm Pharmacol 60:977–985
6. Kwatra D, Venugopal A, Anant S (2013) Nanoparticles in radiation therapy: a summary of various approaches to enhance radiosensitization in cancer. Transl Cancer Res 2:330–342
7. Malam Y, Loizidou M, Seifalian AM (2009) Liposomes and nanoparticles: nanosized vehicles for drug delivery in cancer. Trends Pharmacol Sci 30:592–599
8. Carter JD, Cheng NN, Qu Y, Suarez GD, Guo T (2007) Nanoscale energy deposition by X-ray absorbing nanostructures. J Phys Chem B 111:11622–11625
9. Liu C-J et al (2010) Enhancement of cell radiation sensitivity by pegylated gold nanoparticles. Phys Med Biol 55:931–945
10. Hainfeld JF, Slatkin DN, Smilowitz HM (2004) The use of gold nanoparticles to enhance radiotherapy in mice. Phys Med Biol 49:N309–N315
11. Kobayashi K, Usami N, Porcel E, Lacombe S, Le Sech C (2010) Enhancement of radiation effect by heavy elements. Mutat Res 704:123–131
12. Usami N et al (2008) Mammalian cells loaded with platinum-containing molecules are sensitized to fast atomic ions. Int J Radiat Biol 84:603–611
13. Porcel E, Kobayashi K, Usami N, Remita H, Le Sech C, Lacombe S (2011) Photosensitization of plasmid-DNA loaded with platinum nano-particles and irradiated by low energy X-rays. J Phys Conf Ser 261:012004
14. Asharani PV, Wu, YL, Gong Z, Valiyaveettil S (2008) Toxicity of silver nanoparticles in zebrafish models. Nanotechnology 19:255102
15. Porcel E et al (2014) Gadolinium-based nanoparticles to improve the hadrontherapy performances. Nanomed Nanotechnol 10:1601–1608
16. Alric C et al (2008) Gadolinium chelate coated gold nanoparticles as contrast agents for both X-ray computed tomography and magnetic resonance imaging. J Am Chem Soc 130:5908–5915
17. Barreto JA, O'Malley W, Kubeil M, Graham B, Stephan H, Spiccia L (2011) Nanomaterials: applications in cancer imaging and therapy. Adv Mater 23:H18–H40
18. Alric C et al (2013) The biodistribution of gold nanoparticles designed for renal clearance. Nanoscale 5:5930–5939
19. Sancey L et al (2014) The use of theranostic gadolinium-based nanoprobes to improve radiotherapy efficacy. Br J Radiol 87:20140134
20. Albanese A, Tang PS, Chan WCW (2012) The effect of nanoparticle size, shape, and surface chemistry on biological systems. Annu Rev Biomed Eng 14:1–16
21. Chithrani BD, Ghazani AA, Chan WCW (2006) Determining the size and shape dependence of gold nanoparticle uptake into mammalian cells. Nano Lett 6:662–668
22. Zhang S, Li J, Lykotrafitis G, Bao G, Suresh S (2009) Size-dependent endocytosis of nanoparticles. Adv Mater 21:419–424

23. Perrault SD, Walkey C, Jennings T, Fischer HC, Chan WCW (2009) Mediating tumor targeting efficiency of nanoparticles through design. Nano Lett 9:1909–1915
24. Zhu M, Nie G, Meng H, Xia T (2012) Physicochemical properties determine nanomaterial cellular uptake, transport, and fate. Acc Chem Res 46:622–631
25. Lin Y, McMahon SJ, Paganetti H, Schuemann J (2015) Biological modeling of gold nanoparticle enhanced radiotherapy for proton therapy. Phys Med Biol 60:4149–4168
26. Beddoes CM, Case CP, Briscoe WH (2015) Understanding nanoparticle cellular entry: a physicochemical perspective. Adv Col Interface Sci 218:48–68
27. Hirsch V, Salaklang J, Rothen-Rutishauser B, Petri-Fink A (2013) Influence of serum supplemented cell culture medium on colloidal stability of polymer coated iron oxide and polystyrene nanoparticles with impact on cell interactions in vitro. IEEE Trans Magn 49:402–407
28. Yah CS (2013) The toxicity of gold nanoparticles in relation to their physiochemical properties. Biomed Res 24:400–413
29. Kalay S, Blanchet C, Culha M (2014) Linear assembly and 3D networks of peptide modified gold nanoparticles. Turk J Chem 38:686–700
30. da Rocha EL, Caramori GF, Rambo CR (2013) Nanoparticle translocation through a lipid bilayer tuned by surface chemistry. Phys Chem Chem Phys 15:2282–2290
31. Akhter S, Ahmad MZ, Ahmad FJ, Storm G, Kok RJ (2012) Gold nanoparticles in theranostic oncology: current state-of-the-art. Expert Opin Drug Deliv 9:1225–1243
32. Ranganathan R et al (2012) Nanomedicine: towards development of patient-friendly drug-delivery systems for oncological applications. Int J Nanomed 7:1043–1060
33. Illes E et al (2014) PEGylation of surfacted magnetite core-shell nanoparticles for biomedical application. Colloid Surf A 460:429–440
34. Thierry B, Griesser HJ (2012) Dense PEG layers for efficient immunotargeting of nanoparticles to cancer cells. J Mater Chem 22:8810–8819
35. Otsuka H, Nagasaki Y, Kataoka K (2003) PEGylated nanoparticles for biological and pharmaceutical applications. Adv Drug Deliv Rev 55:403–419
36. Chithrani BD, Stewart J, Allen C, Jaffray DA (2009) Intracellular uptake, transport, and processing of nanostructures in cancer cells. Nanomed Nanotechnol 5:118–127
37. Saptarshi SR, Duschl A, Lopata AL (2013) Interaction of nanoparticles with proteins: relation to bio-reactivity of the nanoparticle. J Nanobiotechnol 11:26
38. Shmeeda H, Tzemach D, Mak L, Gabizon A (2009) Her2-targeted pegylated liposomal doxorubicin: retention of target-specific binding and cytotoxicity after in vivo passage. J Controlled Release 136:155–160
39. Calvaresi EC, Hergenrother PJ (2013) Glucose conjugation for the specific targeting and treatment of cancer. Chem Sci 4:2319–2333
40. Gromnicova R et al (2013) Glucose-coated gold nanoparticles transfer across human brain endothelium and enter astrocytes in vitro. PLoS ONE 8:e81043
41. Hu C, Niestroj M, Yuan D, Chang S, Chen J (2015) Treating cancer stem cells and cancer metastasis using glucose-coated gold nanoparticles. Int J Nanomed 10:2065–2077
42. Dai Q, Walkey C, Chan WC (2014) Polyethylene glycol backfilling mitigates the negative impact of the protein corona on nanoparticle cell targeting. Angew Chem Int Ed 53:5093–5096
43. Miladi I et al (2014) The in vivo radiosensitizing effect of gold nanoparticles based MRI contrast agents. Small 10:1116–1124
44. Zhao P, Li N, Astruc D (2013) State of the art in gold nanoparticle synthesis. Coord Chem Rev 257:638–665
45. Turkevich J, Stevenson PC, Hillier J (1951) A study of the nucleation and growth processes in the synthesis of colloidal gold. Discuss Faraday Soc 11:55–75
46. Frens G (1973) Controlled nucleation for the regulation of the particle size in monodisperse gold suspensions. Nature 241:20–22
47. Brust M, Walker M, Bethell D, Schiffrin DJ, Whyman R (1994) Synthesis of thiol-derivatised gold nanoparticles in a two-phase liquid-liquid system. J Chem Soc Chem Commun 1994:801–802

48. Debouttiere P-J et al (2006) Design of gold nanoparticles for magnetic resonance imaging. Adv Funct Mater 16:2330–2339
49. Perrault SD, Chan WCW (2009) Synthesis and surface modification of highly monodispersed, spherical gold nanoparticles of 50–200 nm. J Am Chem Soc 131:17042–17043
50. Ahmadi T, Wang Z, Green T, Henglein A, El-Sayed M (1996) Shape-controlled synthesis of colloidal platinum nanoparticles. Science 272:1924–1926
51. Stepanov AL, Golubev AN, Nikitin SI, Osin YN (2014) A review on the fabrication and properties of platinum nanoparticles. Rev Adv Mater Sci 38:160–175
52. Miladi I et al (2013) Biodistribution of ultra small gadolinium-based nanoparticles as theranostic agent: application to brain tumors. J Biomater Appl 28:385–394
53. Faucher L, Tremblay M, Lagueux J, Gossuin Y, Fortin M-A (2012) Rapid synthesis of PEGylated ultrasmall gadolinium oxide nanoparticles for cell labeling and tracking with MRI. ACS Appl Mater Interfaces 4:4506–4515
54. Louis C et al (2005) Nanosized hybrid particles with double luminescence for biological labeling. Chem Mater 17:1673–1682
55. Torchilin VP, Papisov MI (1994) Why do polyethylene glycol-coated liposomes circulate so long? J Liposome Res 4:725–739
56. Nicol JR, Dixon D, Coulter JA (2015) Gold nanoparticle surface functionalization: a necessary requirement in the development of novel nanotherapeutics. Nanomedicine 10:1315–1326
57. Chattopadhyay N, Cai Z, Kwon YL, Lechtman E, Pignol J-P, Reilly RM (2013) Molecularly targeted gold nanoparticles enhance the radiation response of breast cancer cells and tumor xenografts to X-radiation. Breast Cancer Res Treat 137:81–91
58. Le Duc G et al (2011) Toward an image-guided microbeam radiation therapy using gadolinium-based nanoparticles. ACS Nano 5:9566–9574
59. Fang J et al (2014) Manipulating the surface coating of ultra-small Gd_2O_3 nanoparticles for improved T1-weighted MR imaging. Biomaterials 35:1636–1642
60. Bregoli L, Movia D, Gavigan-Imedio JD, Lysaght J, Reynolds J, Prina-Mello A (2016) Nanomedicine applied to translational oncology: a future perspective on cancer treatment. Nanomed Nanotechnol 12:81–103
61. Mignot A et al (2013) A top-down synthesis route to ultrasmall multifunctional Gd-based silica nanoparticles for theranostic applications. Chem Eur J 19:6122–6136
62. Tallury P, Payton K, Santra S (2008) Silica-based multimodal/multifunctional nanoparticles for bioimaging and biosensing applications. Nanomedicine 3:579–592
63. Stöber W, Fink A, Bohn E (1968) Controlled growth of monodisperse silica spheres in the micron size range. J Colloid Interface Sci 26:62–69
64. Ma K, Mendoza C, Hanson M, Werner-Zwanziger U, Zwanziger J, Wiesner U (2015) Control of ultrasmall sub-10 nm ligand-functionalized fluorescent core-shell silica nanoparticle growth in water. Chem Mater 27:4119–4133
65. Chi F, Guan B, Yang B, Liu Y, Huo Q (2010) Terminating effects of organosilane in the formation of silica cross-linked micellar core-shell nanoparticles. Langmuir 26:11421–11426
66. Arriagada FJ, Osseo-Asare K (1999) Synthesis of nanosize silica in a nonionic water-in-oil microemulsion: effects of the water/surfactant molar ratio and ammonia concentration. J Colloid Interface Sci 211:210–220
67. Patterson JP, Robin MP, Chassenieux C, Colombani O, O'Reilly RK (2014) The analysis of solution self-assembled polymeric nanomaterials. Chem Soc Rev 43:2412–2425
68. Zetasizer nano series user manual (2004)
69. Lehman SE, Tataurova Y, Mueller PS, Mariappan SVS, Larsen SC (2014) Ligand characterization of covalently functionalized mesoporous silica nanoparticles: an NMR toolbox approach. J Phys Chem C 118:29943–29951
70. Price WS (2005) Applications of pulsed gradient spin-echo NMR diffusion measurements to solution dynamics and organization. Diffus Fundam 2:112
71. Tomaszewska E et al (2013) Detection limits of DLS and UV-Vis spectroscopy in characterization of polydisperse nanoparticles colloids. J Nanomater 2013:313081

72. Pettitt ME, Lead JR (2013) Minimum physicochemical characterisation requirements for nanomaterial regulation. Environ Int 52:41–50
73. Morlieras J et al (2013) Development of gadolinium based nanoparticles having an affinity towards melanin. Nanoscale 5:1603–1615
74. Morlieras J et al (2013) Functionalization of small rigid platforms with cyclic RGD peptides for targeting tumors overexpressing $\alpha_v\beta_3$-integrins. Bioconjug Chem 24:1584–1597
75. Truillet C, Lux F, Tillement O, Dugourd P, Antoine R (2013) Coupling of HPLC with electrospray ionization mass spectrometry for studying the aging of ultrasmall multifunctional gadolinium-based silica nanoparticles. Anal Chem 85:10440–10447
76. Kotb S et al (2016) Gadolinium-based nanoparticles and radiation therapy for multiple brain melanoma metastases: proof of concept before phase I trial. Theranostics 6:418–427
77. Merbach A, Helm L, Toth E (2013) The chemistry of contrast agents in medical magnetic resonance imaging. Wiley
78. Davis ME, Shin DM (2008) Nanoparticle therapeutics: an emerging treatment modality for cancer. Nat Rev Drug Discov 7:771–782
79. Brigger I et al (2012) Nanoparticles in cancer therapy and diagnosis. Adv Drug Deliv Rev 64:24–36
80. Toulany M et al (2014) Cisplatin-mediated radiosensitization of non-small cell lung cancer cells is stimulated by ATM inhibition. Radiother Oncol 111:228–236
81. Liang K et al (2003) Sensitization of breast cancer cells to radiation by trastuzumab. Mol Cancer Ther 2:1113–1120
82. Hermanson GT (2013) Bioconjugate techniques. Academic Press
83. Conde J et al (2014) Revisiting 30 years of biofunctionalization and surface chemistry of inorganic nanoparticles for nanomedicine. Front Chem 2:48
84. Ghosh SS et al (1990) Use of maleimide-thiol coupling chemistry for efficient syntheses of oligonucleotide-enzyme conjugate hybridization probes. Bioconjug Chem 1:71–76
85. Lutz J-F, Zarafshani Z (2008) Efficient construction of therapeutics, bioconjugates, biomaterials and bioactive surfaces using azidealkyne "click" chemistry. Adv Drug Deliv Rev 60:958–970
86. Allen MP, Tildesley DJ (1989) Computer simulation of liquids. Oxford University Press
87. Frenkel D, Smit B (2001) Understanding molecular simulation: from algorithms to applications. Academic Press, San Diego
88. MacKerell AD Jr et al (1998) All-atom empirical potential for molecular modeling and dynamics studies of proteins. J Phys Chem B 102:3586–3616
89. Xiao F et al (2011) On the role of low-energy electrons in the radiosensitization of DNA by gold nanoparticles. Nanotechnology 22:465101
90. Verkhovtsev AV, Korol AV, Solov'yov AV (2015) Revealing the mechanism of the low-energy electron yield enhancement from sensitizing nanoparticles. Phys Rev Lett 114:063401
91. Hohenester U, Trügler A (2012) MNPBEM—A Matlab toolbox for the simulation of plasmonic nanoparticles. Comput Phys Commun 183:370–381
92. Palik ED (1998) Handbook of optical constants of solids. Academic Press
93. Alkilany AM, Murphy CJ (2010) Toxicity and cellular uptake of gold nanoparticles: what we have learned so far? J Nanopart Res 12:2313–2333
94. Boisselier E, Astruc D (2009) Gold nanoparticles in nanomedicine: preparations, imaging, diagnostics, therapies and toxicity. R Soc Chem 38:1759–1782
95. Bianchi A et al (2014) Quantitative biodistribution and pharmacokinetics of multimodal gadolinium-based nanoparticles for lungs using ultrashort TE MRI. Magn Reson Mater Phys Biol Med 27:303–316
96. Pan Y et al (2009) Gold nanoparticles of diameter 1.4 nm trigger necrosis by oxidative stress and mitochondrial damage. Small 5:2067–2076
97. Pan Y et al (2007) Size-dependent cytotoxicity of gold nanoparticles. Small 3:1941–1949
98. Nidome T et al (2006) PEG-modified gold nanorods with a stealth character for in vivo applications. J Control Release 114:343–347

99. Patra HK, Banerjee S, Chaudhuri U, Lahiri P, Dasgupta AK (2007) Cell selective response to gold nanoparticles. Nanomed Nanotechnol 3:111–119
100. Holmes P, Tuckett C (2000) Airborne particles: exposure in the home and health effects. MRC Institute for Environment and Health, Leicester
101. Wallace BA, Janes RW (2001) Synchrotron radiation circular dichroism spectroscopy of proteins: secondary structure, fold recognition and structural genomics. Curr Opin Chem Biol 5:567–571
102. Wallace BA (2000) Synchrotron radiation circular-dichroism spectroscopy as a tool for investigating protein structures. J Synchrotron Radiat 7:289–295
103. World Health Organization. http://www.who.int/mediacentre/factsheets/fs297/en/, http://www.who.int/mediacentre/factsheets/fs334/en/
104. Cancer Research UK. http://www.cancerresearchuk.org/about-cancer/type/
105. American Cancer Society. http://www.cancer.org/cancer/index
106. Wingfield C (2002) Skin cancer: an overview of assessment and management. Primary Health Care 22:28–37
107. Coulter JA et al (2012) Cell type-dependent uptake, localization, and cytotoxicity of 1.9 nm gold nanoparticles. Int J Nanomed 7:2673–2685
108. Lodish H, Berk A, Zipursky SL, Matsudaira P, Baltimore D, Darnell JE (2000) Molecular cell biology. W.H. Freeman, New York
109. Shukla R et al (2005) Biocompatibility of gold nanoparticles and their endocytotic fate inside the cellular compartment: a microscopic overview. Langmuir 21:10644–10654
110. Brust M et al (1995) Synthesis and reactions of functionalized gold nanoparticles. J Chem Soc Chem Commun 1995:1655–1656
111. Chithrani BD et al (2007) Elucidating the mechanism of cellular uptake and removal of protein-coated gold nanoparticles of different sizes and shapes. Nano Lett 7:1542–1550
112. Ivanov AI (2014) Pharmacological inhibitors of exocytosis and endocytosis: novel bullets for old targets. Methods Mol Biol 1174:3–18
113. Gao H et al (2013) Ligand modified nanoparticles increases cell uptake, alters endocytosis and elevates glioma distribution and internalization. Sci Rep 3:2534
114. Rogers DWO (1991) The role of Monte Carlo simulation of electron transport in radiation dosimetry. Appl Radiat Isot 42:965–974
115. Paganetti H et al (2004) Accurate Monte Carlo simulations for nozzle design, commissioning and quality assurance for a proton radiation therapy facility. Med Phys 31:2107–2118
116. Friedland W et al (1998) Monte Carlo simulation of the production of short DNA fragments by low-linear energy transfer radiation using higher-order DNA models. Radiat Res 150:170–182
117. Nikjoo H et al (2002) Modelling of DNA damage induced by energetic electrons (100 eV to 100 keV). Radiat Prot Dosim 99:77–80
118. Pimblott SM, Mozumder A (1991) Structure of electron tracks in water. 2. Distribution of primary ionizations and excitations in water radiolysis. J Phys Chem 95:7291–7300
119. Champion C et al (2012) EPOTRAN: a full-differential Monte Carlo code for electron and positron transport in liquid and gaseous water. Int J Radiat Biol 88:54–61
120. Incerti S et al (2010) Comparison of GEANT4 very low energy cross section models with experimental data in water. Med Phys 37:4692–4708
121. Muñoz A, Pérez JM, García G, Blanco F (2005) An approach to Monte Carlo simulation of low-energy electron and photon interactions in air. Nucl Instr Meth A 536:176–188
122. Krämer M, Durante M (2010) Ion beam transport calculations and treatment plans in particle therapy. Eur Phys J D 60:195–202
123. García Gomez-Tejedor G, Fuss MC (eds) (2012) Radiation damage in biomolecular systems. Springer
124. Nikjoo H et al (2012) Interaction of radiation with matter, CRC Press
125. Muñoz A et al (2008) Single electron tracks in water vapour for energies below 100 eV. Int J Mass Spectrom 277:175–179
126. Wälzlein C et al (2014) Low energy electron transport in non-uniform media. Nucl Instr Meth B 320:75–82

127. Waelzlein C et al (2014) Simulation of dose enhancement for heavy atom nanoparticles irradiated by protons. Phys Med Biol 59:1441–1458
128. Surdutovich E, Solov'yov AV (2014) Multiscale approach to the physics of radiation damage with ions. Eur Phys J D 68:353
129. de Vera P, Garcia-Molina R, Abril I, Solov'yov AV (2013) Semiempirical model for the ion impact ionization of complex biological media. Phys Rev Lett 110:148104
130. de Vera P, Abril I, Garcia-Molina R, Solov'yov AV (2013) Ionization of biomolecular targets by ion impact: input data for radiobiological applications. J Phys Conf Ser 438:012015
131. Surdutovich E, Solov'yov AV (2015) Transport of secondary electrons and reactive species in ion tracks. Eur Phys J D 69:193
132. Toulemonde M, Surdutovich E, Solov'yov AV (2009) Temperature and pressure spikes in ion-beam cancer therapy. Phys Rev E 80:031913
133. Surdutovich E, Solov'yov AV (2010) Shock wave initiated by an ion passing through liquid water. Phys Rev E 82:051915
134. Surdutovich E, Yakubovich AV, Solov'yov AV (2013) DNA damage due to thermomechanical effects caused by heavy ions propagating in tissue. Nucl Instr Meth B 314:63–65
135. de Vera P, Currell FJ, Mason NJ, Solov'yov AV (2016) Molecular dynamics study of accelerated ion-induced shock waves in biological media. Eur Phys J D 70:183
136. Surdutovich E, Yakubovich AV, Solov'yov AV (2013) Biodamage via shock waves initiated by irradiation with ions. Sci Rep 3:1289
137. Yakubovich AV, Surdutovich E, Solov'yov AV (2012) Thermomechanical damage of nucleosome by the shock wave initiated by ion passing through liquid water. Nucl Instr Meth B 279:135–139
138. Yakubovich AV, Surdutovich E, Solov'yov AV (2012) Damage of DNA backbone by nanoscale shock waves. J Phys Conf Ser 373:012014
139. Roots R, Okada S (1972) Protection of DNA molecules of cultured mammalian cells from radiation-induced single-strand scissions by various alcohols and SH compounds. Int J Radiat Biol Relat Stud Phys Chem Med 21:329–342
140. Hirayama R et al (2009) Contributions of direct and indirect actions in cell killing by high-LET radiations. Radiat Res 171:212–218
141. LaVerne JA (2000) Track effects of heavy ions in liquid water. Radiat Res 153:487–496
142. Plante I, Cucinotta F (2008) Ionization and excitation cross sections for the interaction of HZE particles in liquid water and application to Monte Carlo simulation of radiation tracks. New J Phys 10:125020
143. Friedland W, Jacob P, Bernhardt P, Paretzke HG, Dingfelder M (2003) Simulation of DNA damage after proton irradiation. Radiat Res 159:401–410
144. Karamitros M et al (2014) Diffusion-controlled reactions modeling in Geant4-DNA. J Comput Phys 274:841–882
145. Gervais B, Beuve M, Olivera GH, Galassi ME (2006) Numerical simulation of multiple ionization and high LET effects in liquid water radiolysis. Radiat Phys Chem 75:493–513
146. Von Sonntag C (2007) Free-radical-induced DNA damage as approached by quantum-mechanical and Monte Carlo calculations: an overview from the standpoint of an experimentalist. In: Sabin JR, Brändas E (eds) Advances in quantum chemistry, vol 52. Academic Press, pp. 5–20
147. Hirayama R et al (2013) OH radicals from the indirect actions of X-rays induce cell lethality and mediate the majority of the oxygen enhancement effect. Radiat Res 180:514–523
148. Sicard-Roselli C et al (2014) A new mechanism for hydroxyl radical production in irradiated nanoparticle solutions. Small 10:3338–3346
149. Paudel N, Shvydka D, Parsai EI (2015) Comparative study of experimental enhancement in free radical generation against Monte Carlo modeled enhancement in radiation dose deposition due to the presence of high Z materials during irradiation of aqueous media. Int J Med Phys Clin Eng Radiat Oncol 4:300–307
150. Zhang X-D et al (2009) Irradiation stability and cytotoxicity of gold nanoparticles for radiotherapy. Int J Nanomed 4:165–173

151. Sanche L (2008) Low energy electron damage to DNA. In: Shukla M, Leszczynski J (eds) Radiation induced molecular phenomena in nucleic acids, vol 5. Springer, Netherlands, pp 531–575
152. Sanche L (2005) Low energy electron-driven damage in biomolecules. Eur Phys J D 35:367–390
153. Sanche L (2009) Biological chemistry: beyond radical thinking. Nature 461:358–359
154. Lu Q-B (2010) Effects and applications of ultrashort-lived prehydrated electrons in radiation biology and radiotherapy of cancer. Mutat Res 704:190–1999
155. Fuss MC et al (2014) Current prospects on low energy particle track simulation for biomedical applications. Appl Radiat Isot 83B:159–164
156. Elsässer T, Cunrath R, Krämer M, Scholz M (2008) Impact of track structure calculations on biological treatment planning in ion radiotherapy. New J Phys 10:075005
157. Cobut V et al (1998) Monte Carlo simulation of fast electron and proton tracks in liquid water—I. Physical and physicochemical aspects. Radiat Phys Chem 51:229–243
158. Tennyson J (2010) Electron-molecule collision calculations using the R-matrix method. Phys Rep 491:29–76
159. Blanco F, Garcia G (2015) Interference effects in the electron and positron scattering from molecules at intermediate and high energies. Chem Phys Lett 635:321–327
160. Pimblott SM, Laverne JA (2007) Production of low-energy electrons by ionizing radiation. Radiat Phys Chem 76:1244–1247
161. Garrett WR (1975) Molecular scattering: convergence of close-coupling expansions in the presence of many open channels. Phys Rev A 11:1297–1302
162. Blanco F, Ellis-Gibbings L, Garcia G (2016) Screening corrections for the interference contributions to the electron and positron scattering cross sections from polyatomic molecules. Chem Phys Lett 645:71–75
163. Massey HSW, Burhop EHS, Gilbody HB (1970) Electronic and ionic impact phenomena, 2nd edn. In: Electron collisions with molecules and photo-ionization, vol 2. American Association for the Advancement of Science
164. Blanco F, Garcia G (2003) Improvements on the quasifree absorption model for electron scattering. Phys Rev A 67:022701
165. Blanco F, Garcia G (2004) Screening corrections for calculation of electron scattering differential cross sections from polyatomic molecules. Phys Lett A 330:230–237
166. Colmenares R, Sanz AG, Fuss MC, Blanco F, Garcia G (2014) Stopping power for electrons in pyrimidine in the energy range 20–3000 eV. Appl Radiat Isot 83B:91–94
167. Oller JC, Ellis-Gibbings L, da Silva FF, Limao-Vieira P, Garcia G (2015) Novel experimental setup for time-of-flight mass spectrometry ion detection in collisions of anionic species with neutral gas-phase molecular targets. EPJ Tech Instr 2:13
168. Jaffke T, Meinke M, Hashemi R, Christophorou LG, Illenberger E (1992) Dissociative electron attachment to singlet oxygen. Chem Phys Lett 193:62–68
169. Belic DS, Hall RI (1981) Dissociative electron attachment to metastable oxygen ($a^1\Delta g$). J Phys B At Mol Phys 14:365–373
170. Hayashi S, Kuchitsu K (1976) Elastic scattering of electrons by molecules at intermediate energies. Calculation of double scattering effects in N_2 and P_4. Chem Phys Lett 41:575–579
171. Almeida D et al (2012) Mass spectrometry of anions and cations produced in $1-4$ keV H^-, O^-, and OH^- collisions with nitromethane, water, ethanol, and methanol. Int J Mass Spectrom 311:7–16
172. Štefančíková L et al (2014) Cell localisation of gadolinium-based nanoparticles and related radiosensitising efficacy in glioblastoma cells. Cancer Nanotechnol 5:6
173. Harrison KG, Lucas MW (1970) Secondary electron energy spectra from foils under light-ion bombardment. Phys Lett A 33:142
174. Casta R, Champeaux J-P, Sence M, Moretto-Capelle P, Cafarelli P (2015) Comparison between gold nanoparticle and gold plane electron emissions: a way to identify secondary electron emission. Phys Med Biol 60:9095–9106

175. Haberland H, Karrais M, Mall M, Thurner Y (1992) Thin films from energetic cluster impact: a feasibility study. J Vac Sci Technol A 10:3266–3271
176. Kamalou O (2008) PhD Thesis, University of Caen
177. Alpen EL (1998) Radiation biophysics. Academic Press
178. Schardt D, Elsässer T, Schulz-Ertner D (2010) Heavy-ion tumor therapy: physical and radiobiological benefits. Rev Mod Phys 82:383–425
179. Verkhovtsev AV, Korol AV, Solovyov AV (2015) Electron production by sensitizing gold nanoparticles irradiated by fast ions. J Phys Chem C 119:11000–11013
180. Usami N, Kobayashi K, Furusawa Y, Frohlich H, Lacombe S, Le Sech C (2007) Irradiation of DNA loaded with platinum containing molecules by fast atomic ions C^{6+} and Fe^{26+}. Int J Radiat Biol 83:569–576
181. Butterworth KT, McMahon SJ, Currell FJ, Prise KM (2012) Physical basis and biological mechanisms of gold nanoparticle radiosensitization. Nanoscale 4:4830–4838
182. Misawa M, Takahashi J (2011) Generation of reactive oxygen species induced by gold nanoparticles under x-ray and UV irradiations. Nanomed Nanotechnol 7:604–614
183. Jain S et al (2011) Cell-specific radiosensitization by gold nanoparticles at megavoltage radiation energies. Int J Radiat Oncol Biol Phys 79:531–539
184. Chithrani DB et al (2010) Gold nanoparticles as radiation sensitizers in cancer therapy. Radiat Res 173:719–728
185. McMahon SJ, Mendenhall M, Jain S, Currell F (2008) Radiotherapy in the presence of contrast agents: a general figure of merit and its application to gold nanoparticles. Phys Med Biol 53:5635–5651
186. Ando K, Kase Y (2009) Biological characteristics of carbon-ion therapy. Int J Radiat Biol 85:715–728
187. Furusawa Y et al (2000) Inactivation of aerobic and hypoxic cells from three different cell lines by accelerated (3)He-, (12)C- and (20)Ne-ion beams. Radiat Res 154:485–496
188. Hirayama R, Furusawa Y, Fukawa T, Ando K (2005) Repair kinetics of DNA-DSB induced by X-rays or carbon ions under oxic and hypoxic conditions. J Radiat Res 46:325–332
189. Nakano T et al (2006) Carbon beam therapy overcomes the radiation resistance of uterine cervical cancer originating from hypoxia. Clin Cancer Res 12:2185–2190
190. Combs SE et al (2012) PhaseI/II trial evaluating carbon ion radiotherapy for the treatment of recurrent rectal cancer: the PANDORA-01 trial. BMC Cancer 12:137
191. Hirayama R et al (2013) Evaluation of SCCVII tumor cell survival in clamped and non-clamped solidtumors exposed to carbon-ion beams in comparison to X-rays. Mutat Res 756:146–151
192. Rothkamm K, Barnard S, Moquet J, Ellender M, Rana Z, Burdak-Rothkamm S (2015) DNA damage foci: meaning and significance. Environ Mol Mutagen 56:491–504

Biophysics Modeling to Optimize Ion Beam Cancer Therapy

Michaël Beuve

Abstract The optimization of treatments by Ion Beam Cancer Therapy (IBCT) relies on modeling to simulate the transport of the incident ions (and the secondary particles) into patients, and, to predict the biological effects induced by all these particles. Considering the complexity of biological systems, multi-scale approaches seem necessary to build the bridge between the primary physical and chemical events and the consequences for patients both in healthy tissues and tumors. After a brief history of IBCT in France, this chapter presents models used to estimate the probability of tumor control by IBCT, showing the importance of predicting the survival of biological cells to complex irradiation. Then, follows a presentation and analysis of models predicting cell survival to irradiation with ions, including: the procedure developed in Japan for cancer treatments with passive beams; the microdosimetry models TDRA and MKM, and, the MMKM, a modified version of MKM used for active beam in Japan; the Katz models and the LEM, which is presently used by the European centers of therapy with carbon ions. Then, as perspectives, modeling based on nanodosimetry will be addressed with a focus on the NanOx™ model.

1 Brief History of Hadrontherapy in France

The story of IBCT (Ion Beam Cancer Therapy) started in France with proton therapy. The Orsay Protontherapy Center (CPO) was created in 1991 using, as accelerator, a synchrocyclotron, previously dedicated to nuclear research. Then followed the first ophthalmological treatment in 1991 and the first intracranial treatment in 1994. In parallel, in June 1991, the Center Antoine-Lacassagne (CAL) in Nice treated a first patient with a proton beam produced by a 65-MeV cyclotron (Biomedical MEDICYC). The idea of treating patients with carbon ions came soon after, in 1998. Named ETOILE in 2002, the project of creating in Lyon a center combining treatment rooms and a room fully dedicated to research was developed in the European

M. Beuve (✉)
Institute de Physique Nuclaire de Lyon, University of Lyon 1,
4 Enrico Fermi, 69622 Villeurbanne, France
e-mail: michael.beuve@univ-lyon1.fr

A.V. Solov'yov (ed.), *Nanoscale Insights into Ion-Beam Cancer Therapy*,
DOI 10.1007/978-3-319-43030-0_13

context. The development of hadrontherapy, including research and projects of center construction, was increasing in Europe. The ENLIGHT network was thus created in 2002 to coordinate the development of hadrontherapy with light ions. ETOILE members contributed to its creation and dynamics. In parallel, a research community, strongly interdisciplinary, has been created around the Rhône-Alpes Research Program in Hadrontherapy. To facilitate the cooperation between the actors of this research program and the ENLIGHT community, the structuration of this program was decomposed in similar working packages, covering all the aspects of carbon-ion therapy. As a complementary project to ETOILE and based on the experience of GANIL in the construction of accelerators, was developed the idea of constructing in Caen an accelerator for carbon-ion therapy. In 2003, ASCLEPIOS, a project similar to ETOILE was presented to the French government by the Basse-Normandie Region [1] and evolved to ARCHADE project. The aim of ARCHADE was to build a non-clinical center for the development of technologies.

In 2005, the French government decided that the first national carbon-ion center would be built in Lyon and in 2007 agreed to fund ETOILE center. Some years later, it imposed the creation of a public-private consortium. This constrain led to a strong increase in the center cost and constituted an important reason for the end of ETOILE project. Beside, ARCHADE recently signed an agreement to build a proton therapy center. It also supports the project of building a center to host the development, in cooperation with IBA, of a cyclotron to accelerate carbon ions with 400 MeV/n. In case of success, a research room would be set-up to perform research in hadrontherapy with the carbon ions produced by this accelerator. Although ETOILE did not lead to a center of IBCT with carbon ions in Lyon, the project actors successfully initiated and led the creation of the infrastructure France HADRON to enable the consolidation of all medical, scientific and technical teams involved in hadrontherapy in France. The actors, mainly federated by universities and national public research institutions (CNRS, CEA, INSERM, IRSN) have gathered themselves to present the France HADRON project in response to the call for projects "National biology and Health Infrastructure". In April 2012, the project got the label Investments for Future and was financed with 15 M€. The France HADRON objectives are to:

1. coordinate and organize the national program of research and training in hadrontherapy. This program goes from particle physics to clinical research through dosimetry, radiation biology, imaging, control of positioning target, and quality control instruments;
2. organize, facilitate and finance the access of researchers to the particle beams required for experimental work;
3. finance methodological research to increase the research-platform efficiency;
4. finance part of the equipment of the research platforms;
5. and finally ensure participation in international programs such as those already existing and highly invested by the French teams: ENLIGHT, Partner, ULICE, ENVISION, ENTERVISION.

Research is distributed throughout France and concerns over twenty research teams. In France HADRON, nodes are a geographical grouping of research teams

Fig. 1 Geographic localization of the France HADRON nodes

around a therapy or research platform, existing or in project. France HADRON relies on 5 nodes, Lyon, Caen, Orsay, Nice and Toulouse (see Fig. 1). The research program is organized around four Working Packages:

- WP1 – Clinical research
- WP2 – TPS optimisation
- WP3 – Radiobiology
- WP4 – Instrumentation.

2 Biophysics Modeling to Optimize IBCT

Planning and improving cancer treatments by ion beams (IBCT) relies on modeling. First, a faithful description of the transport of the beam through patients is required. Second, the consequences of a treatment on both tumors and healthy tissues need to be predicted in terms of biological effects. A key point of IBCT is that the dose (and dose-rate) cannot possibly be the only physics quantity to consider for the predictions of these biological effects, contrary to conventional radiotherapy. At a given dose, the biological effects depend also on the type of particle and its energy. This fact is on one hand a severe complexity, but on the other hand, it is at the origin of the enhancement of the cell-killing effects induced by light ions in tumors, in addition to the increase of physical dose at the Bragg peak.

Transport models have to accurately describe not only the spacial distribution of dose deposited within patients, but also the distribution in type and energy of all the particles that are at the origin of this dose. Moreover, biophysics models are required to convert these complex distributions into bioeffects, which depends on biological parameters, like for instance, the considered tissues, the cell population that constitutes them or the cell environment. The following sections address this later point, considering is solved the transport problem. Given the complexity of the biological systems, the numerous scales the implied mechanisms cover, both in time (from femtoseconds to years) and space (from atoms to patients), a multi-scale approach is necessary.

2.1 Multi Scale Approach: From Atom to Patients

2.1.1 Cell Survival: A Key Biological Endpoint

The biological endpoint to model is a crucial element to take into consideration at first since the Relative Biological Effectiveness (RBE) clearly depends on it. Damage in DNA, like the induction of mutations, may be a key endpoint if one considers the risk of radiation-induced cancers. However, except for pediatric situations, this risk is not really considered in practice, in particular when dealing with radioresistante tumors: curing patients is a priority when patient survival over five years is low. These tumors are indications for treatments with carbon ions, when they are localized. Instead, the so-called deterministic effects (by opposition to stochastic effects) are considered for both healthy tissues and the tumor volumes. Cell survival is therefore a better biological endpoint, as a quantification of all death pathways that may follow irradiation (necrosis, apoptosis, senescence, mitotic catastrophe, autophagy). Cell survival is even explicit in the modeling of tumor control using Poisson law.

2.1.2 Probability of Tumor Control

In the framework of models aiming at predicting the probability of tumor control (TCP) after treatments, the TCP are directly connected to cell survival through models based on Poisson law. Within these models, the TCP is set to be the probability that not any tumor cell survives. Assuming a Poisson distribution, it reads:

$$\mathrm{TCP} = \exp(-N) \tag{1}$$

where N is the mean number of survival tumor cells.

The simplest expression of these models assumes the tumor contains a given number of "cancer" cells N_0 and that it is irradiated by a uniform dose D of low-LET radiotherapy. The TCP is then given by:

$$\mathrm{TCP} = \exp\left(-N_0 \times S(D)\right) \tag{2}$$

where $S(D)$ is the probability of cancercell survival after a dose D. This survival is generally approximated by a Linear Quadratic (LQ) model:

$$S(D) = \exp(-\alpha.D - \beta.D^2) \tag{3}$$

where α and β are radiobiological constants, which depend on the cell line but also on other biological parameters (cell cycle position, cell environment). This approximation is valid provided the dose does not overcome a certain level above which this approximation is inadequate. The dose needs however to be high enough since at low doses, typically lower than 0.5 Gy, curves of cell survival may be characterized by a structure showing hypersensitivity or increased radioresistance in comparison to the LQ tendency.

The value of N_0 depends on the size of the tumor. Consequently, the dose prescription so should. Its determination is all the more difficult that the definition of cancer cells is not clear. Cancer cells may refer to clonogenic cells or instead steam cells, then the value of N_0 may varies by several orders of magnitude.

Adding a level of complexity allows to take account for the heterogeneity in the spatial distribution of dose, cell population, cell radiosensitivity (due for instance to different level of oxygenation). The previous expression then becomes:

$$\mathrm{TCP} = \exp\left(-\iiint_{\text{tumor}} n(\mathbf{r})\, S(\mathbf{r}, D(\mathbf{r}))\, \mathrm{d}\mathbf{r}^3\right) \tag{4}$$

where $n(\mathbf{r})$ represents the cell density at the position $\mathbf{r}$.

Considering the decomposition of the patient geometry into a regular grid of voxels j of volume υ, the integral is converted into a discrete summation:

$$\mathrm{TCP} = \exp\left(-\upsilon \sum_j n_j S_j\left(D_j\right)\right) \tag{5}$$

While the dose distribution may be reasonably estimated, the other sources of heterogeneity still constitute a limitation to a full application of this expression. Moreover, even at a given position, cell population is heterogeneous and evolves over the whole period of radiotherapy.

Other effects has to be integrated, like for instance the effect of fractionation treatment into n sessions of dose d. Assuming the independence of the response in between sessions leads, for the cell survival to the full treatment, to the expression:

$$S = S(d) \times S(d) \times ...S(d) \tag{6}$$

and then, for a LQ model, to the expression:

$$S = \exp(-D[\alpha + \beta d]) \tag{7}$$

This effect is included for instance in the TCP model by Pedicini et al. [49], who added moreover a doubling time T_{pot}, a time T separating treatment sessions and a delay T_{k} for cell multiplication after irradiation. T_{pot} corresponds to the time necessary by cancer cells to multiply by a factor two their number. TCP became then:

$$\text{TCP} = \exp\left[-N_0 \exp\left(-D(\alpha + \beta d) + \ln(2)\frac{(T - T_{\text{k}})}{T_{\text{pot}}}\right)\right] \tag{8}$$

An important and specific effect to introduce in the planning of IBCT is the effect of the particle type and energy. In principle, this is simply done by replacing, in the TCP expression, the survival to a session dose d of low-LET radiation by the actual survival to a mixed field of ion impacts. Therefore, estimating accurately cell survival to a mixed field of ion impacts is a key point for the optimisation of IBCT. This estimation is a key point for the control of tumors, but as mentioned above, also for the complications induced in healthy tissues.

2.2 *Cell Survival to a Mixed Field*

The approximation proposed by Kanai et al. [33, 34] leads to a simple and fast estimation of cell survival to mixed fields. The authors assumed that cell survival to ion irradiation could be approximated by a LQ model, whatever the irradiation field, mixed or not. Then the couple of α_{mix} and β_{mix} values is estimated by a process of averaging weighted in dose, according to:

$$\alpha_{\text{mix}} = \sum_k f_k \alpha_k \tag{9}$$

and

$$\sqrt{\beta_{\text{mix}}} = \sum_k f_k \sqrt{\beta_k} \tag{10}$$

where α_k and β_k are LQ parameters for the cell survival to irradiation with ions of type T_{k} and energy E_k; f_k is the fraction in dose of these type and energy for the mixed field. Due to the numerous transport processes, including energy loss and nuclear fragmentation, on the one hand, and to the beam field optimisation, on the

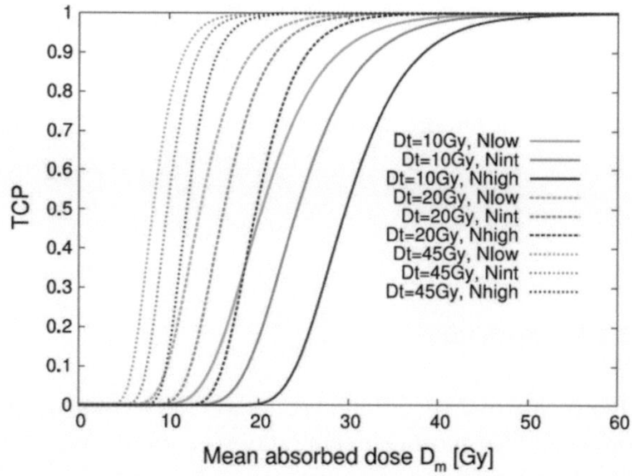

Fig. 2 Example of TCP curves for prostate cancer with different risk (Low, Intermediate, High) treated with carbon ions. The calculation were based on LEM I using different values of threshold dose D_t. Figure extracted from [8]

other hand, the fraction in dose is different from one voxel to another. Finally, within this approximation, the calculation of cell survival to a mixed field simply requires a pre-calculated table gathering the α_k and β_k for a list of couples ion type and energy. This simplification makes possible a clinical application of biophysics models since directly calculating cell survival during a treatment planning would require too much time. Figure 2 illustrates a calculation of TCP curves using this approximation and the LEM I model to calculate the table of (α_k, β_k).

2.3 *The Original Procedure Used in Japan*

The first approach used in clinics was developed at the National Institute of Radiological Sciences (NIRS) in Japan [31]. The list of α_k and β_k was derived empirically from survival of in-vitro human salivary gland (HSG) tumor cells [21]. It was by this approach assumed that the moderate radiosensitivity of HSG cells was a typical response of tumors to carbon beams. This allowed to predict a distribution of 'radiobiological RBE'. However, this 'radiobiological RBE' values were then rescaled to clinical RBE values, using the RBE values observed within clinical studies with fast neutron treatments, to account for the difference between the in-vitro response and clinically required in-vivo response. The radiobiological RBE-weighted dose was referred to as the biological dose, while the clinical RBE-weighted dose was referred to as the clinical dose.

This procedure, pragmatic and simple, was essential to develop IBCT with carbon ions since not any clinical data was available at this time. It was validated later on by determining curves of TCP for non-small-cell lung cancer [35]. But, considering the development of biophysics models, this procedure might be modified or replaced. The renormalization of the biological RBE to estimate the clinical dose should, for instance, depend on the prescription dose. Moreover, all tumors could not be represented by the HSG cell line. Further improvements to take account for effects, like hypoxia, combined effects of radio sensitizers, dose-rate. . . , appear complicated within this procedure and NIRS changed their approach at least for active beam moving to microdosimetry (see Sect. 2.4.3).

2.4 Microdosimetry

2.4.1 Principle

The theory of dual radiation actions (TDRA) [41] was the starting point for the development of the most recent microdosimetry models. The key point of microdosimetry was to consider that the concept of dose needed to be reconsidered when the size of the irradiated objects matches with microscopic scale. Indeed, despite a well-defined macroscopic dose, the energy deposited into such objects may strongly fluctuate. The level of fluctuations depends on the size of the object, on the dose of irradiation and on the particle characteristics, i.e. the type and the energy. Then, the concept of specific energy is introduced according:

$$z = \frac{\sum_i \varepsilon_i}{m} \tag{11}$$

where ε_i is one of the energy-transfer points (energy deposition at a point) deposited into the object of mass m by the ionizing radiation. Another convenient quantity introduced in microdosimetry is the lineal energy y defined by:

$$y = \frac{\sum_i \varepsilon_i}{\bar{l}} \tag{12}$$

where $\bar{l}$ is the mean cord of the object along the beam axis.

While z may strongly vary, it averages over all configurations corresponding to a macroscopic dose D exactly equals this dose. In the same way, the mean of y is related to the LET.

Microdosimetry models are based on the postulate that ionizing radiations may induce sub-lethal lesions, which may interact if close enough, leading then to a lethal lesion. Mathematically a sphere is defined to represent microscopic domains in cells. Its radius sets the critical distance beyond which interaction between sub-lethal lesions cannot occur. The number of created sub-lethal lesions is assumed to scale the specific energy:

$$N_{sub} = c.z \tag{13}$$

and the probability that sub-lethal interaction leads to lethal lesions scales the square of z:

$$n_{\text{lethal}} \propto z^2 \tag{14}$$

Averaging the number of lethal lesions over all irradiation configurations leading to a macroscopic dose D gives a LQ expression:

$$n_{\text{lethal}} = \alpha D + \beta D^2 \tag{15}$$

β is a constant and α a coefficient that depends on the radiation quality (type and energy) according to:

$$\alpha/\beta = \frac{\int \mathrm{d}z_1 f(z_1)\, z_1^2}{\int \mathrm{d}z_1 f(z_1)\, z_1} \tag{16}$$

with $f(z_1)$ the distribution of specific energies by a single-event. For irradiation with charged particles, like ions or electrons, a single event corresponds to the impact of 'one' charged particle that deposits energy into the domain, directly or through the electronic cascades it induces. For neutral particle, like photons or neutrons, a single event corresponds to the interaction of 'one' neutral particle that deposits energy into the domain through the kicked secondary charged particle.

This model was not the first radiobiology model of cell survival. Target theory has already been considered sooner (see for instance [45]), but it was a breakthrough for radiobiology modeling and led to the development of microdosimetry. This model allowed to consider fluctuations of energy deposition at microscopic scale, deducing a LQ expression consistent with experimental evidences for mammalian cells. Moreover, the α coefficient, dependent on the radiation type and energy, could be deduced from a calculation of the specific energy distribution (for instance using Monte Carlo simulation) or from a measurement using small and low-density ionization chamber, like the tissue equivalent proportional counter proposed by Rossi and Failla [51], Rossi and Rosenzweig [50] and Srdo [54].

Beside, some shortcomings of this model required improvements. First, this model relies on a very specific mechanism of cell killing: the interaction of sub-lethal lesions. This mechanism may correspond to the induction of chromosomal aberrations during the process of non-homologous end-joining (NHEJ). Then the sub-lethal lesions would correspond to double-strand break (DSB) DNA damage. But it may not be the only mechanism. Second, the number of sub-lethal lesions was assumed to scale the specific energy, without any explicit dependency with the type of particle. Therefore, this model discards any detail on the spatial distribution of energy at nano-scale despite the fact that the sub-lethal lesions seem to be nano-scale DNA damage. Last, the direct predictions issued from this model is an average number of lethal events per cell and not cell survival. The link to cell survival was found by assuming a Poisson distribution for this number of lethal events. Then, cell survival reads:

$$S(D) = \exp(-n_{\text{lethal}}) \tag{17}$$

However this approximation is not valid for high-LET ions when the ion energy per nucleon is of the order of few MeV/n.

Finally, this model was the starting point for the development of other microdosimetry models, like shortly after, the GTDRA proposed by the same authors [40]. The next section focuses on one of the them, the MK model, since one of its versions is now used in Japan for IBCT with active beam of carbon ions.

2.4.2 Microdosimetric-Kinetic Model (MKM)

The Microdosimetric-Kinetic Model (MKM) was been proposed by Hawking in the 90s [29, 30] as an combination of the Repair-MisRepair model (RMR) [12] and Lethal Potentially Lethal model (LPL) [14, 15]. This model stands within the microdosimetry framework. As in the TDRA, the interaction of sub-lethal lesions may induce a lethal lesion. The lethal lesions correspond to the formation of chromosomal aberrations that prevent cells from proliferation. However is added a kinetic of DNA repair. Some sub-lethal lesions can be repaired preventing from the creation of lethal lesions. Conversely, some of the sub-lethal lesions may be converted into lethal lesions through a first order process. They may also be neither repaired, nor combined with another sub-lethal lesion during a period following creation. If this period exceeds a threshold period, they become lethal.

As in the TDRA, micrometric domains of interaction are defined to calculate the specific energy deposited by irradiation. In the MKM they are represented by a distribution of size. The mathematical formulation of the MKM leads to a set of differential equations, offering the possibility to predict cell survival also as a function of the irradiation dose-rate, or any time structure of irradiation. Considering the specific case of an instantaneous irradiation, which applies to a session of a treatment with IBCT, leads, after approximations, to a simple expression for the mean number of lethal events, which then follows a linear quadratic expression, as for the TDRA. As for the TDRA, the β coefficient is also constant. The expression of the α coefficient is different and reads;

$$\alpha_p = \alpha_0 + \gamma\beta \tag{18}$$

where α_0 is a constant and γ is the dose-weighted average of the dose deposited by a single event, known as z_{1D} in microdosimetry:

$$z_{1D} = \frac{\int \mathrm{d}z_1 f(z_1)\, z_1^2}{\int \mathrm{d}z_1 f(z_1)\, z_1} \tag{19}$$

For a spherical domain of diameter d and mass m, its expression is given by:

$$\gamma = \frac{< E^2 >}{m < E >} + \frac{0.229}{d^2} L \tag{20}$$

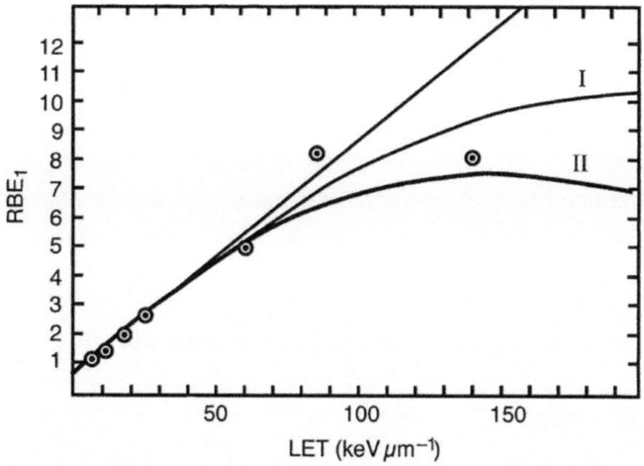

Fig. 3 Impact of the nucleus cross-section (*Curve I* $\sigma = 100$ cm^2, *Curve II* $\sigma = 50$ cm^2) on RBE, calculated by Hawkins as a function of the LET with the MKM. The third line represents a calculation without the non-Poisson correction and the symbols experimental data for human kidney cells. Figure extracted from [28]

where $< E >$ and $< E^2 >$ are respectively, the mean and the quadratic mean of the energy E transferred at each of the points of interaction along the track of an ion traversing a straight-line cord through the domain. L is the LET of this particle.

Finally the MKM brought to microdosimetry a kinetic aspect and a more complete cell death mechanism. On the other hand, this model still relies on a scaling of the number of sub-lethal lesions with the specific energy, disregarding nano-scale, while the formation of DSB are known to depend on the details of tract structure at such a small scale. Moreover, the relation between cell survival and the mean number of lethal lesions is based on the assumption of a Poisson distribution.

To correct this later point, Hawkins [28] introduced a non-Poisson correction in defining a geometry for the cell nucleus. Precisely, the α_p coefficient is multiplied by the factor:

$$\frac{1 - \exp\left(-\alpha_p \overline{Z}_{1Dn}\right)}{\alpha_p \overline{Z_{1Dn}}} \tag{21}$$

where $\overline{Z}_{1Dn}$ is the dose-weighted average of the specific energy deposited by a single event in this nucleus. This quantity is calculated by the equation:

$$\overline{Z}_{1Dn} = 0,16 \frac{\overline{y}_D}{\sigma} \tag{22}$$

Here, σ is the nucleus section and $\overline{y}_D$ is the mean dose lineal energy, i.e. the mean of the lineal energy y (see Eq. (12)) weighted in dose. Figure 3 illustrates the impact of this correction on RBE calculations.

2.4.3 Modified Microdosimetric-Kinetic Model (MMKM)

The original MKM was also modified by Kase et al. [37] to take account for the reduction of RBE due to overkill effect in a region of very high specific energy. With regard to the first version of the MKM, proposed by Hawkins, the survival is still deduced from a Poisson distribution for the number of lethal events:

$$S = \exp\left(-\langle L_n \rangle\right) \tag{23}$$

However, the mean number of lethal events is corrected by substituting z_{1D} by z^*_{1D} in the equation that gives the mean number of lethal events. This equation becomes then:

$$\langle L_n \rangle = \left(\alpha_0 + z^*_{1D}\beta\right) D + \beta D^2 \tag{24}$$

The 'saturation-corrected dose-mean specific energy' z^*_{1D} contains a saturation effect through the introduction of z_{sat}:

$$z^*_{1D} = \frac{\int \mathrm{d}z_1 f\left(z_1\right) z_{\mathrm{sat}} z_1}{\int \mathrm{d}z_1 f\left(z_1\right) z_1} \tag{25}$$

z_{sat} is defined by the two equations:

$$z_{\mathrm{sat}} = \frac{z_0^2}{z}\left(1 - \exp\left(\frac{z^2}{z_0^2}\right)\right) \tag{26}$$

and

$$z_0 = \frac{\left(R_n/r_d\right)^2}{\sqrt{\beta\left(1 + \left(R_n/r_d\right)^2\right)}} \tag{27}$$

Here, R_n and r_d represent the radius of respectively the cell nucleus and the micrometric domains.

Beside, Kase et al. [36] introduced an amorphous track-structure model to calculate the energy deposition instead of using Monte Carlo simulation. For this purpose, the KieferChatterjee track-structure model [9, 42] was used for the radial dose distribution around the trajectory of the ions. This replacement is quite convenient since it simplified the calculation. However, as it will be discussed later on (see Sect. 2.5), this substitution may create severe artifacts. Nevertheless, they found that the MKM calculation was useful for predicting the survival curves of the mammalian cells in vitro for mono-energetic 3He-, 12C- and 20Ne-ion beams. The MMKM model is now implemented in a TPS for IBCT with active beam of carbon ions [32]. Figure 4 shows an example of application of the MMKM for a treatment planning and compares them to experimental data and to calculations performed with historical NIRS procedure. The agreement appears better for the calculations based on MMKM.

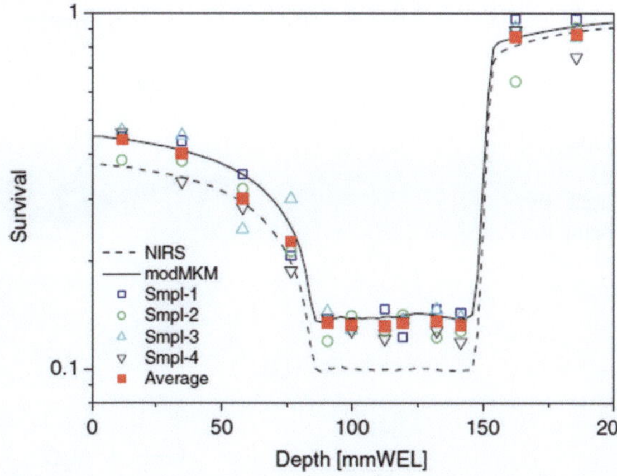

Fig. 4 Measured survival values (*symbols*) compared with the planned survival curve (*dashed curve*) based on the NIRS radio-biological model [33] using the response of the 'old' HSG tumor cells. Recalculated survival curve (*solid curve*) based on the modified MKM in which the current response of the HSG tumor cells was reflected through the value of α_0. Figure extracted from [32]

2.5 *Biophysics Models Based on Amorphous Track Structure*

2.5.1 Radial Dose

This class of models was developed in parallel to the models based on microdosimetry. As microdosimetry models, these models considered that the macroscopic dose was not a sufficient parameter to predict the induced bioeffects. However, instead of using a fully stochastic description of the energy deposition, they introduced the concept of radial dose to represent the capacity of ions to deposit high concentrations of energy along their tracks.

The radial dose is defined as the mean dose deposited at a given distance to one ion trajectory. Due to the cylindrical symmetry, around the ion trajectory, of the ion collisions with matter, the average dose indeed depends only on the distance r to the ion trajectory. Contrary to the specific energy, which cannot be calculated without fixing a target size and geometry, the radial dose can be defined at a point with a point-like volume, because radial dose is a mean quantity. However, to quantify it by Monte-Carlo simulation, a finite target is defined. This target is chosen as small as possible to get good resolution, but large enough to get enough statistics and then an estimation of the mean value with acceptable fluctuations.

In practice, cylindrical shells are defined around the ion trajectory (see Fig. 5). Then, a track is simulated by Monte-Carlo and the energy deposited in each shell is calculated by summing the energy of the transfer points standing in this shell. This energy is normalized to the shell mass and averaged over many track simulations. Generally, the models that were used to performed the Monte-Carlo simulation assumed the matter that is traversed by the ion was made of liquid water. The actual

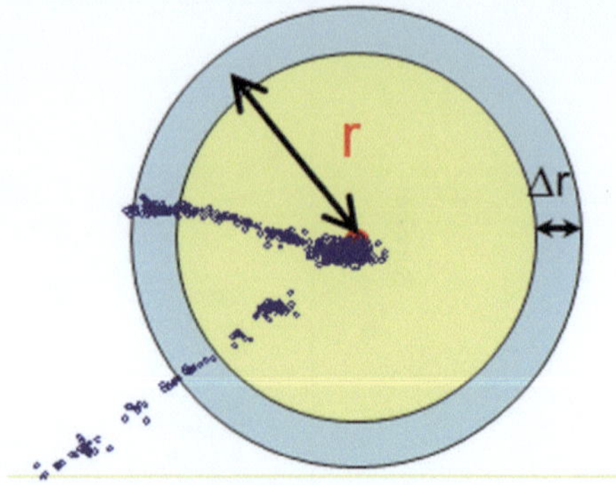

Fig. 5 Scheme of the geometry used to calculate the radial dose by Monte-Carlo simulation. The points correspond to the energy-transfer points generated in an ion track, the trajectory of which being the axis of the cylindrical shell characterized by its radius r and thickness Δr

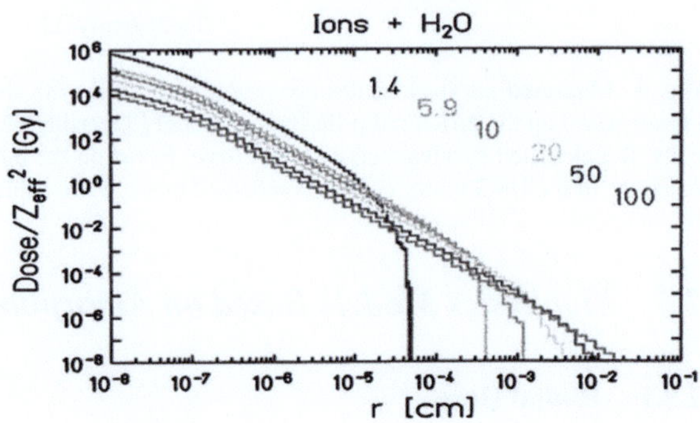

Fig. 6 Radial dose normalised to the square of the effective ion charge Z_{eff} calculated for different values of ion energy. The figure is extracted from [43]

distribution of molecules was not considered and was instead approximated by a uniform statistical distribution. Such models of track are thus known as 'amorphous track structure models'.

Figure 6 shows examples of radial dose calculated by krämer et al. [43]. The radial dose was normalized to the square of the effective charge used by the authors to calculate the cross-section for ion collision with water molecules. Within this approximation, the normalized radial dose did not depend on the projectile charge in their model, while the radial dose did. The radial dose strongly depends on the ion energy per nucleon. At the lowest ion energy, the dose deposition is limited to the ion track core. For higher ion energy, the energy is spread out to higher radial distances because ions can eject fast electrons, which carry out energy far away from the position they are ejected from. The energy transport increases as the probability of ejecting fast primary electrons increases and therefore increases with the ion energy. Finally, the dose deposition in the track core decreases with the ion energy not only because the value of LET does, but also because of the transport to high radial distances of the energy initially transferred by the ions.

The shape of the radial-dose curve depends on the model of track simulation, in particular at short radial distances. A common point of all the radial-dose models is that the dose can be very huge for high LET-ions. It can overcome millions of Grays for carbon ions in the Bragg peak. At higher radial distances, the curve generally shows a decrease according to a $1/r^2$ law. This tendency applies up to a cut-off radius, beyond which radial dose falls down steeply to zero. The cut-off radius corresponds to the range of the highest-energy primary electrons. Thus, the cut-off radius increases with the ion energy.

The $1/r^2$ decrease is due to two effects. One is related to the combined effect of the cylindrical symmetry and the heavy mass of the ions. Due to the cylindrical symmetry, the transferred and transported energy is normalised to the mass of the cylindrical shell around the ion trajectory. This shell mass scales with the radial distance r and leads to a first decrease factor of $1/r$. This factor could be smoothed by the angular diffusion of the ion, but due to its mass, this angular straggling is low for fast ions and the ion trajectory can be approximated by a straight line over micrometers. The second reason for the radial-dose decrease is related to the probability law for the ejection of electrons by ion collision. This latter decreases with the electron energy E, roughly in $1/E$. The energy deposition at high radial distances is strongly related to the electron energy E. Therefore, the probability of energy deposition decreases with the radial distance.

As it was shown by the microdosimetry theories, the fluctuations of deposited energy at microscopical scale is crucial when modeling the biological effects induced by ion irradiation. Although the radial dose is an average quantity, biophysics modeling based on radial dose can contain some effects of fluctuations: the ones related to the distribution of distance between the ion trajectory and the biological targets. For instance, when an ion directly impacts a target, the deposited dose into this target can be large, quite larger than the macroscopic dose of irradiation. The radial dose may be also used to describe effects that are linear with the energy deposited into the targets, or at least linear over a range of energy wider than the range of fluctuations in deposited energy. However, for small targets, for which the energy fluctuations could be huge (see for instance [13] or [4]) and for non-linear functions, radial dose may not be appropriated at all for the prediction of bioeffects induced by ions. This can be illustrated by a look at some orders of magnitude. As can be seen on Fig. 6, the radial dose is generally extremely low in the penumbra (i.e. the track region out of the core). The radial dose appears too low to induce damage in molecules, like DNA. However, this very low value is due to the averaging process mentioned above. Actually, the track penumbra is built by electronic cascades, and thus, contains ionization clusters that may induce molecular damage.

The definition of the radial dose and the limitations that were addressed in this section are useful to well understand the biophysics models that are based on this radial dose.

2.5.2 Katz Model

The model proposed by Katz, developed in the late 60's was the first biophysical model based on the use of an amorphous track structure model [39]. At this period the computing resources were quite limited and the use of the radial dose instead of a full track description by Monte Carlo simulation was convenient. The model of

Katz was strongly related to the modeling of damage induced by ions into matter, like radiosensitive films or track detectors.

Within Katz model, cell inactivation arose from two mechanisms: the ion-kill and γ-kill modes. Cell-surviving fraction is the product of the surviving fraction associated to these modes:

$$S = \Pi_{ion-kill} \cdot \Pi_{\gamma-kill} \tag{28}$$

Ion-kill mode referred to intra-track processes and aimed at representing the high capacity for ion to kill cells by the deposition of high energy concentration, in particular in the track core at high LET. Instead, γ-kill mode referred to inter-track processes and aimed at representing the effect of track superposition. For both these modes, the critical target in mammalian cells was assumed to be a 'substructure' of the cell nucleus. The nucleus contained several of these substructures, 'like beans in a bag', for which the extension a_0 was of the order of one micrometer.

Since the superimposition of tracks corresponds mainly to the superimposition of track penumbras, and therefore of fast-electron tracks, γ-kill mode was modeled by a probability law of cell survival to a dose of D_γ of γ-rays. Using the multi-target single-hit model [45] with an inactivation dose D_0 and a target number m, the survival reads:

$$\Pi_{\gamma-kill} = \left[1 - \exp\left(-\frac{D_\gamma}{D_0}\right)\right]^m \tag{29}$$

Conversely, cell killing by ion-kill mode is induced by intra-track processes and is described by a cross-section σ according to:

$$\Pi_{ion-kill} = \exp\left(-\sigma \cdot F\right) \tag{30}$$

F is the irradiation fluence and σ is given by:

$$\sigma = \int_0^{+\infty} \mathrm{d}t 2\pi t \left[1 - \exp\left(-\frac{D(t)}{D_0}\right)\right]^m \tag{31}$$

In this equation, $D(t)$ is the mean dose deposited by an ion into a target for which the center stands at a distance t from the ion trajectory. This distance is assumed to be the same for all targets. This approximation may appear surprising, in particular for short distances t. $D(t)$ was calculated by using an amorphous track-structure model [59]. For a fixed ion type, σ increases with ion LET to a plateau σ_0 and then may decrease, depending on the target number (see Fig. 7). This decrease is referred as to 'track thinning down'. As explained in the previous section, this is due to the reduction of the fast-electron energy when the ion energy decreases.

Inter- and intra- track processes are both triggered by dose deposition by ions into targets. At this stage, one may suspect a double counting of the energy. However, the total dose D is shared in two contributions through the introduction of a fraction P.

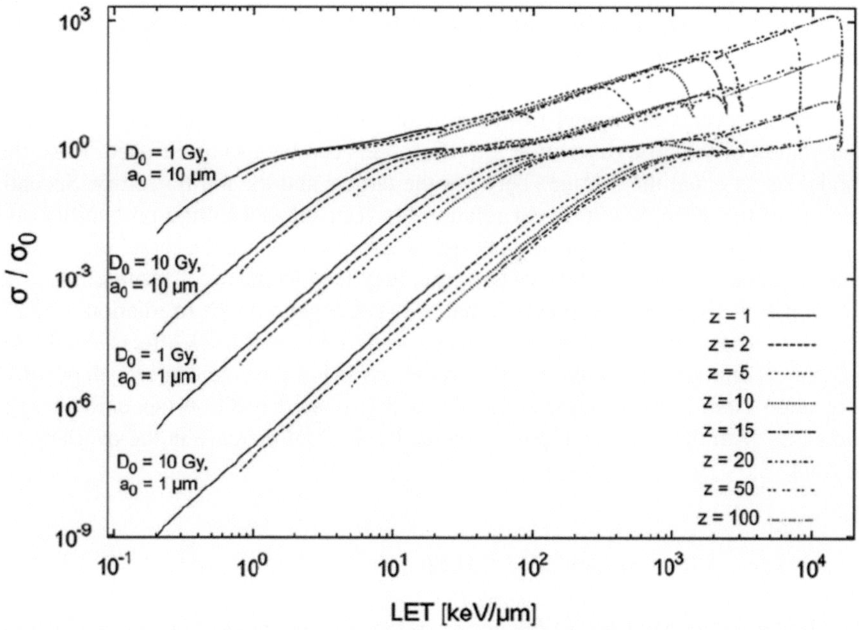

Fig. 7 Calculated values of the average cross sections σ normalised to their saturation or plateau values σ_0 plotted as a function of LET. The family of curves were calculated for m-target detectors ($m = 2, 5$) of $D_0 = 1$ or $10\,\text{Gy}$ and of $a_0 = 1$ or $10\ \mu\text{m}$, irradiated by ions of charges ranging between 1 and 100, and speeds ranging between $\beta = 0.05$ and 0.99. The figure was extracted from [60]

The contribution D_γ to the delivered dose D that corresponds to γ -kill mode is then given by:

$$D_\gamma = (1 - P)\, D \tag{32}$$

The fraction P is expected to be close to zero for fast ions since the inter-track processes dominate. Instead, it may increase to one for thin tracks. Katz proposed an estimation of P with a simple expression that leads to a transition from γ -kill mode to ion-kill mode when the ion LET increases:

$$P = \frac{\sigma}{\sigma_0} \tag{33}$$

Finally, Katz model required four parameters only: m and D_0 to represent the survival to γ rays, a_0, which is crucial for ion-kill mode and σ_0, which controls the transition from a mode to the other. This model was simple to apply and did not require important computer resources. It constituted an important contribution to the modeling of killing effect of ions. With regard to microdosimetry models, it was the first model able to describe the saturation and decrease of cell inactivation cross-sections at high values of LET. The separation into two killing modes was quite innovative and may have been inspired by the observations of damage created by high-LET ion into materials, as compared to the one produced by γ-rays.

On the other hand, using multi-targets single-hit model to represent cell survival to γ-rays prevents the model from predicting a slope in cell survival curves. The calculation of the dose deposition into the targets can be also questioned. First, the model set as equal the distance between the targets and the ion trajectory. Second, despite of the strongly non-linear relation between the cell-killing probability and the dose deposited into targets, this dose is calculated using the radial dose. As it was explained in the previous section, this may lead to incorrect results. Last, the fraction P is given by an expression that does not depend on the irradiation dose D. However, as overlapping of tracks increases with dose, inter-track effects should do. Recent calculations by Monte-Carlo simulations [4] of the actual energy deposition into micrometric targets showed that the actual dose deposition (specific energy) increased with the irradiation dose, as a resulting of an increase in the overlapping of tracks.

2.5.3 Local Effect Model LEM I, II III

The Local Effect Model was developed in the 90s within the context of the pilot project of carbon therapy in the GSI Helmholtzzentrum fr Schwerionenforschung in Germany. The original version [52], known as the LEM I, is the biophysics model presently incorporated into the TPS of the European centers for IBCT with carbon ions.

Like Katz model, the LEM I describes the dose deposition by an ion with the radial dose. However, the principles of the LEM I are quite different. A detailed analysis of this model can be found in [2, 3]. Here are presented the LEM main lines and some important elements of analysis.

The key point of the LEM I was the definition of lethal events. These biological events result from physical or chemical events induced by cell irradiation and one such event is sufficient to kill the irradiated cell. The LEM I assumes that the number of lethal events follows a Poisson distribution. Therefore, the probability of cell survival is related to the mean number of lethal events N_{lethal} according to:

$$S = \exp\left(-N_{\text{lethal}}\right) \tag{34}$$

As a second postulate, the lethal-event extension was assumed to be small enough to allow the definition of local lethal events. Here *local* has to be understood as 'point-like'. This assumption justified then the definition of the probability dP for an irradiation to induce at least one lethal event at the position $\boldsymbol{r}$ in the volume $d\boldsymbol{r}$. According to the authors, these lethal events refereed to the formation of complex damage DNA Complex damage in DNA. The sensitive part of the cell was assumed to be uniformly distributed over the cell nucleus, viewed as an effective nucleus. In the LEM, the nucleus is represented by a cylinder for which the radius is not necessary the radius of the actual nucleus.

Another important postulate of the LEM I was to assume that the density probability $\frac{dP}{dr}$ is a function of the local dose $D(\boldsymbol{r})$ deposited by the irradiation at the position $\boldsymbol{r}$:

$$\frac{dP}{d\boldsymbol{r}^3} = \rho_{\text{lethal}}\left(D(\boldsymbol{r})\right) \tag{35}$$

The local dose was considered by the authors as an expected quantity defined for an irradiation with ions as the superimposition of the radial dose associated to each ion.

The mean number of lethal events is obtained by summing the density probability over the nucleus volume:

$$N_{\text{lethal}} = \iiint_{\boldsymbol{r}\in\text{Nucleus}} \rho_{\text{lethal}}\left(D(\boldsymbol{r})\right) \cdot d\boldsymbol{r}^3 \tag{36}$$

The function ρ_{lethal} contains almost all the biological information on the cell response to radiation. It is determined from the cell survival $S_X(D)$ to an irradiation with a dose D of low-LET X-rays. To derive this expression, the local dose was approximated by the macroscopic dose D for an irradiation with X-rays. It is then easy to show that:

$$\rho_{\text{lethal}}(D) = \frac{-\ln(S_X(D))}{V_{\text{Nucl}}} \tag{37}$$

where V_{Nucl} is the nucleus effective volume. Since the ion trajectories were approximated by straight lines parallel to the axis of the cylinder that represents the cell nucleus, only the effective nucleus radius is required finally.

Gathering all these equations, the cell survival to a configuration of ion irradiation is finally given by:

$$S_{\text{ionconf.}} = \exp\left[\iiint_{\boldsymbol{r}\in\text{Nucleus}} \frac{\ln(S_X\left(D(\boldsymbol{r})\right))}{V_{\text{Nucl}}} \cdot d\boldsymbol{r}^3\right] \tag{38}$$

A configuration gives the number of ion impacts in or close the cell and the position of each impact, which are necessary to calculate the local dose. To be compared to experimental data, this estimate of survival needs to be averaged over many of these configurations leading in average to a macroscopic dose D. Figure 10 shows an example of cell survival calculations. This process of averaging is very important. It is necessary for instance to reproduce the over-killing effect. More recently, for the purpose of fast (α, β) estimations [44], this procedure was simplified by adding some other approximations.

The version I, II and III of the LEM are based on the principles and approximations that are mentioned above (see [16, 17]). Figure 4 gives an illustration of calculations and compares them to results obtained with the MMKM. The only differences between all these versions are the shape of the radial dose and the expression of the cell survival to X-rays S_X. Whatever the version, the cell survival is described by a LQ model for dose D lower than a threshold dose D_t and extrapolated for larger doses. In the LEM I, the extrapolation is linear, assuming a continuity for both the cell survival and its derivative. The extrapolation of the latest versions are more complex. The description of the radial dose is also more complex.

The extrapolation of the cell survival to X-rays is necessary since the local dose $D(r)$, as the radial dose, can reach very huge values (more than millions of Grays). It is impossible in practice to measure cell survival at such high doses. Therefore, as explained in [5, 38] the parameters used to describe the extrapolation need to be fitted to get a good agreement between the experimental estimations of cell survival to ion irradiation and the predictions. For instance, for the LEM I, which is the simplest version, four parameters have to be determined: the effective nucleus radius R_{Nucl}, the α and β parameters for the cell survival to X-rays and the threshold dose D_t.

However, beyond the practical questions, Eq. (38) common to all the three LEM versions raises many questions. The local dose $D(\boldsymbol{r})$ refers to very microscopic mechanisms. It is deposited by each ion within femtoseconds, at least less than microseconds if one includes the chemical processes. It is defined at local scale, therefore the scale of nanometer. The biological mechanisms that are associated to this local dose are the formation of molecular damage, like the formation of damage clusters in DNA. Yet, this microscopic quantity is used as the parameter of the cell survival to X-rays in Eq. (38). The parameter of the cell survival should be a macroscopic dose. A dose lower than few tens of Grays, deposited within minutes and defined at a scale larger than the whole cell. Moreover, the biological mechanisms that are associated are mechanisms that concern the whole cell, like the triggering of apoptosis or necrosis. Last, if it was possible to measure cell survival to millions of Grays, the temperature of the whole cell would be larger than ebullition temperature and these kinds of experiments could not provide any information on the local events at all.

The reasons for these contradictions are largely explained in [2, 3]. Roughly, the main reason is due to the use of an expected quantity, the local dose, to describe nano-scale processes. At such a scale, the fluctuations of deposited energy are huge and cannot be ignored whatever the irradiation source, low-LET X-rays or high-LET ions. The level of fluctuations is visible on Fig. 8, which compares the distribution of specific energy deposited in nanotargets by an irradiation of 1 Gy with photons or ions. Finally, if one considers the actual local dose instead of the expected local dose, some of the equations given above does not hold true.

Another problem that is linked to the use of an expected quantity is related to the prediction of the shoulder in the curves of cell survival when plotted as a function of the irradiation dose D. In the microdosimetry models, the shoulders are the same whatever the irradiation particles since the β coefficient is predicted to be constant. In the model by Katz, the shoulder is reduced for high-LET ions because the fraction P of dose D associated to ion-kill mode increases when the inactivation cross-section increases. In the LEM, the shoulder also decreases for high-LET ions because of the following reason. The shoulder in the LEM is due to the superimposition of the

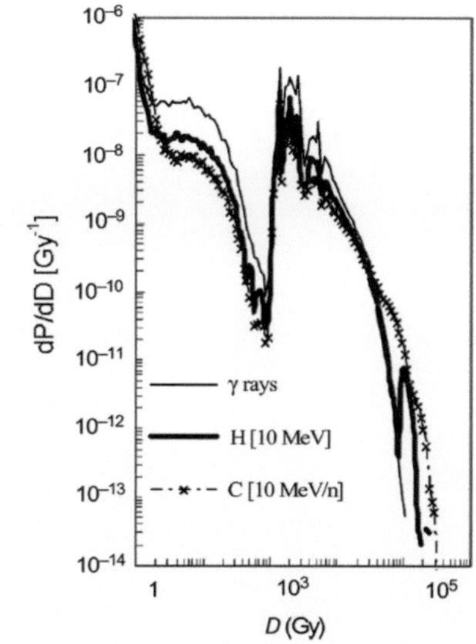

Fig. 8 Distribution of local dose calculated for a water sample of 50 × 50 × 10 μm^3 irradiated at a dose of 1 Gray with a beam of Co^{60} γ-rays, H[10 MeV] and C[10 MeV/n]. The mesh resolution was 10 nm. Figure extracted from [4]

radial doses associated to the impacting ions. Indeed, the density of lethal events ρ_{lethal} is a nonlinear increasing function of the dose. Therefore, for an irradiation with several ions, the effects induced by a superimposition of radial doses is larger than the superimposition of the effects induced by the radial dose of each ion taken separately. The later superimposition leads to a curve of cell survival that decreases as a pure exponential. The former superimposition instead leads to non linear terms and therefore to a shoulder. When the ion LET increases, the radial dose profile becomes thinner, the superimposition decreases and the shoulder as well.

However, when considering the actual local dose instead of the expected dose, the probability that two tracks overlapped at nano-scale resolution is extremely low for the dose that is used in IBCT. This is illustrated by Fig. 9 and proved in [4]. Therefore, the generation of lethal events, as defined in the LEM, cannot actually lead to shoulders in curves of cell survival.

2.5.4 LEM IV

The last version of the LEM, as described in [20], strongly differs from the three precious versions. The notion of local events, these lethal point-like events that are very specific to the LEM, disappeared in this new version. Instead, as for the MKM, the cell nucleus is divided into micrometric domains. In the LEM IV, these domains correspond to the 'giant loops' of DNA for which the size was set to 0.54 μm. The irradiation may create double strand breaks (DSBs) in these DNA loops. As for the sub-lethal lesions in the MKM, the DSB are not necessary lethal. The mean number of such events is deduced from the energy deposited in the domains. As in the MMKM or in the Katz model, the energy is calculated using the radial dose.

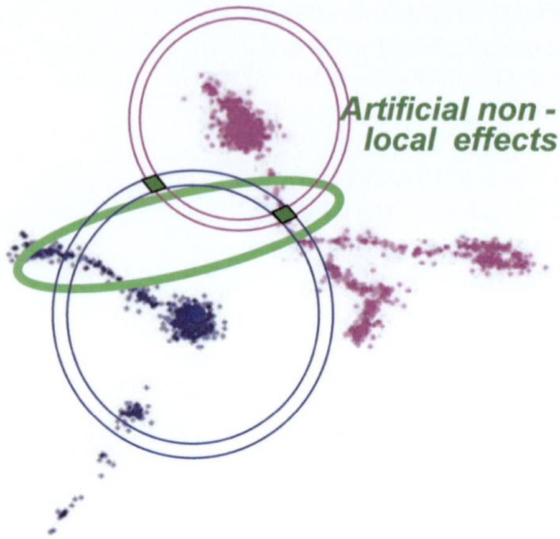

Fig. 9 Scheme of track overlapping. The points represent the energy-transfer points generated by two ions. The cercles illustrate the superimposition of radial dose

However, in the LEM, 'a potential enhancement of the DSB induction results from a combination of SSB in close vicinity as a consequence of the extremely high local doses in the track center'. As in the microdosimetry model, the creation of several sub-lethal events in the domains increases the possibility for cell killing, but the link between the lethality and the number of sub-lethal events is different. In the LEM IV, a coefficient C of damage complexity is defined by:

$$C = \frac{N_{cDSB}}{N_{cDSB} + N_{iDSB}} \tag{39}$$

where N_{iDSB} (resp. N_{cDSB}) is the number of cell domains that suffered one (resp. two or more) DSBs. This complexity coefficient is determined by a Monte-Carlo process, which generates the number of DSBs in a grid of domains, calculates N_{iDSB} and N_{cDSB} and finally determines a mean value of complexity C over many configurations. The cell survival to ion irradiation is derived from the cell survival to an irradiation with photons through an equivalent photon dose inducing an equal complexity.

Regarding, the problem of track overlapping pointed out by [2–4] for the previous versions of the LEM, it is now better treated since the domain extension is micrometric: two ion tracks can contribute to the number of DSB in a micrometric domains.

The LEM IV is very different from the previous versions. In particular, the notion of local effects has been given up. This model is closer to Katz model and the MMKM. Target size is clearly defined and matches micrometric scale. The energy transferred to the domains is calculated using a radial dose. The notion of sub-lethal lesions is also introduced (Fig. 10).

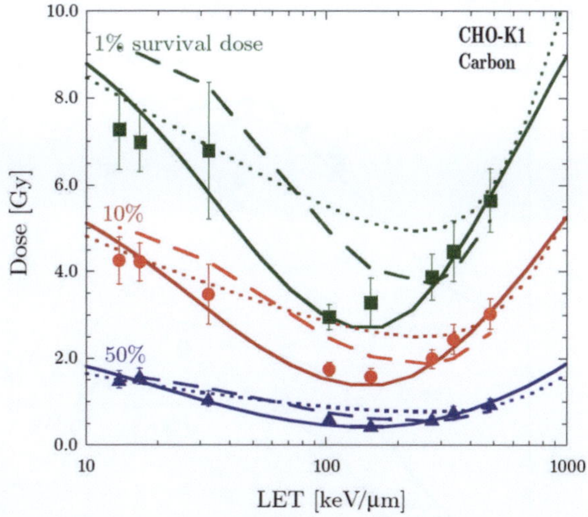

Fig. 10 1, 10 and 50 % survival doses of CHO cells as a function of LET for 12C-ion beams. The *closed square*, *circle* and *triangle symbols* respectively show the 1, 10 and 50 % survival doses of the experimental results. The *solid lines* indicate the 1, 10 and 50 % survival doses calculated by the MKM using the KieferChatterjee track structure model. The *dotted lines* show the LEM calculation with the LEM I track structure model and the dashed lines with the LEM II track structure model. Figure extracted from [36]

2.6 Nanodosimetry

The interest for considering nano-scale to model biophysics effects induced by ion irradiations is not new. For instance, Goodhead *et al.* supported the idea that clusters of DNA damage could play an important role in biological effects, by facing experimental evidences and Monte-Carlo simulations of the energy deposited into nano-targets [25, 26, 48]. Beside, Chetioui et al. [10] pointed out a correlation between the cross-section for cell inactivation and the cross-section for core ionization when plotted as a function of LET (see Fig. 11). They proposed a scenario, known as model K, in which core ionizations could efficiently induce lethal events and quantified it in comparing cell killing induced by soft X-rays with photon energy below and above the binding energy of core electrons (see [57] for details and [6] for an analysis).

More recently, the interest in nano-scale mechanisms as a cause of many induced biological effects has increased probably because of theoretical and experimental evidences. Computing resources give nowadays the possibility to perform more exhaustive and realistic simulations for the deposition of energy, the formation of damage and the modeling of more complex biological structures. The most popular code is the PARTRAC code, which was used to estimate DNA damage [18] and chromosomal aberrations [19]. More recently the GEANT4-DNA collaboration undertook the development of an open source code similar to PARTRAC.

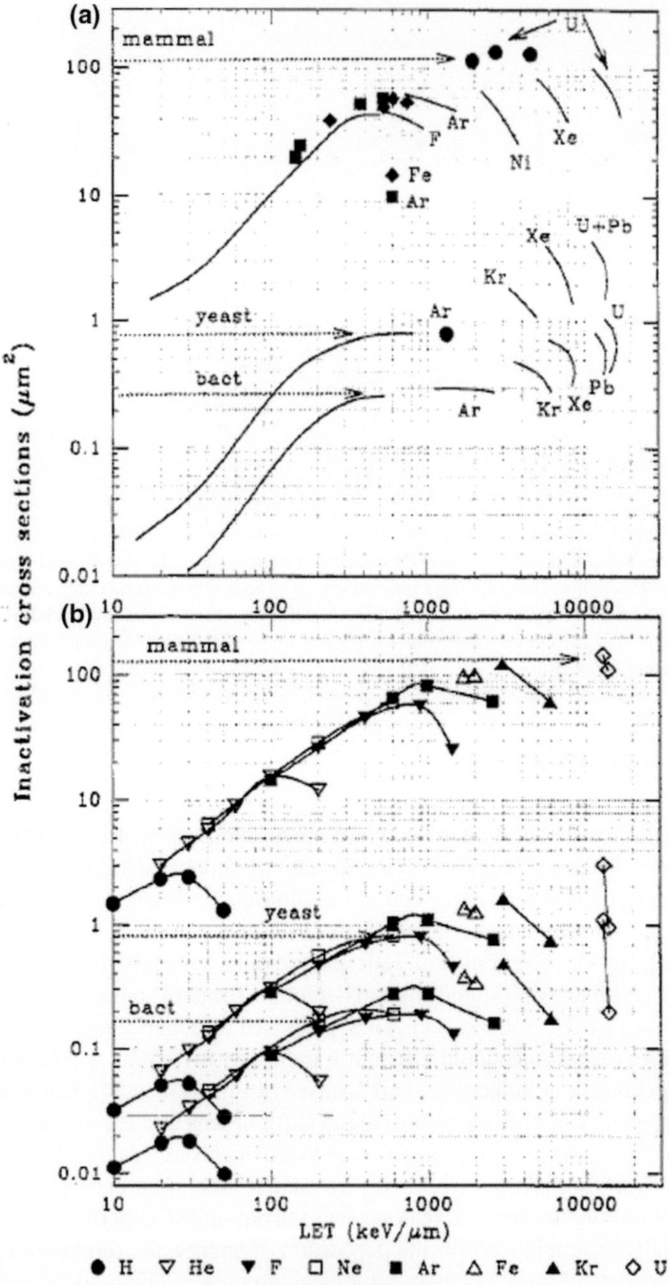

Fig. 11 **a** Experimental cross section for cell inactivation by heavy ions of various LET: mammal cells, yeast, bacillus subtilis; *arrows* indicate the geometrical cross section of the cell nucleus or of the DNA concentration. **b** Calculated cross sections for the induction of at least one efficient K-vacancy in cell nucleus (from [56])

Another reason for the growing interest in nanodosimetry may be attributed to the development of nano-detectors [27]. The notion of experimental nanodosimetry was extended to quantities other than the specific energy. It is indeed possible to estimate experimentally the size of ionization clusters ν, by collecting either the electrons [11, 47] or the ions [23] that are generated by the penetration of an ion beam into a small volume of low-pressure gas. New quantities were then introduced to characterize the level of ionization: the mean cluster size M_1 and, more generally, the high-order moments M_ξ; the cumulative probabilities F_k of forming a cluster size $\nu \geq k$; the cluster yield Y that gives the number of clusters for which the size stands in a defined interval.

These nanodosimetric indicators have been expected to be closely related to the formation of DNA damage. For instance, [22] proposed a simple model, based on combinatorial analysis, to convert distributions of ionization cluster size into predictions of the yield of DSB and non-DSB.

However, the link to more complex biological endpoints like cell survival is less direct. Schulte et al. proposed an estimation for the α and β coefficients of LQ model in introducing the cluster yield Y for the cluster-size intervals 2–5 and 6–10 to represent respectively the repairable and irreparable DNA lesions [53]. They successfully applied their model for V79 cells irradiated with low-energy protons and helium ions, but a more exhaustive benchmark should be undertaken to estimate the potential of this model for IBCT. Beside, the model presented in [61] is based on energy deposition at two nano-scales, 5 and 25 nm. They also introduced the definition of three functions F_1, F_2, F_3 corresponding to the probability of depositing energy into these nano-targets within three different intervals. Then, a prediction of cell survival was derived, in a similar way as in the MKM. This model has been tested only for low-energy protons and helium ions, since the field of applications was related to irradiation with neutrons.

Regarding, treatment planning, not any nanodosimetry model has been incorporated in a clinical TPS to now. Casiraghi et al. recently proposed a novel treatment plan optimization strategy based on measurable nanodosimetry quantities instead of biophysical models [7]. In particular, a biological indicator M_1^{bio} was defined for evaluating the radiation quality of the mixed radiation field in terms of clusters with 2–10 ionizations. In the context of Biologically Effective Dose (BED), a quantity that was introduced to better manage the fractionating effects in clinical treatments, Lindborg *et al.* proposed to extend the framework of microdosimetry to nanodosimetry. Precisely, they reported that the lineal energy deposited into a volume of about 10–15 nm was approximately proportional to the α-parameter of the linear-quadratic relation used in fractionated radiotherapy in both low- and high-LET beams. This correlation was explored for a very limited number of particle types and energies [46].

Finally, despite of the research dynamics in the field of nanodosimetry, an important effort is required to take advantage of nanodosimetry for the optimization of IBCT. While LEM I, II and III were constructed to predict cell survival from a description at nano-scale of the energy deposition by ions, the use of expected quantities has introduced many artifacts (see Sect. 2.5.3). Considering the relevance of predicting cell survival for clinical applications, models based on nanodosimetry still need to be developed to make the bridge between the descriptions of initial radiation events at nano-scale and biological endpoints useful for IBCT. The development of the NanOx™ model has been undertaken in this context.

2.6.1 NanOx™

As it has been addressed in some of the previous sections, the fluctuations in deposited energy are very important and cannot be ignored when looking at microscopic scale. They need to be even more considered when nano-scale processes are involved. One important motivation to set up NanOx™ model was then to propose a mathematical framework to develop modeling consistently with the stochastic nature of energy deposition by radiations.

A second motivation for developing NanOx™ was to suggest other quantities than the specific energy as relevant parameters to describe the induced biological effects. For instance, some authors [58] suggested to take account for the thermal spike generated by high-LET ions. Parameters related to the shock wave that follows this thermal spike could also be considered [55]. Distributions in ionization clusters size, as discussed in the previous section, or even descriptors of damage in DNA could be quite relevant as well. NanOx™ is the acronym of NANodosimetry and OXydative stress to suggest that mechanisms occurring at nano-scale may play a crucial role in the specificity of the biological effects with regard to the incident particle, and, that the production of radicals may stand within the important parameters. The effects of radiations on living systems are complex to model. At this state of knowledge and computing resources, it seems impossible to predict complex biological endpoints like cell survival without adding some empirical ingredients and phenomenological descriptions. However, scientific progresses, in particular on the initial physical and chemical processes, and the inclusion of new knowledge into biophysics models should help in reducing the empiricism for better and more sophisticated predictions.

A third purpose of NanOx™ is not only to provide a framework, but also to propose a predictive model reasonable in terms of computing requirements and able to produce data for IBCT TPS.

NanOx™ is based on a rigorous mathematical formulation, which is long and cannot be summarized easily. Therefore, the following presentation avoids this rigorous but a bit heavy formulation. Instead, the following main lines aim at giving a view of the basic NanOx™ principles, without any demonstration.

Considering a number N of cells irradiated with a source of ionising radiation, a number n among them will survive to this irradiation configuration. The purpose of NanOx™ is not to predict this actual number, but instead to predict a mean number of surviving cells. This mean number corresponds to the average of n over the space of cell and irradiation configurations that match the protocol of the experiments for which one wants to predict the issues.

The cell configurations should take account for the spatial distribution of cells, the heterogeneity in the population and, for each cell, its particular state. In this first version of NanOx™, a representative cell is considered and, in particular, the communication between cells is assumed to be implicitly incorporated in the representative cell response. Although this approximation is done in the other models and in experimental work related to cell survival, it is not explicitely mentioned.

An irradiation configuration c_K corresponds, for ion beams, to the number K of ions that impact the representative cell (directly or through the electronic cascades) and, for each ion, a track configuration c_k, made of physical and chemical events described at nano-scale. For non-charged particles, the number K refers to the number of particle interactions that lead to energy deposition into the cell. Thus, the mean cell survival reads:

$$\overline{S} = \left\langle {}^{c_K} S \right\rangle_{c_K} \tag{40}$$

In this version of the model, the irradiation dose-rate is assumed to be high enough to ignore, for instance, the description of the repair kinetics. Then, cell killing is assumed to be independently induced by local and non-local events. The survival fraction is the product of the cell survival to local (S_L) and non-local (S_{NL}) events according to:

$${}^{c_K} S = S_L \times S_{NL} \tag{41}$$

The local events are defined as directly lethal events produced by physical and chemical events at local scale. Local scale is defined as a scale small enough so that only one track significantly contributes to energy deposition into a local volume when the dose value of the global irradiation corresponds to clinical doses. Typical clinical doses delivered during a treatment session are lower than tens of Grays and local scale matches therefore nano-scale. The non-local events correspond by definition to any event other than local events.

The definition of an irradiation configuration is largely inspired by the framework of microdosimetry. It is nevertheless generalized to include chemical events. Moreover, these configurations are not only used to derive a number of lethal lesions but directly the cell survival. The expression of the cell survival as a product of the cell survival to two independent mechanisms is inspired by Katz model. It is moreover possible to prove that the cell survival to local effects can be described by a cross-section, as for the 'ion-kill' mode. However, the target size are not the same, being micrometric for Katz model and nanometric for NanOx™.

The local lethal events can be triggered by various events at nano-scale, like energy deposition, chemical events, thermal spikes or shock waves. The creation of lethal events is represented by the knock-out of nano targets. In the current version of NanOx™ (NanOx™I), these targets are uniformly distributed over a sensitive volume and the probability of knock-out a target is a function of the specific energy deposited into this target. These nano targets might represent groups of DNA base pairs. Thus, the sensitive volume is taken to be the cell nucleus and the nano targets are approximated by cylinders of size 10-nm.

Since, one target knock-out is sufficient to create one lethal event, then the cell survival to lethal events obeys:

$$S_L = \exp(-n_L^*) \tag{42}$$

where n_L^* is an effective number of lethal events derived from an effective local function F and from the distribution of energy deposited into the nano-targets. F contains all the information on the cell response to radiations in terms of local lethal events. Several arguments led us to test the form:

$$F(z) = \frac{h}{2}\left[1 + \operatorname{erf}\left(\frac{z - z_0}{\sigma}\right)\right] \tag{43}$$

for which z is the specific energy deposited into a nano target and h, z_0 and σ parameters that are fitted to get a good predictions for the cell survival to irradiation measured for different particle types and energies. The results obtained for this example of function form were satisfactory for both V79 and HSG cell lines, and, already competitive with regard to LEM and MKM predictions. The calculation of the energy deposition into targets were achieved by using the Monte-Carlo code LQD [24], but it could be calculated by any code able to simulate tracks with a nano-scale resolution.

Regarding non-local events, the interaction of sub-lethal lesions, as defined in the framework of microdosimetry, matches the requirement of NanOx™. However, as an attempt, it is proposed in NanOx™ I to introduce the notion of global events. These global events are at the extreme opposite of the local events since the target volume associated to these events is closer to the volume of the cell rather than to the local targets. For the sake of simplicity, this volume is set equal to the volume containing the local targets. A LQ form is taken for the cell survival to global events but the parameter of this function is not the specific energy deposited into the global sensitive volume. Instead, a level of oxidative stress induced by irradiation is chosen as parameter. A chemical dose is even defined for convenience in the same way as the biological dose but for the production of radicals. For the first implementation of NanOx™, the index of oxidative stress was represented by the production in hydroxyl radicals, calculated at 10 ps using the Monte-Carlo code LQD [24].

The model NanOx™ is relatively recent and needs to be challenged extensively with available experimental data. Whatever the accuracy of the predictions, it aims at evolving with the state of knowledge. Although it was not possible to detail the dosimetric aspects in this very brief presentation, NanOx™ has been built as a framework for modeling cell survival based on dosimetry at micro- and nano-scales. It may be used to test scenarios that may be involved in the triggering of cell death by ionizing radiations. This model may also evolve to predict other outcomes useful for IBCT.

Acknowledgements Some of the sections refer to work performed within the framework of the LABEX PRIMES (ANR-11-LABX-0063) of Université de Lyon, within the program 'Investissements d'Avenir' (ANR-11-IDEX-0007) operated by the French National Research Agency (ANR).

The author acknowledges the financial support by ITMO Cancer in the framework of Plan Cancer 2009–2013 and of the project no. PC201312, designated 'Domaine de la physique, des mathématiques ou des sciences de l'ingénieur appliqués au Cancer'. He is also grateful to his collaborators, in particular M. Cunha, D. Dauvergne, C. Monini, J. Remillieux and E. Testa, for fruitful discussions.

References

1. Amaldi U, Kraft G (2005) Radiotherapy with beams of carbon ions. Rep Progr Phys 68(8):1861
2. Beuve M (2009) Formalization and theoretical analysis of the local effect model. Radiat Res 172(3):394–402
3. Beuve M (2010) In response to the comment by T. Elssser and M. Scholz. Radiat Res 173(6):856–858
4. Beuve M, Colliaux A, Dabli D, Dauvergne D, Gervais B, Montarou G, Testa E (2009) Statistical effects of dose deposition in track-structure modelling of radiobiology efficiency. Nucl Instrum Methods Phys Resh Sect B: Beam Interact Mater Atoms 267(6):983–988 (Proceedings of the seventh international symposium on swift heavy ions in matter)
5. Beuve M, Gersende A, Mira M, Colliaux A, Battiston-Montagne P, Jalade P, Balanzat E, Demeyer A, Bajard M, Rodriguez-Lafrasse C (2008) Radiobiologic parameters and local effect model predictions for head-and-neck squamous cell carcinomas exposed to high linear energy transfer ions. Int J Radiat Oncol Biol Phys 71(2):635–642
6. Beuve M, Moreau J-M, Rodriguez C, Testa E (2015) Biological systems: from water radiolysis to carbon ion radiotherapy. J Phys: Conf Ser 629(1)
7. Casiraghi M, Schulte RW (2015) Nanodosimetry-based plan optimization for particle therapy. Comput Math Methods Med
8. Chanrion M-A, Sauerwein W, Jelen U, Wittig A, Engenhart-Cabillic R, Beuve M (2014) The influence of the local effect model parameters on the prediction of the tumor control probability for prostate cancer. Phys Med Biol 59(12):3019
9. Chatterjee A, Schaefer HJ (1976) Microdosimetric structure of heavy ion tracks in tissue. Radiat Environ Biophys M13:215–227
10. Chetioui A, Despiney I, Guiraud L, AdouiL L, Sabatier L, Dutrillaux B (1994) Possible role of inner-shell ionization phenomena in cell inactivation by heavy ions. Int J Radiat Biol 655(5):511–522
11. Conte V, Colautti P, Moro D, Grosswendt B (2014) Track structure of carbon ions: measurements and simulations. Radiat Prot Dosim 161(1–4):445448
12. Tobias C, Blakel E, Ngo FQH, Yang TCH (1980) The repair-misrepair model of cell survival. Radiat Biol Cancer Res 195–230
13. Cunha M, Testa E, Komova O, Nasonova E, Melnikova L, Shmakova N, Beuve M (2015) Modeling cell response to low doses of photon irradiationpart 1: on the origin of fluctuations. Radiat Environ Biophys 1–12
14. Curtis S (1986) Lethal and potentially lethal lesions induced by radiation: a unified repair model. Radiat Res 106(2):252–270
15. Curtis S (1989) Erratum: Lethal and potentially lethal lesions induced by radiation: a unified repair model. Radiat Res 119(3):584
16. Elsasser T, Scholz M (2007) Cluster effects within the local effect model. Rad Res 167(3):319–329
17. Elssser T, Krmer M, Scholz M (2008) Accuracy of the local effect model for the prediction of biologic effects of carbon ion beams in vitro and in vivo. Int J Radiat Oncol Biol Phys 71(3):866–872
18. Friedland W, Jacob P, Bernhardt P, Paretzke HG, Dingfelder M (2003) Simulation of dna damage after proton irradiation. Radiat Res 159:40110

19. Friedland W, Kundrt P (2015) Chromosome aberration model combining radiation tracks, chromatin structure, dsb repair and chromatin mobility. Radiat Protect Dosim
20. Friedrich T, Scholz U, Elssser T, Durante M, Scholz M (2012) Calculation of the biological effects of ion beams based on the microscopic spatial damage distribution pattern. Int J Radiat Biol 88(1–2):103–107
21. Furusawa Y, Fukutsu K, Aoki M, Itsukaichi H, Eguchi-Kasai K, Ohara H, Yatagai F, Kanai T, Ando K (2000) Inactivation of aerobic and hypoxic cells from three different cell lines by accelerated he-, c- and ne-ion beams. Radiat Res 154:48596
22. Garty G, Schulte R, Shchemelinin S, Leloup C, Assaf G, Breskin A, Chechik R, Bashkirov V, Milligan J, Grosswendt B (2010) A nanodosimetric model of radiation-induced clustered dna damage yields. Phys Med Biol 55(3):761
23. Garty G, Shchemelinin S, Breskin A, Chechik R, Assaf G, Orion I, Bashkirov V, Schulte RW, Grosswendt B (2002) The performance of a novel ion-counting nanodosimeter. Nucl Instrum Methods A 491:21235
24. Gervais B, Beuve M, Olivera G, Galassi M (2006) Numerical simulation of multiple ionization and high LET effects in liquid water radiolysis. Radiat Phys Chem 75(4):493–513
25. Goodhead D (1994) Initial events in the cellular effects of ionizing radiations: clustered damage in dna. Int J Radiat Biol 65(1):7–17
26. Goodhead D (1995) Molecular and cell models of biological effects of heavy ion radiation. Radiat Environ Biophys 34(2):67–72
27. Grosswendt B (2004) Recent advances of nanodosimetry. Radiat Prot Dosim 110(1–4):789–799
28. Hawkins R (2003) A microdosimetric-kinetic model for the effect of non-poisson distribution of lethal lesions on the variation of rbe with let. Rad Res 160(160):61–69
29. Hawkins RB (1994) A statistical theory of cell killing by radiation of varying linear energy transfer. Radiat Res 140:36647
30. Hawkins RB (1996) A microdosimetric-kinetic model of cell death from exposure to ionizing radiation of any let, with experimental and clinical applications. Int J Radiat Biol 69(6):739–755
31. Hirao Y, Ogawa H, Yamada S, Sato Y, Yamada T, Sato K, Itano A, Kanazawa M, Noda K, Kawachi K, Endo M, Kanai T, Kohno T, Sudou M, Minohara S, Kitagawa A, Soga F, Takada E, Watanabe S, Endo K, Kumada M, Matsumoto S (1992) Heavy ion synchrotron for medical use himac project at nirs -japan. Nucl Phys A 538:541–550
32. Inaniwa T, Furukawa T, Kase Y, Matsufuji N, Toshito T, Matsumoto Y, Furusawa Y, Noda K (2010) Treatment planning for a scanned carbon beam with a modified microdosimetric kinetic model. Phys Med Biol 55:67216737
33. Kanai T, Endo M, Minohara S, Miyahara N, Ito HK, Tomura H, Matsufuji N, Futami Y, Fukumura A, Hiraoka T, Furusawa Y, Ando K, Suzuki M, Soga F, Kawachi K (1999) Biophysical characteristics of himac clinical irradiation system for heavy-ion radiation therapy. Int J Radiat Oncol Biol Phys, 44(1):201–210
34. Kanai T, Furusawa Y, Fukutsu K, Itsukaichi H, Kasai KE, Ohara H (1997) Irradiation of mixed beam and design of spread-out bragg peak for heavy-ion radiotherapy. Radiat Res 147(1):78–85
35. Kanai T, Matsufuji N, Miyamoto T, Mizoe J, Kamada T, Tsuji H, Kato H, Baba M, Tsujii H (2006) Examination of gye system for himac carbon therapy. Int J Radiat Oncol Biol Phys 61:65056
36. Kase Y, Kanai T, Matsufuji N, Furusawa Y, Elsasser T, Scholz M (2008) Biophysical calculation of cell survival probabilities using amorphous track structure models for heavy-ion irradiation. Phys Med Biol 53:3759
37. Kase Y, Kanai T, Matsumoto Y, Furusawa Y, Okamoto H, Asaba T, Sakama M, Shinoda H (2006) Microdosimetric measurements and estimation of human cell survival for heavy-ion beams. Radiat Res 166:62938
38. Katz R (2003) The parameter-free track structure model of scholz and kraft for heavy-ion cross sections. Radiat Res 160(6):724–728
39. Katz R, Ackerson B, Homayoonfar M, Sharma SC (1971) Inactivation of cells by heavy ion bombardment. Radiat Res 47(2):402425

40. Kellerer A, Rossi H (1978) A generalized formulation of dual radiation action. Radiat Res 75:471–488
41. Kellerer A, Rossi HH (1972) The theory of dual radiation action. Curr Topics Radiat Res Q 8(85):158
42. Kiefer J, Straaten H (1986) A model of ion track structure based on classical collision dynamics. Phys Med Biol 31:1201–1209
43. Krämer M, Kraft G (1994) Calculations of heavy-ion track structure. Radiat Environ Biophys 33(3):91–109
44. Krämer M, Scholz M (2006) Rapid calculation of biological effects in ion radiotherapy. Phys Med Biol 51:1959–1970
45. Lea DE (1955) Actions of radiation on living cells. Cambridge University Press, Cambridge
46. Lindborg L, Hultqvist M, Tedgren AC, Nikjoo H (2015) Nanodosimetry and rbe values in radiotherapy. Radiat Prot Dosim 166(1–4):339–342
47. Nardo LD, Colautti P, Conte V, Baek WY, Grosswendt B, Tornielli G (2002) Ionization-cluster distributions of a-particles in nanometric volumes of propane: measurement and calculation. Radiat Environ Biophys 41:235256
48. Nikjoo H, Charlton D, Goodhead D (1994) Monte carlo track structure studies of energy deposition and calculation of initial dsb and rbe. Adv Space Res 14(10):161–180
49. Pedicini P, Strigari L, Benassi M (2013) Estimation of a self-consistent set of radiobiological parameters from hypofractionated versus standard radiation therapy of prostate cancer. Int J Radiat Oncol Biol Phys 85(5):231–237
50. Rossi HH, Failla G (1956) Tissue-equivalent ionization chambers. Nucleonics 14(2):32–37
51. Rossi HH, Rosenzweig W (1955) A device for the measurement of dose as a function of specific ionization. Radiology 64(3):404–411
52. Scholz M, Kellerer A, Kraft-Weyrather W, Kraft G (1997) Computation of cell survival in heavy ion beams for therapy the model and its approximation. Radiat Environ Biophys 36:59–66
53. Schulte R, Bashkirov V, Shchemelinin S, Garty G, Chechik R, Breskin A (2001) Modeling of radiation action based on nanodosimetric event spectra. Phys Med 17(Suppl 1):177–180
54. Srdo D (1970) Experimental technique of measurement of microscopic energy distribution in irradiated matter using rossi counters. Radiat Res 43(2):302–319
55. Surdutovich E, Yakubovich A, Solovyov A (2013). Biodamage via shock waves initiated by irradiation with ions. Sci Rep 3(3):1289 (1–6)
56. Touati A, du Penhoat MAH, Bailly-Despiney I, Gobert F, Champion C, Fayard B, Abel F, L'Hoir A, Moulin J, Sabatier L, Chetioui A (1997) Biological implication of atomic collisions at the molecular level. Nucl Instrum Methods Phys Res Sect B: Beam Interact Mater Atoms 132(2):276–279
57. Touati A, du Penhoat MH, Bailly-Despiney I, Gobert F, Champion C, Fayard B, Abel F, L'Hoir A, Moulin J, Sabatier L, Chetioui A (2002) Biological effects induced by k photo-ionisation in and near constituent atoms of DNA. Radiat Protect Dosim 99(1–4):83–84
58. Toulemonde M, Surdutovich E, Solovyov A (2009) Temperature and pressure spikes in ion-beam cancer therapy. Phys Rev E 80(3):031913(1–9)
59. Waligorski MPR, Hamm R, Katz R (1986) The radial distribution of dose around the path of a heavy ion in liquid water. Tracks Radiat Meas 11(6):309
60. Waligrski MPR, Grzanka L, Korcyl M (2015) The principles of katz's cellular track structure radiobiological model. Radiat Protect Dosim 166(1–4):49–55
61. Wang C-KC, Zhang X (2006) A nanodosimetry-based linear-quadratic model of cell survival for mixed-let radiations. Phys Med Biol 51(23):6087

Treatment Planning Systems and Hadron Therapy Practice in France

Ludovic De Marzi, Annalisa Patriarca, Alejandro Mazal and Jean-Louis Habrand

1 Introduction

Radiotherapy, particularly particle therapy, is one of the main techniques used for the treatment of cancer today. The medical use of protons, with energies ranging from 60 to 250 MeV, is expanding all over the world, including in France. Since 1980, thousands of patients in France have been treated for a variety of different diseases by particle therapy. One neutron (1980–2007) and two proton (1991) facilities were opened in Orleans, Nice and Orsay, respectively. Early applications of protons concerned ophthalmological treatments (1991), followed by adult and paediatric intracranial or skull base tumours (1993), and then spine and sacrum sites (2012), while the use of neutron therapy stopped in about 2007. This experience gave rise to several carbon research or therapy projects in France (Lyon, Caen) as well as several other proton therapy projects. The first accelerators used for particle therapy were derived from facilities dedicated to nuclear physics research, mainly cyclotrons, synchrotrons or synchrocyclotrons. A clinical environment was then added to complement these installations: passive beam lines, treatment planning systems and dose algorithms were all developed in-house, before being commercially available. Scanning techniques will likely become the dominant proton or ion therapy modalities in the near future, and considerable attention has recently been paid to the development and improvement of physical models in dose calculation algorithms (still widely based on analytical approximations such as ray-tracing and pencil-beams), as well as in dosimetry equipment for quality assurance or commissioning. Due to their ballistic properties, small angular diffusion and precise depth dose distributions, protons and light ions allow highly conformational dose deposition and fairly good protec-

L. De Marzi (✉) · A. Patriarca · A. Mazal · J.-L. Habrand
Institut Curie – Centre de Protonthérapie d'Orsay, Paris, France
e-mail: ludovic.demarzi@curie.fr

J.-L. Habrand
Centre François Baclesse, Caen, France

A.V. Solov'yov (ed.), *Nanoscale Insights into Ion-Beam Cancer Therapy*,
DOI 10.1007/978-3-319-43030-0_14

tion of organ at risks. In turn, this type of therapy requires a more accurate planning system (submillimetric) for calculation of geometries and 3D dose distributions, for example using Monte Carlo methods.

This chapter will discuss the technical and clinical aspects of proton beam treatment planning, as many similarities exist between proton and ion therapy. This chapter includes a summary of the physics and approximations used in proton dose algorithms, including the impact of accelerator and nozzle modelling, a description of conventional delivery approaches such as passive scattering or pencil beam scanning, immobilization specificities and the need for accurate imaging of patient geometry. The issues of neutron generation, risk of second cancers, and radiobiological effectiveness (RBE) of protons will also be discussed. As several of these aspects are common to proton and ion therapy, one section of this chapter will be devoted to the differences between these techniques, especially the biological effects of radiation. Finally, recent developments and perspectives in the planning process will be presented.

2 Treatment Planning Process

The aim of treatment planning is to optimize and simulate dose distributions to a target volume and the surrounding normal tissues. This step is generally performed before the patient's first irradiation session, and allows prediction of treatment outcome based on the knowledge of specific dosimetric or biological parameters representative of local tumour control or probabilities of normal tissue complications. The treatment planning system (TPS) is software designed to perform these simulations and manipulate physical doses in order to deliver a uniform biological equivalent dose to the target volume, sparing the surrounding tissues as much as possible.

Several tasks are mandatory and strongly dependent in the treatment planning process. First of all, the patient's anatomy must be reconstructed in 3D.

2.1 Conversion of CT Information for Dose Calculation

Ion beam treatment planning is based on computed tomography (CT) imaging, acquired at kilovoltage X-ray energies. The information provided by CT images is used to account for attenuation and scattering of particle beams in dose calculations, and also to delineate target volumes and normal tissues. The relative ion stopping powers (stopping power ratio or SPR) in human tissues are deduced from Hounsfield units (HU) or CT numbers.

2.1.1 Principle

A commonly applied methodology to obtain an accurate relationship between HU values and SPR has been proposed for charged particle therapy [49, 63, 65]. This multistep stoichiometric calibration procedure has been described in the literature and is used to determine a tissue substitute calibration curve:

- A set of materials with known elemental composition and mass density close to tissue samples is scanned in the CT imaging system and the corresponding HU values are measured. Tissue samples are scanned individually in the centre of a water-equivalent phantom in order to ensure the same photon spectrum for each sample [63].
- The parameterization used for calculation of photon total attenuation coefficient ([36], Eq. 1) is defined from the known chemical composition and HU measurements of the tissue samples.
- This parameterization is used to compute the HU values of a selection of tissue samples and the SPR are calculated, for example from the Bethe Bloch equation for these tissue samples (Eq. 4).
- A final calibration curve (HU values vs. SPR) is fitted to the data: linear fits are calculated separately for adipose tissue, organ and muscle and bone categories, in order to divide the final curve into several segments (Fig. 1).

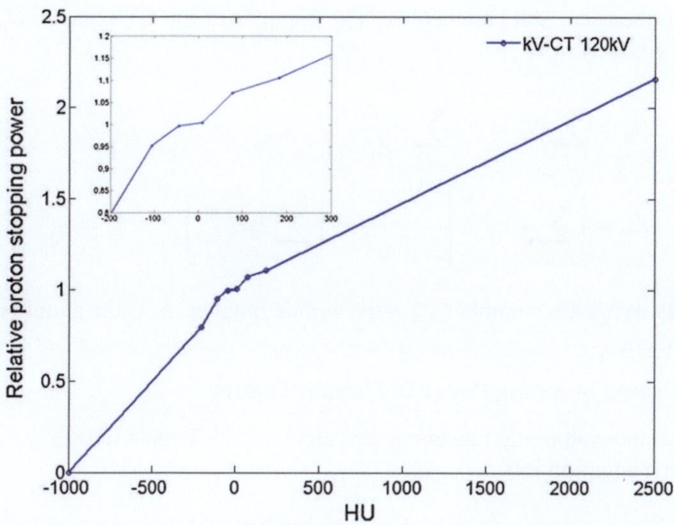

Fig. 1 Variation of relative proton stopping power as a function of HU values, for a 120 kV CT scan with a phantom diameter of 200 and 2 mm slice thickness. The *different lines* correspond to the linear fits to the biological tissues grouped into categories (adipose tissue, organ, muscle and bone)

Depending on the dose calculation model, the calibration curves (HU values vs. SPR or mass density) are then stored in the treatment planning system database for each CT scanner protocol. An internal material list, typically composed of a few dozen materials, mass densities, elemental composition and mean ionization energy, is used to associate HU values with material properties for each voxel of CT images. CT conversion methods have also been applied to Monte Carlo (MC) dose calculations, essentially based on conversion to mass density and elemental composition assigned to materials rather than water equivalent properties [40, 66, 76].

2.1.2 Basic Equations

Total Attenuation Coefficient

For a mixture of elements, the total attenuation coefficient μ can be obtained in the "Jackson and Hawkes" form [36], using Rutherford's parameterized cross-section of scattering processes:

$$\mu = \rho\, N_g(Z, A) \left\{ K^{ph} Z_1^{3.62} + K^{coh} Z_2^{1.86} + K^{KN} \right\} \tag{1}$$

where ρN_g is the electron density and K^{ph}, K^{coh} and K^{KN} are constants that characterize the cross-sections of the photoelectric effect, coherent scattering and the cross-section of Klein Nishina, respectively. These constants are dependent on the scan technique used, and parameterize the response of the CT scanner. N_g, Z_1, Z_2, and λ_i are given by:

$$N_g = \sum_i N_g^i = N_A \sum_i \frac{w_i Z_i}{A_i} \tag{2}$$

$$Z_1 = \left[\sum_i \lambda_i Z_i^{3.62} \right]^{1/3.62} \quad Z_2 = \left[\sum_i \lambda_i Z_i^{1,86} \right]^{1/1,86} \quad \lambda_i = \frac{N_g^i}{N_g} \tag{3}$$

where N_A is Avogadro's number, Z_i is the atomic number, A_i is the atomic weight of the *i*th element

Relative Proton Stopping Power in Human Tissues

The relative mass stopping power can be calculated for human tissues, based on the Bethe-Bloch equation [65]:

$$SP_w^m = \frac{\rho_e^m \rho^w \left\{ \ln\left[2m_e c^2 \beta^2 / I^m (1-\beta^2) \right] - \beta^2 \right\}}{\rho_e^w \rho^m \left\{ \ln\left[2m_e c^2 \beta^2 / I^w (1-\beta^2) \right] - \beta^2 \right\}} \tag{4}$$

where $\beta = v/c$ is the 'normalized' velocity (v the projectile's velocity, c the speed of light), m_e is the electron mass and $I^{m,w}$ are the mean ionization energies of atoms for medium and water. The value of $\ln(I^m)$ for a mixture can be calculated using the Bragg additivity rule:

$$\ln(I^m) = \left(\sum_i \frac{w_i Z_i}{A_i} \ln I_i\right)\left(\sum_i \frac{w_i Z_i}{A_i}\right)^{-1} \quad (5)$$

with Z_i, A_i Ii and w_i atomic number, atomic weight, ionization energy and percentage contribution by weight of *i*th element i in tissue, respectively. The relative electron density of medium to water can be calculated using:

$$\frac{\rho_e^m \rho^w}{\rho_e^w \rho^m} = \frac{N_g^m}{N_g^{eau}} \quad (6)$$

where ρ is the mass density and N_g is the number of electrons per unit volume defined in Eq. (2).

2.1.3 Uncertainties and Perspectives

Potential errors in the prediction of beam range in patients, derived from the conversion process, noise and partial volume effects in images, metallic implants or beam hardening artifacts, were estimated to be between 1–3 mm [48]. The approximation of relative biological effectiveness values in clinical practice and underestimation of its value at the end of the Bragg peak (also see an example Fig. 7) can generate an extension of the biological range of the order of 1 mm [53]. Moreover, different values have been obtained for the ionization energy of water, typically estimated to be around 75 eV that can lead to uncertainties in the depth of the Bragg peak by up to several millimetres [1].

Consequently, in practice, the gradient at the distal end of particle dose distributions is also rarely used (or used for a small fraction of the total dose) to spare critical normal tissues due to uncertainties about their exact position, and safety margins proportional to a few % of the range are added to the planning target volume (PTV). Recently, dual energy (DE) or megavoltage (MV) computed tomography has been investigated in order to improve elemental mass fraction predictions, potentially mitigate beam hardening and metal artifacts which limits the accuracy of kV-CT [13]. Photon attenuation is strongly dependent on the energy spectrum used and can vary with the size of the phantom and the position of the heterogeneity, and metallic implants induce artifacts leading to inaccuracies in the calculated ion beam range.

Moreover, several alternatives are also being explored to manage range uncertainty: proton tomography is being investigated by several teams as a means to improve treatment planning in terms of range and dose deposition predictions. The use of proton imaging could provide supplementary information on the stoichiometric composition of the tissues and cross-sections for nuclear interactions and could reduce uncertainties in the final proton range [4, 12]. Proton radiography is also expected to achieve submillimetric spatial resolution with low imaging dose deposition to the patient, for example when using a filtered-backprojection reconstruction algorithm with estimation of the most likely path of protons [57]. Finally, evaluation of tissue activation or detection of prompt gamma rays produced by nuclear interactions [24, 39, 43] could lead to an in vivo estimation of the position of each individual Bragg peak delivered during the treatment session. Visualization and quantification (offline and/or online) of carbon-11, oxygen-15 isotopes or prompt gamma emission produced in the patient before the beam stops has been made possible by the use of new PET, single- or multiple-slit cameras in or adjacent to the treatment room.

2.2 Segmentation

Structure segmentation constitutes a second step in the treatment planning process. In this task, the tumour and volumes (Planning Target Volume—PTV), organs at risk (OAR) and other structures are defined, for example according to the published recommendations for prescribing, recording and reporting radiation treatments (for example ICRU report 78, addressing proton beam therapy). This task, similar to segmentation in standard 3D conformal radiation therapy, is generally performed with conventional segmentation software and has no specific characteristics in particle therapy, one of the several treatment modalities that will benefit from recent software developments in terms of deformable registration and dose accumulation.

2.3 Dose Calculation Algorithms

The next steps in the treatment planning process consist of definition and optimization of treatment parameters and calculation of the resulting dose distributions. A few definitions, useful for a better understanding of dose calculation algorithms without discussing the details of analytical models, are presented below.

2.3.1 Proton Beam Algorithms

Range-Energy Relationship, Bragg Peak Model

Assuming a continuous slowing down approximation (CDSA), i.e. the range of a particle is given by integrating the total stopping power from 0 to the initial energy, a power law relationship known as the Bragg-Kleeman rule describes the range-energy relationship of the particle. For therapeutic protons in water with E_0 <200 MeV (see Eq. 7), the power p and the factor α have been determined based on ICRU 49 by p = 1.77 and $\alpha = 2.2 \times 10^{-3}$ [5].

$$R_0 = \alpha E_0^p \tag{7}$$

This equation can be used to derive an analytical model of the Bragg curve [5], which includes an empirical model of nuclear fragmentation (primary fluence reduction due to nuclear interactions) based on data fitting, energy spread of polyenergetic beams (Eq. 8), which accounts for range straggling distribution with depth z caused by the statistical fluctuation in the energy loss process (Eq. 9).

$$\hat{D}(z) = \phi_0 \frac{(R_0 - z)^{1/p-1} + (\beta + \gamma\beta p)(R_0 - z)^{1/p}}{\varsigma p \alpha^{1/p}(1 + \beta R_0)} + \varepsilon\phi_0 \frac{(R_0 - z)^{1/p}}{\varsigma \alpha^{1/p} R_0 (1 + \beta R_0)} \quad \text{for } z \leq R_0 - 10\sigma \tag{8}$$

$$D(z) = \frac{1}{\sqrt{2\pi}\sigma} \int_0^{R_0} \hat{D}(z) \times e^{\frac{-(z-\bar{z})^2}{2\sigma^2}} d\bar{z} \qquad \text{for } R_0 - 10\sigma < z < R_0 + 5\sigma \tag{9}$$

where R_0, ϕ_0, σ and ε are nominal range, primary fluence, standard deviation of the Gaussian distribution and fraction of the primary fluence contributing to the 'tail' of the energy spectrum, respectively. This type of model, sometimes modified with a combination of parabolic functions of R_0 [82], can easily be simultaneously fitted to experimental data with sufficient accuracy to allow interpolation from data and to be used in dose calculation algorithms for routine treatment planning.

Ray-Tracing Algorithm

Accurate dose algorithms must include models for the sharp lateral penumbra and the rapid fall-off of the dose of protons or light ion beams. Fast broad-beam versions of such algorithms, giving accurate results for penumbra calculation, were first developed and based on ray-tracing techniques applied to CT images. The dose to a point of interest P with coordinates (X, Y, Z) is obtained from water-equivalent depth calculations (based on CT images information) and linear interpolation of reference depth dose curves measured in a water phantom. However, ray-tracing algorithms usually do not fully take into account straggling effects emerging from complex inhomogeneities, body surface irregularities, lateral spreading of the beam from beam-modifying devices upstream of the patient such as apertures and compensators. To achieve better modelling of the collimator edge effects, a lateral dose profile of a broad beam was therefore defined as the product of a central axis dose and a beam profile function [42]. This lateral dose profile function can model the radial distribution of protons induced by multiple Coulomb scattering through the media along the beam path, particularly scattering of the beamline elements and within the patient. A semi-experimental method was subsequently developed [51] to determine the beamline contribution from lateral penumbra measurements in air, and to take into account complex apertures using a sector integration method for irregular fields based on the Clarkson and Cunningham algorithm.

Pencil Beam Algorithm

As broad beam algorithms do not fully take into account internal inhomogeneities in the patient, the most commonly applied dose calculation algorithm for particle therapy is pencil beam. For example, in most models, the proton pencil beam algorithm factorizes into a depth-dependent term, proportional to the central axis depth dose curve for a broad beam, and a lateral fluence distribution that includes the effect of multiple Coulomb scattering. The lateral fluence distribution is usually decomposed into a large number of small pencil beams, whose parameters are scaled to model the effects of media heterogeneities. In the case of pencil beam scanning (i.e. beam delivery system), the same algorithm is used but modified as a discrete summation.

The central axis depth-dose curve can be described by analytical approximations ([5], Eqs. 8, 9), taken from measurements or from theoretical calculations [74]. The lateral fluence distribution for any pencil beam can, at first approximation, be analytically described by a Gaussian shape with a standard deviation σ or small angles, but deviates from the Gaussian shape for larger angles (an accurate multiple Coulomb scattering theory was developed by Molière in 1948). Empirical formulae and parameterizations based on a theoretical background were then described to predict σ as a function of beam energy and depth in media [19, 32, 33]. An experimental procedure for the determination and verification of the parameters used in a proton pencil beam algorithm has been presented [71], as well as approximations that may be used for the lateral fluence distribution, beam energy modulation and lateral penumbra in the presence of compensators. An improvement of the algorithm for heterogeneous slab geometries and based on an additional 2D scaling of the lateral proton fluence was then described [72], and was further generalized to any heterogeneous geometry and scattering power model [80].

The first component of the lateral fluence distribution is then described analytically, based on a two-dimensional symmetric Gaussian function, given by Eqs. (10) and (11):

$$D_1(X, Y, Z) = D(Z) \times \frac{1}{\left(2\pi\sigma(Z)^2\right)} \exp\left(\frac{-\left[(X - X_0)^2 + (Y - Y_0)^2\right]}{2\sigma(Z)^2}\right) \tag{10}$$

$$\sigma(z)^2 = \sigma_{beamline}(z)^2 + \sigma_{tissue}(z)^2 \tag{11}$$

where (X_0, Y_0) are the coordinates of the pencil beam axis, σ escribes the total lateral spread of the beam, $\sigma_{beamline}$ and σ_{tissue} account for the beamline dependent scattering and multiple Coulomb scattering within the tissue at depth z, respectively.

Dose Algorithms for Scanned Pencil Beams

In high-energy particle beams, nuclear reactions are responsible for removal of primary particles from the incident beam, as well as production of fragment particles, resulting in a tail of the beam lateral dose distribution at more than three standard widths away from the central axis. As shown in Fig. 2, the main reaction products for a proton pencil beam are secondary protons, including alphas and deuterons. The most important dose contributions are those from secondary protons (as much as 10 % of the total dose).

The effects of these large-angle scattered fragments (produced by nuclear interactions) or large angle Coulomb scattering of primary particles on the physical dose distribution are accounted for in treatment planning systems by a sum of Gaussians fitted with measured or simulated data. As proposed by several authors [55, 69] for protons, the transverse dose profile of a scanned particle beam is modelled as the superimposition of at least two Gaussian distributions. In this model (see Eq. 12),

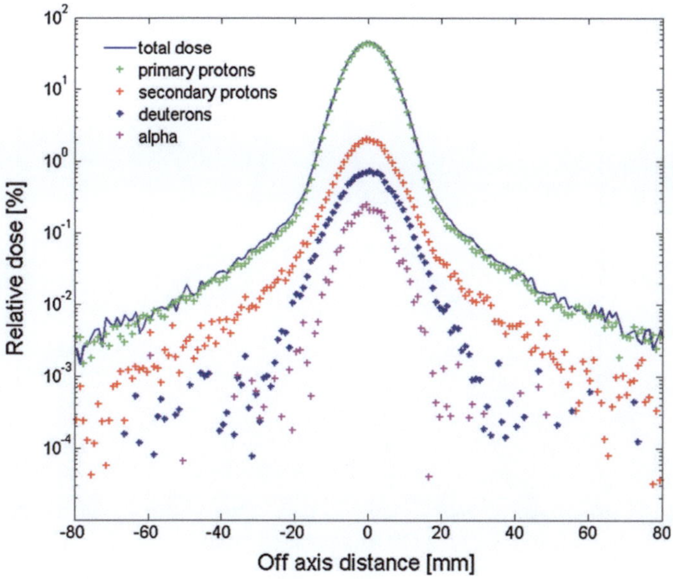

Fig. 2 Lateral profile of a pristine 160 MeV proton beam (sigma in air of 5 mm) in water at mid-range (at depth = 8.5 cm) showing the transverse dose distribution for different secondary components. Data are from GEANT4.9.3/GATE6.2 simulations

the first Gaussian component (G_1) describes the primary particles, whereas the second component (G_2) describes the beam halo from large angle scattered particles. Figure 3 presents an example of comparison between Monte Carlo simulated data and calculations based on a three Gaussians parameterization (from TPS ISOgray, Dosisoft).

As the exact extent of nuclear contribution at large scattering angles is not properly validated by calculation models nor separately accessible by measurements, several analytical approximations or fitting procedures have been described to determine the weight w and functional form of the second component in Eq. (12) for proton beams [69] or ions [34, 68]. Different parameterizations based on accurate measurement for a 177 MeV proton beam and description of the distinct components from a physics point of view have also recently been proposed [20].

$$D(X, Y, Z) \equiv D(Z) \times [(1 - w)\, G_1(X, Y, Z) + wG_2(X, Y, Z)] \qquad (12)$$

Monte Carlo Simulations

However, fast analytical algorithms present several limitations in terms of accuracy of calculation and more accurate dose calculation codes are needed, for example in very heterogeneous geometries where the effects of diffusion might be excessively smoothed and hot or cold spots may be underestimated. Monte Carlo dose calculations are considered to be the most accurate method to compute doses in radi-

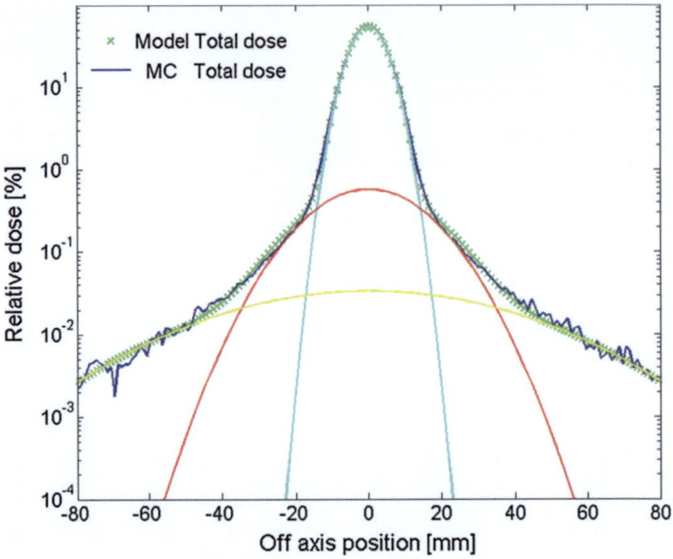

Fig. 3 Lateral profile of a pristine 180 MeV proton beam (sigma in air of 4 mm) in water at mid-range (at depth = 10 cm) showing the transverse dose distribution: simulated data from GEANT4.9.3/GATE6.2 (*solid line*), model with a superimposition of three Gaussians (*crosses*). The contributions of the different terms in the fit are also shown

ation therapy, as Monte Carlo simulations take into account the physics of particle interactions on a particle-by-particle basis using theoretical models or experimental cross-section data for electromagnetic as well as nuclear interactions. Monte Carlo dose calculations also consider tissue inhomogeneities by using material properties, atomic elemental composition, electron density, mass density or ionization potential, and secondary particle tracking. The main advantages of these codes in radiation therapy are that they can be used as references for validation purposes, for simulations of the components of the treatment head and to extract parameterized phase spaces for complex beam delivery systems. They can also validate or be used for the commissioning of beam delivery systems, and the quality assurance of clinical beam delivery. Therefore, although the main well-known Monte Carlo codes (MCNPX, GEANT4, and FLUKA) were initially designed for simulations in particle and nuclear physics, all of them have been successfully used in the field of particle therapy, as illustrated by the following examples.

In the framework of research studies, the Centre Antoine Lacassagne installation has been modelled in MCNPX [30, 31] to provide absolute dosimetry and independent monitor unit calculations for ophthalmological proton beam therapy. Also with MCNPX, a partnership between several French teams (Institut Curie, CEA/IRFU, IRSN) has led to the modelling of all Institut Curie passive beam lines [61, 70], and extensive comparisons of proton and neutron dose calculations versus experimental measurements have been performed. The MCS algorithm of MCNPX was modified

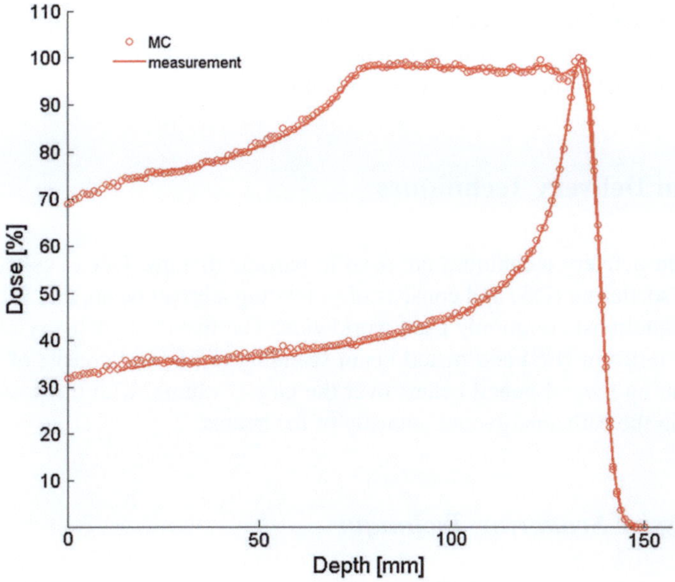

Fig. 4 Comparison of central axis depth dose in water (120 mm diameter field, 140 MeV SOBP and pristine Bragg peak). Data are from CC13 ionization chamber measurements (*solid line*) and GEANT4/GATE MC simulation (*circle*)

in order to improve the modelling of multiple Coulomb scattering in the case of thin foils [70].

GATE (Geant4 Application for Emission Tomography) is an advanced and versatile open source software, which contains tools dedicated to radiotherapy applications [37, 38, 60]. A Monte Carlo simulation of the IBA active scanning system with this platform, a reference physics list and a list of optimized parameters have been proposed for proton therapy [21–23]. An example of calculation using this platform is shown in Fig. 4 for a spread-out Bragg peak (SOBP) of a passive beam line.

In the framework of nuclear imaging for hadron therapy monitoring, nuclear models implemented in GEANT4/GATE and FLUKA were compared for monoenergetic protons and carbon ions, showing discrepancies between the two codes in terms of the spatial and time distributions of secondary particles [58].

Until very recently, Monte Carlo dose calculation, including treatment head simulation and dose calculation for passive scattering or beam scanning delivery were not commercially available. The main challenges for fast and reliable MC codes are to decrease the computer calculation time, while maintaining reliability, and to provide users with tools for automatic conversion of HU values, DICOM RT-ION and 3D dose or phase space outputs. Consequently in collaboration with several French teams (CEA/LIST, Institut Curie, DOSisoft, Centre Antoine Lacassagne, CEA/IRFU, INSA Lyon), a code (called PROUESSE) based on the Monte Carlo code PENELOPE [59] has been developed for proton-induced dose calculations and is currently in the process of validation. Furthermore, parallel architecture, for example based on graphical processing units (GPU), is now very frequently addressed in new versions of Monte Carlo codes [3] or dose calculation engines in TPS, opening the way for four-dimensional or real-time treatment planning.

3 Beam Delivery Techniques

Three main delivery techniques are used in particle therapy. One is called passive or double scattering (DS) and consists of producing a broad beam, and is the delivery technique most commonly used worldwide. The two other delivery techniques are called uniform (US) and pencil beam scanning (PBS) and consist of magnetically scanning several pencil beams over the target volume, with the possibility of modulating the path, energy and intensity of the beams.

3.1 Passive Scattering Technique

To achieve the necessary uniformity of the radiation field required for clinical treatments (within a few percent), many methods for lateral spreading of particle beams have been investigated. The passive systems, composed of single or double scatterers, are one of the main techniques used to achieve conformal beams [50]. In a double scattering beam line, the dose is delivered with good conformity to the lateral and distal edges, but not to the proximal edge of the tumour. For this purpose, the range of different Bragg Peaks is adjusted with a range-shifter, while a second scatterer laterally enlarges the beam size. A dynamic wheel (e.g. aluminium propeller), that can be synchronized to the beam source in each case, is used to adjust the modulation value (a characteristic spread-out Bragg peak is shown in Fig. 4). Sets of collimators along the beam line limit secondaries, and divergent brass collimators, personalized to each patient, are used to obtain a smaller lateral penumbra at the end of the beamline. A range-compensator is used to achieve accurate distal target dose conformation. For example, at Institut Curie, compensators are made of Lucite blocks, manufactured with a computer-controlled drilling machine according to the target volume and each individual patient's anatomical data. Patients are usually preferably treated in the horizontal position, immobilized on a patient table, but sometimes in supine, prone or seated positions. To irradiate patients with all degrees of freedom, patient supports are usually mounted on robotic arms, which allow rotations and translations with few angular restrictions [47].

3.2 Uniform Scanning

The Uniform Scanning systems (for example those of IBA for proton beams) can deliver uniform transverse dose distributions up to 40x30 cm^2 treatment area. The beam is scanned continuously according to a predefined scanning pattern. Two perpendicular dipole magnets scan a large spot along a fixed pattern at a constant frequency. The optimal beam spot size is determined by the system to achieve the required field size and uniformity for all layers. The scanning amplitude is also related to the field size and beam diameter: the scanning area is defined as the uncollimated proton field projected at the isocentre plane. Patient-specific apertures and range compensators are used to laterally and distally shape the radiation field appropriate for treatment. Minor, if any, modifications of the treatment planning system dose calculation algorithms are assumed from double scattering to uniform scanning. Indeed, the main differences between US and DS concern the transverse characteristics and absolute values of lateral penumbras, usually customizable in the TPS beam data library. The uniform scanning method also has a few advantages over passive beam delivery systems, mainly due to a smaller thickness of scattering material, as the transverse and distal penumbra widths show a slight improvement in comparison to those achieved with scattered beams for an equivalent field size. The maximum field size, the maximal range in water as well as the modulation width of a spread-out Bragg peak (SOBP) can also be slightly increased at a given accelerator energy.

3.3 Pencil Beam Scanning

With intensity-modulated particle therapy (IMPT) techniques, a narrow pencil beam is scanned magnetically over the target volume, while both the energy and the intensity of the beam are modulated: pristine Bragg peaks of a few millimetres are obtained and can be scanned transversally at average speed of a few m/s. Between two spot irradiations, the beam is usually turned off during magnet and energy modifications. Just as IMRT with photons led to vast improvements in conventional radiotherapy, simulations show that IMPT can provide significant improvements and much more conformal dose distributions. For IMPT, a discrete fluence map is optimized and converted into a set of continuous time-functions describing the beam position and current required for continuous beam scanning.

Two methods can be used for intensity-modulated proton therapy: the first is called single field uniform dose (SFUD) and the second one is called intensity-modulated proton therapy (IMPT). In SFUD, the extent of the spread-out Bragg peak is matched to the thickness of the target volume on each pencil beam axis. The modulation of the pencil beams is limited in the transverse plane in order to minimize the proximal dose to the target volume and to maintain a uniform dose distribution in the tumour. With the IMPT technique, Bragg peaks are distributed in three dimensions throughout the volume with free optimization of the intensity of each individual Bragg peak. As shown in Fig. 5, the definition of spot geometry and assignment of initial weights depend on the optimization procedure and must take into account accurate beam modelling to ensure correct coverage of the target.

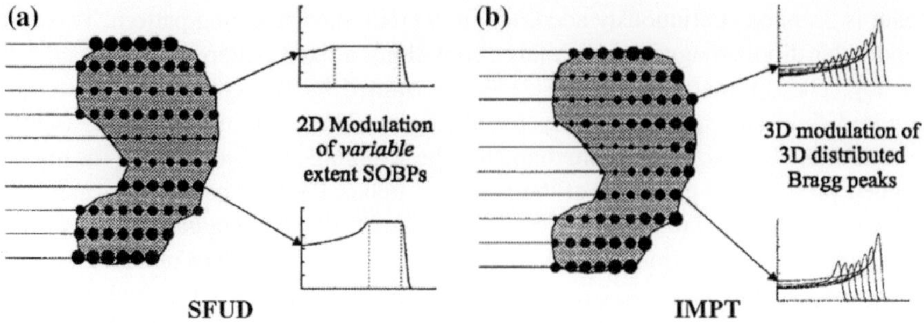

Fig. 5 Diagrams and representative depth-dose curves for **a** SFUD and **b** IMPT approaches to pencil beam scanning proton therapy (from [44])

4 Treatment Plan Optimization

Many special issues in treatment planning optimization for particle therapy are currently being addressed: multi-criteria optimization, sensitivity to the delivery uncertainties for inhomogeneous fields, radiobiological effects that can be included in pencil-beam based inverse optimization (through linear energy transfer (LET) calculations). The degeneracy of solutions in treatment planning can be used to incorporate management of uncertainties, especially in the optimization process. However, optimization techniques differ between the main beam delivery methods (i.e. passively scattered and scanned beams). With the double-scattering technique, optimization is based on manual operations and a set of practices and tools. For example, some of the following parameters can be adjusted: the compensator can be enlarged to guarantee target coverage even in the presence of small misalignments, narrow target extensions, or internal organ motion, the beam angles can be optimized to minimize or avoid consequences from high-gradient changes in density (bone-air) and reduce integral dose to healthy tissues, beams that pass through metal implants can be avoided because of range uncertainties associated with artefact in images especially near critical structures, and field overlapping on skin is usually also minimized. Treatment angles are also selected manually (only a few beams are usually used in pencil beam scanning or ion therapy, as dose conformation is excellent), sometimes with the help of accurate 3D visualizations available in TPS. Intensity-modulated particle therapy, made possible by pencil beam scanning, is based on inverse planning and can lead to significant improvements in particle therapy [44]. A discrete fluence map is optimized by the treatment planning system, and is converted into a set of beam weights. The optimization problem is based on the minimization of the cost function that expresses the difference between the planned dose D_0 and the calculated dose D_j (Eq. 13) for the set of beam weights w (Eq. 14). Iterative algorithms are used to find the solution of this problem, which reflects the physician's requirements by the use of dose-volume constraints and objectives.

There is a linear relationship between the dose D_j to the point of calculation *j* and the beam weights w_i:

$$D_j(x,y,z) = \sum_i w_i \times D_{ij}(x,y,z) \tag{13}$$

where D_{ij} is the dose contribution of pencil beam *i* to the point of calculation *j*.

$$w = [w_i] = \min_w \left\{ \sum_{j \in structure} g_{j^2} \times \left(D_j(x,y,z) - D_0\right)^2 \right\} \tag{14}$$

where g_j is the importance factor to the point of calculation *j*.

The problem in Eq. (14) can be reduced to solving a quadratic equation. Boundary conditions have to be added to take into account the non-negativity or minimum constraints for beam weights (for example, depending on the machine, there is a minimum value for monitor units and beam-on time), and possibly constraints on calculated doses for certain points. The projected conjugate gradient with penalties is a relevant algorithm for such problems and is implemented in some TPS (ISOgray, Dosisoft).

4.1 Robustness

Range uncertainties constitute the most significant risk in particle therapy, and may be initially managed by applying margins that expand the distal target volume (the current standard is to use linear scaling of stopping power and apply a range-uncertainty margin of ≈3 % + 1 mm). However, this approach does not guarantee robustness of the treatment plan, as uncertainties may combine and distort the dose in a non-linear way.

Table 1 Indicative overview of strengths (+) and weaknesses (−) of the various beam delivery techniques, as these parameters may vary according to acceleration type and optimization algorithms

	DS	US	SFUD	3D-IMPT
Optimization	No	No	Single-field	Multi-field
Homogeneity	Yes	Yes	Yes	no
Integrated boost	No	No	No	yes
Robustness (setup errors)	−	−	+	+
Robustness (motion errors)	+	+	−	−
Target size	−	+	+	+
Treatment time	++	+	−	-
Neutron contamination	−	+	++	++
Conformity	−	+	+	++

Various other approaches to ensure robust IMPT have been investigated by different teams worldwide: the "worst case scenario" [45] selects the best plan, while considering minimum and maximum doses in the target and healthy volumes for a limited number of range errors and shifted positioning scenarios. This strategy was subsequently refined assuming prior knowledge of the probability distribution of the uncertainty [75]. A multi-criteria optimization (MCO) framework has also been introduced to investigate the trade-offs between conformity and robustness when selecting clinically achievable plans [9] (Table 1).

5 Clinical Aspects and Status

The clinical experience acquired worldwide in hadron therapy at the end of 2014 exceeds 118,000 patients and 15,000 patients for proton therapy and carbon therapy, respectively [41]. For example, current indications for protons and carbon ions include unresected or incompletely resected locally aggressive tumours located close to healthy critical structures, while paediatric malignancies that require improved tolerability of radiation are considered to be a priority for proton therapy. Detailed information can be found in recent articles and reviews [26].

5.1 Particle Type Selection

Although selection of the optimal type of ion remains a relevant topic of discussion, new facilities are now offering proton and carbon ions (for example, a manufacturer such as IBA, is developing hybrid systems), usually close to the photon therapy facility, raising the question of the preference of one type of particle over another. On the one hand, limited data are available concerning the toxicity of particle-based treatments and there is insufficient evidence to recommend the use of particles in many disease sites, making further clinical research programmes essential. On the other hand, reduction of the integral dose and increased RBE in the distal part of the physical dose distribution makes ion therapy highly attractive in challenging or radioresistant tumours. For example, in the case of skull base tumours, the overall conformality of combined photon-proton treatments has been discussed [18] to assess the clinical benefit of exclusive particle therapy. In the case of spinal locations, 3D conformal, IMRT, VMAT, tomotherapy and proton therapy have been compared in terms of dose escalation possibilities [77]. As discussed in Habrand 2009, the major advantage of proton therapy might concern paediatrics due to the potential for sparing normal tissues, but more clinical data and careful assessment of long-term side effects in children are needed. Prospective studies will inevitably be conducted, in parallel with optimization of dose delivery over time (dose escalations, hypofractionation), integration of particles into multimodal treatment and their implications on treatment planning.

5.2 *Matching and Patching Fields*

To improve conformality and reduce the dose to organs at risk, a dose delivery technique called "patching", specific to passive scattering in particle therapy, can be used. A multiple beam ballistic (usually two or three beams) is used to partially cover a complex shaped target volume while avoiding critical structures. In two-field patching, the beams are combined so that the distal edge of the first patch field is designed to stop on the lateral penumbra of the second through field (two typical examples are shown in Fig. 6). The distal fall-off of the first field and the lateral penumbra of the other field are matched to the 50 hot spots. However, the dose distribution along the patch junction is generally non-uniform because of tissue heterogeneities, heterogeneous dose gradients and a lack of analytical tools in treatment planning systems to take the scattering effects into account in the design of compensators. Therefore, patching and matching are always located within the target volumes, sometimes with small overshoots, as hot spots may not be detrimental if they are located within the target areas. To reduce the uncertainties, several patching fields are usually combined in order to create different shifted junctions that can be alternated every day and to minimize the dose delivered by each field.

5.3 *Plan Review*

Treatment planning review and assessment of doses distributions in proton or ion therapy use the same tools as in 3D conformal radiation therapy. For instance, many important parameters of an organ at risk or target volume dose distribution can

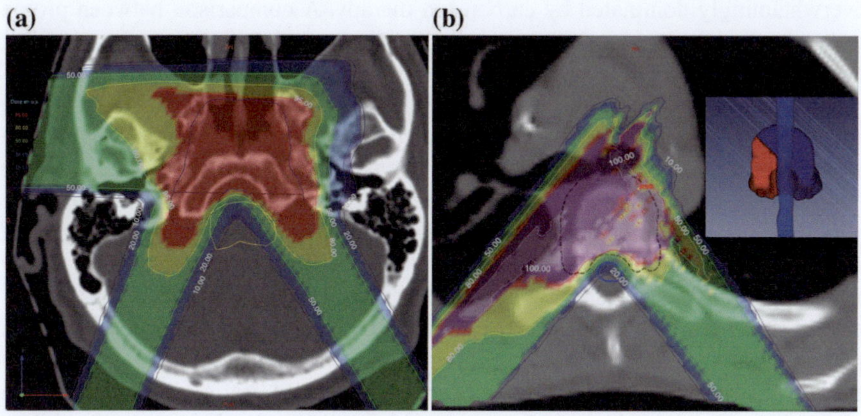

Fig. 6 Axial CT images with dose display in percent for typical chordoma cases, with a combination of patch/through fields: **a** three fields which irradiate anterior and posterior portions of the target while avoiding brainstem, **b** two fields avoiding spinal cord

be evaluated by inspection of the cumulative dose-volume histogram (DVH). The probability of achieving tumour control (called tumour control probability TCP), or complications for healthy structures (called normal tissue complication probability NTCP) can be used to describe the dose–response relationships for normal tissues or targets, using sigmoid-shaped formulas. In particular, generalized equivalent uniform dose (gEUD) can be used as an input parameter for such probability models (Eq. 15). The gEUD, which measures the physical dose for non-uniformly irradiated structures, while taking into account a biological dose response, is given by:

$$gEUD = \left(\sum_i \left(v_i D_i^a \right) \right)^{\frac{1}{a}} \tag{15}$$

where v_i is the volume fraction of the dose bin corresponding to the dose D_i and a characterizes the dose response behaviour of the OAR.

gEUD, with a value expressed in Gray, provides a simple method to compare treatment plans from different optimization results, can be a better prognostic predictor for late effects than the mean or maximum doses usually considered and can be easily used to formulate objective functions for IMRT or IMPT optimization. For example, this concept has been applied to determine optimal parameters of a gEUD-based NTCP for a group of patients with long-term follow-up data after skull base proton therapy [12].

6 Protons and Ions: Similarities and Comparisons

Clinical experience with the various types of ions is limited and is now becoming overwhelmingly dominated by carbon ion therapy. A comparison between proton and carbon ion therapy has been proposed [78] that includes many aspects of therapy in practice. A few aspects are highlighted here.

For example, proton beams have a lower RBE (1.1) than carbon ions (1.5–4): the uncertainties in the RBE and actual dose calculations are smaller for protons, but the increased RBE in the Bragg peak of carbon ions may increase TCP for challenging tumours. The angular spread of charged particles and the lateral penumbras increase in both beam line and patient. However, the angular spread for heavy ions is much smaller than for protons or X-rays (particularly for high beam energies for which scattering can even be neglected in some dose calculations) and constitutes a particular advantage of heavy ions compared to other radiations. In addition, as multiple scattering in air is an important contributor to lateral penumbra, especially at low energies, the use of patient apertures (even with scanned beams) and minimization of the air-gap between the aperture (or if necessary the range-shifter) and the patient may be required to achieve best penumbras. The dose of secondary lighter fragments created by nuclear fragmentation in the case of heavy ions produces a tail (which does not exist for protons) in the depth dose curve at large depths: this nuclear fragmentation must be modelled to obtain a sufficiently accurate model of RBE and dose distribution for spread-out Bragg peaks. Range straggling with depth is markedly reduced for carbon ions compared to protons: the distal fall-off is much smaller for carbons, thus potentially requiring the use of a ridge-filter to improve the flatness in the spread-out Bragg peak, taking into account the biologically effective dose.

7 Radioprotection, Neutron Contamination

The high energies of the primary beam (from several dozen to several hundred MeV) commonly used in particle therapy are able to generate secondary neutrons through nuclear reactions within the different beamline components and the patient himself. In addition, the TPS used in routine clinical practice do not take into account secondary neutrons when calculating doses from primary particles and do not consider the elevated biological effectiveness of secondary and scattered neutrons from the treatment head.

Dosimetric studies on neutron doses have shown that particle therapy, especially with scanning techniques, could possibly lead to a further reduction in second malignancies because of the following factors: higher conformality because of the much lower entrance dose and no exit dose, less neutron scatter and, in many cases, fewer beams are needed to achieve good dose conformation. However, these findings may vary according to the beam delivery technique. As reported by Zheng et al. [83], a similar behaviour of neutron dose equivalent dependence on patient-specific beam parameters is expected between passive scattering and uniform scanning proton delivery systems. The neutron dose equivalent per absorbed dose is also expected to be slightly lower for uniform scanning beams than for passive scattering beams. The contribution of secondary neutron is expected to be the lowest for pencil beam scanning techniques, due to the absence of scattering devices in the treatment nozzle, patient apertures or compensators [16]. In reality, these findings could be mitigated by the possibility of using a range-shifter and the recent reintroduction of shaping devices to improve lateral penumbra. After having precisely characterized the secondary neutron doses received by paediatric patients treated for intracranial tumours [62], some authors are now developing facility-specific analytical models that could replace time-consuming MC calculations to assess accurate doses to healthy organs [17]. Furthermore, comparisons of the neutron contamination between photon, proton and ion therapy (for passive and scanning techniques) have shown that the out-of-field dose from secondary neutrons was lowest for ions (protons followed by heavier ions) delivered by scanning, followed by passive modulation, and finally by high-energy IMRT photons [73].

8 Biological Modelling

The biological effect of a given dose distribution differs between protons and other light ions. A constant relative biological effectiveness (RBE) equal to 1.1 is usually assumed for protons, while the RBE for other light ions varies substantially and must be estimated by mathematical models. Consequently, the radiobiological response of cells to particle deliveries has been extensively studied and its dependence on dose/fraction, position in the irradiated volume, beam energy and tissue has been characterized. The fraction of surviving cells as a function of dose is usually described by the linear-quadratic model (Eq. 16), where α and β are the parameters of the model, associated with initial slope and curvature for a single dose D. As experimental observations indicate a linear trend at high doses, the model has been modified and extrapolated by a straight line at doses higher than a threshold D_t [2].

$$S = \begin{cases} \exp\left(-\alpha D - \beta D^2\right) & \text{for } D < D_t \\ \exp\left(-\alpha D_t - \beta D_t^2 - (\alpha + 2\beta D_t) \cdot (D - D_t)\right) & \text{for } D \geq D_t \end{cases} \tag{16}$$

The RBE is defined as the ratio of the doses required by two radiations to cause the same level of effect. At a certain particle dose D_P, the RBE can then be expressed (Eq. 17) as a function of the parameters α_X, β_X and α_P, β_P (X refers to the reference radiation while P refers to the ion radiation).

$$RBE(\alpha_X, \beta_X, \alpha_P, \beta_P, D_P) = \frac{\sqrt{\alpha_X^2 + 4\beta_X D_P(\alpha_P + \beta_P D_P)} - \alpha_X}{2\beta_X D_P} \tag{17}$$

8.1 Microdosimetry

Microdosimetry can be defined as the estimation of the energy probability distribution imparted in an irradiated volume of matter, whereas dosimetry is the estimation of the mean energy imparted at one point of an irradiated volume of matter. In particle therapy, the spatial distribution of energy depositions (linear energy transfer analogous) and the amount of energy deposited in a volume (dose analogous) influence the relative biological effectiveness of radiation and must be defined experimentally or theoretically. Depending on the application area (microscopic or macroscopic), different physical quantities (stochastic or mean values) are then essential to define the biophysical models of radiation effects. For example, the LET concept, based on a mean value, has been developed as an approximation of the energy transfer by charged particles. The lineal energy y and the (frequency or dose) mean lineal energy, being stochastic values, are microdosimetric quantities that are commonly used to estimate biological effectiveness on scales similar to a mammalian cell nucleus (see section *RBE modelling for ions*). As a detailed review of the radiobiological modelling based on these concepts is beyond the scope of this chapter, only a few definitions and a brief summary of some recent improvements in the main models used in particle therapy will be presented in the next sections.

8.2 *RBE Modelling for Protons*

The use of a generic RBE of 1.1 at 2 Gy in clinical proton therapy appears to be reasonable in view of the lack of experimental data to define accurate RBE models and the lack of clear clinical evidence for RBE variations [54]. Experimental in vivo and clinical data have also shown that this generic RBE value seems to be appropriate. However, many authors have reported variations of the RBE with depth in the spread-out Bragg peak proton beams used for treatment [6, 11], as well as significant differences in the final range and linear energy transfer, which can modulate biological effectiveness [53]. Indeed, wide-angle proton-proton scattering, nuclear interactions, and internally or externally produced neutrons are part of the physical processes involved in proton therapy. Some of these physical processes are not completely modelled in dose calculations by treatment planning systems, mostly based on measured data and water equivalent approximations. To account for this variable RBE in treatment planning, a few RBE models have been specifically developed for proton beams.

8.2.1 Definitions

Based on the observations that the maximum RBE for proton beams is observed at LET values around 30 keV/μm and that high LET values are of little practical relevance in clinical proton therapy, most biophysical models for the prediction of the RBE have been developed by assuming a linear relationship between RBE and LET. A few common definitions are provided below.

The dose-averaged LET (LET_d) distributions can easily be obtained from Monte Carlo simulations (Eq. 18) as a function of local particle spectrum and stopping power S taking into account primary particles, or from analytical calculations (Eq. 19) as a function of the mean stopping power:

$$LET_d(z) = \frac{\int_0^\infty \varphi_E(z) S^2(E) dE}{\int_0^\infty \varphi_E(z) S(E) dE} \quad (18)$$

where φ_E is the local particle spectrum at depth z with energies ranging between E et $E + dE$.

$$LET_d(z) = \frac{\langle S(z)^2 \rangle}{\langle S(z) \rangle} \quad (19)$$

where $\langle S(z) \rangle$ is the mean stopping power at depth z.

The absorbed dose D (or energy imparted per unit mass) can be expressed as a function of the beam fluency and LET (Eq. 20):

$$D = 1.6 \times 10^{-5} \times \Phi \times \frac{LET}{\rho} \quad (20)$$

with D in Gy, σn cm^{-2}, LET in keV/μm, and ρ the mass density in g/cm^3.

For example, assuming a 0.5 keV/μm LET value for a 200 MeV proton beam, the fluency to deposit 1 Gy in water would be 1.25 Gp/cm^2, (i.e. 12 $p/\mu m^2$). This means that a cell nucleus (≈5 μm diameter) would be crossed by 235 particles, each 0.25 μm apart. For alpha particles or gamma rays, this value would vary between a few tracks and a thousand tracks, respectively.

8.2.2 LET-RBE Models

The RBE can then be obtained by several parameterization as a function of LET, dose and tissue-specific parameters α/β, s described below:

- For one selected type of cells and a LET less than several dozen keV/μm (Eq. 21, [81], Eq. 22, [10]), the following two approximations have been proposed:

$$\alpha_P(LET) = \alpha_0 + \lambda LET \quad (21)$$

or

$$\alpha_P(LET) = \alpha_0 + \frac{1 - e^{-\lambda_1 . LET^2}}{\lambda_2} \quad (22)$$

(α_0 and λ_i have to be fitted to the experimental data for the tissue considered). To ensure equality between low LET protons (≈0.5 keV/μm) and photons, as frequently observed experimentally, the following assumption can be adopted (Eq. 23):

$$\alpha_0 = \alpha_X - 0.5.\lambda \quad (23)$$

- Fits to a dozen cell lines result in the following two parameterizations of the previous linear relationship (Eq. 24, [79], Eq. 25, [8]):

$$\alpha_P(LET) = \alpha_X(1 + 0.434 \times \frac{\beta_X}{\alpha_X} \times LET) \quad (24)$$

or

$$\alpha_P(LET) = \alpha_X \left(0.843 + 0.154 \times 2.686 \times \frac{\beta_X}{\alpha_X} \times LET\right) \quad (25)$$

- In most of the literature, the quadratic parameter β is assumed to be constant ($\beta_P = \beta_X$) and LET-independent. However, several authors have shown experimentally that the β values for some cell lines appear to be LET-dependent. Therefore, one model (Eq. 26) explicitly assumes a dependency of β on LETand proposes a fit to experimental data (V79 Chinese hamster cells with mean $(\alpha/\beta)_X$ = 2.686 Gy) of the behaviour of $\surd\,(\beta_P/\beta_X)$ as a function of LET [7, 8]:

$$\beta_P(LET) = \beta_X \left(1.09 + 0.006 \times 2.686 \times \frac{\beta_X}{\alpha_X} \times LET\right)^2 \tag{26}$$

An example of calculation of the RBE-weighted dose with the "Wedenberg" parameterization for a spread-out Bragg peak is given in Fig. 7.

8.3 *RBE Modelling for Ions*

The complex dependencies of the RBE for ions (with depth in tissue, applied dose, cell type, biological endpoint, particle type) have led to the development of several models to estimate RBE. Two main radiobiological models have been proposed, based on different physical and biological models already integrated into existing TPS (not yet used in France except for research purposes): the first one is used in

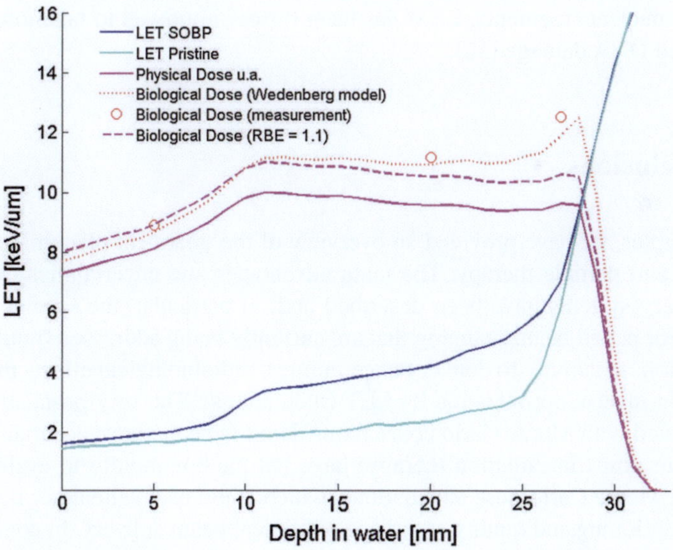

Fig. 7 Comparison of simulated LET_d, measured (*circle*) and calculated (*dotted* and *dashed lines*) RBE-weighted dose (for D37 endpoint, HeLa cells) of 58 MeV proton beams

Japan (HIMAC, NIRS) and is based on the microdosimetric kinetic model (MKM) developed by Hawkins [27–29], whereas the second model is used in Germany (GSI) and is based on the local effect model (LEM), now in its 5th version [67].

- The MKM combines assumptions from microdosimetry with kinetic relations for lethal lesions and sublethal lesions that are not repaired. Cell survival is then correlated with stochastic dose deposits in the volume of a small sensitive site (called domain with diameter $d < 1\,\mu m$) within the cell nucleus.

$$\alpha_P(E) = \alpha_0 + \beta_X \times \frac{4\bar{y}_d(E)}{\pi d^2} \tag{27}$$

  In Eq. 27, the term $\frac{4\bar{y}_d(E)}{\pi d^2}$ stands for an approximation of the mean specific energy deposited by a single event in a spherical domain composed of water, and is calculated from the mean lineal energy y_d. For high LET (above 100 keV/μm), the mean lineal energy is usually corrected for a saturation effect of the RBE. In practice, and according to the *Wilkens* model (Eq. 21), α_0 and β_x are independent of the radiation quality and equal to the cell-line parameters in the limit of zero LET.
- The LEM relates the response of biological systems after ion irradiation to the response after X-ray irradiation, and uses a parameterized dose-response curve derived from experimental photon data. It assumes that the biological effect of irradiation is determined by the spatial dose distribution inside the volume of a small sensitive site within the cell nucleus (nm scale as compared to the μm scale of MKM). By doing so, it also takes into account the biological effectiveness of the various nuclear fragments. LEM has been further improved to take into account clustered DNA damage [15].

9 Conclusions

In this chapter, we have provided an overview of the treatment planning procedure in proton and particle therapy. The main advantages and uncertainties of the current delivery systems have been described and, in particular, the special issues in planning for pencil beam scanning that are currently being addressed (multi-criteria optimization, sensitivity to delivery uncertainties, radiobiological effects that can be included in inverse optimization by LET calculations). The very promising results demonstrated with Monte Carlo codes (considered to be the most accurate methods to compute doses in radiation therapy) have led the community to evaluate fully integrated Monte Carlo dose calculations, which could also validate or be used for the commissioning and quality assurance of clinical beam delivery. In contrast with photons, protons and ions present the advantage of stopping at a given depth (at the site of the tumour) with reduced straggling, implying a significant reduction of integral dose to surrounding tissues. However, range uncertainties still represent one of the major limitations to clinical application of the full potential of hadron therapy, and accurate modelling of relative biological effectiveness also remains highly challenging.

References

1. Andreo P (2009) On the clinical spatial resolution achievable with protons and heavier charged particle radiotherapy beams. Phys Med Biol 54(11):N205–N215
2. Astrahan M (2008) Some implications of linear-quadratic-linear radiation dose-response with regard to hypofractionation. Med Phys 35(9):4161–4173
3. Bert J et al (2013) Geant4-based Monte Carlo simulations on GPU for medical applications. Phys Med Biol 21;58(16):5593–5611
4. Bopp C et al (2015) Quantitative proton imaging from multiple physics processes: a proof of concept. Phys Med Biol 60:5325–5341
5. Bortfeld T (1997) An analytical approximation of the Bragg curve for therapeutic proton beams. Med Phys 24:2024–2033
6. Calugaru V et al (2011) Radiobiological characterization of two therapeutic proton beams with different initial energy spectra used at the Institut Curie Proton Therapy Center in Orsay. Int J Radiat Oncol Biol Phys 81(4):1136–1143
7. Carabe A et al (2007) The incorporation of the concept of minimum RBE (RBEmin) into the linear-quadratic model and the potential for improved radiobiological analysis of high-LET treatments. Int J Radiat Biol 83(1):27–39
8. Carabe A et al (2012) Range uncertainty in proton therapy due to variable biological effectiveness. Phys Med Biol 57:1159–1172
9. Chen W et al (2012) Including robustness in multi-criteria optimization for intensity-modulated proton therapy. Phys Med Biol 57:591–608
10. Chen Y et al (2012) Empirical model estimation of relative biological effectiveness for proton beam therapy. Radiat Protect Dosim 149(2):116–123
11. Courdi A et al (1994) The depth-dependent radiation response of human melanoma cells exposed to 65 MeV protons. Br J Radiol 67:800–804
12. Dedes G et al (2015) Monte Carlo study of the precision and accuracy of proton CT reconstructed relative stopping power maps. Med Phys 42:3298
13. De Marzi L et al (2013) Calibration of CT Hounsfield units for proton therapy treatment planning: use of kilovoltage and megavoltage images and comparison of parameterized methods. Phys Med Biol 58:4255–4276
14. De Marzi L et al (2015) Use of gEUD for predicting ear and pituitary gland damage following proton and photon radiation therapy. Br J Radiol 88:20140413
15. Elsässer T et al (2007) Cluster effects within the local effect model. Radiat Res 167:319–329
16. Farah J et al (2015) Measurement of stray radiation within a scanning proton therapy facility: EURADOS WG9 intercomparison exercise of active dosimetry systems. Med Phys 42(5):2572–84
17. Farah J et al (2015) Configuration and validation of an analytical model predicting secondary neutron radiation in proton therapy using Monte Carlo simulations and experimental measurements. Phys Med 31(3):248–256
18. Feuvret L et al (2007) A treatment planning comparison of combined photon-proton beams versus proton beams-only for the treatment of skull base tumors. Int J Radiat Oncol Biol Phys 69:944–954
19. Gottschalk B et al (1993) Multiple Coulomb scattering of 160 MeV protons. Nucl Instrum Methods Phys Res B 74:467–490
20. Gottschalk B et al (2015) On the nuclear halo of a proton pencil beam stopping in water. Phys Med Biol 60(2015):5627–5654

21. Grevillot L et al (2010) Optimization of GEANT4 settings for proton pencil beam scanning simulations using GATE. Nucl Instr Meth Phys Res B 268(20) 3295-3305
22. Grevillot L et al (2011) A Monte Carlo pencil beam scanning model for proton treatment plan simulation using GATE/GEANT4. Phys Med Biol 56:5203–5219
23. Grevillot L et al (2012) GATE as a GEANT4-based Monte Carlo platform for the evaluation of proton pencil beam scanning treatment plans. Phys Med Biol 57:4223–4244
24. Gueth P et al (2013) Machine learning-based patient specific prompt-gamma dose monitoring in proton therapy. Phys Med Biol 58(13):4563–4577
25. Habrand JL et al (2009) Proton beam therapy in pediatric radiotherapy. Cancer Radiother 13:550–5
26. Habrand JL et al (2013) Update of clinical programs using hadrontherapy 2008–2012. Cancer Radiother 17(5-6):400–406
27. Hawkins RB (1994) A statistical theory of cell killing by radiation of varying linear energy transfer. Radiat Res 140(3):366–374
28. Hawkins RB (1996) A microdosimetric-kinetic model of cell death from exposure to ionizing radiation of any LET, with experimental and clinical applications Int. J. Radiat. Biol. 69(6):739–755
29. Hawkins RB (2003) A microdosimetric-kinetic model for the effect of non-poisson distribution of lethal lesions on the variation of RBE with LET. Radiat Res 60(1):61–69
30. Hérault J et al (2005) Monte Carlo simulation of a protontherapy platform devoted to ocular melanoma. Med Phys 32(4):910–919
31. Hérault J et al (2007) Spread-out Bragg peak and monitor units calculation with the Monte Carlo code MCNPX. Med Phys 34(2):680–8
32. Highland VL (1975) Some practical remarks on multiple scattering. Nucl Instrum Methods 129:497–499
33. Hong L et al (1996) A pencil beam algorithm for proton dose calculations. Phys Med Biol 41:1305–1330
34. Inaniwa T (2014) Implementation of a triple Gaussian beam model with subdivision and redefinition against density heterogeneities in treatment planning for scanned carbon-ion radiotherapy. Phys Med Biol 59(18):5361–5386
35. Journal of the ICRU 2007 Vol. 7 No 2 Report 78
36. Jackson DF et al (1981) X-ray coefficients of elements and mixture. Phys Rep 70:169–223
37. Jan S et al (2004) GATE: a simulation toolkit for PET and SPECT. Phys Med Biol 49:4543–4561
38. Jan S et al (2011) GATE V6: a major enhancement of the GATE simulation platform enabling modelling of CT and radiotherapy. Phys Med Biol 56:881–901
39. Jan S et al (2013) GATE simulation of 12C hadrontherapy treatment combined with a PET imaging system for dose monitoring: a feasibility study. IEEE Trans Nucl Sci 60(1 part2):423–429
40. Jiang HY et al (2007) Effects of Hounsfield number conversion on CT based proton Monte Carlo dose calculations. Med Phys 34(4):1439–1449
41. Jermann M (2015) Particle therapy statistics in 2014. Int J Particle Ther 2(1):50–54
42. Lee M (1993) An empirical method to build up a model of proton dose distribution for a radiotherapy treatment-planning package. Phys Med Biol 38:989–998
43. Lestand L et al (2012) In-beam quality assurance using induced $\beta+$ activity in hadrontherapy: a preliminary physical requirements study using Geant4. Phys Med Biol 57:6497–6518
44. Lomax AJ (1999) Intensity modulated methods for proton therapy. Phys Med Biol 44:185–205
45. Lomax AJ et al (2004) The clinical potential of intensity modulated proton therapy. Z Med Phys 14:147–52
46. Martínez-Rovira I et al (2015) Implementation of biological washout processes within GATE/Geant4-A Monte Carlo study in the case of carbon therapy treatments. Med Phys 42:1773
47. Mazal A et al (1997) Robots in high precision patient positioning for conformal radiotherapy World Congress on Medical Physics and Biomedical Engineering. Med Biol Eng Comput 35:824

48. McGowan SE et al (2013) Treatment planning optimisation in proton therapy. Br J Radiol 86:20120288
49. Mustapha AA et al (1983) The relation between X-ray CT numbers and charged particles stopping powers and its significance for radiotherapy treatment planning. Phys Med Biol 28:169–176
50. Nauraye C et al (1995) An experimental approach to the design of a scattering system for a proton therapy beam line dedicated to ophthalmological applications. Int J Radiat Oncol Biol Phys 32(4):1177–1183
51. Oozeer R (1997) A model for the lateral penumbra in water of a 200-MeV proton beam devoted to clinical applications. Med Phys 24(10):1599–1604
52. Paganetti H (2005) Interpretation of proton relative biological effectiveness using lesion induction, lesion repair and cellular dose distribution. Med Phys 32:2548–2556
53. Paganetti H (2012) Range uncertainties in proton therapy and the role of Monte Carlo simulations. Phys Med Biol 57:R99–117
54. Paganetti H (2014) Relative biological effectiveness (RBE) values for proton beam therapy. Variations as a function of biological endpoint, dose, and linear energy transfer. Phys Med Biol 59:R419–R472
55. Pedroni E et al (2005) Experimental characterization and physical modelling of the dose distribution of scanned proton pencil beams. Phys Med Biol 50:541–561
56. Peucelle C et al (2015) Exploring new paths in particle therapy: a Monte Carlo study on the use of spatial fractionation of the dose. Physica Medica 30(Supplement 1):e27
57. Rit S et al (2013) Filtered backprojection proton CT reconstruction along most likely paths. Med Phys 40:031103
58. Robert C et al (2013) Distributions of secondary particles in proton and carbon-ion therapy: a comparison between GATE/Geant4 and FLUKA Monte Carlo codes. Phys Med Biol 58:2879–2900
59. Salvat F et al (2006) PENELOPE-2006, a code system for Monte Carlo simulation of electron and photon transport OECD, Spain. ISBN 92-64-02301-1
60. Sarrut D et al (2014) A review of the use and potential of the GATE Monte Carlo code for radiation therapy and dosimetry applications. Med Phys 41:064301
61. Sayah R et al (2013) Monte Carlo simulation of a proton therapy beamline for intracranial treatments. Radioprotection 48:317–339
62. Sayah R et al (2014) Secondary neutron doses received by paediatric patients during intracranial proton therapy treatments. J Radiol Prot 34(2):279–296
63. Schaffner B et al (1998) The precision of proton range calculations in proton radiotherapy treatment planning: experimental verification of the relation between ct-hu and proton stopping power. Phys Med Biol 43(6):1579–1592
64. Schaffner B (2008) Proton dose calculation based on in-air fluence measurements. Phys Med Biol 53:1545–1562
65. Schneider U et al (1996) The calibration of CT Hounsfield units for radiotherapy treatment planning. Phys Med Biol 41(1):111–124
66. Schneider W et al (2000) Correlation between CT numbers and tissue parameters needed for Monte Carlo simulations of clinical dose distributions. Phys Med Biol 45:459–478
67. Scholz M et al (1996) Track structure and the calculation of biological effects of heavy charged particles. Adv Space Res 18:5–14
68. Schwaab J et al (2011) Experimental characterization of lateral profiles of scanned proton and carbon ion pencil beams for improved beam models in ion therapy treatment planning. Phys Med Biol 56:7813–7827
69. Soukup M et al (2005) A pencil beam algorithm for intensity modulated proton therapy derived from Monte Carlo simulations. Phys Med Biol 50:5089–5104
70. Stankovskiy A et al (2009) Monte Carlo modelling of the treatment line of the Proton Therapy Center in Orsay. Phys Med Biol 54(8):2377–2394
71. Szymanowski H et al (2001) Experimental determination and verification of the parameters used in a proton pencil beam algorithm. Med Phys 28(6):975–987

72. Szymanowski H et al (2002) Two-dimensional pencil beam scaling: an improved proton dose algorithm for heterogeneous media Phys. Med Biol 47:3313–3330
73. Tessa CL et al (2014) Characterization of the secondary neutron field produced during treatment of an anthropomorphic phantom with x-rays, protons and carbon ions. Phys Med Biol 59(8):2111–2125
74. Ulmer W (2007) Theoretical aspects of energy-range relations, stopping power and energy straggling of protons. Radiat Phys Chem 76:1089–1107
75. Unkelbach J et al (2009) Reducing the sensitivity of IMPT treatment plans to setup errors and range uncertainties via probabilistic treatment planning. Med Phys 36:149–163
76. Vanderstraeten B et al (2007) Conversion of CT numbers into tissue parameters for Monte Carlo dose calculations: a multi-centre study. Phys Med Biol 52(3):539–562
77. Vogin G et al (2013) Localized Ewing sarcoma of the spine: a preliminary dose-escalation study comparing innovative radiation techniques in a single patient. Cancer Radiother 17(1):26–33
78. Weber U (2009) Comparison of carbon ions versus protons. Cancer J 15(4):325–332
79. Wedenberg M et al (2013) A model for the relative biological effectiveness of protons: The tissue specific parameter α/β of photons is a predictor for the sensitivity to LET changes. Acta Oncologica 52:580–588
80. Westerly DC et al (2013) A generalized 2D pencil beam scaling algorithm for proton dose calculation in heterogeneous slab geometries. Med Phys 40(6):061706
81. Wilkens JJ et al (2004) A phenomenological model for the relative biological effectiveness in therapeutic proton beams. Phys Med Biol 49:2811–2825
82. Zhang X et al (2011) Parameterization of multiple Bragg curves for scanning proton beams using simultaneous fitting of multiple curves. Phys Med Biol 56:7725–7735
83. Zheng Y et al (2012) Measurements of neutron dose equivalent for a proton therapy center using uniform scanning proton beams. Med Phys 39(6):3484–3492

Author Index

A.V. Solov'yov (ed.), *Nanoscale Insights into Ion-Beam Cancer Therapy*,
DOI 10.1007/978-3-319-43030-0

Subject Index

A.V. Solov'yov (ed.), *Nanoscale Insights into Ion-Beam Cancer Therapy*,
DOI 10.1007/978-3-319-43030-0